AF358669

Formation and Evolution of the Continental Crust in North China Craton during Precambrian

Formation and Evolution of the Continental Crust in North China Craton during Precambrian

Jin Liu
Jiahui Qian
Xiaoguang Liu

Basel • Beijing • Wuhan • Barcelona • Belgrade • Novi Sad • Cluj • Manchester

Jin Liu
College of Earth Sciences
Jilin University
Changchun
China

Jiahui Qian
School of Earth Sciences
and Engineering
Sun Yat-sen University
Zhuhai
China

Xiaoguang Liu
College of Earth Science
and Engineering
Shandong University of
Science and Technology
Qingdao
China

Editorial Office
MDPI AG
Grosspeteranlage 5
4052 Basel, Switzerland

This is a reprint of articles from the Special Issue published online in the open access journal *Minerals* (ISSN 2075-163X) (available at: www.mdpi.com/journal/minerals/special_issues/Q0BL2D34YP).

For citation purposes, cite each article independently as indicated on the article page online and using the guide below:

Lastname, A.A.; Lastname, B.B. Article Title. *Journal Name* **Year**, *Volume Number*, Page Range.

ISBN 978-3-7258-1950-8 (Hbk)
ISBN 978-3-7258-1949-2 (PDF)
https://doi.org/10.3390/books978-3-7258-1949-2

Contents

 *minerals*

MDPI

Editorial

Editorial for the Special Issue "Formation and Evolution of the Continental Crust in North China Craton during Precambrian"

Jin Liu [1,*], Jiahui Qian [2,*] and Xiaoguang Liu [3,*]

1 College of Earth Sciences, Jilin University, Changchun 130012, China
2 School of Earth Sciences and Engineering, Sun Yat-sen University, Zhuhai 519082, China
3 Shandong Provincial Key Laboratory of Depositional Mineralization & Sedimentary Minerals, Shandong University of Science and Technology, Qingdao 266590, China
* Correspondence: liujin@jlu.edu.cn (J.L.); qianjh5@mail.sysu.edu.cn (J.Q.); liuxiaoguang@ouc.edu.cn (X.L.)

Citation: Liu, J.; Qian, J.; Liu, X. Editorial for the Special Issue "Formation and Evolution of the Continental Crust in North China Craton during Precambrian". *Minerals* 2024, 14, 800. https://doi.org/10.3390/min14080800

Received: 1 August 2024
Accepted: 5 August 2024
Published: 6 August 2024

The North China Craton (NCC) preserves ancient rocks dating back to ca. 3.8 Ga, and has witnessed multiple magmatic–metamorphic events during the Archean–Proterozoic era [1–3]. This critical period may have marked the transformation of the geodynamic regime on early Earth. As a result, the NCC serves as an ideal natural laboratory for deciphering the growth and evolution of continental crust, the onset of plate tectonics, and the architecture of the Archean geodynamic regime. To better understand the geological evolution of the NCC during Precambrian, we collected 11 contributions for this Special Issue.

A significant component within the NCC is the Khondalite Belt, a major Paleoproterozoic orogenic belt. In this Special Issue, three contributions focus on the tectonic evolution of the Khondalite Belt, examining it from the perspectives of magmatism and deformation. Zhu et al. [4] proposed that the Sanchakou gabbro in the Khondalite Belt formed during the late Paleoproterozoic, based on zircon U-Pb dating. A comprehensive analysis of elemental and Hf-O-Sr-Nd isotopes further revealed a complex process involving assimilation and fractional crystallization within a post-collision tectonic environment. Qiao et al. [5] provided geochronological constraints for the Paleoproterozoic gneiss domes in the Qianlishan region. They identified two phases of deformation during the late Paleoproterozoic: D1 deformation, characterized by isoclinal folds and penetrative transposed foliations/gneissosities, occurring at ~1.95 Ga; and D2 deformation, featuring doubly plunging upright folds, formed between 1.93 and 1.90 Ga. The authors suggested that the collision between the Yinshan and Ordos Blocks led to the formation of the Qilishan gneiss domes. Qiao et al. [6] presented time and structural constraints for the Qianlishan ductile shear zones within the Khondalite Belt. The zircon U-Pb dating of mylonites is indicative of the metamorphic ages of 1.90–1.88 Ga, interpreted as the activity timing of Qianlishan ductile shear zones. Notably, their findings revealed three phases of deformation during Orosirian, suggesting that the Khondalite Belt underwent an orogenesis lasting over 100 Ma.

In this Special Issue, three contributions specifically focused on the Paleoproterozoic granitic dykes, felsic metavolcanic rocks, and meta-sedimentary rocks within the Jiao–Liao–Ji Belt. Zhao et al. [7] investigated newly identified late Paleoproterozoic granitic dykes from the Liaodong Peninsula. Zircon U-Pb dating indicated crystallization ages of 1859–1852 Ma for these dykes. Geochemical and zircon Hf isotopic analyses suggested that these granitic dykes formed through the partial melting of early Paleoproterozoic granitoids and meta-sedimentary rocks in a post-collisional environment. Cheng et al. [8] conducted systematic analyses of zircon geochronology, geochemistry, and zircon Hf isotope on the 2.2–2.1 Ga felsic metavolcanic rocks from the Li'eryu Formation in the Liaodong Peninsula. They proposed that these metavolcanic rocks originated from the partial melting of Archean TTG rocks, with some contributions from lower crustal materials. Additionally, they suggested an intra-continental rift setting for the Jiao–Liao–Ji Belt during 2.2–2.1 Ga. Gao et al. [9] reported new detrital zircon ages for the Paleoproterozoic Langzishan and Li'eryu Formations within the Jiao–Liao–Ji Belt. The geochronological results indicated deposition ages of 2136 Ma for Langzishan and 1974 Ma for Li'eryu. Combining those findings with previous magmatic studies, they proposed a back-arc basin environment for these sedimentary rocks.

Two contributions explore the affinity between the NCC and the Alxa Block based on magmatism and metamorphism. Niu et al. [10] conducted systematic zircon geochronological analyses, revealing magmatic activities at ~2.84, ~2.76, ~2.54, and ~2.49 Ga within the Alxa Block, previously considered part of the NCC. Geochemical and zircon Hf isotopic data suggest a distinct crustal evolutionary history for the Alxa Block during the Archean. Zhou et al. [11] used zircon U-Pb dating, mineral geochemistry, detailed petrological observations, and phase equilibrium modeling to constrain the P–T conditions of amphibolites in the Diebusige and Bayanwulashan complexes of the Alxa Block. These amphibolites experienced amphibolite–facies metamorphism at temperatures ranging from 800 to 910 °C and pressures of 7.0–10.8 kbar during 1901–1817 Ma. A post-collision extensional setting played a critical role in this metamorphic event.

In this Special Issue, researchers also discussed three key aspects: the oldest fuchsite quartzite from the eastern Hebei, the oldest clastic strata of the Xiong'er Group, and newly identified Paleoproterozoic mafic igneous rocks in the northern Liaoning. Zhao et al. [12] presented new zircon ages for the oldest fuchsite quartzite found in the Lulong area of the eastern Hebei terrane. Numerous 3.8–3.4 Ga detrital zircons were identified, suggesting a depositional age of 3.3–3.1 Ga for this fuchsite quartzite. Zircon Hf isotopes further implied significant crustal growth during the Eoarchean–Paleoarchean in the NCC. Zhang et al. [13] reported new detrital zircon ages for the earliest clastic strata of the Xiong'er Group from the southern margin of the NCC. Zircon U-Pb dating revealed four distinct age groups: 1905–1925, 2154–2295, 2529–2536, and 2713–2720 Ma, indicating a provenance link to the NCC basement. They proposed that these sedimentary rocks of the Dagushi Formtion formed through continuous crust fracturing. Chen et al. [14] identified four stages (~2.21, ~2.15, ~2.06, and ~2.02 Ga) of mafic magmatism in northern Liaoning. These mafic rocks constrain their surrounding strata to be Paleoproterozoic rather than Neoproterozoic. These mafic rocks likely formed in island arc or oceanic island environments associated with the oceanic subduction and extension during the Paleoproterozoic.

In conclusion, this Special Issue significantly contributes to our understanding of the Archean and Paleoproterozoic evolution of the NCC. However, further investigations into magmatism, sedimentary processes, metamorphism, and structural analyses remain essential.

Conflicts of Interest: The authors declare no conflicts of interest.

References

1. Wan, Y.; Dong, C.; Xie, H.; Wilde, S.A.; Liu, S.; Li, P.; Ma, M.; Li, Y.; Wang, Y.; Wang, K.; et al. Hadean to early Mesoarchean rocks and zircons in the North China Craton: A review. *Earth-Sci. Rev.* **2023**, *243*, 104489. [CrossRef]
2. Liu, J.; Palin, R.M.; Mitchell, R.N.; Liu, Z.; Zhang, J.; Li, Z.; Cheng, C.; Zhang, H. Archaean multi-stage magmatic underplating drove formation of continental nuclei in the North China Craton. *Nat. Commun.* **2024**, *15*, 6231. [CrossRef] [PubMed]
3. Zhao, G.; Cawood, P.A.; Li, S.; Wilde, S.A.; Sun, M.; Zhang, J.; He, Y.; Yin, C. Amalgamation of the North China Craton: Key issues and discussion. *Precambrian Res.* **2012**, *222–223*, 55–76. [CrossRef]
4. Zhu, W.P.; Tian, W.; Wang, B.; Zhang, Y.H.; Wei, C.J. Paleoproterozoic Crust–Mantle Interaction in the Khondalite Belt, North China Craton: Constraints from Geochronology, Elements, and Hf-O-Sr-Nd Isotopes of the Layered Complex in the Jining Terrane. *Minerals* **2023**, *13*, 426. [CrossRef]
5. Qiao, H.; Deng, P.; Li, J. Geochronological Constraints on the Origin of the Paleoproterozoic Qianlishan Gneiss Domes in the Khondalite Belt of the North China Craton and Their Tectonic implications. *Minerals* **2023**, *13*, 1361. [CrossRef]
6. Qiao, H.; Liu, M.; Dai, C. Timing and Tectonic Implications of the Development of the Orosirian Qianlishan Ductile Shear Zones in the Khondalite Belt, North China Craton. *Minerals* **2024**, *14*, 561. [CrossRef]
7. Zhao, Y.; Lyu, J.; Han, X.; Lin, S.; Zhang, P.; Yang, X.; Chen, C. Geochronology and Geological Implications of Paleoproterozoic Post-Collisional Monzogranitic Dykes in the Ne Jiao-Liao-Ji Belt, North China Craton. *Minerals* **2023**, *13*, 928. [CrossRef]

8. Cheng, C.; Liu, J.; Zhang, J.; Zhang, H.; Chen, Y.; Wang, X.; Li, Z.; Yu, H. Petrogenesis of the Early Paleoproterozoic Felsic Metavolcanic Rocks from the Liaodong Peninsula, NE China: Implications for the Tectonic Evolution of the Jiao-Liao-Ji Belt, North China Craton. *Minerals* **2023**, *13*, 1168. [CrossRef]
9. Gao, J.; Li, W.; Liu, Y.; Zhao, Y.; Liu, T.; Wen, Q. Detrital Zircon LA-ICP-MS U-Pb Ages of the North Liaohe Group from the Lianshanguan Area, NE China: Implications for the Tectonic Evolution of the Paleoproterozoic Jiao-Liao-Ji Belt. *Minerals* **2023**, *13*, 708. [CrossRef]
10. Niu, P.; Qu, J.; Zhang, J.; Zhang, B.; Zhao, H. Archean Crustal Evolution of the Alxa Block, Western North China Craton: Constraints from Zircon U-Pb Ages and the Hf Isotopic Composition. *Minerals* **2023**, *13*, 685. [CrossRef]
11. Zhou, F.; Gou, L.; Xu, X.; Tian, Z. Metamorphic Ages and PT Conditions of Amphibolites in the Diebusige and Bayanwulashan Complexes of the Alxa Block, North China Craton. *Minerals* **2023**, *13*, 1426. [CrossRef]
12. Zhao, C.; Zhang, J.; Wang, X.; Zhang, C.; Chen, G.; Zhang, S.; Guo, M. Tracing the Early Crustal Evolution of the North China Craton: New Constraints from the Geochronology and Hf Isotopes of Fuchsite Quartzite in the Lulong Area, Eastern Hebei Province. *Minerals* **2023**, *13*, 1174. [CrossRef]
13. Zhang, Y.; Zhang, G.; Sun, F. The Earliest Clastic Sediments of the Xiong'er Group: Implications for the Early Mesoproterozoic Sediment Source System of the Southern North China Craton. *Minerals* **2023**, *13*, 971. [CrossRef]
14. Chen, J.; Tian, Y.; Gao, Z.; Li, B.; Zhao, C.; Li, W.; Zhang, C.; Wang, Y. Geochronology and Geochemistry of Paleoproterozoic Mafic Rocks in Northern Liaoning and Their Geological Significance. *Minerals* **2024**, *14*, 717. [CrossRef]

Article

Paleoproterozoic Crust–Mantle Interaction in the Khondalite Belt, North China Craton: Constraints from Geochronology, Elements, and Hf-O-Sr-Nd Isotopes of the Layered Complex in the Jining Terrane

Wei-Peng Zhu [1,2], Wei Tian [1,*], Bin Wang [1], Ying-Hui Zhang [2] and Chun-Jing Wei [1]

[1] MOE Key Laboratory of Orogenic Belts and Crustal Evolution, School of Earth and Space Sciences, Peking University, Beijing 100871, China
[2] Key Laboratory of Deep-Earth Dynamics of Ministry of Natural Resources, Institute of Geology, Chinese Academy of Geological Sciences, Beijing 100037, China
* Correspondence: davidtian@pku.edu.cn

Abstract: The Paleoproterozoic Khondalite Belt, located in the northwestern segment of North China Craton (NCC), is characterized by widespread high-temperature/ultrahigh-temperature (UHT) granulite/gneiss and large-scale magmatic activity. The tectonic evolution is still controversial. Here, we report new geochronological, elemental, and Hf-O-Sr-Nd isotopic data for a Paleoproterozoic layered complex in the Jining terrane to constrain the tectonic evolution of the Khondalite Belt. In situ zircon U-Pb dating indicates that the Sanchakou gabbros were emplaced between ~1.94 Ga and ~1.82 Ga, which might be the heat source of UHT metamorphism. The elemental and Hf-O-Sr-Nd isotopic analysis shows that the formation of Sanchakou gabbros is consistent with the assimilation and fractional crystallization (AFC) process. The magma originates from the 10%~20% partial melting of the spinel + garnet lherzolite mantle. The Sanchakou gabbros are magmatic crystallization products mixed with crustal wallrocks in the magma chamber. We have established a tectonic evolution model involving asthenosphere upwelling after the amalgamation of the Ordos and Yinshan Blocks at ~1.95 Ga.

Keywords: NCC; Paleoproterozoic Khondalite Belt; UHT granulite/gneiss; layered complex; AFC; tectonic evolution

Citation: Zhu, W.-P.; Tian, W.; Wang, B.; Zhang, Y.-H.; Wei, C.-J. Paleoproterozoic Crust–Mantle Interaction in the Khondalite Belt, North China Craton: Constraints from Geochronology, Elements, and Hf-O-Sr-Nd Isotopes of the Layered Complex in the Jining Terrane. *Minerals* **2023**, *13*, 462. https://doi.org/10.3390/min13040462

Academic Editors: Jin Liu, Jiahui Qian, Xiaoguang Liu and Dominique Gasquet

Received: 23 February 2023
Revised: 18 March 2023
Accepted: 21 March 2023
Published: 24 March 2023

1. Introduction

The concept of "layered complex" was first recorded in *Layered Igneous Rocks* [1]. Most of the layered complex is close to gabbros, mainly composed of basic or ultrabasic rocks. The most notable feature is the well-developed layered rhythmic structure. The layered complexes formed mainly in Archean and Proterozoic [2–6], with a lesser amount in Phanerozoic [7–9]. The tectonic environment is related to mantle plume or intracontinental rifting [10–13], controlled by regional fractures. Crustal contamination plays an important role in the formation of layered complexes, which can greatly change the composition of magma and result in the difference in products [14,15]. Therefore, the compositional changes in layered complexes in the open magmatic system are significant for revealing crust-mantle interaction [16–19].

The North China Craton (NCC) is a fundamental geological unit of the early Precambrian in China [20]. The Jining terrane, located in the eastern Khondalite Belt of the NCC, has been widely studied by researchers in past decades. A large number of ultrahigh-temperature (UHT) metamorphic rocks have been reported in this area [21–32]. In addition, there are small amounts of basic intrusive rocks dominated by gabbros, some of which occurred as multiple sets of layered complexes [33]. The episodic crystallization ages are 2.45~2.10 Ga, 1.97~1.92 Ga, and 1.85~1.84 Ga [34–37]. The magma may have originated from a deep mantle of ~3.0 GPa, and ~1550 °C [35]. However, the genesis of these basic intrusive rocks and tectonic environment have not been determined yet.

In this paper, we focus on a Paleoproterozoic layered complex in the Jining area, northwestern margin of the NCC, which is accompanied by extensive granulites/gneisses. Through the chronological, elemental, and Hf-O-Sr-Nd isotopic analysis methods, we have found that the layered complex is a product of crust–mantle interaction, which reflects the large-scale thermal fluctuation under the influence of the upwelling asthenosphere.

2. Geological Setting

The NCC is one of the rare ancient continental blocks with ~3.8 Ga crustal rock in the world [38–40]. There are different views about the formation and evolution of the NCC [20,41,42]. The Khondalite Belt is in the western part of the NCC (Figure 1a), considered to be formed via a collision between the Ordos and Yinshan Blocks at ~1.95 Ga [20,43]. From west to east, there are Qianlishan-Helanshan terrane, Daqingshan-Helanshan terrane, and Jining terrane. The Jining terrane is located in the eastern segment of the Khondalite Belt (Figure 1b), where UHT granulites/gneisses, S-type granites, and a small amount of basic intrusive rocks are mainly exposed (Figure 1c).

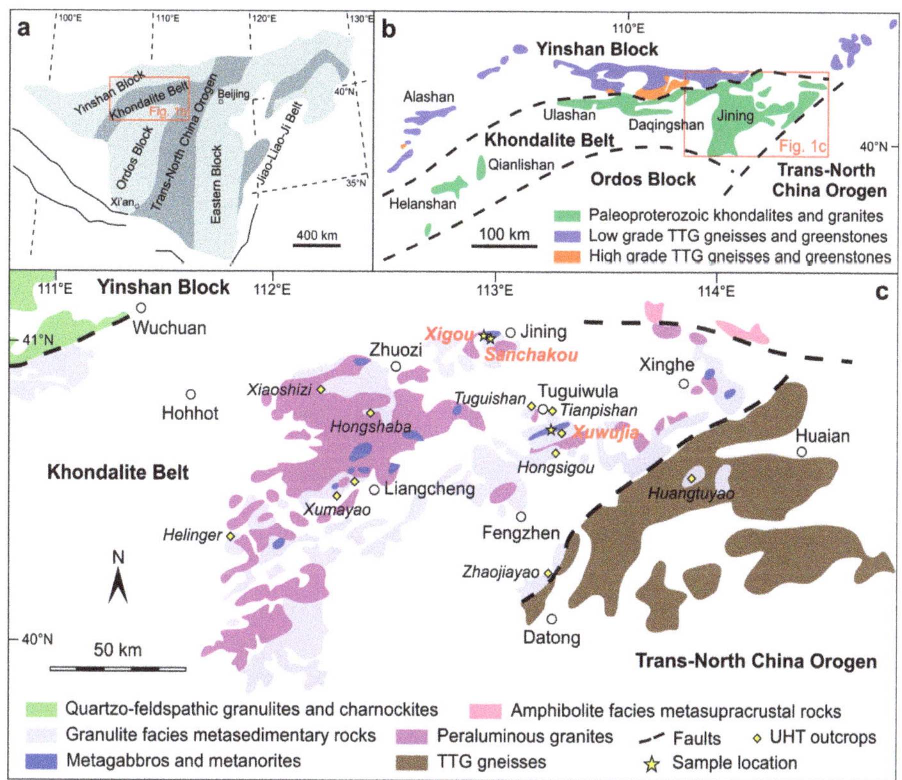

Figure 1. The regional geological map of the Jining terrane in the Khondalite Belt of the NCC. (**a**) The structural sketch of the NCC (modified from [20]); (**b**) The distribution map of structural units contained in the Khondalite Belt and adjacent structural units of the Khondalite Belt (modified from [20]); (**c**) The lithology distribution map of the Jining terrane (modified from [34]).

The UHT granulites/gneisses are scattered in Tuguishan, Dajing, Tianpishan, Xuwujia, Xumayao, Helinger, Hongsigou, Zhaojiayao, Liangcheng, and Hongshaba [23–26,28–31,44–50], continuously forming a UHT metamorphic zone of about 250 × 150 square kilometers. Most of these UHT granulites/gneisses are sillimanite-garnet gneisses, and only a few contain spinel/sapphirine + quartz assemblages. They have average metamorphic ages of 1.92~1.91 Ga [30,31,44,50] and ~1.88 Ga [28]. The peak temperature of UHT metamorphism is mainly concentrated at 950~1050 °C [23–26,28–31,44–50].

The vast majority of the S-type granites are garnet granites, which intruded into the khondalite and covered more than 40% of the Jining terrane. The crystallization age is about 1.94~1.90 Ga [51,52]. One view [51] believed that garnet granite is a mixture of the mantle-derived basic magma and the melt produced by the anatexis of metasedimentary rocks, formed in the process of UHT metamorphism (1.93~1.92 Ga). The other view [52] believed that the garnet granite was formed before UHT metamorphism (1.94~1.93 Ga) and underwent UHT metamorphism (~1.92 Ga) with metasedimentary rocks.

A few basic intrusive rocks are mainly gabbros with an average crystallization age of ~1.93 Ga, believed to originate from the mantle plume or mantle upwelling under the background of the mid-ocean ridge [35].

3. Samples and Methods

3.1. Sample Description

The studied sample is from a drilling core of the layered complex, taken from Sanchakou town, Jining District, Ulanqab City, Inner Mongolia Autonomous Region (Figure 1c). We selected a section in which the overall lithology is gabbro, with obvious changes in feldspar content in the vertical direction (Figure 2a). Therefore, the layered complex was named the Sanchakou gabbros. We chose typical positions of the drilling core, cut into rock thin slices, ground the rock into powder, and numbered SCK-1 to SCK-6, respectively.

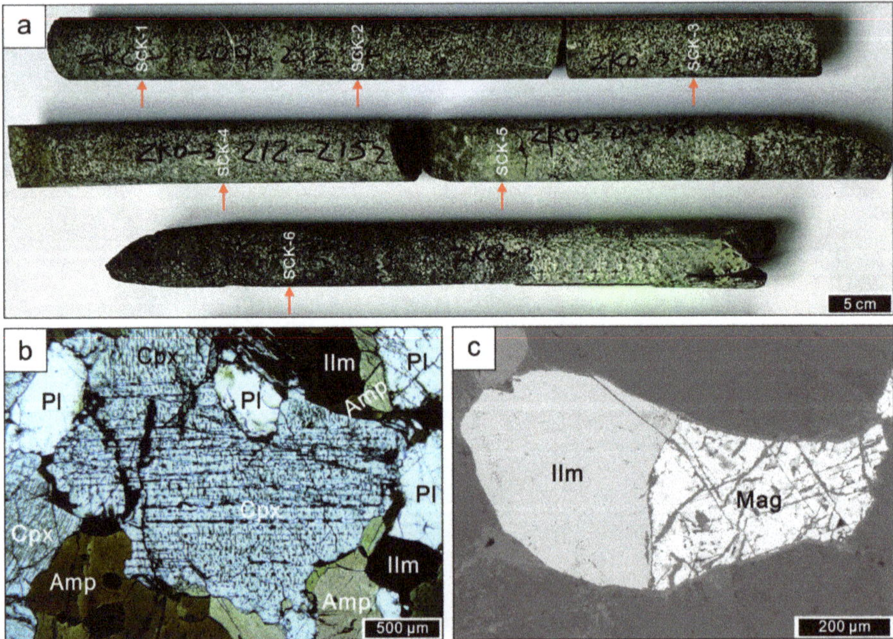

Figure 2. Petrological characteristics of the Sanchakou gabbros from the Jining terrane. (**a**) The Sanchakou gabbro drilling core (The arrows indicate the slicing and sampling positions); (**b**) The main rock-forming minerals, including the diopside, pargasite, and plagioclase; (**c**) The equilibrium assemblage of magnetite and ilmenite. Mineral abbreviations: Amp, amphibole; Cpx, clinopyroxene; Ilm, ilmenite; Mag, magnetite; Pl, plagioclase.

The rock has a full-crystalline unequal granular texture. The main minerals include clinopyroxene, amphibole, and plagioclase. The altered minerals mainly include zoisite and epidote. The representative accessory minerals include ilmenite, magnetite, apatite, and zircon. The clinopyroxene is mostly irregular prismatic, widely developing a group of parallel cleavage (Figure 2b). The amphibole has obvious pleochroism, and its interference color is lower than that of the clinopyroxene. The primary amphibole is generally subhedral plate-prismatic. The secondary amphibole is in a xenomorphic long-prismatic shape, growing along the edge of other minerals. The plagioclase has undergone intense zoisitization and idolization alteration. The opaque minerals mainly include ilmenite and magnetite, some of which form the equilibrium assemblages (Figure 2c). The apatite is transparent and round. The zircon is subhedral long-prismatic, developing a higher white interference color.

3.2. Analytical Methods

3.2.1. Mineral Major Elements Analysis

Mineral major elemental analysis was conducted by EPMA at the MOE Key Laboratory of Orogenic Belts and Crustal Evolution, School of Earth and Space Sciences, Peking University, Beijing, China. We used JEOL JXA-8230 for in situ analysis, equipped with four-channel spectrometers (CH_1-CH_4). PETJ crystal (CH_1) was used to determine K, Ca, and Ti. Two TAP crystals (CH_2, CH_4) were used to determine Na, Si, Mg, and Al. LIFH crystal (CH_3) was used to determine Cr, Mn, Fe, and Ni. The acceleration voltage, the beam current, and the beam spot were set to 15 kV, 10 nA, and 2 μm. The counting times for the background and peak values of Ca, Mg, Al, and Fe are 5 s and 20 s, and those of other elements are 5 s and 10 s. The standard samples are 53 kinds of minerals from American Structure Probe Inc. Supplies^{TM} Company. We used the stoichiometric method to calculate and the PRZ method to correct. Detailed information for the configuration of crystal spectrometers and the mineral standard samples of related elements was the same as described by [53].

3.2.2. In Situ Zircon U-Pb Isotopic Analysis

In situ zircon U-Pb isotopic analysis was conducted by LA-ICP-MS at the Wuhan SampleSolution Analytical Technology Company Limited, Hubei, China. Detailed operating conditions for the laser ablation system and the ICP-MS instrument were the same as described by [54]. Laser sampling was performed using a GeolasPro laser ablation system that consists of a COMPexPro 102 ArF excimer laser (wavelength of 193 nm and maximum energy of 200 mJ) and a MicroLas optical system. We used an Agilent 7900 ICP-MS instrument to acquire ion-signal intensities. Helium was applied as a carrier gas. Argon was used as the make-up gas and mixed with the carrier gas via a T-connector before entering the ICP. A "wire" signal smoothing device is included in this laser ablation system [55]. The spot size and frequency of the laser were set to 24 μm and 5 Hz. International zircon standard 91500 and glass NIST610 were used as external standards for U-Pb dating and trace element calibration, respectively. TEMORA-2 was treated as a blind sample to monitor the working state of the instrument. Each analysis incorporated a background acquisition of approximately 20~30 s followed by 50 s of data acquisition from the sample. We used ICPMSDataCal to perform offline selection and integration of background and analyze signals, time-drift correction, and quantitative calibration [56,57]. Concordia diagrams, probability density diagrams, and weighted mean calculations were made using Isoplot 4.15 [58].

3.2.3. Whole-Rock Major and Trace Elements Analysis

Whole-rock chemical pretreatment and major and trace element analysis were conducted at Nanjing FocuMS Technology Company Limited, Jiangsu, China. For major elements, we adopt the acid dissolution method to dissolve the sample powder to analyze the major elements except for Si and the alkali fusion method to liquate the sample powder to analyze Si. The analytical instrument was Agilent 5110 ICP-OES. The analysis accuracy of SiO_2 is better than 1%; that of Al_2O_3, TFeO, MgO, K_2O, Na_2O, and CaO is better than 3%; and that of TiO_2, MnO, and P_2O_5 is better than 5%. For trace elements, we adopt the acid dissolution method to dissolve the sample powder. The analytical instrument was Agilent 7700x ICP-MS. Among them, the analysis accuracy of trace elements with content more than 50×10^{-6} is better than 5%; that of trace elements with content between 5×10^{-6} and 50×10^{-6} is better than 10%; and that of trace elements with content between 0.5×10^{-6} and 5×10^{-6} is better than 20%. Geochemical reference materials of BHVO-2 and AGV-2 were treated as blind samples for quality assurance of measurement, the measured values of which were compared with *Geological and Environmental Reference Materials* (GeoReM) [59]. The detailed operating process was the same as described by [60].

3.2.4. In Situ Zircon Hf-O Isotopic Analysis

In situ zircon Hf isotope ratio analysis was conducted using a Neptune Plus MC-ICP-MS in combination with a Geolas HD excimer ArF laser ablation system that was hosted at the Wuhan SampleSolution Analytical Technology Company Limited, Hubei, China. A "wire" signal smoothing device is included in this laser ablation system, by which smooth signals are produced even at very low laser repetition rates down to 1 Hz [55]. Helium was used as the carrier gas within the ablation cell and was merged with argon after the ablation cell. Small amounts of nitrogen were added to the argon makeup gas flow for sensitivity improvement [61]. The single spot size and energy density of the laser were set to 32 μm and ~7.0 J cm^{-2}. Each measurement included 20 s of background signal acquisition and 50 s of ablation signal acquisition. Detailed operating conditions for the laser ablation system, the MC-ICP-MS instrument, and the analytical method were the same as described by [61]. The interference of ^{176}Lu and ^{176}Yb on ^{176}Hf was corrected according to [62]. Off-line selection and integration of analytical signal and mass bias calibrations were performed using ICPMSDataCal [57]. To ensure the data reliability, three international zircon standards of Plešovice, 91500, and GJ-1 were analyzed simultaneously with the actual samples, the Hf isotopic compositions of which have been reported by [63]. The external accuracies (2σ) of Plešovice, 91500, and GJ-1 were better than 0.000020. The test values were consistent with the recommended value within the error range.

In situ zircon O isotopic analysis was conducted using a SHRIMP IIe-MC that was hosted at the Beijing SHRIMP Center, Institute of Geology, Chinese Academy of Geological Sciences, Beijing, China. Detailed operating conditions were the same as described by [64,65]. Two groups of scans were set for each data analysis. Each group was scanned 6 times, and the integral time of each scan was 10 s. Between the two groups of scans, the instrument can automatically readjust the parameters of the primary ion current and the secondary ion current to achieve the best result. During the analysis process, the ratio of sample zircon data points to standard zircon data points was 1:2~1:4. The internal accuracy of single analysis data was generally better than ±0.3‰ (2σ). We used TEMORA-2 ($\delta^{18}O = 8.20‰$) [66] and 91500 ($\delta^{18}O = 9.86‰$) [67] to monitor the working state of the instrument and correct for the mass fractionation. *Vienna standard mean ocean water* (V-SMOW) was adopted to standardize the O isotopic analysis results.

In particular, the zircon number and position of in situ Hf-O isotopic analysis were consistent with those of in situ U-Pb isotopic analysis.

3.2.5. Whole-Rock Sr-Nd Isotopic Analysis

Whole-rock chemical pretreatment and Sr-Nd isotopic analysis were conducted at Nanjing FocuMS Technology Company Limited, Jiangsu, China. We adopt the acid dissolution method to dissolve the sample powder, followed by extraction and chromatographic separation according to the method of [68]. The detailed operating process was the same as described by [69]. The analytical instrument was Nu Plasma II MC-ICP-MS. Raw data of isotope ratios were internally corrected for the mass fractionation by normalizing $^{86}Sr/^{88}Sr = 0.1194$ for Sr, $^{146}Nd/^{144}Nd = 0.7219$ for Nd with exponential law. International isotopic standards (NIST SRM 987 for Sr, JNdi-1 for Nd) were periodically analyzed to correct instrumental drift. Geochemical reference materials of BCR-2 and BHVO-2 were treated as blind samples for quality assurance of measurement. These isotopic results agreed with previous publications regarding analytical uncertainty [70].

4. Results

4.1. Mineral Chemistry

The EPMA analytical results of clinopyroxene, amphibole, plagioclase, magnetite, and ilmenite in the Sanchakou gabbros are listed in Table 1.

Table 1. The major elemental data (wt.%) of representative minerals in the Sanchakou gabbros.

Mineral	SiO_2	TiO_2	Al_2O_3	Cr_2O_3	TFeO	MnO	NiO	MgO	CaO	Na_2O	K_2O	Total
Cpx	52.42	0.25	1.99	0.01	8.09	0.36	0.01	14.05	23.87	0.39	0.01	101.45
Amp	43.89	2.01	10.65	0.03	13.03	0.21	0.02	12.98	12.38	1.46	1.21	97.87
Pl	55.91	0.01	28.08	0.01	0.15	0.01	0.03	0	11.17	5.11	0.25	100.73
Mag	0.31	0.10	0.17	0.04	93.65	0.02	0.01	0.14	0.11	0.06	0.01	94.62
Ilm	0.03	50.58	0.03	0.02	46.29	2.63	0.01	0.05	0.01	0.03	0.00	99.68

Notes: Amp, amphibole; Cpx, clinopyroxene; Ilm, ilmenite; Mag, magnetite; Pl, plagioclase.

4.1.1. Clinopyroxene

According to the classification of [71], the clinopyroxene of Sanchakou gabbros corresponds to the Ca-Mg-Fe pyroxenes (Figure 3a). Its composition is $Wo_{46.71\sim47.48}En_{36.26\sim39.35}Fs_{12.17\sim14.74}$, belonging to the diopside (Figure 3b). The $Mg^{\#}$ of the clinopyroxene ranges from 72.11 to 77.20, with an average value of 75.58. The $Cr^{\#}$ is mostly 0. The clinopyroxene has relatively low contents of TiO_2 (0.14~0.56 wt.%), Al_2O_3 (1.42~3.61 wt.%), TFeO (7.57~8.97 wt.%), and Na_2O (0.30~0.55 wt.%).

4.1.2. Amphibole

The amphibole of Sanchakou gabbros has relatively high contents of Al_2O_3 (10.14~10.97 wt.%), TFeO (12.12~14.12 wt.%), MgO (12.30~13.84 wt.%), and CaO (11.76~12.86 wt.%) and relatively low contents of SiO_2 (42.15~45.53 wt.%), Na_2O (1.21~1.61 wt.%), and K_2O (0.96~1.32 wt.%). In general, its composition has little change. According to the classification of [72], the amphibole belongs to the pargasite and edenite (Figure 3c).

4.1.3. Plagioclase

The plagioclase of Sanchakou gabbros has relatively high contents of Al_2O_3 (27.46~29.03 wt.%) and CaO (10.49~11.94 wt.%) and a relatively low content of K_2O (0.14~0.35 wt.%). Its composition is $An_{50.20\sim57.16}Ab_{41.42\sim48.12}Or_{0.80\sim1.99}$. According to the classification of [73], the plagioclase belongs to the labradorite (Figure 3d).

4.2. In Situ Zircon Isotopic Characteristics

4.2.1. Zircon U-Pb Dating

The in situ zircon U-Pb isotopic data of the Sanchakou gabbros are listed in Table 2. The zircons selected from the Sanchakou gabbros are irregularly rounded crystals, ranging from 70 μm to 180 μm in diameter. Almost all zircons have bright growth edges in CL images, which vary in width (Figure 4a). The core of zircons is generally dark. A few of the cores have obvious growth stripes inside. Small amounts of zircons have 2225~2375 Ma inherited cores. All characteristics reflect the crystallization history of zircons and the superposition of multi-stage magmatism. The zircon Th/U ratios range from 0.11 to 7.53, with an average value of 1.44. Based on these features, the zircon belongs to the magmatic zircon.

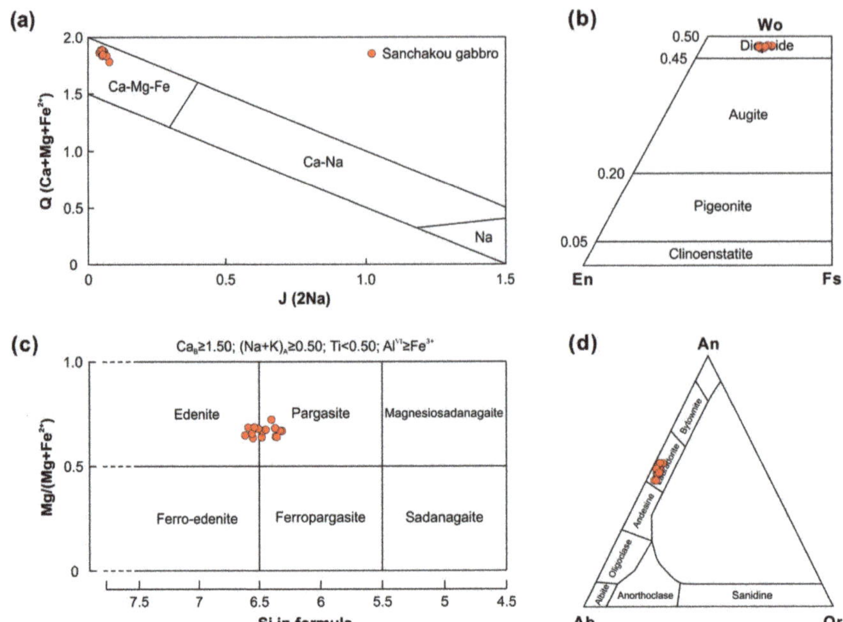

Figure 3. Mineral compositions of the Sanchakou gabbros from the Jining terrane. (**a**) The Q-J diagram of pyroxene according to the nomenclature of [71]; (**b**) The classification diagram of pyroxene according to the nomenclature of [71]; (**c**) The classification diagram of amphibole according to the nomenclature of [72]; (**d**) The classification diagram of feldspar according to the nomenclature of [73]. Mineral abbreviations: Ab, albite; An, anorthite; En, enstatite; Fs, ferrosilite; Or, potassium feldspar; Wo, wollastonite.

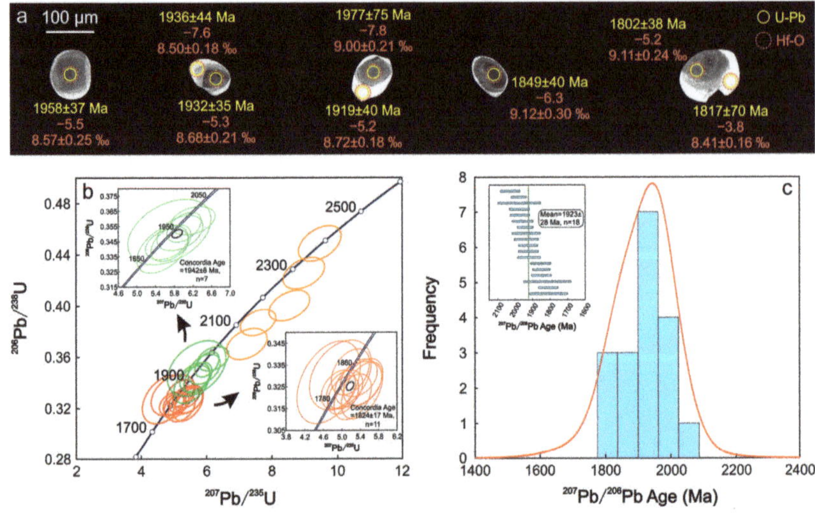

Figure 4. Zircon U-Pb isotopic analytical results of the Sanchakou gabbros from the Jining terrane. (**a**) Typical zircon CL images, marked with spot positions of in situ zircon U-Pb and Hf-O isotopic analysis; (**b**) The zircon U-Pb concordia diagram; (**c**) The ^{207}Pb/^{206}Pb age probability density histogram together with the ^{207}Pb/^{206}Pb weighted average age distribution.

Table 2. The in situ zircon U-Pb isotopic data of the Sanchakou gabbros.

Sample	Th ($\times 10^{-6}$)	U ($\times 10^{-6}$)	Th/U	$^{207}Pb/^{206}Pb$	$\pm 1\sigma$	$^{207}Pb/^{235}U$	$\pm 1\sigma$	$^{206}Pb/^{238}U$	$\pm 1\sigma$	$^{207}Pb/^{206}Pb$ Age (Ma)	$\pm 1\sigma$	Corrected $^{207}Pb/^{206}Pb$ Age (Ma)	$\pm 1\sigma$
SCK-1	93.74	146.87	0.67	0.1201	0.0025	5.4627	0.1757	0.3271	0.0044	1958	60	1958	37
SCK-2.1	31.93	280.56	0.11	0.1184	0.0024	5.5976	0.1810	0.3429	0.0050	1943	45	1932	35
SCK-2.2	99.94	42.62	2.34	0.1187	0.0030	5.5456	0.2253	0.3400	0.0065	1936	81	1936	44
SCK-3.1	286.49	529.80	0.54	0.1196	0.0026	5.8190	0.1880	0.3529	0.0043	1979	43	1950	38
SCK-3.2	84.66	110.80	0.76	0.1212	0.0025	5.4406	0.1839	0.3233	0.0053	1976	57	1974	36
SCK-4.1	79.50	149.93	0.53	0.1399	0.0033	7.2351	0.2538	0.3703	0.0050	2226	56	2225	41
SCK-4.2	112.83	82.50	1.37	0.1187	0.0033	5.3588	0.2084	0.3262	0.0043	1936	71	1936	49
SCK-5.1	175.25	349.17	0.50	0.1526	0.0029	8.5282	0.2523	0.4003	0.0049	2376	46	2375	32
SCK-5.2	70.99	95.85	0.74	0.1136	0.0037	4.9834	0.2247	0.3182	0.0047	1924	58	1924	57
SCK-6	97.22	265.73	0.37	0.1505	0.0032	8.8052	0.2818	0.4242	0.0053	2355	48	2352	36
SCK-8.1	91.48	13.31	6.87	0.1214	0.0052	5.6946	0.3648	0.3514	0.0088	1977	123	1977	75
SCK-8.2	65.32	164.50	0.40	0.1175	0.0027	5.6357	0.1866	0.3434	0.0042	1920	66	1919	40
SCK-9.1	126.56	162.62	0.78	0.1405	0.0028	7.5763	0.2212	0.3870	0.0042	2235	50	2234	34
SCK-9.2	68.40	113.10	0.60	0.1212	0.0027	5.9214	0.2012	0.3507	0.0048	1974	56	1974	40
SCK-10.1	127.14	248.59	0.51	0.1516	0.0027	9.4564	0.2680	0.4488	0.0057	2365	46	2364	30
SCK-10.2	155.94	183.12	0.85	0.1180	0.0025	5.3711	0.1641	0.3291	0.0035	1926	55	1926	38
SCK-11	78.17	140.98	0.55	0.1131	0.0025	5.0615	0.1605	0.3238	0.0036	1850	57	1849	40
SCK-12.1	110.85	271.20	0.41	0.1250	0.0028	6.1952	0.1955	0.3589	0.0038	2029	56	2029	39
SCK-12.2	172.81	136.29	1.27	0.1133	0.0025	5.1886	0.1703	0.3318	0.0043	1854	58	1853	39
SCK-12.3	94.04	21.42	4.39	0.1107	0.0042	4.9726	0.2740	0.3304	0.0066	1810	109	1810	68
SCK-13.1	95.48	133.61	0.71	0.1101	0.0023	4.9251	0.1510	0.3237	0.0037	1811	57	1802	38
SCK-13.2	104.55	13.89	7.53	0.1111	0.0044	4.7167	0.2796	0.3271	0.0077	1817	136	1817	70
SCK-14	62.50	193.00	0.32	0.1130	0.0020	5.0281	0.1381	0.3212	0.0036	1850	50	1848	32

Except for the data of inheritance cores, the rest zircon grains show a "beaded" distribution along the concordia line (Figure 4b). Based on the principle of statistics, the data can be divided into two groups: the concordia age of the first group is 1942 ± 8 Ma ($n = 7$) and the concordia age of the second group is 1824 ± 17 Ma ($n = 11$). The distribution of corrected ^{207}Pb/^{206}Pb apparent ages ranges from 1802 ± 38 Ma to 2029 ± 39 Ma. The ^{207}Pb/^{206}Pb weighted average age is concentrated at 1923 ± 28 Ma ($n = 18$). The ^{207}Pb/^{206}Pb age probability density histogram shows that the data distribution is continuous (Figure 4c), and the peak value is 1941 ± 22 Ma ($n = 18$).

4.2.2. Zircon Hf-O Isotopes

The in situ zircon Hf-O isotopic data of the Sanchakou gabbros are listed in Table 3.

Table 3. The in situ zircon Hf-O isotopic data of the Sanchakou gabbros.

Sample	^{207}Pb/^{206}Pb Age (Ma)	^{176}Lu/^{177}Hf	±2σ	^{176}Hf/^{177}Hf	±2σ	εHf (t)	εHf (0)	±2σ	TDM (Ma)	δ18O (‰)	±2σ
SCK-1	1958	0.000770	0.000020	0.281411	0.000012	−5.5	−48.1	0.4	2558	8.57	0.25
SCK-2.1	1943	0.000738	0.000006	0.281396	0.000012	−5.3	−47.1	0.4	2526	8.68	0.21
SCK-2.2	1936	0.000383	0.000004	0.281424	0.000012	−7.6	−47.8	0.4	2546	8.50	0.18
SCK-3.1	1979	0.000995	0.000026	0.281385	0.000013	−3.9	−48.7	0.4	2553	8.29	0.17
SCK-3.2	1976	0.000563	0.000013	0.281403	0.000012	−7.1	−48.2	0.4	2520	9.06	0.28
SCK-4.2	1936	0.000609	0.000047	0.281388	0.000013	−8.4	−48.5	0.5	2531	8.87	0.20
SCK-5.2	1924	0.000355	0.000016	0.281407	0.000012	−7.4	−47.7	0.4	2493	9.04	0.26
SCK-8.1	1977	0.000839	0.000001	0.281446	0.000012	−7.8	−48.2	0.5	2512	9.00	0.21
SCK-8.2	1920	0.000042	0.000001	0.281418	0.000012	−5.2	−46.1	0.5	2444	8.72	0.18
SCK-9.2	1974	0.000744	0.000007	0.281383	0.000013	−6.3	−48.6	0.4	2576	8.38	0.27
SCK-10.2	1926	0.000858	0.000005	0.281439	0.000012	−5.0	−47.7	0.4	2516	8.84	0.24
SCK-11	1850	0.000761	0.000004	0.281420	0.000012	−6.3	−49.1	0.5	2609	9.12	0.30
SCK-12.1	2029	0.000329	0.000023	0.281394	0.000012	−5.1	−48.4	0.4	2555	8.08	0.19
SCK-12.2	1854	0.000139	0.000002	0.281409	0.000012	−6.6	−48.9	0.5	2578	8.78	0.25
SCK-12.3	1810	0.000152	0.000002	0.281402	0.000015	−5.9	−48.3	0.4	2536	9.22	0.16
SCK-13.1	1811	0.000031	0.000001	0.281424	0.000011	−5.2	−46.9	0.4	2515	9.11	0.24
SCK-13.2	1817	0.000011	0.000000	0.281408	0.000013	−3.8	−47.9	0.4	2501	8.41	0.16
SCK-14	1850	0.000187	0.000012	0.281468	0.000013	−6.1	−49.1	0.5	2595	9.26	0.21

The zircon has a high Hf content and a very low Lu content, resulting in a very low ^{176}Lu/^{177}Hf ratio and a very low content of ^{176}Hf formed by Lu decay. So, there is no obvious radioactive accumulation after the formation of zircon. The ^{176}Hf/^{177}Hf measured in the samples can represent the Hf isotopic composition of the system when the zircons formed. The initial ^{176}Hf/^{177}Hf value of zircon in the samples is low (0.281383~0.281468), and the weighted average value is 0.281413 ± 0.000011. The $\varepsilon_{Hf}(t)$ values are all negative, ranging from −8.4 to −3.8 (Figure 5a). The Lu-Hf T_{DM} age ranges from 2444 Ma to 2609 Ma, which belongs to the Paleoproterozoic and Archean, indicating that the magma may have originated from the enriched mantle or suffered from crustal contamination [74].

The zircon δ^{18}O value of the samples ranges from 8.08 to 9.26, with a weighted average value of 8.75 ± 0.16‰ (Figure 5b), which is much higher than that of the mantle zircon (5.3 ± 0.6‰) [75]. Its genesis may be that the parent magmatic source of the Sanchakou gabbros was added with the high-δ^{18}O crustal materials, or the parent magma suffered from contamination by high-δ^{18}O crustal wallrocks during emplacement.

4.3. Whole-Rock Geochemical Characteristics

4.3.1. Major Elements

The whole-rock major elemental data of the Sanchakou gabbros are listed in Table 4. The loss on ignition (LOI) of the Sanchakou gabbros is negative, indicating that the samples have suffered from a low degree of alteration. It is consistent with the relatively fresh characteristics observed in the field. The content of SiO_2 is 43.45~45.01 wt.%, indicating that the samples belong to the basic rocks. The samples contain 12.73~13.64 wt.% Al_2O_3, 0.14~0.23 wt.% K_2O, and 2.05~2.25 wt.% Na_2O. The content of $Na_2O + K_2O$ is 2.21~2.47 wt.%, and the ratio of K_2O/Na_2O is 0.06~0.10, indicating that the samples belong to calc-alkaline gabbros and have not suffered obvious potassium metasomatism. The samples fall into the peridot-gabbro field in the (Na$_2$O + K$_2$O)-SiO$_2$ diagram (Figure 6a) and the medium-K calc-alkaline series field in the K$_2$O-SiO$_2$ diagram (Figure 6b). The samples contain 7.44~7.77 wt.% MgO and 17.17~20.27 wt.% TFeO. The Mg$^{\#}$ of the samples ranges from 40.08 to 44.64, with an average value of 41.78. The Mg$^{\#}$ is relatively low, indicating that the Sanchakou gabbros were not formed in primary magma. The content of TiO$_2$ is 2.06~2.65 wt.%, which is higher than that of the high-K alkaline and shoshonitic rocks, indicating that the Sanchakou gabbros should originate from the mantle.

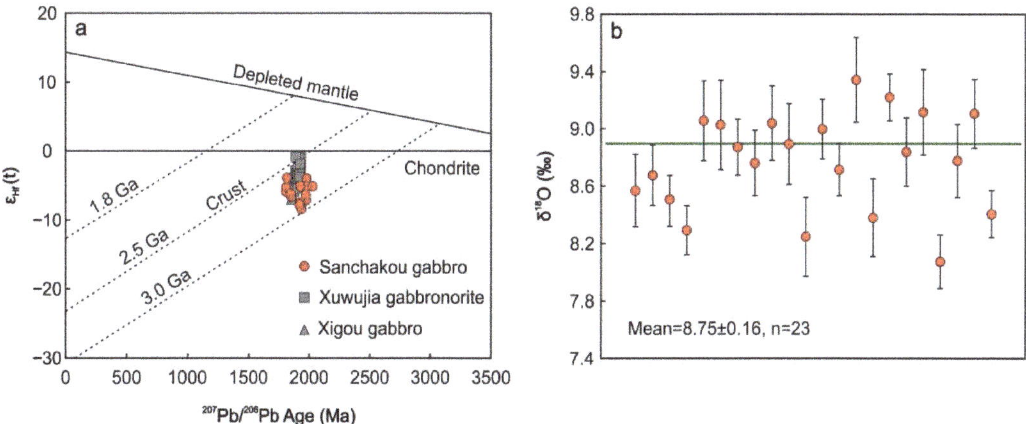

Figure 5. Zircon Hf-O isotopic compositions for the Sanchakou gabbros. (**a**) The ε_{Hf} (t) versus ^{207}Pb/^{206}Pb age diagram; (**b**) The δ^{18}O weighted average value distribution diagram. The Hf isotopic data of Xuwujia gabbronorites and Xigou gabbros are from [35,76].

Table 4. The whole-rock major elemental data (wt.%) of the Sanchakou gabbros.

Sample	SCK-1	SCK-2	SCK-3	SCK-4	SCK-5	SCK-6
SiO$_2$	45.01	44.87	43.89	43.45	43.75	43.50
TiO$_2$	2.06	2.29	2.46	2.65	2.52	2.59
Al$_2$O$_3$	13.64	13.50	13.11	12.73	13.04	13.15
TFeO	17.17	17.87	19.18	20.27	19.20	19.96
MnO	0.26	0.27	0.27	0.28	0.26	0.28
MgO	7.77	7.61	7.68	7.61	7.44	7.56
CaO	11.41	11.09	11.17	10.79	11.38	10.46
Na$_2$O	2.25	2.24	2.12	2.05	2.13	2.17
K$_2$O	0.23	0.17	0.21	0.16	0.19	0.14
P$_2$O$_5$	0.21	0.23	0.25	0.27	0.26	0.27
LOI	−0.14	−0.24	−0.24	−0.33	−0.14	−0.01
Total	99.86	99.90	100.08	99.94	100.02	100.08
Na$_2$O + K$_2$O	2.47	2.41	2.33	2.21	2.32	2.31
K$_2$O/Na$_2$O	0.10	0.08	0.10	0.08	0.09	0.06
Mg$^{\#}$	44.64	43.16	41.66	40.08	40.84	40.32
CaO/Al$_2$O$_3$	0.84	0.82	0.85	0.85	0.87	0.80

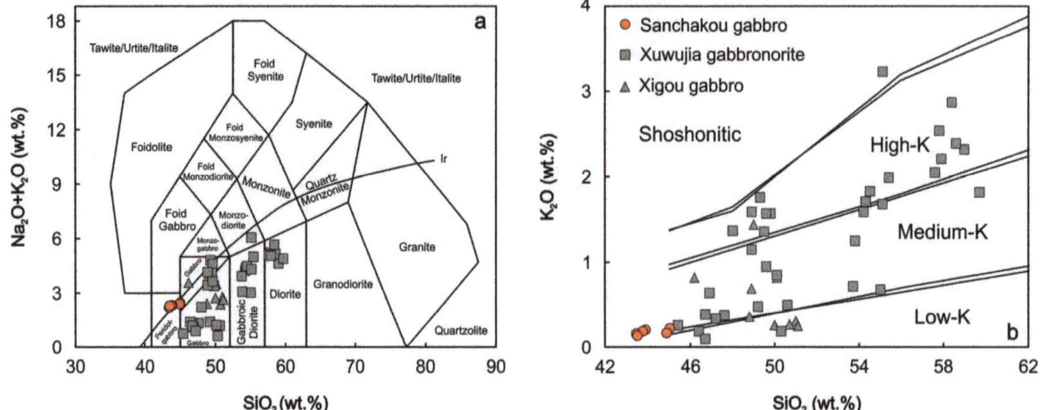

Figure 6. Binary classification diagrams for the Sanchakou gabbros. (**a**) The (Na$_2$O + K$_2$O) versus SiO$_2$ diagram [77]; (**b**) The K$_2$O versus SiO$_2$ diagram [78]. The major elemental data of Xuwujia gabbronorites and Xigou gabbros are from [35].

4.3.2. Trace Elements

The whole-rock trace elemental data of the Sanchakou gabbros are listed in Table 5. In the primitive mantle-normalized trace elements patterns of the Sanchakou gabbros, it is generally shown that Ba, K, Nb, Ta, Sr, and Ti are enriched, while Th and U are depleted (Figure 7a). In the chondrite-normalized rare earth elements (REE) patterns of the Sanchakou gabbros, the curve has a slightly right-leaning trend (Figure 7b). The total amount of REE (ΣREE) is not high, ranging from 58.6×10^{-6} to 77.40×10^{-6}, with an average value of 68.51×10^{-6}. The LREE/HREE ratio is 2.30~2.84, and the (La/Yb)$_N$ ratio is 1.37~1.88, indicating that the LREE and the HREE of the samples are not differentiated. The δEu value is 0.95~1.08, indicating that the samples have no obvious Eu anomaly.

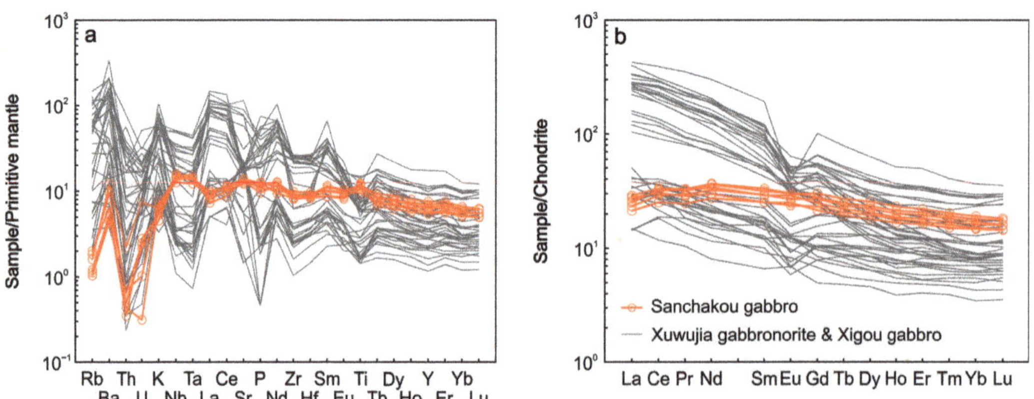

Figure 7. Chondrite-normalized REE (**a**) and primitive mantle-normalized trace element (**b**) patterns for the Sanchakou gabbros. The normalization values are from [79]. The trace elemental data of Xuwujia gabbronorites and Xigou gabbros are from [35].

Table 5. The whole-rock trace elemental data ($\times 10^{-6}$) of the Sanchakou gabbros.

Sample	SCK-1	SCK-2	SCK-3	SCK-4	SCK-5	SCK-6
Rb	1.28	1.17	0.77	0.72	1.02	0.66
Sr	271.97	282.96	258.95	257.62	289.98	287.52
Ba	45.74	67.76	37.23	35.06	84.33	31.27
Th	0.06	0.05	0.04	0.03	0.19	0.04
U	0.02	0.05	0.01	0.06	0.16	0.05
Zr	90.73	101.63	95.27	109.80	101.95	92.60
Hf	2.59	2.67	2.71	2.99	2.79	2.56
Nb	9.02	9.90	10.74	11.30	10.74	10.41
Ta	0.50	0.56	0.58	0.61	0.57	0.58
Sc	47.28	46.05	51.06	50.66	54.06	49.49
La	6.17	6.59	5.95	5.52	6.85	5.03
Ce	19.61	18.62	20.04	17.80	20.48	15.44
Pr	3.06	2.72	3.11	2.80	3.15	2.35
Nd	15.94	14.00	16.83	15.17	17.13	12.80
Sm	4.44	3.85	4.86	4.49	5.09	3.87
Eu	1.68	1.44	1.75	1.56	1.74	1.37
Gd	5.11	4.42	5.78	5.60	6.22	4.79
Tb	0.83	0.73	0.93	0.88	1.01	0.77
Dy	5.24	4.60	5.94	5.65	6.52	4.99
Ho	1.06	0.95	1.21	1.14	1.32	1.02
Er	3.01	2.70	3.42	3.24	3.65	2.82
Tm	0.44	0.39	0.49	0.47	0.52	0.41
Yb	2.75	2.51	3.06	2.90	3.25	2.57
Lu	0.41	0.37	0.45	0.41	0.46	0.37
Y	28.98	24.92	31.89	29.86	34.37	26.19
ΣREE	69.73	63.88	73.81	67.63	77.40	58.60
LREE/HREE	2.70	2.84	2.47	2.33	2.37	2.30
La$_N$/Yb$_N$	1.61	1.88	1.39	1.37	1.51	1.41
δEu	1.08	1.07	1.01	0.95	0.95	0.97
Sc/Y	1.63	1.85	1.60	1.70	1.57	1.89
Eu*/Eu	2.85	2.87	3.04	3.24	3.25	3.16
Sm/Yb	1.62	1.53	1.59	1.55	1.57	1.51
La/Yb	2.24	2.63	1.94	1.90	2.11	1.96
Th/Yb	0.02	0.02	0.01	0.01	0.06	0.02
TiO2/Yb	0.75	0.91	0.80	0.92	0.77	1.01
La/Nb	0.68	0.67	0.55	0.49	0.64	0.48
La/Ba	0.13	0.10	0.16	0.16	0.08	0.16
Th/Yb	0.02	0.02	0.01	0.01	0.06	0.02
Ba/La	7.41	10.28	6.26	6.35	12.31	6.21
Ba/Th	782.96	1243.80	1008.25	1185.70	447.10	739.64
Th/Nb	0.01	0.01	0.00	0.00	0.02	0.00

4.3.3. Sr-Nd Isotopes

The whole-rock Sr-Nd isotopic data of the Sanchakou gabbros are listed in Table 6. The ^{87}Sr/^{86}Sr ratios range from 0.704313 to 0.704882, slightly higher than the mantle value (0.704) and lower than the average value of the continental crust (0.719). The calculated (^{87}Sr/^{86}Sr)$_i$ ratios range from 0.703942 to 0.704702, lower than that of basaltic magma formed by mantle (0.706) and granite formed by partial melting of crust (0.718). The ^{143}Nd/^{144}Nd ratios range from 0.512364 to 0.512573, lower than the modern value of the primitive mantle (0.512638). The $\varepsilon_{Nd}(t)$ values are all positive, ranging from 2.12 to 2.39. The Sm-Nd T_{DM} age ranges from 2535 Ma to 2834 Ma. The Sm-Nd T_{DM2} age ranges from 2204 Ma to 2226 Ma. In the $\varepsilon_{Nd}(t)$-(^{87}Sr/^{86}Sr)$_i$ diagram [80], the Sanchakou gabbros fall into the continental basalt or oceanic island basalt (OIB) field (Figure 8a).

Table 6. The whole-rock Sr-Nd isotopic data of the Sanchakou gabbros.

Sample	87Rb/86Sr	87Sr/86Sr	$\pm 1\sigma$	(87Sr/86Sr)$_i$	147Sm/144Nd	143Nd/144Nd	$\pm 1\sigma$	εNd (t)	TDM (Ma)	TDM2 (Ma)
SCK-1	0.0136	0.704313	0.000005	0.703942	0.1684	0.512389	0.000002	2.12	2563	2226
SCK-2	0.0120	0.704346	0.000004	0.704019	0.1664	0.512364	0.000002	2.13	2535	2225
SCK-3	0.0086	0.704425	0.000005	0.704191	0.1743	0.512474	0.000002	2.33	2619	2209
SCK-4	0.0080	0.704634	0.000004	0.704415	0.1787	0.512528	0.000003	2.28	2716	2213
SCK-5	0.0102	0.704731	0.000005	0.704454	0.1797	0.512545	0.000002	2.39	2713	2204
SCK-6	0.0066	0.704882	0.000004	0.704702	0.1826	0.512573	0.000003	2.21	2834	2219

5. Discussion

5.1. Duration of Magmatism and UHT Metamorphism

The Paleoproterozoic Khondalite Belt is characterized by widespread UHT granulites/gneisses and large-scale magmatic events. Combined with previous studies in the Jining terrane, the results of zircon U-Pb geochronology analysis in UHT granulites/gneisses show the extensive and continuous "beaded" distribution along the concordia line, with an average age of ~1.92 Ga [30,31,44,50]. The occurrence time of UHT metamorphism is still controversial. One view is that it occurred at approximately ~1.92 Ga [52]. The other view is that it occurred before ~1.94 Ga, and then the rock experienced a slow cooling process of ~40 Myr [32]. However, it is generally believed that the cooling duration of UHT metamorphism varies with the tectonic environment. The slow cooling is usually >30 Myr, while the rapid cooling is mostly <10 Myr [81,82]. The "beaded" concordia diagram is more likely to reflect the slow cooling process [29,83–85]. Therefore, the UHT metamorphism in the Jining terrane belongs to the slow cooling and long-term persistent type, which is similar to that in the Rogaland area of Norway [84] and the Napier complex in East Antarctica [83]. Through the zircon U-Pb geochronology analysis of the Sanchakou gabbros, we found that the magmatism occurred from ~1.94 Ga to ~1.82 Ga, with a duration over 100 Ma and a peak time of ~1.94 Ga. It is coupled with the peak period and slow cooling process of UHT metamorphism in the Jining terrane. Therefore, the magma of these basic intrusive rocks may be the heat source for the UHT metamorphism [30,35,52,86].

5.2. Genesis of Layered Complex

In previous studies, the layered complex is always considered to be the magmatic product undergone a special process of fractional crystallization. In the closed magmatic system, the layered complex can be formed by the magmatic differentiation with only one large-scale magma injection and little or no magma replenishment, such as the Skaergaard complex in Greenland [87]. In the open magmatic system, the layered complex can be formed by magmatic differentiation after the multiple magma injection, such as the Muskox complex in Canada [88], the Bushveld complex in South Africa [89], the Stillwater complex in America [90], and the Rum complex in Britain [91]. However, the fractional crystallization process is usually accompanied by the contamination of crustal wallrocks in many cases [92,93]. Crustal contamination can change the compositions of magmatic melts [94,95]. Here, we discuss the genesis of the Sanchakou gabbros.

5.2.1. Crustal Contamination and Fractional Crystallization

We often regard "Assimilation and Fractional Crystallization (AFC)" as a significant process during the magmatic evolution, which can modify the geochemical compositions of the initial magma [94–96]. It does not probably occur in the shallow crust, because there is not enough energy accumulation [97]. The binary diagrams of $\varepsilon_{Nd}(t)$, $Mg^{\#}$ versus SiO_2 show roughly negative correlations, indicating the Sanchakou gabbros are formed in the AFC process and similar to experimental results from peridotite melts (Figure 8b,c). The occurrence of the AFC process is supported by the 2225~2375 Ma inherited zircon in CL images. The $Mg^{\#}$ is often taken to reveal the process of fractional crystallization. There is a positive correlation between CaO/Al_2O_3 ratio and $Mg^{\#}$, further suggesting that the clinopyroxene is probably the main fractionating mineral phase (Figure 8d). The Sc/Y ratio is usually controlled by clinopyroxene crystallization [98]. The decreasing Sc/Y ratio with the decreasing $Mg^\#$ also indicates the crystallization of clinopyroxene (Figure 8e). The plagioclase does not play a vital role during the magmatic evolution, as shown by the nearly constant Eu^*/Eu with the decreasing $Mg^\#$ (Figure 8f).

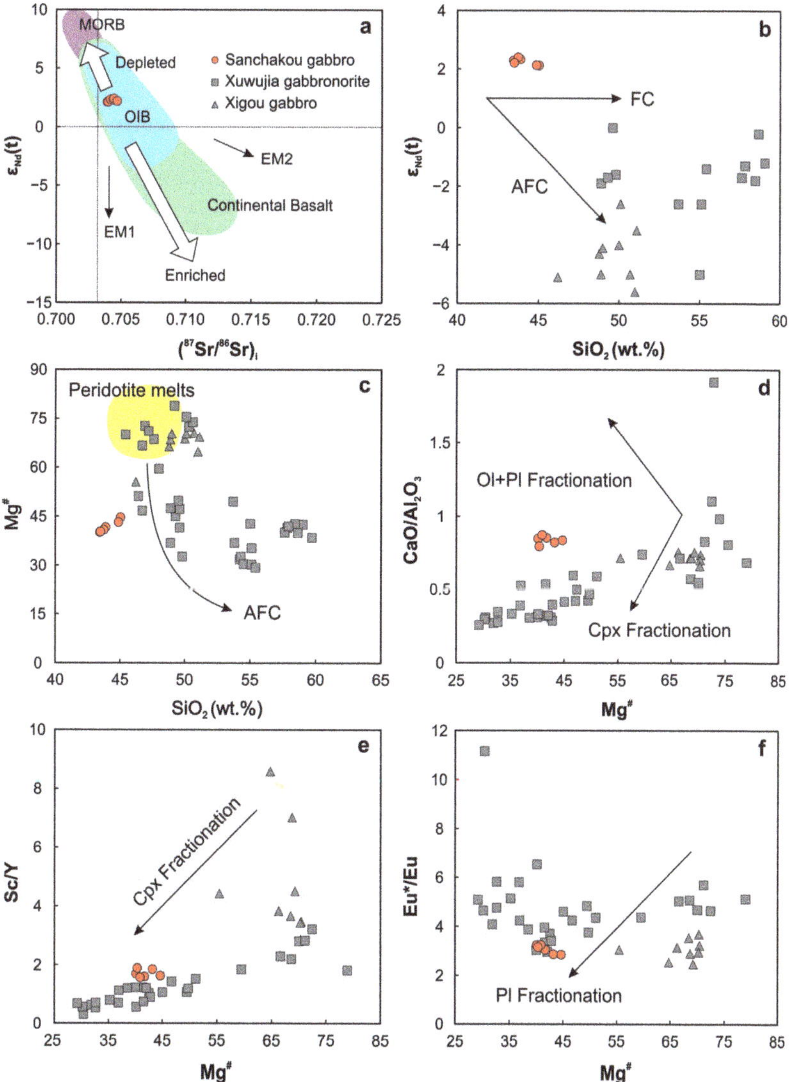

Figure 8. Geochemical diagrams of the Sanchakou gabbros. (**a**) The $\varepsilon_{Nd}(t)$ versus I_{Sr} diagram (after [80]). (**b**) The $\varepsilon_{Nd}(t)$ versus SiO_2 diagram. (**c**) The $Mg^\#$ versus SiO_2 diagram. Supposed peridotite melts and crust AFC curve are after [99]. (**d**) The CaO/Al_2O_3 versus $Mg^\#$ diagram. (**e**) The Sc/Y versus $Mg^\#$ diagram. (**f**) The Eu^*/Eu versus $Mg^\#$ diagram. Mineral abbreviations: Cpx, clinopyroxene; Ol, olivine; Pl, plagioclase. The major and trace elemental data of Xuwujia gabbronorites and Xigou gabbros are from [35].

5.2.2. Magmatic Source and Fluid Metasomatism

The Sr-Nd and Hf-O isotopic systems are used to determine the source of magmatic rocks. The Sanchakou gabbros have a positive whole-rock $\varepsilon_{Nd}(t)$ value (2.12~2.39) and a negative zircon $\varepsilon_{Hf}(t)$ value (-8.4~-3.8), reflecting the characteristics of Nd-Hf isotopic decoupling. The $\varepsilon_{Nd}(t)$-$(^{87}Sr/^{86}Sr)_i$ diagram [80] of Sanchakou gabbros indicates that the magma source may be a depleted mantle (Figure 8a). In the $\delta^{18}O$-εHf (Paleoproterozoic) diagram [100], most concordant zircons fall in the mixing line between the 3.5 Ga supracrustal component and the depleted mantle with Hf concentration ratios (Hf_{pm}/Hf_c) of 0.7 and 1.5 (Figure 9), indicating that the magma source may be enriched mantle or suffered from the crustal contamination [74]. In the process of mantle evolution, Nd and Hf are more likely to enter the melts than their parent isotopes, but the mobility of Nd is higher than that of Hf [101]. Lu-Hf isotope system is mainly controlled by Hf-rich minerals such as zircon, apatite, and garnet, and the Nd-Hf isotopic decoupling can occur during partial melting [102,103]. Due to the difference in partial melting conditions and the duration of magmatic events, it is difficult to achieve the isotopic equilibrium, resulting in the Nd-Hf isotopic decoupling between the melt and the source region [104]. However, the Sm-Nd T_{DM} age (2535~2834 Ma) and the Lu-Hf T_{DM} age (2444~2609 Ma) are in a similar range, which is older than the formation age of Sanchakou gabbros. Therefore, it is very likely that the Nd-Hf isotopic decoupling occurred during mantle-derived magma mixing with crustal wallrocks in the magma chamber.

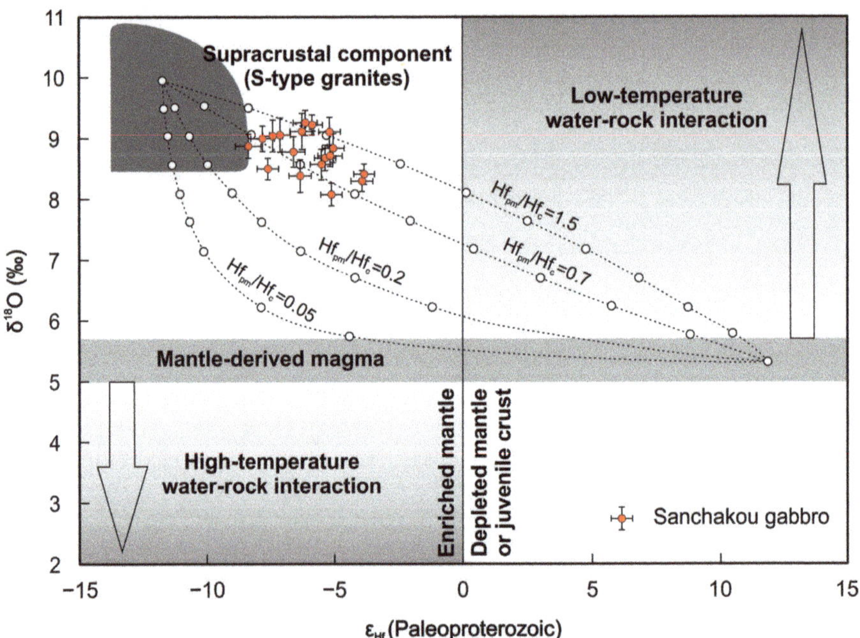

Figure 9. The zircon $\delta^{18}O$ versus $\varepsilon_{Hf}(t)$ diagram of the Sanchakou gabbros. The dotted lines indicate the two-component mixing trends between the depleted mantle and the supracrust-derived magma. We assume that the supracrustal zircons have $\varepsilon_{Hf}(t) = -12$ and $\delta^{18}O = 10‰$, the depleted mantle zircons have $\varepsilon_{Hf}(t) = 12$ and $\delta^{18}O = 5.3‰$ [105]. Hf_{pm}/Hf_c is the Hf concentration ratio between the parental mantle-derived magma and supracrustal components indicated for each curve. The small circles on the curves indicate 10% mixing increments. The zircon Hf-O isotope compositions of the supracrustal component (S-type granites) are from [105], and those of the depleted mantle are from [75].

The melting depths of the layered complex can be modeled with related trace elements [106,107]. REE ratios and abundances (e.g., Sm/Yb, and La/Yb and Sm) are widely used to determine the origin of magma and the melting degree of mantle [107–110]. The Sanchakou gabbros plot near the spinel + garnet lherzolite melting curves with primitive mantle (PM) starting compositions. The Sm/Yb ratios are lower than the garnet lherzolite melting curves and higher than the spinel lherzolite melting curves (Figure 10a,b). The parental magma is likely to be derived from a mantle source consisting of spinel + garnet lherzolite. Additionally, approximately 10%~20% partial melting of the lherzolites is required.

The Sanchakou gabbros have a relatively lower Th/Yb ratio and higher TiO_2/Yb ratio, and all of the points fall outside the MORB-OIB array (Figure 10c,d), similar to subduction-related enrichment [111]. All of the samples exhibit a relatively high La/Nb ratio (0.48~33.67) and low La/Ba ratio (0.01~0.16), derived from a similar modified continental lithospheric mantle (CLM) source [112]. The basic magma formed by the partial melting of mantle peridotite that has interacted with fluids usually has relatively high Na_2O and P_2O_5 contents, positive to weakly negative Nb anomalies, and non-negative Ti anomalies relative to the PM [113,114]. It is similar to the characteristics of Sanchakou gabbros, thus supporting the interaction between the CLM and the fluids.

The trace element ratios (e.g., Th/Yb, Ba/La, Ba/Th, and Th/Nb) are widely used to distinguish whether metasomatic agents belong to fluids or sediments [115,116]. All the samples have variable Ba/La (6.21~76.85) and Ba/Th (39.50~18,537.50) ratios but relatively constant Th/Yb (0.01~6.59) and Th/Nb (0.00~2.51) ratios (Figure 10e,f), which can be considered as the addition of fluids into the mantle source [116–118]. The highly variable Sr/Nd (2.40~127.39) ratio and relatively low Th/Yb (0.01~6.59) ratio further suggest that the fluids could be derived from the large-scale melting crust [119].

In summary, we believe that crustal contamination plays an important role in the formation of Sanchakou gabbros. The magma originates from the 10%~20% partial melting of the spinel + garnet lherzolite mantle. The Sanchakou gabbros are magmatic crystallization products mixed with crustal wallrocks in the magma chamber.

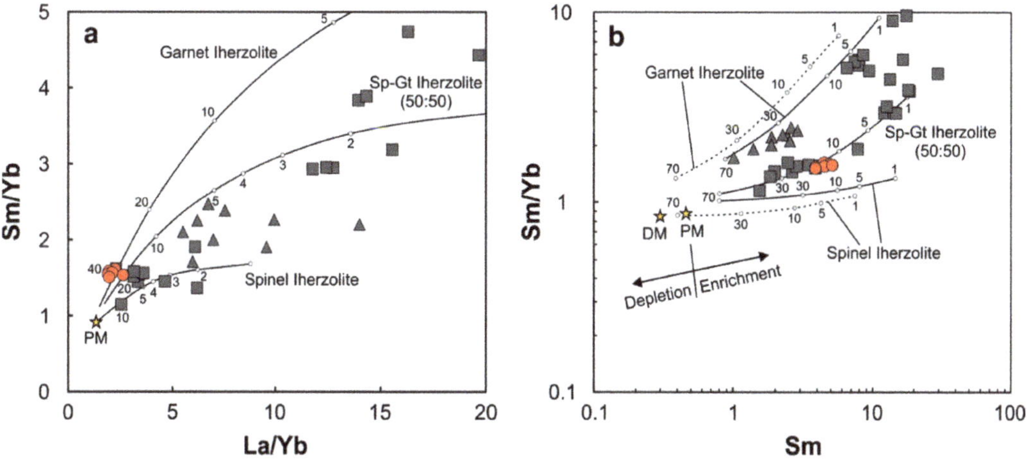

Figure 10. *Cont.*

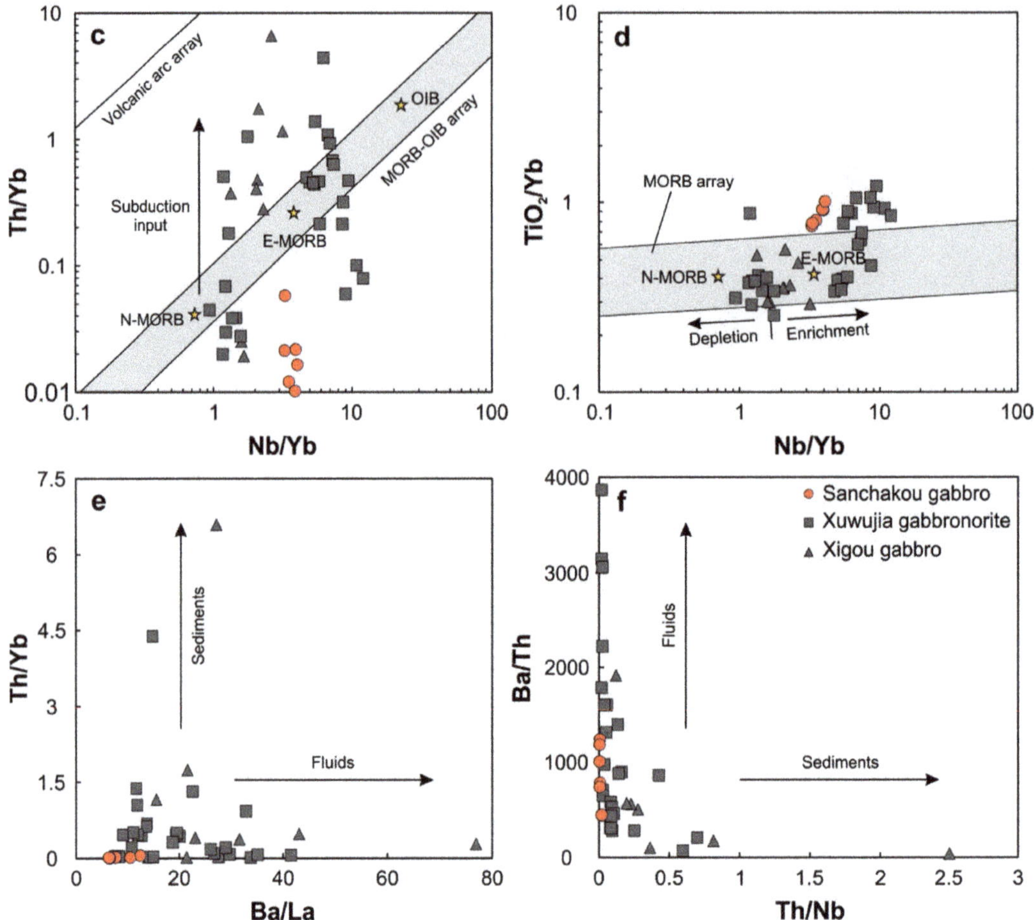

Figure 10. Geochemical diagrams of the Sanchakou gabbros. (**a**) The Sm/Yb versus La/Yb diagram (after [120]). (**b**) The Sm/Yb versus Sm diagram (after [107]). Solid and dashed curves are the melting trends from DM (depleted MORB) and PM (primitive mantle). (**c**) The Th/Yb versus Nb/Yb diagram (after [121]). (**d**) The TiO₂/Yb versus Nb/Yb diagram (after [121]). (**e**) The Th/Yb versus Ba/La diagram (after [116]). (**f**) The Ba/Th versus Th/Nb diagram (after [115]). The major and trace elemental data of Xuwujia gabbronorites and Xigou gabbros are from [35].

5.3. Tectonic Implications

It is generally believed that the Khondalite Belt is the tectonic amalgamation belt of the Ordos and Yinshan Blocks at ~1.95 Ga, which is a typical Paleoproterozoic continent-continent collisional belt [20,43]. This view is supported by the Precambrian granulites/gneisses, but it is still unclear whether it is consistent with the magmatic rocks. A few studies have focused on 2.45~2.10 Ga, 1.97~1.92 Ga, and 1.85~1.84 Ga basic intrusive rocks in the Jining terrane, indicating that they were involved in the subduction and collision processes during the formation of the Khondalite Belt [34–37]. However, the detailed tectonic environment is still controversial.

For the tectonic environment of Jining terrane at 1.95~1.82 Ga, researchers have proposed a variety of models, including mantle plume events [21], mid-ocean ridge subduction [22,35,86], and post-collision mantle upwelling [30,122]. The mid-ocean ridge subduction usually forms double metamorphic zones and adakitic rocks [123,124], which have not been found in the Jining terrane. The post-collision mantle upwelling model also seems to be insufficient, because the duration of UHT metamorphism formed under this condition is generally within ~30 Myr [81,82], and the upper-temperature limit is usually less than 1000 °C [125,126], which contradicts the fact that extremely high-temperature metamorphic rocks are exposed in the Jining terrane [32]. In addition, the back-arc basin has twice the heat flow value compared with the normal craton, which is also an ideal environment for UHT metamorphism [127,128]. The geothermal gradient is only 20~25 °C/km [127,129], and high-grade metamorphic rocks usually have anticlockwise P-T paths [82,130]. However, it is contrary to the fact that most UHT granulites/gneisses reported in the Jining terrane have clockwise P-T paths [20,43]. Combined with previous studies, the layered complex often forms in the tectonic environment associated with mantle plume or intracontinental rifting, such as the Stillwater complex, the Duluth complex in America, and the Muskox complex in Canada [10–13]. Only a few form in the late-orogenic or post-orogenic extensional environment, such as the Bjerkreim-Sokndal complex and the Fongen-Hyllingen complex in Norway [131,132]. For basic intrusive rocks in the Jining terrane, the magma emplacement temperature is as high as ~1400 °C, and the mantle potential temperature is about ~1550 °C [35] which is slightly higher than that of the Paleoproterozoic mantle (~1500 °C) [133–135]. It is probably caused by the upwelling asthenosphere. The duration of UHT metamorphism depends on the duration of asthenosphere upwelling.

Therefore, we produce a hypothetical tectonic framework for the Khondalite Belt to explain the tectonic evolution at 1.95~1.82 Ga. With the end of the amalgamation between the Ordos and Yinshan Blocks at ~1.95 Ga [20,43], the asthenosphere upwelling resulted in large-scale crustal melting and long-term magmatism. The magma assimilated crustal wallrocks, and then fractional crystallized to form the layered complex (Figure 11), in which several fluids of granulites/gneisses were mixed. Meanwhile, the surrounding granulites/gneisses were heated to form the UHT metamorphism.

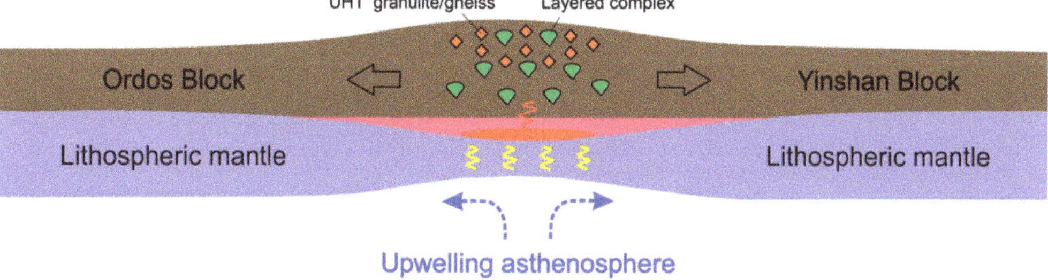

Figure 11. Cartoons showing the tectonic evolution of the Khondalite Belt between ~1.95 Ga and ~1.82 Ga.

6. Concluding Remarks

Based on the in situ zircon U-Pb isotopic analysis of the Sanchakou gabbros, we have found that they have experienced a slow cooling process from ~1.94 Ga to ~1.82 Ga, with a weighted average age of 1923 ± 28 Ma. Combined with the study of elemental and Hf-O-Sr-Nd isotopic analysis, we believe that they are the crystallization products of assimilating crustal wallrocks after 10%~20% partial melting of spinel + garnet lherzolite mantle, probably formed by the asthenosphere upwelling after the amalgamation of the Ordos and Yinshan Blocks.

Author Contributions: Conceptualization, W.-P.Z. and W.T.; methodology, W.-P.Z.; software, W.-P.Z.; validation, W.-P.Z., W.T. and B.W.; formal analysis, W.-P.Z.; investigation, W.-P.Z., Y.-H.Z. and C.-J.W.; resources, Y.-H.Z.; data curation, W.-P.Z. and W.T.; writing—original draft preparation, W.-P.Z.; writing—review and editing, W.T. and C.-J.W.; visualization, B.W.; supervision, W.T.; project administration, W.T.; funding acquisition, W.T., Y.-H.Z. and C.-J.W. All authors have read and agreed to the published version of the manuscript.

Funding: This research was funded by the National Natural Science Foundation of China (Grant No. 42030304, 42202047), the National Key Research & Development Program of China (Grant No. 2018YFE0204202, 2017YFC0601302), and the China Geological Survey (Grant No. DD20221649, DD20221647).

Data Availability Statement: Not applicable.

Acknowledgments: Constructive suggestions and comments from anonymous reviewers led to great improvements in the quality of this article.

Conflicts of Interest: The authors declare no conflict of interest.

References

1. Wager, L.R.; Brown, G.M. *Layered Igneous Rocks*; Oliver and Boyd: Edinburgh, UK; London, UK, 1968.
2. Hoatson, D.M.; Sun, S.S. Archean layered mafic-ultramafic intrusions in the West Pilbara Craton, Western Australia: A synthesis of some of the oldest orthomagmatic mineralizing systems in the world. *Econ. Geol.* **2002**, *97*, 847–872. [CrossRef]
3. Ripley, E.M.; Shafer, P.; Li, C.; Hauck, S.A. Re-Os and O isotopic variations in magnetite from the contact zone of the Duluth Complex and the Biwabik Iron Formation, northeastern Minnesota. *Chem. Geol.* **2008**, *249*, 213–226. [CrossRef]
4. Roelofse, F.; Ashwal, L.D. The Lower Main Zone in the Northern Limb of the Bushveld Complex—A >1·3 km thick sequence of intruded and variably contaminated crystal mushes. *J. Petrol.* **2012**, *53*, 1449–1476. [CrossRef]
5. Maier, W.D.; Halkoaho, T.; Huhma, H.; Hanski, E.; Barnes, S.J. The Penikat intrusion, Finland: Geochemistry, geochronology, and origin of platinum-palladium reefs. *J. Petrol.* **2018**, *59*, 967–1006. [CrossRef]
6. Wall, C.J.; Scoates, J.S.; Weis, D.; Friedman, R.M.; Amini, M.; Meurer, W.P. The Stillwater Complex: Integrating zircon geochronological and geochemical constraints on the age, emplacement history and crystallization of a large, open-system layered intrusion. *J. Petrol.* **2018**, *59*, 153–190. [CrossRef]
7. Zhou, M.F.; Arndt, N.T.; Malpas, J.; Wang, C.Y.; Kennedy, A.K. Two magma series and associated ore deposit types in the Permian Emeishan large igneous province, SW China. *Lithos* **2008**, *103*, 352–368. [CrossRef]
8. Zhong, H.; Campbell, I.H.; Zhu, W.G.; Allen, C.M.; Hu, R.Z.; Xie, L.W.; He, D.F. Timing and source constraints on the relationship between mafic and felsic intrusions in the Emeishan large igneous province. *Geochim. Cosmochim. Acta* **2011**, *75*, 1374–1395. [CrossRef]
9. Wotzlaw, J.F.; Bindeman, I.N.; Schaltegger, U.; Brooks, C.K.; Naslund, H.R. High-resolution insights into episodes of crystallization, hydrothermal alteration and remelting in the Skaergaard intrusive complex. *Earth Planet. Sci. Lett.* **2012**, *355*, 199–212. [CrossRef]
10. Nicholson, S.W.; Shirey, S.B. Midcontinent rift volcanism in the Lake Superior region: Sr, Nd, and Pb isotopic evidence for a mantle plume origin. *J. Geophys. Res. Solid Earth* **1990**, *95*, 10851–10868. [CrossRef]
11. Paces, J.B.; Miller, J.D., Jr. Precise U-Pb ages of Duluth complex and related mafic intrusions, northeastern Minnesota: Geochronological insights to physical, petrogenetic, paleomagnetic, and tectonomagmatic processes associated with the 1.1 Ga midcontinent rift system. *J. Geophys. Res. Solid Earth* **1993**, *98*, 13997–14013. [CrossRef]
12. Mackie, R.A.; Scoates, J.S.; Weis, D. Age and Nd-Hf isotopic constraints on the origin of marginal rocks from the Muskox layered intrusion (Nunavut, Canada) and implications for the evolution of the 1.27 Ga Mackenzie large igneous province. *Precambrian Res.* **2009**, *172*, 46–66. [CrossRef]
13. Bai, Y.; Su, B.X.; Xiao, Y.; Chen, C.; Cui, M.M.; He, X.Q.; Qin, L.P.; Charlier, B. Diffusion-driven chromium isotope fractionation in ultramafic cumulate minerals: Elemental and isotopic evidence from the Stillwater Complex. *Geochim. Cosmochim. Acta* **2019**, *263*, 167–181. [CrossRef]
14. Bowen, N.L. *The Evolution of Igneous Rocks*; Princeton University Press: Princeton, NJ, USA, 1928.
15. Reiners, P.W.; Nelson, B.K.; Ghiorso, M.S. Assimilation of felsic crust by basaltic magma: Thermal limits and extents of crustal contamination of mantle-derived magmas. *Geology* **1995**, *23*, 563–566. [CrossRef]
16. Anderson, A.T., Jr. Parental basalts in subduction zones: Implications for continental evolution. *J. Geophys. Res. Solid Earth* **1982**, *87*, 7047–7060. [CrossRef]
17. Taylor, S.R.; McLennan, S.M. The geochemical evolution of the continental crust. *Rev. Geophys.* **1995**, *33*, 241–265. [CrossRef]
18. Kuşcu, G.G.; Floyd, P.A. Mineral compositional and textural evidence for magma mingling in the Saraykent volcanics. *Lithos* **2001**, *56*, 207–230. [CrossRef]
19. Arvin, M.; Dargahi, S.; Babaei, A.A. Mafic microgranular enclave swarms in the Chenar granitoid stock, NW of Kerman, Iran: Evidence for magma mingling. *J. Asian Earth Sci.* **2004**, *24*, 105–113. [CrossRef]

20. Zhao, G.C.; Sun, M.; Wilde, S.A.; Li, S.Z. Late Archean to Paleoproterozoic evolution of the North China Craton: Key issues revisited. *Precambrian Res.* **2005**, *136*, 177–202. [CrossRef]

21. Santosh, M.; Tsunogae, T.; Ohyama, H.; Sato, K.; Li, J.H.; Liu, S.J. Carbonic metamorphism at ultrahigh-temperatures: Evidence from North China Craton. *Earth Planet. Sci. Lett.* **2008**, *266*, 149–165. [CrossRef]

22. Santosh, M.; Liu, S.J.; Tsunogae, T.; Li, J.H. Paleoproterozoic ultrahigh-temperature granulites in the North China Craton: Implications for tectonic models on extreme crustal metamorphism. *Precambrian Res.* **2012**, *223*, 77–106. [CrossRef]

23. Liu, S.J.; Li, J.H.; Santosh, M. First application of the revised Ti-in-zircon geothermometer to Paleoproterozoic ultrahigh-temperature granulites of Tuguiwula, Inner Mongolia, North China Craton. *Contrib. Mineral. Petrol.* **2010**, *159*, 225–235. [CrossRef]

24. Liu, S.J.; Xiang, B.; Li, J.H.; Santosh, M. Retrograde metamorphism of ultrahigh-temperature granulites from the Khondalite belt in Inner Mongolia, North China Craton: Evidence from aluminous orthopyroxenes. *Geol. J.* **2011**, *46*, 263–275. [CrossRef]

25. Liu, S.J.; Tsunogae, T.; Li, W.S.; Shimizu, H.; Santosh, M.; Wan, Y.S.; Li, J.H. Paleoproterozoic granulites from Heling'er: Implications for regional ultrahigh-temperature metamorphism in the North China Craton. *Lithos* **2012**, *148*, 54–70. [CrossRef]

26. Zhang, H.T.; Li, J.H.; Liu, S.J.; Li, W.S.; Santosh, M.; Wang, H.H. Spinel + quartz-bearing ultrahigh-temperature granulites from Xumayao, Inner Mongolia Suture Zone, North China Craton: Petrology, phase equilibria and counterclockwise P-T path. *Geosci. Front.* **2012**, *3*, 603–611. [CrossRef]

27. Cai, J.; Liu, F.L.; Liu, P.H.; Shi, J.R. Metamorphic P-T conditions and U-Pb dating of the sillimanite-cordierite-garnet paragneisses in Sanchakou, Jining area, Inner Mongolia. *Acta Petrol. Sin.* **2014**, *30*, 472–490. (In Chinese with English Abstract)

28. Yang, Q.Y.; Santosh, M.; Tsunogae, T. Ultrahigh-temperature metamorphism under isobaric heating: New evidence from the North China Craton. *J. Asian Earth Sci.* **2014**, *95*, 2–16. [CrossRef]

29. Li, X.W.; Wei, C.J. Phase equilibria modelling and zircon age dating of pelitic granulites in Zhaojiayao, from the Jining Group of the Khondalite Belt, North China Craton. *J. Metamorph. Geol.* **2016**, *34*, 595–615. [CrossRef]

30. Li, X.W.; Wei, C.J. Ultrahigh-temperature metamorphism in the Tuguiwula area, Khondalite Belt, North China Craton. *J. Metamorph. Geol.* **2018**, *36*, 489–509. [CrossRef]

31. Li, X.W.; White, R.W.; Wei, C.J. Can we extract ultrahigh-temperature conditions from Fe-rich metapelites? An example from the Khondalite Belt, North China Craton. *Lithos* **2019**, *328*, 228–243. [CrossRef]

32. Wang, B.; Wei, C.J.; Tian, W.; Fu, B. UHT metamorphism peaking above 1100 °C with slow cooling: Insights from pelitic granulites in the Jining complex, North China Craton. *J. Petrol.* **2020**, *61*, egaa070. [CrossRef]

33. Cawthorn, R.G. *Layered Intrusions*; Elsevier Science: Amsterdam, The Netherlands, 1996.

34. Guo, J.H.; Shi, X.; Bian, A.G.; Xu, R.H.; Zhai, M.G.; Li, Y.G. Pb isotopic composition of feldspar and U-Pb age of zircon from early Proterozoic granite in Sanggan area, North China Craton: Metamorphism, crustal melting and tectono-thermal event. *Acta Petrol. Sin.* **1999**, *15*, 199–207. (In Chinese with English Abstract)

35. Peng, P.; Guo, J.H.; Zhai, M.G.; Bleeker, W. Paleoproterozoic gabbronoritic and granitic magmatism in the northern margin of the North China Craton: Evidence of crust-mantle interaction. *Precambrian Res.* **2010**, *183*, 635–659. [CrossRef]

36. Wan, Y.S.; Xu, Z.Y.; Dong, C.Y.; Nutman, A.P.; Ma, M.; Xie, H.; Cu, H. Episodic Paleoproterozoic (~2.45, ~1.95 and ~1.85 Ga) mafic magmatism and associated high temperature metamorphism in the Daqingshan area, North China Craton: SHRIMP zircon U-Pb dating and whole-rock geochemistry. *Precambrian Res.* **2013**, *224*, 71–93. [CrossRef]

37. Liu, P.H.; Liu, F.L.; Liu, C.H.; Liu, J.H.; Wang, F.; Xiao, L.; Shi, J. Multiple mafic magmatic and high-grade metamorphic events revealed by zircons from meta-mafic rocks in the Daqingshan-Wulashan Complex of the Khondalite Belt, North China Craton. *Precambrian Res.* **2014**, *246*, 334–357. [CrossRef]

38. Liu, D.Y.; Nutman, A.P.; Compston, W.; Wu, J.S.; Shen, Q.H. Remnants of ≥3800 Ma crust in the Chinese part of the Sino-Korean craton. *Geology* **1992**, *20*, 339–342. [CrossRef]

39. Song, B.; Nutman, A.P.; Liu, D.; Wu, J. 3800 to 2500 Ma crustal evolution in the Anshan area of Liaoning Province, northeastern China. *Precambrian Res.* **1996**, *78*, 79–94. [CrossRef]

40. Wu, F.Y.; Zhang, Y.B.; Yang, J.H.; Xie, L.W.; Yang, Y.H. Zircon U-Pb and Hf isotopic constraints on the Early Archean crustal evolution in Anshan of the North China Craton. *Precambrian Res.* **2008**, *167*, 339–362. [CrossRef]

41. Zhai, M.G.; Santosh, M. The early Precambrian odyssey of the North China Craton: A synoptic overview. *Gondwana Res.* **2011**, *20*, 6–25. [CrossRef]

42. Kusky, T.M.; Polat, A.; Windley, B.F.; Burke, K.C.; Dewey, J.F.; Kidd, W.S.F.; Maruyama, S.; Wang, J.P.; Deng, H.; Wang, Z.S.; et al. Insights into the tectonic evolution of the North China Craton through comparative tectonic analysis: A record of outward growth of Precambrian continents. *Earth-Sci. Rev.* **2016**, *162*, 387–432. [CrossRef]

43. Zhao, G.C.; Cawood, P.A.; Li, S.Z.; Wilde, S.A.; Sun, M.; Zhang, J.; He, Y.H.; Yin, C.Q. Amalgamation of the North China Craton: Key issues and discussion. *Precambrian Res.* **2012**, *222*, 55–76. [CrossRef]

44. Santosh, M.; Wilde, S.A.; Li, J.H. Timing of Paleoproterozoic ultrahigh-temperature metamorphism in the North China Craton: Evidence from SHRIMP U-Pb zircon geochronology. *Precambrian Res.* **2007**, *159*, 178–196. [CrossRef]

45. Santosh, M.; Tsunogae, T.; Li, J.H.; Liu, S.J. Discovery of sapphirine-bearing Mg-Al granulites in the North China Craton: Implications for Paleoproterozoic ultrahigh temperature metamorphism. *Gondwana Res.* **2007**, *11*, 263–285. [CrossRef]

46. Santosh, M.; Sajeev, K.; Li, J.H.; Liu, S.J.; Itaya, T. Counterclockwise exhumation of a hot orogen: The Paleoproterozoic ultrahigh-temperature granulites in the North China Craton. *Lithos* **2009**, *110*, 140–152. [CrossRef]

47. Jiao, S.J.; Guo, J.H. Application of the two-feldspar geothermometer to ultrahigh-temperature (UHT) rocks in the Khondalite Belt, North China Craton and its implications. *Am. Mineral.* **2011**, *96*, 250–260. [CrossRef]

48. Jiao, S.J.; Guo, J.H.; Qian, M.; Zhao, R. Application of Zr-in-rutile thermometry: A case study from ultrahigh-temperature granulites of the Khondalite belt, North China Craton. *Contrib. Mineral. Petrol.* **2011**, *162*, 379–393. [CrossRef]

49. Shimizu, H.; Tsunogae, T.; Santosh, M.; Liu, S.J.; Li, J.H. Phase equilibrium modelling of Palaeoproterozoic ultrahigh temperature sapphirine granulite from the Inner Mongolia suture zone, North China Craton: Implications for counterclockwise P-T path. *Geol. J.* **2013**, *48*, 456–466. [CrossRef]

50. Lobjoie, C.; Wei, L.; Trap, P.; Goncalves, P.; Li, Q.; Marquer, D.; Devoir, A. Ultra-high temperature metamorphism recorded in Fe-rich olivine-bearing migmatite from the Khondalite belt, North China Craton. *J. Metamorph. Geol.* **2018**, *36*, 343–368. [CrossRef]

51. Wang, L.J.; Guo, J.H.; Yin, C.Q.; Peng, P.; Zhang, J.; Spencer, C.J.; Qian, J.H. High-temperature S-type granitoids (charnockites) in the Jining complex, North China Craton: Restite entrainment and hybridization with mafic magma. *Lithos* **2018**, *320*, 435–453. [CrossRef]

52. Huang, G.Y.; Guo, J.H.; Jiao, S.J.; Palin, R.M. What drives the continental crust to be extremely hot so quickly? *J. Geophys. Res. Solid Earth* **2019**, *124*, 11218–11231. [CrossRef]

53. Li, X.L.; Zhang, L.F.; Wei, C.J.; Slabunov, A.I.; Bader, T. Quartz and orthopyroxene exsolution lamellae in clinopyroxene and the metamorphic P-T path of Belomorian eclogites. *J. Metamorph. Geol.* **2018**, *36*, 1–12. [CrossRef]

54. Zong, K.Q.; Klemd, R.; Yuan, Y.; He, Z.Y.; Guo, J.L.; Shi, X.L.; Liu, Y.S.; Hu, Z.C.; Zhang, Z.M. The assembly of Rodinia: The correlation of early Neoproterozoic (ca. 900 Ma) high-grade metamorphism and continental arc formation in the southern Beishan Orogen, southern Central Asian Orogenic Belt (CAOB). *Precambrian Res.* **2017**, *290*, 32–48. [CrossRef]

55. Hu, Z.C.; Zhang, W.; Liu, Y.S.; Gao, S.; Li, M.; Zong, K.Q.; Chen, H.H.; Hu, S.H. "Wave" signal-smoothing and mercury-removing device for laser ablation quadrupole and multiple collector ICPMS analysis: Application to lead isotope analysis. *Anal. Chem.* **2015**, *87*, 1152–1157. [CrossRef]

56. Liu, Y.S.; Hu, Z.C.; Gao, S.; Günther, D.; Xu, J.; Gao, C.G.; Chen, H.H. In situ analysis of major and trace elements of anhydrous minerals by LA-ICP-MS without applying an internal standard. *Chem. Geol.* **2008**, *257*, 34–43. [CrossRef]

57. Liu, Y.S.; Gao, S.; Hu, Z.C.; Gao, C.G.; Zong, K.Q.; Wang, D.B. Continental and oceanic crust recycling-induced melt-peridotite interactions in the Trans-North China Orogen: U-Pb dating, Hf isotopes and trace elements in zircons of mantle xenoliths. *J. Petrol.* **2010**, *51*, 537–571. [CrossRef]

58. Ludwig, K.R. *ISOPLOT 3.00: A Geochronological Toolkit for Microsoft Excel*; Berkeley Geochronology Center: Berkeley, CA, USA, 2003; p. 39.

59. Jochum, K.P.; Nohl, U. Reference materials in geochemistry and environmental research and the GeoReM database. *Chem. Geol.* **2008**, *253*, 50–53. [CrossRef]

60. Chen, S.; Wang, X.H.; Niu, Y.L.; Sun, P.; Duan, M.; Xiao, Y.Y.; Guo, P.Y.; Gong, H.M.; Wang, G.D.; Xue, Q.Q. Simple and cost-effective methods for precise analysis of trace element abundances in geological materials with ICP-MS. *Sci. Bull.* **2017**, *62*, 277–289. [CrossRef] [PubMed]

61. Hu, Z.C.; Liu, Y.S.; Gao, S.; Liu, W.; Yang, L.; Zhang, W.; Tong, X.; Lin, L.; Zong, K.Q.; Li, M.; et al. Improved in situ Hf isotope ratio analysis of zircon using newly designed X skimmer cone and Jet sample cone in combination with the addition of nitrogen by laser ablation multiple collector ICP-MS. *J. Anal. At. Spectrom.* **2012**, *27*, 1391–1399. [CrossRef]

62. Woodhead, J.; Hergt, J.; Shelley, M.; Eggins, S.; Kemp, R. Zircon Hf-isotope analysis with an excimer laser, depth profiling, ablation of complex geometries, and concomitant age estimation. *Chem. Geol.* **2004**, *209*, 121–135. [CrossRef]

63. Zhang, W.; Hu, Z.C. Estimation of isotopic reference values for pure materials and geological reference materials. *At. Spectrosc.* **2020**, *41*, 93–102. [CrossRef]

64. Ickert, R.B.; Hiess, J.; Williams, I.S.; Holden, P.; Ireland, T.R.; Lanc, P.; Schram, N.; Foster, J.J.; Clement, S.W. Determining high precision, in situ, oxygen isotope ratios with a SHRIMP II: Analyses of MPI-DING silicate-glass reference materials and zircon from contrasting granites. *Chem. Geol.* **2008**, *257*, 114–128. [CrossRef]

65. Dong, C.Y.; Wan, Y.S.; Long, T.; Zhang, Y.H.; Liu, J.H.; Ma, M.Z.; Xie, H.Q.; Liu, D.Y. Oxygen isotopic compositions of zircons from Paleoproterozoic metasedimentary rocks in the Daqingshan-Jining area, North China Craton: In situ SHRIMP analysis. *Acta Petrol. Sin.* **2016**, *32*, 659–681. (In Chinese with English Abstract)

66. Black, L.P.; Kamo, S.L.; Allen, C.M.; Davis, D.W.; Aleinikoff, J.N.; Valley, J.W.; Mundil, R.; Campbell, I.H.; Korsch, R.J.; Williams, I.S.; et al. Improved 206Pb/238U microprobe geochronology by the monitoring of a trace-element-related matrix effect; SHRIMP, ID-TIMS, ELA-ICP-MS and oxygen isotope documentation for a series of zircon standards. *Chem. Geol.* **2004**, *205*, 115–140. [CrossRef]

67. Wiedenbeck, M.; Hanchar, J.M.; Peck, W.H.; Sylvester, P.; Valley, J.; Whitehouse, M.; Kronz, A.; Morishita, Y.; Nasdala, L.; Fiebig, J.; et al. Further Characterisation of the 91500 Zircon Crystal. *Geostand. Geoanal. Res.* **2004**, *28*, 9–39. [CrossRef]

68. Pin, C.; Gannoun, A.; Dupont, A. Rapid, simultaneous separation of Sr, Pb, and Nd by extraction chromatography prior to isotope ratios determination by TIMS and MC-ICP-MS. *J. Anal. At. Spectrom.* **2014**, *29*, 1858–1870. [CrossRef]

69. Pu, W.; Gao, J.F.; Zhao, K.D.; Ling, H.F.; Jiang, S.Y. Separation method of Rb-Sr, Sm-Nd using DCTA and HIBA. *J. Nanjing Univ. (Nat. Sci.)* **2005**, *41*, 445–450. (In Chinese with English Abstract)

70. Weis, D.; Kieffer, B.; Maerschalk, C.; Barling, J.; De Jong, J.; Williams, G.A.; Hanano, D.; Pretorius, W.; Mattielli, N.; Scoates, J.S.; et al. High-precision isotopic characterization of USGS reference materials by TIMS and MC-ICP-MS. *Geochem. Geophys. Geosystems* **2006**, *7*. [CrossRef]

71. Morimoto, N. Nomenclature of pyroxenes. *Mineral. Petrol.* **1988**, *39*, 55–76. [CrossRef]

72. Leake, B.E.; Woolley, A.R.; Birch, W.D.; Burke, E.A.; Ferraris, G.; Grice, J.D.; Hawthorne, F.C.; Kato, A.; Kisch, H.J.; Krivovichev, V.G.; et al. Nomenclature of amphiboles: Additions and revisions to the International Mineralogical Association's amphibole nomenclature. *Mineral. Mag.* **2004**, *68*, 209–215. [CrossRef]

73. Smith, J.V. I. Crystal Structure and Physical Properties. II. Chemical and Textural Properties. In *Feldspar Minerals*; Springer: Berlin/Heidelberg, Germany; New York, NY, USA, 1974; pp. 627, 690.

74. Wu, F.Y.; Li, X.H.; Zheng, Y.F.; Gao, S. Lu-Hf isotopic systematics and their applications in petrology. *Acta Petrol. Sin.* **2007**, *23*, 185–220. (In Chinese with English Abstract)

75. Valley, J.W.; Lackey, J.S.; Cavosie, A.J.; Clechenko, C.C.; Spicuzza, M.J.; Basei, M.A.S.; Bindeman, I.N.; Ferreira, V.P.; Sial, A.N.; King, E.M. 4.4 billion years of crustal maturation: Oxygen isotope ratios of magmatic zircon. *Contrib. Mineral. Petrol.* **2005**, *150*, 561–580. [CrossRef]

76. Wang, Z.; Wang, H.C.; Shi, J.R.; Chang, Q.S.; Zhang, J.H.; Ren, Y.W.; Xiang, Z.Q. Tectonic setting and geological significance of Xuwujia metagabbro in Jining area, Inner Mongolia. *Geol. Surv. Res.* **2020**, *43*, 97–113. (In Chinese with English Abstract)

77. Middlemost, E.A. Naming materials in the magma/igneous rock system. *Earth-Sci. Rev.* **1994**, *37*, 215–224. [CrossRef]

78. Rickwood, P.C. Boundary lines within petrologic diagrams which use oxides of major and minor elements. *Lithos* **1989**, *22*, 247–263. [CrossRef]

79. Sun, S.S.; McDonough, W.F. Chemical and isotopic systematics of oceanic basalts: Implications for mantle composition and processes. *Geol. Soc. Lond. Spec. Publ.* **1989**, *42*, 313–345. [CrossRef]

80. Zindler, A.; Hart, S. Chemical geodynamics. *Annu. Rev. Earth Planet. Sci.* **1986**, *14*, 493–571. [CrossRef]

81. Kelsey, D.E.; Hand, M. On ultrahigh temperature crustal metamorphism: Phase equilibria, trace element thermometry, bulk composition, heat sources, timescales and tectonic settings. *Geosci. Front.* **2015**, *6*, 311–356. [CrossRef]

82. Harley, S.L. A matter of time: The importance of the duration of UHT metamorphism. *J. Mineral. Petrol.* **2016**, *111*, 50–72. [CrossRef]

83. Clark, C.; Taylor, R.J.M.; Kylander-Clark, A.R.C.; Hacker, B.R. Prolonged (>100 Ma) ultrahigh temperature metamorphism in the Napier Complex, East Antarctica: A petrochronological investigation of Earth's hottest crust. *J. Metamorph. Geol.* **2018**, *36*, 1117–1139. [CrossRef]

84. Laurent, A.T.; Bingen, B.; Duchene, S.; Whitehouse, M.J.; Seydoux-Guillaume, A.M.; Bosse, V. Decoding a protracted zircon geochronological record in ultrahigh temperature granulite, and persistence of partial melting in the crust, Rogaland, Norway. *Contrib. Mineral. Petrol.* **2018**, *173*, 29. [CrossRef]

85. Jiao, S.J.; Guo, J.H.; Evans, N.J.; Mcdonald, B.J.; Liu, P.; Ouyang, D.J.; Fitzsimons, I.C.W. The timing and duration of high-temperature to ultrahigh-temperature metamorphism constrained by zircon U-Pb-Hf and trace element signatures in the Khondalite Belt, North China Craton. *Contrib. Mineral. Petrol.* **2020**, *175*, 66. [CrossRef]

86. Guo, J.H.; Peng, P.; Chen, Y.; Jiao, S.J.; Windley, B.F. UHT sapphirine granulite metamorphism at 1.93–1.92 Ga caused by gabbronorite intrusions: Implications for tectonic evolution of the northern margin of the North China Craton. *Precambrian Res.* **2012**, *222*, 124–142. [CrossRef]

87. Stewart, B.W.; DePaolo, D.J. Isotopic studies of processes in mafic magma chambers: II. The Skaergaard Intrusion, East Greenland. *Contrib. Mineral. Petrol.* **1990**, *104*, 125–141. [CrossRef]

88. Irvine, T.N. Crystallization sequences in the Muskox intrusion and other layered intrusions—II. Origin of chromitite layers and similar deposits of other magmatic ores. *Geochim. Cosmochim. Acta* **1976**, *39*, 991–1020. [CrossRef]

89. Kruger, F.J.; Marsh, J.S. Significance of 87Sr/86Sr ratios in the Merensky cyclic unit of the Bushveld Complex. *Nature* **1982**, *298*, 53–55. [CrossRef]

90. Irvine, T.N.; Keith, D.W.; Todd, S.G. The JM platinum-palladium reef of the Stillwater Complex, Montana; II, Origin by double-diffusive convective magma mixing and implications for the Bushveld Complex. *Econ. Geol.* **1983**, *78*, 1287–1334. [CrossRef]

91. Palacz, Z.A. Isotopic and geochemical evidence for the evolution of a cyclic unit in the Rhum intrusion, north-west Scotland. *Nature* **1984**, *307*, 618–620. [CrossRef]

92. Palacz, Z.A. Sr-Nd-Pb isotopic evidence for crustal contamination in the Rhum intrusion. *Earth Planet. Sci. Lett.* **1985**, *74*, 35–44. [CrossRef]

93. Ripley, E.M.; Lambert, D.D.; Frick, L.R. Re-Os, Sm-Nd, and Pb isotopic constraints on mantle and crustal contributions to magmatic sulfide mineralization in the Duluth Complex. *Geochim. Cosmochim. Acta* **1998**, *62*, 3349–3365. [CrossRef]

94. DePaolo, D.J. Trace element and isotopic effects of combined wallrock assimilation and fractional crystallization. *Earth Planet. Sci. Lett.* **1981**, *53*, 189–202. [CrossRef]

95. Halama, R.; Marks, M.; Brügmann, G.; Siebel, W.; Wenzel, T.; Markl, G. Crustal contamination of mafic magmas: Evidence from a petrological, geochemical and Sr-Nd-Os-O isotopic study of the Proterozoic Isortoq dike swarm, South Greenland. *Lithos* **2004**, *74*, 199–232. [CrossRef]

96. Mir, A.R.; Alvi, S.H.; Balaram, V. Geochemistry of the mafic dykes in parts of the Singhbhum granitoid complex: Petrogenesis and tectonic setting. *Arab. J. Geosci.* **2011**, *4*, 933–943. [CrossRef]

97. Scarrow, J.H.; Leat, P.T.; Wareham, C.D.; Millar, I.L. Geochemistry of mafic dykes in the Antarctic Peninsula continental-margin batholith: A record of arc evolution. *Contrib. Mineral. Petrol.* **1998**, *131*, 289–305.

98. Naumann, T.R.; Geist, D.J. Generation of alkalic basalt by crystal fractionation of tholeiitic magma. *Geology* **1999**, *27*, 423–426. [CrossRef]

99. Stern, C.R.; Kilian, R. Role of the subducted slab, mantle wedge and continental crust in the generation of adakites from the Andean Austral Volcanic Zone. *Contrib. Mineral. Petrol.* **1996**, *123*, 263–281. [CrossRef]

100. Zhang, H.F.; Wang, J.L.; Zhou, D.W.; Yang, Y.H.; Zhang, G.W.; Santosh, M.; Yu, H.; Zhang, J. Hadean to Neoarchean episodic crustal growth: Detrital zircon records in Paleoproterozoic quartzites from the southern North China Craton. *Precambrian Res.* **2014**, *254*, 245–257. [CrossRef]

101. Yu, Y.; Huang, X.L.; Sun, M.; Yuan, C. Missing Sr-Nd isotopic decoupling in subduction zone: Decoding the multi-stage dehydration and melting of subducted slab in the Chinese Altai. *Lithos* **2020**, *362*, 105465. [CrossRef]

102. White, W.M. *Geochemistry*; John Wiley & Sons Inc.: Hoboken, NJ, USA, 2013.

103. Chen, Y.X.; Gao, P.; Zheng, Y.F. The anatectic effect on the zircon Hf isotope composition of migmatites and associated granites. *Lithos* **2015**, *238*, 174–184. [CrossRef]

104. Davies, G.R.; Tommasini, S. Isotopic disequilibrium during rapid crustal anatexis: Implications for petrogenetic studies of magmatic processes. *Chem. Geol.* **2000**, *162*, 169–191. [CrossRef]

105. Li, X.H.; Li, W.X.; Wang, X.C.; Li, Q.L.; Liu, Y.; Tang, G.Q. Role of mantle-derived magma in genesis of early Yanshanian granites in the Nanling Range, South China: In situ zircon Hf-O isotopic constraints. *Sci. China (Ser. D) Earth Sci.* **2009**, *52*, 1262–1278. [CrossRef]

106. Ellam, R.M. Lithospheric thickness as a control on basalt geochemistry. *Geology* **1992**, *20*, 153–156. [CrossRef]

107. Aldanmaz, E.R.C.A.N.; Pearce, J.A.; Thirlwall, M.F.; Mitchell, J.G. Petrogenetic evolution of late Cenozoic, post-collision volcanism in western Anatolia, Turkey. *J. Volcanol. Geotherm. Res.* **2000**, *102*, 67–95. [CrossRef]

108. Mckenzie, D.A.N.; Bickle, M.J. The volume and composition of melt generated by extension of the lithosphere. *J. Petrol.* **1988**, *29*, 625–679. [CrossRef]

109. McKenzie, D.A.N.; O'Nions, R.K. Partial melt distributions from inversion of rare earth element concentrations. *J. Petrol.* **1991**, *32*, 1021–1091. [CrossRef]

110. Su, Y.; Zheng, J.; Griffin, W.L.; Zhao, J.; Tang, H.; Ma, Q.; Lin, X. Geochemistry and geochronology of Carboniferous volcanic rocks in the eastern Junggar terrane, NW China: Implication for a tectonic transition. *Gondwana Res.* **2012**, *22*, 1009–1029. [CrossRef]

111. Pearce, J.A.; Peate, D.W. Tectonic implications of the composition of volcanic arc magmas. *Annu. Rev. Earth Planet. Sci.* **1995**, *23*, 251–285. [CrossRef]

112. Saunders, A.D.; Storey, M.; Kent, R.W.; Norry, M.J. Consequences of plume-lithosphere interactions. *Geol. Soc. Lond. Spec. Publ.* **1992**, *68*, 41–60. [CrossRef]

113. Sajona, F.G.; Maury, R.C.; Pubellier, M.; Leterrier, J.; Bellon, H.; Cotten, J. Magmatic source enrichment by slab-derived melts in a young post-collision setting, central Mindanao (Philippines). *Lithos* **2000**, *54*, 173–206. [CrossRef]

114. Wang, Q.; Zhao, Z.H.; Bai, Z.H.; Bao, Z.W.; Xiong, X.L.; Mei, H.J.; Xu, J.F.; Wang, Y.X. Carboniferous adakites and Nb-enriched arc basaltic rocks association in the Alataw Mountains, north Xinjiang: Interactions between slab melt and mantle peridotite and implications for crustal growth. *Chin. Sci. Bull.* **2003**, *48*, 2108–2115. [CrossRef]

115. Hawkesworth, C.J.; Turner, S.P.; McDermott, F.; Peate, D.W.; Van Calsteren, P. U-Th isotopes in arc magmas: Implications for element transfer from the subducted crust. *Science* **1997**, *276*, 551–555. [CrossRef]

116. Woodhead, J.D.; Hergt, J.M.; Davidson, J.P.; Eggins, S.M. Hafnium isotope evidence for 'conservative' element mobility during subduction zone processes. *Earth Planet. Sci. Lett.* **2001**, *192*, 331–346. [CrossRef]

117. Hanyu, T.; Tatsumi, Y.; Nakai, S.I.; Chang, Q.; Miyazaki, T.; Sato, K.; Tani, K.; Shibata, T.; Yoshida, T. Contribution of slab melting and slab dehydration to magmatism in the NE Japan arc for the last 25 Myr: Constraints from geochemistry. *Geochem. Geophys. Geosystems* **2006**, *7*. [CrossRef]

118. Tian, L.Y.; Castillo, P.R.; Hilton, D.R.; Hawkins, J.W.; Hanan, B.B.; Pietruszka, A.J. Major and trace element and Sr-Nd isotope signatures of the northern Lau Basin lavas: Implications for the composition and dynamics of the back-arc basin mantle. *J. Geophys. Res. Solid Earth* **2011**, *116*. [CrossRef]

119. Woodhead, J.D.; Eggins, S.M.; Johnson, R.W. Magma genesis in the New Britain island arc: Further insights into melting and mass transfer processes. *J. Petrol.* **1998**, *39*, 1641–1668. [CrossRef]

120. Zhao, J.H.; Zhou, M.F. Secular evolution of the Neoproterozoic lithospheric mantle underneath the northern margin of the Yangtze Block, South China. *Lithos* **2009**, *107*, 152–168. [CrossRef]

121. Pearce, J.A. Geochemical fingerprinting of oceanic basalts with applications to ophiolite classification and the search for Archean oceanic crust. *Lithos* **2008**, *100*, 14–48. [CrossRef]

122. Zhao, G.C. Metamorphic evolution of major tectonic units in the basement of the North China Craton: Key issues and discussion. *Acta Petrol. Sin.* **2009**, *25*, 1772–1792. (In Chinese with English Abstract)

123. Iwamori, H. Thermal effects of ridge subduction and its implications for the origin of granitic batholiths and paired metamorphic belts. *Earth Planet. Sci. Lett.* **2000**, *181*, 131–144. [CrossRef]

124. Santosh, M.; Kusky, T. Origin of paired high pressure-ultrahigh-temperature orogens: A ridge subduction and slab window model. *Terra Nova* **2010**, *22*, 35–42. [CrossRef]

125. Harley, S.L. Extending our understanding of ultrahigh temperature crustal metamorphism. *J. Mineral. Petrol. Sci.* **2004**, *99*, 140–158. [CrossRef]
126. Stüwe, K. *Geodynamics of the Lithosphere: Quantitative Description of Geological Problems*; Springer: Berlin/Heidelberg, Germany; Dordrecht, The Netherlands, 2007.
127. Hyndman, R.D.; Currie, C.A.; Mazzotti, S.P. Subduction zone back-arcs, mobile belts, and orogenic heat. *GSA Today* **2005**, *15*, 4–10. [CrossRef]
128. Brown, M. Duality of thermal regimes is the distinctive characteristic of plate tectonics since the Neoarchaean. *Geology* **2006**, *34*, 961–964. [CrossRef]
129. Currie, C.A.; Hyndman, R.D. The thermal structure of subduction zone back arcs. *J. Geophys. Res.* **2006**, *111*, 1–22. [CrossRef]
130. Collins, W.J. Hot orogens, tectonic switching, and creation of continental crust. *Geology* **2002**, *30*, 535–538. [CrossRef]
131. Wilson, J.R.; Hansen, B.T.; Pedersen, S. Zircon U-Pb evidence for the age of the Fongen-Hyllingen complex, Trondheim region, Norway. *Geol. Fören. I Stockh. Förh.* **1983**, *105*, 68–70. [CrossRef]
132. Schärer, U.; Wilmart, E.; Duchesne, J.C. The short duration and anorogenic character of anorthosite magmatism: U-Pb dating of the Rogaland complex, Norway. *Earth Planet. Sci. Lett.* **1996**, *139*, 335–350. [CrossRef]
133. Pollack, H.N. Thermal characteristics of the Archaean. *Oxf. Monogr. Geol. Geophys.* **1997**, *35*, 223–232.
134. Herzberg, C.; O'hara, M.J. Plume-associated ultramafic magmas of Phanerozoic age. *J. Petrol.* **2002**, *43*, 1857–1883. [CrossRef]
135. Herzberg, C.; Condie, K.; Korenaga, J. Thermal history of the Earth and its petrological expression. *Earth Planet. Sci. Lett.* **2010**, *292*, 79–88. [CrossRef]

Article

Geochronological Constraints on the Origin of the Paleoproterozoic Qianlishan Gneiss Domes in the Khondalite Belt of the North China Craton and Their Tectonic implications

Hengzhong Qiao [1,2,*], Peipei Deng [2] and Jiawei Li [2]

[1] Department of Geology, Northwest University, Xi'an 710069, China
[2] College of Tourism and Geographical Science, Leshan Normal University, Leshan 614000, China
* Correspondence: qiaohzh@lsnu.edu.cn

Abstract: The Paleoproterozoic gneiss domes are important structures of the Khondalite Belt in the northwestern North China Craton. However, less attention has been paid to their formation and evolution, and it thus hampers a better understanding of the deformation history of the Khondalite Belt. In this paper, we conducted structural and geochronological studies on the Qianlishan gneiss domes of the Khondalite Belt. The field observations and zircon U–Pb dating results show that the Qianlishan gneiss domes consist of 2.06–2.01 Ga granitoid plutons in the core, rimmed by granulite facies metasedimentary rocks (khondalites) of the Qianlishan Group. Both of them were subjected to two major phases of deformation (D1–D2) in the late Paleoproterozoic. Of these, D1 deformation mainly generated overturned to recumbent isoclinal folds F1 and penetrative transposed foliations/gneissosities S1 at ~1.95 Ga. Subsequently, D2 deformation produced the NW(W)–SE(E)-trending doubly plunging upright folds F2 at 1.93–1.90 Ga, and they have strongly re-oriented S1 gneissosities, giving rise to the Qianlishan gneiss domes. Combined with previous studies, we argue that the Qianlishan gneiss domes were the products of the Paleoproterozoic collisional orogenesis between the Yinshan and Ordos Blocks. Additionally, the development of doubly plunging antiforms is considered an important dome-forming mechanism in the Khondalite Belt.

Keywords: gneiss dome; zircon U–Pb dating; Paleoproterozoic; Qianlishan Complex; Khondalite Belt; North China Craton

Citation: Qiao, H.; Deng, P.; Li, J. Geochronological Constraints on the Origin of the Paleoproterozoic Qianlishan Gneiss Domes in the Khondalite Belt of the North China Craton and Their Tectonic implications. *Minerals* **2023**, *13*, 1361. https://doi.org/10.3390/min13111361

Academic Editor: Jaroslav Dostal

Received: 5 September 2023
Revised: 20 October 2023
Accepted: 22 October 2023
Published: 25 October 2023

1. Introduction

Gneiss domes are important features of many orogenic belts around the world, characterized by domal structures that typically comprise granitoid plutons or high-grade metamorphic rocks in the core, with outward-dipping gneissic foliations [1–5]. Most gneiss domes tend to display elliptical or elongated shapes in map view, and their long axes are commonly parallel to the strike of orogens [3,6]. Although gneiss domes show some similarities in geometry, petrology, and structure, a variety of dome-forming mechanisms have been proposed, mainly including diapirism driven by density inversion, buckling under horizontal contraction, extension-controlled exhumation, superposition of multiple deformations, duplex-induced folding, or some combination of these processes [7–11]. The potential link between doming and fundamental orogenic processes (e.g., crustal melting, flow, and exhumation) highlights the significance of the anatomy of gneiss domes in comprehensively understanding the geodynamics of orogens [3]. Of these, a prerequisite for unraveling the origin of gneiss domes is to determine temporal and spatial relationships between major structures, magmatic intrusions, and gneiss domes [2,4].

The Qianlishan Complex is a high-grade gneiss terrain located in the western segment of the Khondalite Belt that has been widely accepted as a Paleoproterozoic collisional orogen in the northwestern North China Craton (Figure 1) [12–17]. Of particular interest is the occurrence of a series of nearly orogen-parallel elongated gneiss domes in this region, together called the Qianlishan gneiss domes (Figure 2) [18–20]. These gneiss domes are dominated by granitoid plutons in the core and medium- to high-pressure granulite facies metasedimentary rocks of the Qianlishan Group in the rim, respectively (Figure 2) [18]. Notably, previous geochronological, metamorphic, and structural studies have primarily focused on the Qianlishan Group [14–21], but comparatively less attention was paid to the plutonic cores and domal structures in the Qianlishan Complex. Consequently, the origin of the Qianlishan gneiss domes still remains enigmatic, which hinders a comprehensive understanding of the deformation history of the Khondalite Belt. To resolve these key issues, we carried out detailed structural investigations on the Qianlishan gneiss domes and conducted LA-ICP-MS zircon U–Pb dating on granitoid plutons and leucocratic dykes that were variably involved in these domal structures. Integrated with previous studies, field observations and geochronological data in this paper will provide important insights into the development of the Qianlishan gneiss domes and further help to understand the tectonic evolution of the Khondalite Belt.

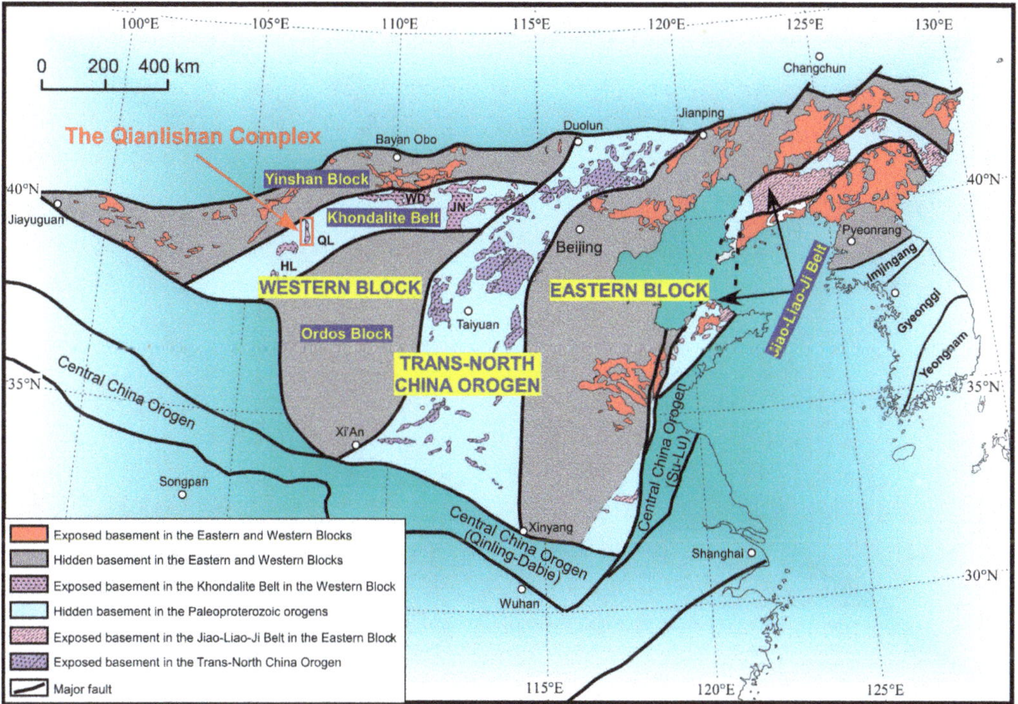

Figure 1. Tectonic subdivision of the North China Craton (modified after [12]). Abbreviations for the exposed high-grade metamorphic complexes in the Khondalite Belt: HL, Helanshan Complex; QL, Qianlishan Complex; WD, Wulashan and Daqingshan complexes; JN, Jining Complex.

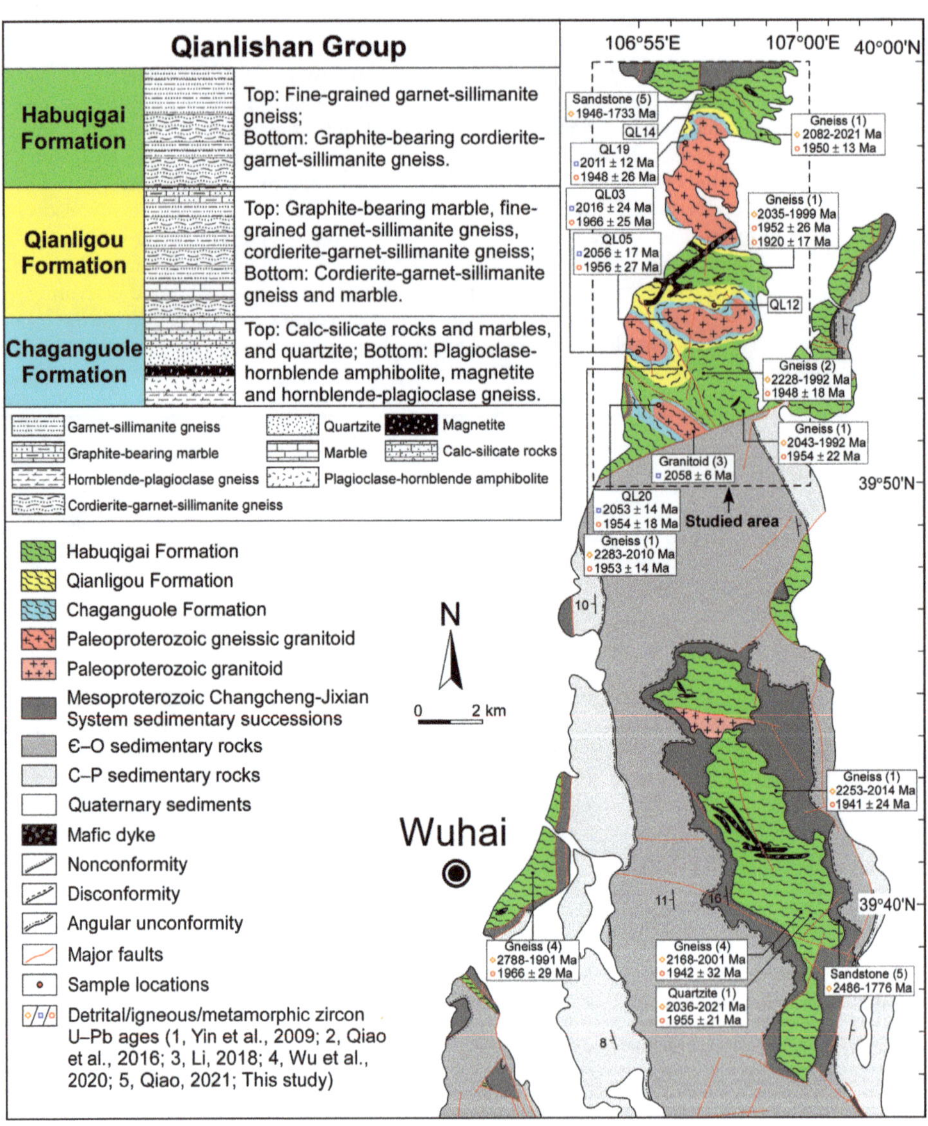

Figure 2. Simplified geological map of the Qianlishan Complex (modified after [14,18–20]).

2. Geological Setting

The North China Craton is amongst the oldest cratons in the world, and it has been regarded to result from the assembly of several Archean to Paleoproterozoic micro-blocks along linear structural belts [12,13,22–25]. The Khondalite Belt is a nearly ~1000 km, E-W-trending collisional orogen, where the northern Yinshan Block has been considered to amalgamate with the southern Ordos Block to form the Western Block at ~1.95 Ga [12–14,26–28]. From west to east, the Khondalite Belt well exposes the Helanshan, Qianlishan, Wulashan, Daqingshan, and Jining Complexes (Figure 1) [12]. These complexes are dominated by upper amphibolite to granulite facies metasedimentary rocks (i.e., the khondalites), mainly including felsic paragneisses, graphite-bearing pelitic gneisses, garnet-bearing quartzites, marbles, and calc-silicate rocks [18,29–31]. The khondalites are spatially juxtaposed with dioritic–granitic gneisses, S-type granites, minor charnockites, and mafic granulites [18,31]. Traditionally, the protoliths of the khondalites were inferred to form on a stable continental margin [12,18,32,33], but recently an active continental margin has also been proposed [34–38]. Extensive geochronological data revealed that the protoliths of khondalites were primarily sourced from a 2.2–2.0 Ga provenance, subsequently deposited at 2.0–1.95 Ga, and experienced regional high-grade metamorphism at 1.95–1.85 Ga [14–17,26,27,39–53]. Ultrahigh temperature (UHT) metamorphism was also regarded to coevally appear throughout the Khondalite Belt [28,54–61]. Moreover, syn- and post-collisional S-type granites that resulted from partial melting of the khondalites are mostly dated at ~1.95 Ga, 1.93–1.90 Ga, and 1.88–1.84 Ga [14,35,36,46,52,62–65].

The Qianlishan Complex is one of the most representative litho-tectonic units in the Khondalite Belt and unconformably overlain by (sub-)horizontal to gently-dipping unmetamorphosed Mesoproterozoic sedimentary sequences of the Changcheng–Jixian System (Figure 2) [18,66]. This complex mainly consists of the Paleoproterozoic granitoid plutons and the khondalites that are termed the Qianlishan Group [18]. The Qianlishan Group is subdivided into the Chaganguole, Qianligou, and Habuqigai formations. Their typical rock assemblages and distributions are shown in Figure 2. Metamorphic studies demonstrate that pelitic and felsic granulites recorded similar clockwise P-T paths, of which pelitic granulites were characterized by peak high-pressure metamorphism and post-peak near-isothermal decompression processes, with their P-T conditions constrained at 11–15 kbar/800–850 °C and 5.7–6.2 kbar/800–815 °C [15,17], respectively. Available U–Pb data show that detrital zircons from the Qianlishan Group mainly gave apparent $^{207}Pb/^{206}Pb$ ages (discordance degree < 10%) ranging from 2788 Ma to 1991 Ma [14,16,17]. Metamorphic zircons dominantly yielded a major age group at ~1.95 Ga that was interpreted as the timing of granulite facies metamorphism in the Qianlishan Complex [14,16,17]. Minor ~1.92 Ga metamorphic zircons were also reported, which were related to post-peak decompression [14]. Meanwhile, a crystallization mean age of 2058 ± 6 Ma was obtained from a granitoid pluton in the Qianligou quarry [37,67], but other plutons in the studied area lacked age constraints. In addition, previous structural investigations revealed that the Qianlishan Group underwent two major stages of deformation (D1–D2) in the Paleoproterozoic [19,20]. D1 deformation was regarded to has occurred at 1976–1936 Ma, characterized by small-scale overturned to recumbent isoclinal folds F1, transposition foliations S1 with NNE-SSW mineral lineations L1 [19,20]. D2 deformation mainly produced NW(W)–SE(E)-trending doubly plunging upright folds F2, and it is inferred to broadly take place at 1936–1854 Ma [19,20].

3. Samples and Methods

Field-based structural investigations were carried out to document the geometry of the Qianlishan gneiss domes. Particularly, domal structures (e.g., foliations, fold hinges) have been analyzed and measured in four representative domains of the studied area (Figure 3a). In order to put age constraints on the development of these gneiss domes, we conducted zircon U–Pb dating on six critical rock samples. Of these, four samples were from gneissic granitoid plutons (Samples QL03, 05, 19, and 20), and the other two samples were from deformation-related leucocratic dykes (Samples QL14 and 12). Zircon U–Pb analyses were performed by LA-ICP-MS at the Guangzhou Tuoyan Analytical Technology Co., Ltd., Guangzhou, China. Detailed analytical procedures were similar to those described in [68]. Laser sampling was conducted using an NWR 193 nm ArF excimer laser ablation system, and an iCAP RQ quadrupole ICP-MS instrument was used to acquire ion-signal intensities. The frequency and spot size of the laser were set to 5 Hz and 30 µm. Zircon 91500 [69] and glass NIST610 [70] were used as external standards for U–Pb dating and trace element calibration, respectively. Zircon standard Plešovice [71] was used as an unknown sample to monitor the working state of the instrument. Each analysis incorporated a background acquisition of approximately 30 s, followed by 40 s of data acquisition from the sample. An Excel-based software, ICPMSDataCal [72], was used to perform offline selection and integration of background and analyzed signals, time-drift correction, and quantitative calibration for U–Pb dating and trace element analysis. Concordia diagrams and age calculations were made using Isoplot/Ex_ver4.15 [73]. Individual analyses have been presented at 1σ level, and uncertainties on the weighted mean age, lower and upper intercept age were quoted at the 95% confidence level (2σ). Zircon U–Pb data in this study were provided as Supplementary Materials (Table S1).

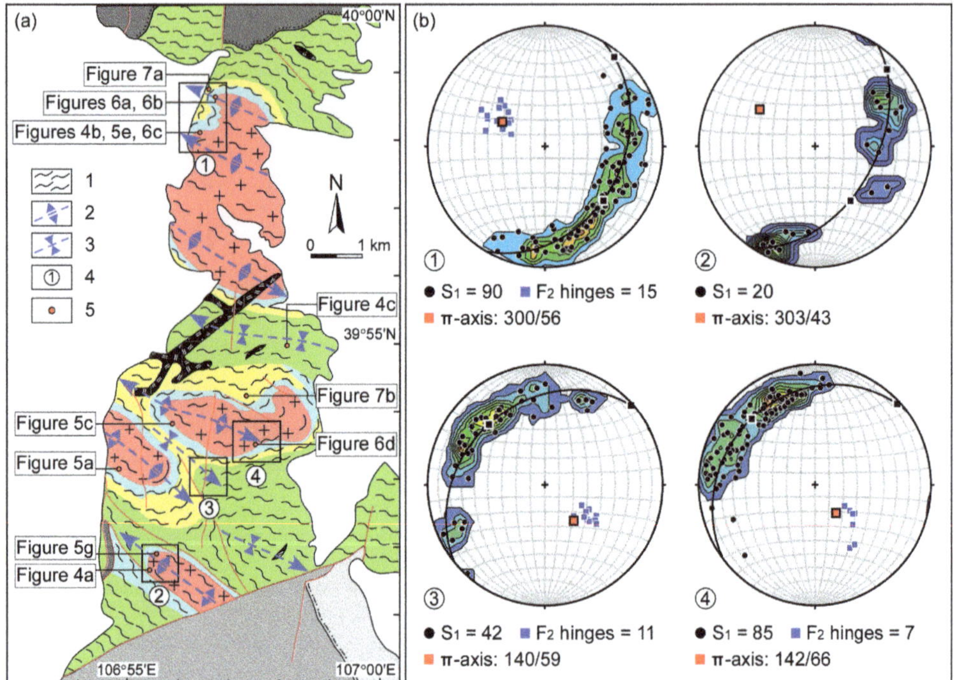

Figure 3. Simplified structural map and stereonet diagrams of the Qianlishan gneiss domes. (**a**) Litho-tectonic map mainly showing structural features of the regional-scale NW(W)–SE(E)-trending doubly plunging upright folds F2 and the penetrative gneissosities S1 (modified after [14,19,20]). 1, S1 gneissosity; 2, F2 antiform; 3, F2 synform; 4, domain for stereonet projection; 5, photo location. (**b**) Stereonet diagrams (lower hemisphere equal-area projections) exhibiting the orientations of S1 gneissosities and F2 fold hinges in four representative areas (domains 1–4 in Figure 3a).

4. Results

4.1. Field Structural Observations

The Qianlishan gneiss domes are exposed in the northern part of the Qianlishan Complex, locally covered by Quaternary sediments. In map view, they are unevenly spaced and roughly display NW(W)–SE(E)-trending elliptical shapes with outward-dipping gneissic foliations (Figure 3a). These domes consist of granitoid plutons in the core, rimmed by high-grade metasedimentary rocks from the Chaganguole, Qianligou, and Habuqigai formations of the Qianlishan Group (Figures 2 and 3), but their metamorphic grades did not significantly vary. There is no large-scale ductile shear zone observed along the contacts between the core and rim of these domes. Remarkably, the penetrative gneissosity in the plutonic cores and the supracrustal rocks are consistent, showing parallelism with the core/rim contacts (Figures 3a and 4a). Similar to the structural features of the Qianlishan Group [19,20], the granitoid plutons in the core were also subjected to D1–D2 deformation (Figures 4–7). In the field, overturned to recumbent isoclinal folds F1 are infrequent and difficult to discern in the plutons (Figure 4b), though they commonly appear in the Qianlishan Group (Figure 4c). Whereas these granitoid plutons have obviously developed penetrative gneissosities S1 that are mainly defined by the strong alignment of quartz + plagioclase ± K-feldspar ± biotite ± hornblende (Figure 5).

Figure 4. Typical photographs of D1 structures in the Qianlishan gneiss domes. (**a**) The contact between the granitoid pluton in the core and the Qianlishan Group (khondalites) in the rim of the Qianlishan gneiss domes. (**b**) Recumbent tight to isoclinal folds F1 in the granitoid pluton. (**c**) SSW-vergent overturned isoclinal folds F1 that have intensively transposed sedimentary bedding S0 to newly-developed S1 foliations in marbles and calc–silicate rocks.

D2 deformation is dominantly represented by doubly plunging upright open folds F2. The F2 structures commonly occurred as a series of micro- to macroscopic antiforms and synforms with NW(W)–SE(E)-striking sub-vertical fold axial surfaces (Figures 6 and 7). These F2 folds have strongly reworked the previous D1 structures in the metasedimentary rocks and the plutonic cores of the Qianlishan gneiss domes (Figures 6 and 7). Regionally, the F2 folds show doubly plunging geometry, and their fold hinges plunge to either NW(W) or SE(E) (Figure 3). In this study, the attitudes of S1 gneissosities from four representative F2 fold hinge zones (domains 1–4 in Figure 3a) were plotted on the stereonet. In domains 1 and 2, the poles to S1 gneissosities mostly appear along great circle girdles (Figure 3b), the π-axes of which are located at 300°/56° and 303°/43° (azimuth/plunge), respectively. Whereas the orientations of those π-axes in domains 3 and 4 are plotted at 140°/59° and 142°/66° (Figure 3b). These results are in good agreement with the measurements of F2 fold hinges in the field (Figures 3b, 6 and 7). This also confirms the overprinting relationships between the NW(W)–SE(E)-trending doubly plunging F2 folds and the penetrative gneissosities S1.

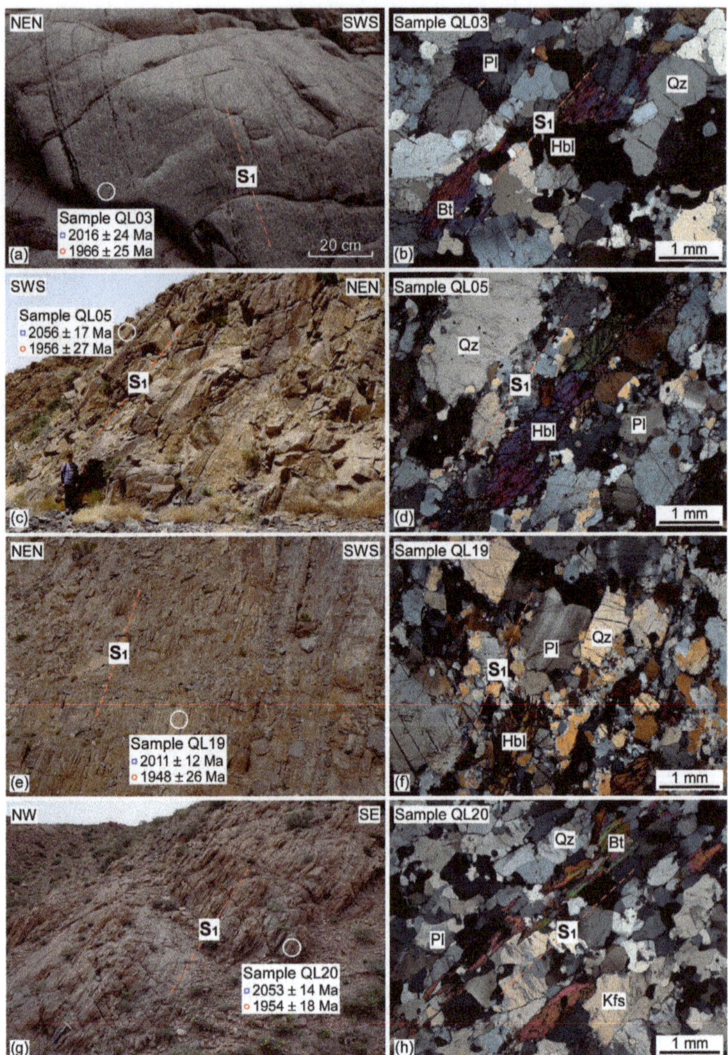

Figure 5. Representative photos of the S1-foliated granitoid plutons in the core of the Qianlishan gneiss domes. (**a**,**b**) The granitoid pluton (Sample QL03) underwent D1 deformation and developed ubiquitous gneissosities S1 that are mainly defined by well-oriented aggregates of quartz, plagioclase, and hornblende (cross-polarized light). (**c**,**d**) The gneissic granitoid plutons (Sample QL05) with pervasive S1 foliations, featured by the preferred alignment of quartz, plagioclase, and hornblende (cross-polarized light). (**e**,**f**) The pluton (Sample QL19) has apparently been foliated during D1 deformation, mainly composed of aligned aggregates of quartz, plagioclase, and hornblende (cross-polarized light). (**g**,**h**) The penetratively S1-foliated granitoid pluton (Sample QL20) is characterized by the preferential orientation of quartz, K-feldspar, plagioclase, and biotite (cross-polarized light). Blue squares and red circles indicate U–Pb ages of magmatic and metamorphic zircons, respectively. Qz, quartz; Pl, plagioclase; Hbl, hornblende; Kfs, K-feldspar; Bt, biotite.

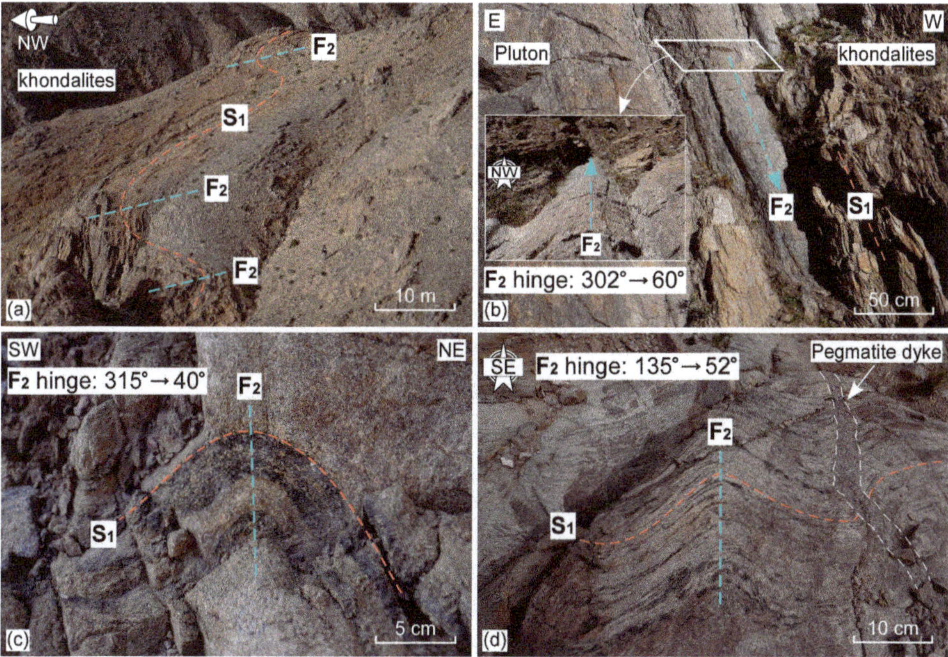

Figure 6. Typical photos of NW(W)–SE(E)-trending doubly plunging folds F2 in the Qianlishan gneiss domes. (**a**) Mesoscopic upright folds F2 that mostly plunge to NW(W) in the Qianlishan Group. (**b**) Open upright folds F2 with moderately NW(W)-plunging fold hinges formed along the core/rim contact of the Qianlishan gneiss domes. (**c**,**d**) Moderately NW- and SE-plunging open folds F2 with vertical fold axial surfaces have reworked the S1 gneissosities in the granitoid pluton.

Figure 7. Field photos of two representative leucocratic dykes that displayed different structural features in the Qianlishan gneiss domes. (**a**) A fine-grained leucocratic dyke (Sample QL14) along the pervasive S1 foliations was deflected by moderately NW-plunging open upright folds F2, suggesting that it most likely formed earlier than D2 deformation. (**b**) An undeformed coarse-grained pegmatite dyke (Sample QL12) obliquely truncated the limbs of gently NWW-plunging open upright folds F2, probably indicative of its post-D2 intrusion.

4.2. Zircon U–Pb Geochronology

4.2.1. S1-Foliated Granitoid Plutons

Four representative granitoid gneisses (Samples QL03, 05, 19, and 20) were collected from four plutons in the core of the Qianlishan gneiss domes (39°53.0′ N/106°55.3′ E, 39°54.0′ N/106°55.9′ E, 39°58.3′ N/106°56.8′ E, and 39°51.9′ N/106°55.7′ E; Figure 2), respectively. Zircons separated from these samples are mostly euhedral to subhedral and vary from 100–250 μm in grain size. Cathodoluminescence (CL) images illustrate that they are characterized by typical core–rim textures (Figure 8a–d), of which the zircon cores are commonly bright and oscillatory-zoned, interpreted to be of magmatic origin. These zircon cores are generally surrounded by relatively dark and structureless overgrowth rims (Figure 8a–d), typical of metamorphic origin.

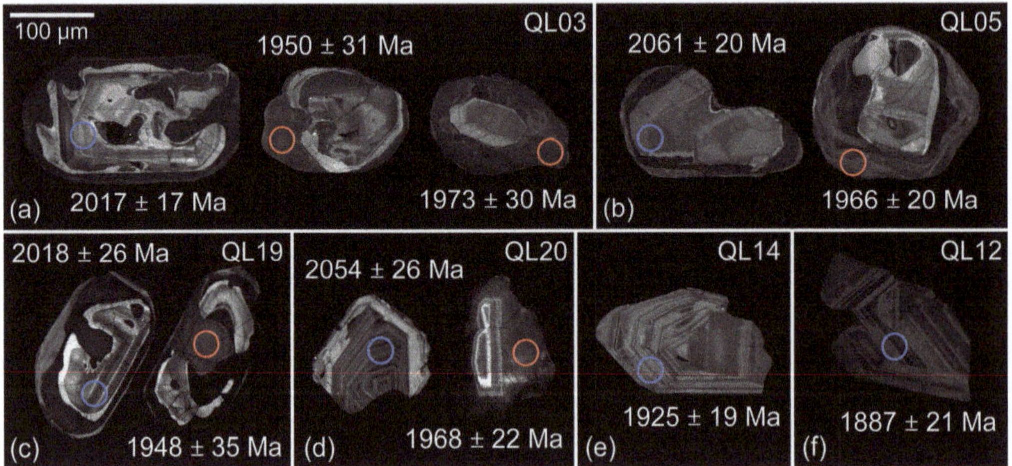

Figure 8. Representative cathodoluminescence (CL) zircon images with their LA-ICP-MS U–Pb ages of dating samples from the Qianlishan gneiss domes. (**a–d**) The S1-foliated granitoid plutons (Samples QL03, 05, 19, and 20). (**e**) The pre-D2 F2-folded leucocratic dyke (Sample QL14). (**f**) The post-D2 F2-cutting undeformed pegmatite dyke (Sample QL12). Blue circle, magmatic zircon; red circle, metamorphic overgrowth rim.

(1)　Sample QL03

A total of 22 zircons have been analyzed in this sample, and the dating results are presented in Figure 9a and Table S1. Of these, five spots (discordance degree $\leq$ 5%) were made on the zircon cores of igneous origin (Figure 8a), with high Th/U values of 0.51–1.8. They yielded a weighted mean ^{207}Pb/^{206}Pb age of 2016 ± 24 Ma (n = 5, MSWD = 3.4; Figure 9a). The remaining 17 spots on metamorphic overgrowth rims (Figure 8a) have lower Th/U values of 0.02–0.25 (Table S1), and they were variably discordant, probably due to the loss of Pb. On the concordia diagram (Figure 9a), these data define a discordia line with the lower and upper intercept age of 567 ± 29 Ma and 1966 ± 25 Ma (n = 17, MSWD = 8.0), respectively.

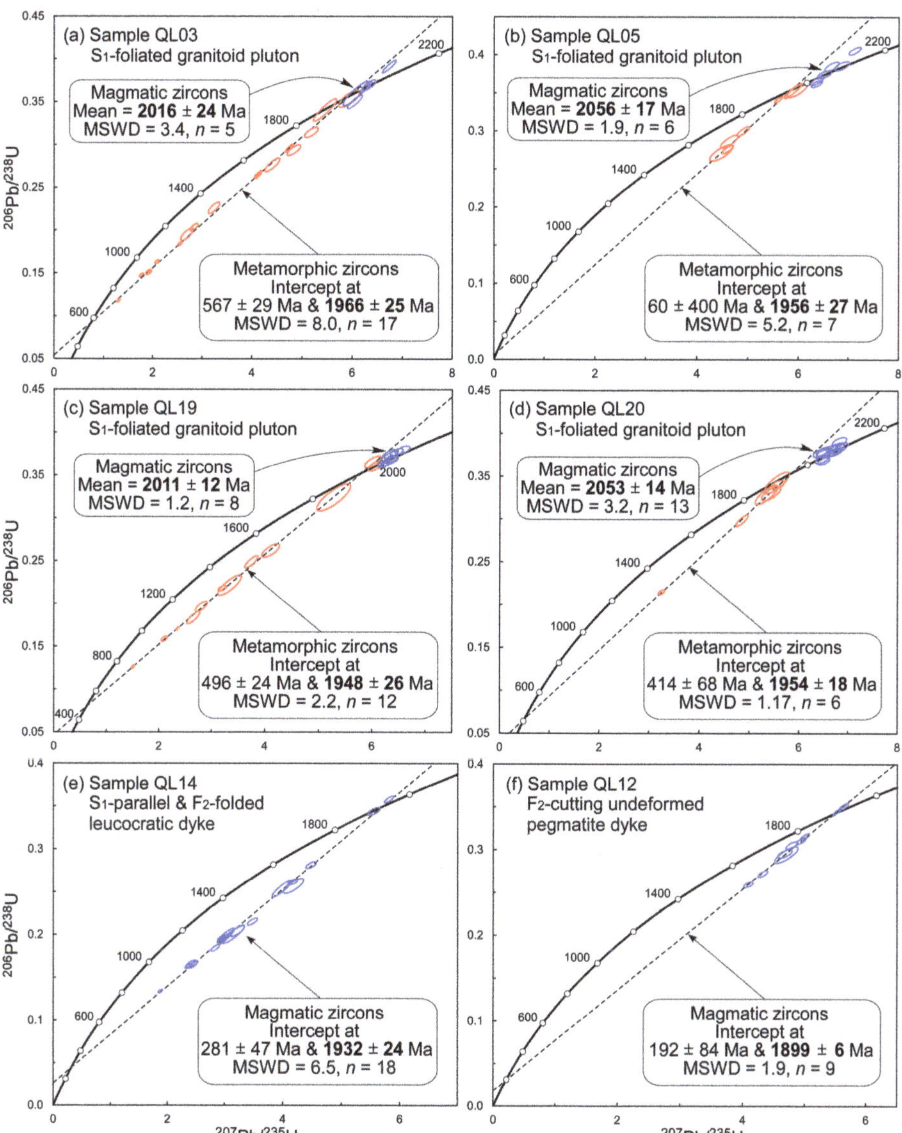

Figure 9. Concordia diagram of LA-ICP-MS zircon U–Pb dating results from the Qianlishan gneiss domes. (**a–d**) The S1-foliated granitoid plutons (Samples QL03, 05, 19, and 20). (**e**) The pre-D2 F2-folded leucocratic dyke (Sample QL14). (**f**) The post-D2 F2-cutting undeformed pegmatite dyke (Sample QL12). Blue ellipse, magmatic zircon; red ellipse, metamorphic overgrowth rim. Uncertainties on weighted mean age, lower and upper intercept age were quoted at the 95% confidence level (2σ).

(2)   Sample QL05

Six spots (discordance degree $\leq$ 3.7%; Table S1) were conducted on the magmatic zircon cores in Sample QL05, and they gave a weighted mean ^{207}Pb/^{206}Pb age of 2056 $\pm$ 17 Ma (n = 6, MSWD = 1.9; Figures 8b and 9b). Meanwhile, seven spots were analyzed on metamorphic overgrowth rims, and they yielded the lower and upper intercept age of 60 $\pm$ 400 Ma and 1956 $\pm$ 27 Ma (n = 7, MSWD = 5.2; Figures 8b and 9b), respectively.

(3)   Sample QL19

A total of 20 zircons were analyzed in this sample, of which eight spots on the igneous zircon cores were concordant (discordance degree $\leq$ 3.1%; Table S1), yielding a weighted mean ^{207}Pb/^{206}Pb age of 2011 $\pm$ 12 Ma (n = 8, MSWD = 1.2; Figures 8c and 9c). The remaining 12 spots were undertaken on zircon overgrowth rims of metamorphic origin (Figure 8c) and defined the lower and upper intercept age of 496 $\pm$ 24 Ma and 1948 $\pm$ 26 Ma (n = 12, MSWD = 2.2; Figure 9c), respectively.

(4)   Sample QL20

Thirteen spots (discordance degree $\leq$ 4.1%; Table S1) were carried out on the magmatic zircon cores in Sample QL20 and gave a weighted mean ^{207}Pb/^{206}Pb age of 2053 $\pm$ 14 Ma (n = 13, MSWD = 3.2; Figures 8d and 9d). Another six analyses were obtained from metamorphic overgrowth rims and yielded the lower and upper intercept age of 414 $\pm$ 68 Ma and 1954 $\pm$ 18 Ma (n = 6, MSWD = 1.17; Figures 8d and 9d), respectively.

4.2.2. F2-Folded Leucocratic Dyke

Sample QL14 was collected from a fine-grained leucocratic dyke that intruded the Qianligou Formation in the northern part of the Qianlishan Complex (Figure 2; 39°58.8′ N/106°56.9′ E). In the field, this dyke appeared along the S1 foliations of the host metasedimentary rocks (Figure 7a), which were together deflected by the moderately NWW-plunging upright open folds F2. The dyke mainly exhibited granitic textures and did not develop any D1 deformational fabrics. These features indicate that the dyke most likely formed after D1 deformation but prior to D2 deformation. Zircons from Sample QL14 are euhedral, prismatic, and 150–200 μm in grain size. CL images reveal that they are generally single grains with bright, patchy, or oscillatory zoning (Figure 8e). Meanwhile, most of these zircons have high Th/U values of 0.5–0.9 (Table S1), indicative of magmatic origin [74]. A total of 18 spots were conducted on 18 igneous zircons in this sample. On the concordia diagram (Figure 9e), these data were plotted as a discordia line with the lower and upper intercept age of 281 $\pm$ 47 Ma and 1932 $\pm$ 24 Ma (n = 18, MSWD = 6.5).

4.2.3. F2-Cutting Undeformed Pegmatite Dyke

Sample QL12 was collected from a coarse-grained pegmatite dyke without deformational fabrics (Figure 2; 39°54.2′ N, 106°57.6′ E). This dyke obliquely cut the limb of a gently NWW-plunging upright open fold F2 (Figure 6b), suggesting that its emplacement postdated D2 deformation. Zircons separated from this sample are 100–200 μm, euhedral to subhedral single grains. In CL images, they are commonly patchy and oscillatory-zoned (Figure 8f). Combined with their Th/U values of 0.33–0.76 (Table S1), these zircons are considered to be of igneous origin [74]. A total of nine zircons were analyzed in the sample, which defined the lower and upper intercept age of 192 $\pm$ 84 Ma and 1899 $\pm$ 6 Ma (n = 9, MSWD = 1.9; Figure 9f), respectively.

5. Discussions

5.1. Significance of Zircon U–Pb Ages

A critical aspect of understanding the development of gneiss domes is temporal relations among magmatism, deformation, and metamorphism involved in the dome-forming processes [3,4,9]. The granitoid plutons in the core of the Qianlishan gneiss domes have long been inferred to have formed in the Paleoproterozoic [18], but their crystallization ages were poorly understood. In this study, magmatic zircons from four gneissic granitoid plutons (Samples QL03, 05, 19, and 20; Figures 2 and 5) yielded weighted mean ages of 2016 ± 24 Ma, 2056 ± 17 Ma, 2011 ± 12 Ma, and 2053 ± 14 Ma, respectively (Figures 8a–d, 9a–d and 10). Similarly, previous dating results have revealed a crystallization age of 2058 ± 6 Ma from the same pluton with Sample QL20 [37,67]. These ages suggested that the granitoid plutons were broadly emplaced at 2.06–2.01 Ga. Noticeably, the compilation of available data demonstrates that detrital zircons from the metasedimentary rocks of the Qianlishan Group predominantly gave ages of 2.1–2.0 Ga (discordance degree < 10%) [14,16,17], with a prominent single peak at ~2.02 Ga. This supports that the 2.06–2.01 Ga granitoid plutons have probably provided important clastic sediments for the protoliths of the Qianlishan Group [14,16]. Subsequently, the 2.06–2.01 Ga granitoid plutons have been highly S1-foliated during D1 deformation (Figure 5). Metamorphic zircon overgrowth rims from the plutons (Samples QL03, 05, 19, and 20) yielded upper intercept ages of 1966 ± 25 Ma, 1956 ± 27 Ma, 1948 ± 26 Ma, and 1954 ± 18 Ma (Figures 8a–d, 9a–d and 10), which coincide with those ~1.95 Ga metamorphic ages of pelitic/felsic granulites from the Qianlishan Group [14,16,17], interpreted as the timing of syn-D1 regional metamorphic event. Taken together, we consider that the plutonic cores and khondalites of the Qianlishan gneiss domes underwent regional high-grade metamorphism and coevally developed penetrative S1 gneissosities at ~1.95 Ga (Figure 10). This interpretation is also supported by the assumption that D1 structures in the Qianlishan Group roughly occurred at some time between ~1976 Ma and ~1936 Ma [19,20].

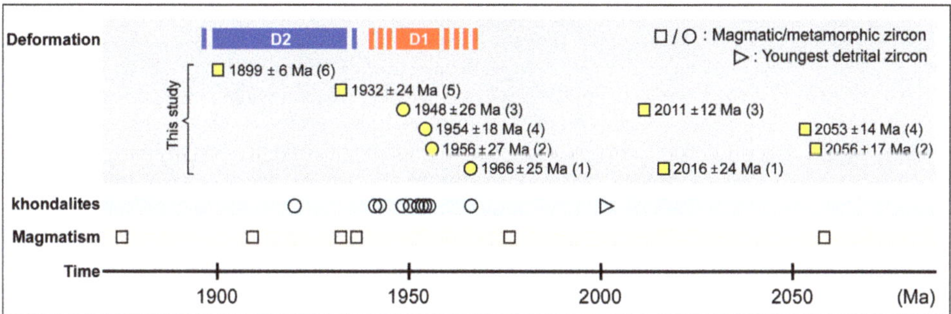

Figure 10. Schematic diagram illustrating temporal relations among major magmatic, deformational, and metamorphic events in the Paleoproterozoic Qianlishan gneiss domes. Dating samples are from this study (1–4, the S1-foliated granitoid plutons, Samples QL03, 05, 19, and 20; 5, the pre-D2 F2-folded leucocratic dyke, Sample QL14; 6, the post-D2 F2-cutting undeformed pegmatite dyke, Sample QL12) and previous studies [14,16,17,19,20,37,67]. See the text for details.

Abundant leucocratic dykes were involved in D2 structures in the Qianlishan gneiss domes, and they provided good opportunities to put age constraints on D2 deformation. Of these, an S1-parallel fine-grained leucocratic dyke (Sample QL14; Figure 7a) has been reworked by the upright open fold F2, interpreted to predate D2 deformation. This pre-D2 dyke yielded an upper intercept age of 1932 ± 24 Ma that was considered as its approximate crystallization age (Figures 8e and 9e), indicating that D2 deformation took place later than ~1.93 Ga. Meanwhile, a similar formation age of 1932 ± 47 Ma was obtained from a syn-D2 dyke that both truncated the S1 gneissosity and appeared along the F2 fold axial surface [19,20]. Later, a series of randomly oriented pegmatite dykes intruded into the Qianlishan Complex, which exhibited clear crosscutting relationships with all the D1 and D2 structures (e.g., Figures 4b, 6d and 7). Based on our observations, an undeformed pegmatite dyke (Sample QL12; Figure 7b) obliquely cut the limb of the F2 fold, indicative of post-D2 intrusion. The dyke gave an upper intercept age of 1899 ± 6 Ma (Figures 8f and 9f), regarded as its crystallization age. This suggests that D2 deformation occurred approximately earlier than ~1.90 Ga. Consistently, an 1854 ± 68 Ma post-D2 undeformed pegmatite dyke has also been found in this region, implying that D2 deformation was broadly constrained before ~1854 Ma [19,20]. Thus, we further proposed that D2 deformation has most likely happened in the period of 1.93–1.90 Ga (Figure 10). In addition, we note that magmatic zircons from Sample QL14 and 12 yielded lower intercept ages of 281 ± 47 Ma and 192 ± 84 Ma (Figure 9e,f), respectively. Similarly, metamorphic zircons from Samples QL03, 05, 19, and 20 gave lower intercept ages of 567 ± 29 Ma, 60 ± 400 Ma, 496 ± 24 Ma and 414 ± 68 Ma (Figure 9a–d). But these ages are difficult to relate to a known late-stage tectonothermal event in the Khondalite Belt.

5.2. Structural Evolution of the Qianlishan Gneiss Domes

The map-view NW(W)–SE(E)-trending elliptical Qianlishan gneiss domes occupy the northern part of the Qianlishan Complex (Figures 2 and 3a) [18–20], and the cores of these domes are dominated by the 2.06–2.01 Ga granitoid plutons (Figures 3, 5, 9 and 10). Available structural and geochronological data demonstrate that the Qianlishan gneiss domes underwent two major phases of deformation (D1–D2) in the late Paleoproterozoic (Figures 4–7 and 10). Of these, D1 deformation has approximately taken place at ~1.95 Ga, mainly manifested by overturned to recumbent isoclinal folds F1, transposed foliations/gneissosities S1 and NNE-SSW mineral lineations L1 (Figures 4, 5 and 10) [19,20]. The penetrative gneissosities S1 in the core and rim of the Qianlishan gneiss domes are continuous and parallel to the core/rim contacts (Figures 3a and 4a). We interpret these contacts as deformed unconformities between the granitoid plutons and the overlying Qianlishan Group. They can also serve as important markers to trace the Qianlishan gneiss domes in map view (Figures 2 and 3a). Subsequently, D2 deformation generated NW(W)–SE(E)-trending doubly plunging open upright folds F2 at 1.93–1.90 Ga (Figures 6, 7 and 10). D2 deformation apparently lacked the development of ubiquitous foliations, implying that it perhaps happened at a relatively shallower level than D1 deformation. Noticeably, the superposition of D2 on D1 deformation played an important role in shaping the general structural framework of the Qianlishan Complex (Figures 3, 6, 7 and 10). We consider that the doubly plunging open upright antiforms F2 have strongly re-oriented the penetrative gneissosities S1, giving rise to the Qianlishan gneiss domes. This is supported by the fact that a series of regional-scale antiforms and synforms F2 alternatively occurred in the northern Qianlishan Complex (Figures 3, 6 and 7). Meanwhile, the absence of syn-doming large-scale plutonism and radial lineation pattern in this region suggests that the origin of Qianlishan gneiss domes might not be primarily controlled by magmatic diapirism. This interpretation corresponds to the viewpoint that gneiss domes without syn-kinematic plutons can be created by the formation of doubly plunging antiforms [4,9,75].

5.3. Tectonic Implications

As mentioned earlier, high-pressure pelitic/felsic granulites in the Qianlishan Complex recorded clockwise P-T paths with peak conditions of 11–15 kbar/800–850 °C, indicating that they were once buried at lower crustal depths of 45–50 km [15,17]. Previous structural investigations also unraveled that the contractional deformation D1 was associated with peak granulite facies metamorphism at ~1.95 Ga, linked to the continent–continent collision [12–17,19,20]. Later, D2 deformation and post-peak decompression simultaneously occurred, related to tectonic exhumation after crustal thickening [15,17,19,20]. The NW(W)–SE(E)-trending doubly plunging open upright antiforms F2 were responsible for the development of the Qianlishan gneiss domes at 1.93–1.90 Ga. Meanwhile, extensive post-collisional partial melting happened in the Qianlishan Complex, as partly indicated by the presence of ~1.93 Ga and 1.91–1.90 Ga leucocratic dykes (Figure 10) [14,19,20,37,67], which might facilitate dome-forming processes during D2 deformation. Moreover, we regard that D2 doming deformation played an important role in the upward exhumation of high-pressure granulites in the Qianlishan Complex. Taken together, the Qianlishan gneiss domes are considered to have witnessed the collisional orogenesis between the Yinshan and Ordos Blocks in the late Paleoproterozoic.

Remarkably, dome-like structures that are dominated by 2.3–2.0 Ga gneissic granitoid plutons in the core have also been found in other segments of the Khondalite Belt (e.g., the Daqingshan Complex [76]; the Helanshan Complex [77,78]), but their origin still remains enigmatic. Meanwhile, the D1 and D2 deformational fabrics documented in this study are ubiquitous throughout the Khondalite Belt [18,78–82]. Some authors assumed that D1 structures resulted from sub-horizontal extensional detachment deformation [83–85]. This interpretation is not supported by structural observations in the Qianlishan Complex, and we attribute D1 deformational fabrics to the nearly NNE–SSW collision between the Yinshan and Ordos Blocks [19,20]. Moreover, the superposition of doubly plunging open upright folds F2 on the penetrative gneissosities S1 is common in the Daqingshan Complex [79], similar to the case of the Qianlishan gneiss domes in this study. We consider that the development of doubly plunging antiforms was an important dome-forming mechanism in the Khondalite Belt. Additionally, it is worth noting that the Qianlishan gneiss domes are well-preserved, but domal structures in other complexes were variably reworked by post-D2 (ca. 1.89–1.87 Ga) large-scale NE- to E-striking ductile shear zones in the Khondalite Belt [19,61,86].

6. Concluding Remarks

Based on our new structural and geochronological data, the Qianlishan gneiss domes consist of 2.06–2.01 Ga granitoid plutons in the core, rimmed by high-grade metasedimentary rocks of the Qianlishan Group. They were subjected to regional high-grade metamorphism and D1 deformation at ~1.95 Ga, mainly characterized by overturned to recumbent isoclinal folds F1 and penetrative transposed gneissosities S1. Subsequently, D2 deformation produced the NW(W)–SE(E)-trending doubly plunging upright folds F2 that have strongly re-oriented S1 gneissosities at 1.93–1.90 Ga, responsible for the origin of the Qianlishan gneiss domes. Combined with previous studies, we consider that the Paleoproterozoic Qianlishan gneiss domes have resulted from the collisional orogenesis between the Yinshan and Ordos Blocks. In addition, the development of doubly plunging antiforms is regarded as an important dome-forming mechanism in the Khondalite Belt.

Supplementary Materials: The following supporting information can be downloaded at https://www.mdpi.com/article/10.3390/min13111361/s1, Table S1: LA-ICP-MS zircon U–Pb data for dating samples from the Qianlishan gneiss domes.

Author Contributions: Conceptualization, H.Q.; investigation, H.Q. and J.L.; methodology, H.Q. and P.D.; data curation, H.Q. and P.D.; writing—original draft preparation, H.Q., P.D. and J.L.; writing—review and editing, H.Q.; supervision, H.Q.; funding acquisition, H.Q.; All authors have read and agreed to the published version of the manuscript.

Funding: This research was funded by National Natural Science Foundation of China (Grant No. 42002221), China Postdoctoral Science Foundation (Grant No. 2022M712569), Science & Technology Department of Sichuan Province (Grant No. MZGC20230103) and Leshan Normal University (Grant No. KYCXTD2023-2, RC202009, LZD031).

Data Availability Statement: Not applicable.

Conflicts of Interest: The authors declare no conflict of interest.

References

1.   Eskola, P.E. The problem of mantled gneiss domes. *Q. J. Geol. Soc.* **1948**, *104*, 461–476. [CrossRef]
2.   Hobbs, B.E.; Means, W.D.; Williams, P.F. *An outline of Structural Geology*; Wiley: New York, NY, USA, 1976; pp. 428–430
3.   Whitney, D.L.; Teyssier, C.; Vanderhaeghe, O. Gneiss domes and crustal flow. In *Gneiss Domes in Orogeny*; Whitney, D.L., Teyssier, C., Siddoway, C.S., Eds.; Geological Society of America: Boulder, CO, USA, 2004; Volume 380, pp. 15–34.
4.   Yin, A. Gneiss domes and gneiss dome systems. In *Gneiss Domes in Orogeny*; Whitney, D.L., Teyssier, C., Siddoway, C.S., Eds.; Geological Society of America Special Paper; Geological Society of America: Boulder, CO, USA, 2004; Volume 380, pp. 1–14.

5. Xu, Z.Q.; Li, Y.; Ji, S.C.; Li, G.W.; Pei, X.Z.; Ma, X.X.; Xiang, H.; Wang, R.R. Qinling gneiss domes and implications for tectonic evolution of the Early Paleozoic Orogen in Central China. *J. Asian Earth Sci.* **2020**, *188*, 104052. [CrossRef]

6. Brun, J.P. The cluster-ridge pattern of mantled gneiss domes in eastern Finland: Evidence for large-scale gravitational instability of the Proterozoic crust. *Earth Planet. Sci. Lett.* **1980**, *47*, 441–449. [CrossRef]

7. Yin, A. Mechanisms for the formation of domal and basinal detachment faults: A three-dimensional analysis. *J. Geophys. Res. Solid Earth* **1991**, *96*, 14577–14594. [CrossRef]

8. Fletcher, R.C. Three-dimensional folding and necking of a power-law layer: Are folds cylindrical, and, if so, do we understand why? *Tectonophysics* **1995**, *247*, 65–83. [CrossRef]

9. Bell, T.H.; Ham, A.P.; Hayward, N.; Hickey, K.A. On the development of gneiss domes. *Aust. J. Earth Sci.* **2005**, *52*, 183–204. [CrossRef]

10. Lee, J.; Hacker, B.; Wang, Y. Evolution of North Himalayan gneiss domes: Structural and metamorphic studies in Mabja Dome, southern Tibet. *J. Struct. Geol.* **2004**, *26*, 2297–2316. [CrossRef]

11. Zhang, B.; Chai, Z.; Yin, C.Y.; Huang, W.T.; Wang, Y.; Zhang, J.J.; Wang, X.X.; Cao, K. Intra-continental transpression and gneiss doming in an obliquely convergent regime in SE Asia. *J. Struct. Geol.* **2017**, *97*, 48–70. [CrossRef]

12. Zhao, G.C.; Sun, M.; Wilde, S.A.; Li, S.Z. Late Archean to Paleoproterozoic evolution of the North China Craton: Key issues revisited. *Precambrian Res.* **2005**, *136*, 177–202. [CrossRef]

13. Zhao, G.C.; Cawood, P.A.; Li, S.Z.; Wilde, S.A.; Sun, M.; Zhang, J.; He, Y.H.; Yin, C.Q. Amalgamation of the North China Craton: Key issues and discussion. *Precambrian Res.* **2012**, *222*, 55–76. [CrossRef]

14. Yin, C.Q.; Zhao, G.C.; Sun, M.; Xia, X.P.; Wei, C.J.; Zhou, X.W.; Leung, W.H. LA-ICP-MS U–Pb zircon ages of the Qianlishan Complex: Constrains on the evolution of the Khondalite Belt in the Western Block of the North China Craton. *Precambrian Res.* **2009**, *174*, 78–94. [CrossRef]

15. Yin, C.Q.; Zhao, G.C.; Wei, C.J.; Sun, M.; Guo, J.H.; Zhou, X.W. Metamorphism and partial melting of high-pressure pelitic granulites from the Qianlishan Complex: Constraints on the tectonic evolution of the Khondalite Belt in the North China Craton. *Precambrian Res.* **2014**, *242*, 172–186. [CrossRef]

16. Qiao, H.Z.; Yin, C.Q.; Li, Q.L.; He, X.L.; Qian, J.H.; Li, W.J. Application of the revised Ti-in-zircon thermometer and SIMS zircon U–Pb dating of high-pressure pelitic granulites from the Qianlishan-Helanshan Complex of the Khondalite Belt, North China Craton. *Precambrian Res.* **2016**, *276*, 1–13. [CrossRef]

17. Wu, S.J.; Yin, C.Q.; Davis, D.W.; Zhang, J.; Qian, J.H.; Qiao, H.Z.; Xia, Y.F.; Liu, J.N. Metamorphic evolution of high-pressure felsic and pelitic granulites from the Qianlishan Complex and tectonic implications for the Khondalite Belt, North China Craton. *Geol. Soc. Am. Bull.* **2020**, *132*, 2253–2266. [CrossRef]

18. Lu, L.Z.; Xu, X.C.; Liu, F.L. *Early Precambrian Khondalites in North China*; Changchun Publishing House: Changchun, China, 1996; pp. 1–277. (In Chinese)

19. Qiao, H.Z. Structural and Geochronological Studies of the Qianlishan-Helanshan Complex in North China. Ph.D. Dissertation, Sun Yat-sen University, Guangzhou, China, 2019; pp. 1–163, (In Chinese with English Abstract).

20. Yin, C.Q.; Qiao, H.Z.; Lin, S.F.; Li, C.C. Zhang, J.; Qian, J.H.; Wu, S.J. Deformation history of the Qianlishan complex, Khondalite belt, North China: Structures, ages and tectonic implications. *J. Struct. Geol.* **2020**, *141*, 104176. [CrossRef]

21. He, X.L.; Yin, C.Q.; Long, X.P.; Qian, J.H.; Wang, L.J.; Qiao, H.Z. Archean to Paleoproterozoic continental crust growth in the Western Block of North China: Constraints from zircon Hf isotopic and whole-rock Nd isotopic data. *Precambrian Res.* **2017**, *303*, 105–116. [CrossRef]

22. Zhai, M.G.; Peng, P. Paleoproterozoic events in North China Craton. *Acta Petrol. Sin.* **2007**, *23*, 2665–2682. (In Chinese with English Abstract)

23. Zhai, M.G.; Santosh, M. The early Precambrian odyssey of the North China Craton: A synoptic overview. *Gondwana Res.* **2011**, *20*, 6–25. [CrossRef]

24. Kusky, T.M.; Polat, A.; Windley, B.F.; Burke, K.C.; Dewey, J.F.; Kidd, W.S.F.; Maruyama, S.; Wang, J.P.; Deng, H.; Wang, Z.S.; et al. Insights into the tectonic evolution of the North China Craton through comparative tectonic analysis: A record of outward growth of Precambrian continents. *Earth-Sci. Rev.* **2016**, *162*, 387–432. [CrossRef]

25. Wei, C.; Zhai, M.; Wang, B. Four phases of Orosirian metamorphism in the north North China Craton (NNCC): Insights into the regional tectonic framework and evolution. *Earth-Sci. Rev.* **2023**, *241*, 104449. [CrossRef]

26. Wan, Y.S.; Liu, D.Y.; Xu, Z.Y.; Dong, C.Y.; Wang, Z.J.; Zhou, H.Y.; Yang, Z.S.; Liu, Z.H.; Wu, J.S. Paleoproterozoic crustally derived carbonate-rich magmatic rocks from the Daqinshan area, North China Craton: Geological, petrographical, geochronological and geochemical (Hf, Nd, O and C) evidence. *Am. J. Sci.* **2008**, *308*, 351–378. [CrossRef]

27. Zhou, X.W.; Geng, Y.S. Metamorphic age of the khondalites in the Helanshan region: Constraints on the evolution of the Western block in the North China Craton. *Acta Petrol. Sin.* **2009**, *25*, 1843–1852, (In Chinese with English Abstract).

28. Guo, J.H.; Peng, P.; Chen, Y.; Jiao, S.J.; Windley, B.F. UHT sapphirine granulite metamorphism at 1.93–1.92 Ga caused by gabbronorite intrusions: Implications for tectonic evolution of the northern margin of the North China Craton. *Precambrian Res.* **2012**, *222*, 124–142. [CrossRef]

29. Qian, X.L.; Cui, W.Y.; Wang, S.Q. Evolution of the Inner Mongolia-eastern Hebei Archean Granulite Belt in the North China Craton. In *The Records of Geological Research*; Department of Geology, Beijing University Press: Beijing, China, 1985.

30. Shen, Q.H.; Xu, H.F.; Zhang, Z.Q.; Gao, J.F.; Wu, J.S.; Ji, C.L. *Precambrian Granulites in China*; Geological Publishing House: Beijing, China, 1992.
31. Zhai, M. Khondalite revisited-record of special geological processes on Earth. *Acta Petrol. Sin.* **2022**, *96*, 2967–2997, (In Chinese with English Abstract).
32. Condie, K.C.; Boryta, M.D.; Liu, J.Z. The origin of khondalites: Geochemical evidence from the Archean to Early Proterozoic granulite belt in the North China Craton. *Precambrian Res.* **1992**, *59*, 207–223. [CrossRef]
33. Lu, L.Z.; Jin, S.Q. P-T-t paths and tectonic history of an early Precambrian granulite-facies terrane, Jining District, South-East Inner-Mongolia, China. *J. Metamorph. Geol.* **1993**, *11*, 483–498.
34. Wan, Y.S.; Liu, D.Y.; Dong, C.Y.; Xu, Z.Y.; Wang, Z.J.; Wilde, S.A.; Yang, Y.H.; Liu, Z.H.; Zhou, H.Y. The Precambrian Khondalite Belt in the Daqingshan area, North China Craton: Evidence for multiple metamorphic events in the Palaeoproterozoic era. *Geol. Soc. Lond. Spec. Publ.* **2009**, *323*, 73–97. [CrossRef]
35. Dan, W.; Li, X.H.; Guo, J.H.; Liu, Y.; Wang, X.C. Integrated in situ zircon U–Pb age and Hf–O isotopes for the Helanshan khondalites in North China Craton: Juvenile crustal materials deposited in active or passive continental margin? *Precambrian Res.* **2012**, *222*, 143–158. [CrossRef]
36. Wang, L.J.; Guo, J.H.; Yin, C.Q.; Peng, P. Petrogenesis of ca. 1.95 Ga meta-leucogranites from the Jining Complex in the Khondalite Belt, North China Craton: Water-fluxed melting of metasedimentary rocks. *Precambrian Res.* **2017**, *303*, 355–371. [CrossRef]
37. Li, W.X.; Yin, C.Q.; Lin, S.F.; Li, W.J.; Gao, P.; Zhang, J.; Qian, J.H.; Qiao, H.Z. Paleoproterozoic tectonic evolution from subduction to collision of the Khondalite Belt in North China: Evidence from multiple magmatism in the Qianlishan Complex. *Precambrian Res.* **2022**, *368*, 106471. [CrossRef]
38. Yin, C.Q.; Zhao, G.C.; Xiao, W.J.; Lin, S.F.; Gao, R.; Zhang, J.; Qian, J.H.; Gao, P.; Qiao, H.Z.; Li, W.X. Paleoproterozoic accretion and assembly of the Western Block of North China: A new model. *Earth-Sci. Rev.* **2023**, *241*, 104448. [CrossRef]
39. Wan, Y.S.; Song, B.; Liu, D.Y.; Wilde, S.A.; Wu, J.S.; Shi, Y.R.; Yin, X.Y.; Zhou, H.Y. SHRIMP U–Pb zircon geochronology of Palaeoproterozoic metasedimentary rocks in the North China Craton: Evidence for a major Late Palaeoproterozoic tectonothermal event. *Precambrian Res.* **2006**, *149*, 249–271. [CrossRef]
40. Wan, Y.S.; Xu, Z.Y.; Dong, C.Y.; Nutman, A.; Ma, M.Z.; Xie, H.Q.; Liu, S.J.; Liu, D.Y.; Wang, H.C.; Cu, H. Episodic paleoproterozoic (~2.45, ~1.95 and ~1.85 Ga) mafic magmatism and associated high temperature metamorphism in the Daqingshan area, North China Craton: SHRIMP zircon U–Pb dating and whole-rock geochemistry. *Precambrian Res.* **2013**, *224*, 71–93. [CrossRef]
41. Xia, X.P.; Sun, M.; Zhao, G.C.; Luo, Y. LA-ICP-MS U–Pb geochronology of detrital zircons from the Jining Complex, North China Craton and its tectonic significance. *Precambrian Res.* **2006**, *144*, 199–212. [CrossRef]
42. Xia, X.P.; Sun, M.; Zhao, G.C.; Wu, F.Y.; Xu, P.; Zhang, J.H.; Luo, Y. U–Pb and Hf isotopic study of detrital zircons from the Wulashan khondalites: Constraints on the evolution of the Ordos Terrane, Western Block of the North China Craton. *Earth Planet. Sci. Lett.* **2006**, *241*, 581–593. [CrossRef]
43. Xia, X.P.; Sun, M.; Zhao, G.C.; Wu, F.Y.; Xu, P.; Zhang, J.; He, Y.H. Paleoproterozoic crustal growth in the Western Block of the North China Craton: Evidence from detrital zircon Hf and whole rock Sr-Nd isotopic compositions of the Khondalites from the Jining Complex. *Am. J. Sci.* **2008**, *308*, 304–327. [CrossRef]
44. Dong, C.Y.; Liu, D.Y.; Li, J.J.; Wan, Y.S.; Zhou, H.Y.; Li, C.D.; Yang, Y.H.; Xie, L.W. Palaeoproterozoic Khondalite Belt in the western North China Craton: New evidence from SHRIMP dating and Hf isotope composition of zircons from metamorphic rocks in the Bayan Ul-Helan Mountains area. *Chin. Sci. Bull.* **2007**, *52*, 2984–2994, (In Chinese with English Abstract). [CrossRef]
45. Dong, C.Y.; Wan, Y.S.; Xu, Z.Y.; Liu, D.Y.; Yang, Z.S.; Ma, M.Z.; Xie, H.Q. SHRIMP zircon U–Pb dating of late Paleoproterozoic kondalites in the Daqing Mountains area on the North China Craton. *Sci. China Earth Sci.* **2013**, *56*, 115–125. [CrossRef]
46. Yin, C.Q.; Zhao, G.C.; Guo, J.H.; Sun, M.; Xia, X.P.; Zhou, X.W.; Liu, C.H. U–Pb and Hf isotopic study of zircons of the Helanshan Complex: Constrains on the evolution of the Khondalite Belt in the Western Block of the North China Craton. *Lithos* **2011**, *122*, 25–38. [CrossRef]
47. Jiao, S.J.; Guo, J.H.; Harley, S.L.; Peng, P. Geochronology and trace element geochemistry of zircon, monazite and garnet from the garnetite and/or associated other high-grade rocks: Implications for Palaeoproterozoic tectonothermal evolution of the Khondalite Belt, North China Craton. *Precambrian Res.* **2013**, *237*, 78–100. [CrossRef]
48. Liu, P.H.; Liu, F.L.; Liu, C.H.; Liu, J.H.; Wang, F.; Xiao, L.L.; Cai, J.; Shi, J.R. Multiple mafic magmatic and high-grade metamorphic events revealed by zircons from meta-mafic rocks in the Daqingshan–Wulashan Complex of the Khondalite Belt, North China Craton. *Precambrian Res.* **2014**, *246*, 334–357. [CrossRef]
49. Liu, P.H.; Liu, F.L.; Cai, J.; Liu, C.H.; Liu, J.H.; Wang, F.; Xiao, L.L.; Shi, J.R. Spatial distribution, P–T–t paths, and tectonic significance of high-pressure mafic granulites from the Daqingshan–Wulashan Complex in the Khondalite Belt, North China Craton. *Precambrian Res.* **2017**, *303*, 687–708. [CrossRef]
50. Gou, L.L.; Zhang, C.L.; Brown, M.; Piccoli, P.M.; Lin, H.B.; Wei, X.S. P–T–t evolution of pelitic gneiss from the basement underlying the Northwestern Ordos Basin, North China Craton, and the tectonic implications. *Precambrian Res.* **2016**, *276*, 67–84. [CrossRef]
51. Cai, J.; Liu, F.L.; Liu, P.H. Paleoproterozoic multistage metamorphic events in Jining metapelitic rocks from the Khondalite Belt in the North China Craton: Evidence from petrology, phase equilibria modelling and U–Pb geochronology. *J. Asian Earth Sci.* **2017**, *138*, 515–534. [CrossRef]
52. Cai, J.; Liu, F.L.; Liu, P.H.; Wang, F.; Liu, C.H.; Shi, J.R. Anatectic record and P–T path evolution of metapelites from the Wulashan Complex, Khondalite Belt, North China Craton. *Precambrian Res.* **2017**, *303*, 10–29. [CrossRef]

53. Xu, X.F.; Gou, L.L.; Long, X.P.; Dong, Y.P.; Liu, X.M.; Zi, J.W.; Li, Z.H.; Zhang, C.L.; Liu, L.; Zhao, J. Phase equilibrium modelling and SHRIMP zircon U–Pb dating of medium-pressure pelitic granulites in the Helanshan complex of the Khondalite Belt, North China Craton, and their tectonic implications. *Precambrian Res.* **2018**, *314*, 62–75. [CrossRef]

54. Santosh, M.; Tsunogae, T.; Li, J.H.; Liu, S.J. Discovery of sapphirine-bearing Mg–Al granulites in the North China Craton: Implications for Paleoproterozoic ultrahigh temperature metamorphism. *Gondwana Res.* **2007**, *11*, 263–285. [CrossRef]

55. Jiao, S.J.; Fitzsimons, I.C.; Guo, J.H. Paleoproterozoic UHT metamorphism in the Daqingshan Terrane, North China Craton: New constraints from phase equilibria modeling and SIMS U–Pb zircon dating. *Precambrian Res.* **2017**, *303*, 208–227. [CrossRef]

56. Jiao, S.J.; Fitzsimons, I.C.; Zi, J.W.; Evans, N.J.; Mcdonald, B.J.; Guo, J.H. Texturally controlled U–Th–Pb monazite geochronology reveals Paleoproterozoic UHT metamorphic evolution in the Khondalite belt, North China craton. *J. Petrol.* **2020**, *61*, egaa023. [CrossRef]

57. Jiao, S.J.; Evans, N.J.; Guo, J.H.; Fitzsimons, I.C.W.; Zi, J.W.; McDonald, B.J. Establishing the PT path of UHT granulites by geochemically distinguishing peritectic from retrograde garnet. *Am. Mineral.* **2021**, *106*, 1640–1653. [CrossRef]

58. Gou, L.L.; Li, Z.H.; Liu, X.M.; Dong, Y.P.; Zhao, J.; Zhang, C.L.; Liu, L.; Long, X.P. Ultrahigh-temperature metamorphism in the Helanshan complex of the Khondalite Belt, North China Craton: Petrology and phase equilibria of spinel-bearing pelitic granulites. *J. Metamorph. Geol.* **2018**, *36*, 1199–1220. [CrossRef]

59. Gou, L.L.; Zi, J.W.; Dong, Y.P.; Liu, X.M.; Li, Z.H.; Xu, X.F.; Zhang, C.L.; Liu, L.; Long, X.P.; Zhao, Y.H. Timing of two separate granulite-facies metamorphic events in the Helanshan complex, North China Craton: Constraints from monazite and zircon U–Pb dating of pelitic granulites. *Lithos* **2019**, *350*, 105216. [CrossRef]

60. Jiao, S.J.; Guo, J.H. Paleoproterozoic UHT metamorphism with isobaric cooling (IBC) followed by decompression–heating in the Khondalite Belt (North China Craton): New evidence from two sapphirine formation processes. *J. Metamorph. Geol.* **2020**, *38*, 357–378. [CrossRef]

61. Zheng, Y.Y.; Qi, Y.; Zhang, D.; Jiao, S.J.; Huang, G.Y.; Guo, J.H. New insight from the first application of Ti-in-Quartz (TitaniQ) thermometry mapping in the eastern Khondalite Belt, North China Craton. *Front. Earth Sci.* **2022**, *10*, 860057. [CrossRef]

62. Wang, L.J.; Guo, J.H.; Yin, C.Q.; Peng, P.; Zhang, J.; Spencer, C.J.; Qian, J.H. High-temperature S-type granitoids (charnockites) in the Jining complex, North China Craton: Restite entrainment and hybridization with mafic magma. *Lithos* **2018**, *320–321*, 435–453. [CrossRef]

63. Wang, L.J.; Guo, J.H.; Peng, P. Petrogenesis of Paleoproterozoic Liangcheng garnet granitoids in the Khondalite Belt, North China Craton. *Acta Petrol. Sin.* **2021**, *37*, 375–390, (In Chinese with English Abstract).

64. Shi, Q.; Ding, D.; Xu, Z.Y.; Li, W.Q.; Li, G.; Li, C.X.; Zhao, Z.H.; Zhang, G.B.; Jiang, X.Y.; Yang, R.B.; et al. Metamorphic evolution of Daqingshan supracrustal rocks and garnet granite from the North China Craton: Constraints from phase equilibria modelling, geochemistry, and SHRIMP U–Pb geochronology. *Gondwana Res.* **2021**, *97*, 101–120. [CrossRef]

65. Jiang, X.Z.; Yu, S.Y.; Liu, Y.J.; Li, S.Z.; Lv, P.; Peng, Y.B.; Gao, X.Y.; Ji, W.T.; Li, C.Z.; Xie, W.M. Episodic metamorphism and anatexis within the Khondalite Belt, North China Craton: Constraint from Late-Paleoproterozoic fluid-fluxed melting of the Daqingshan Complex. *Precambrian Res.* **2022**, *369*, 106504. [CrossRef]

66. Qiao, H.Z. Detrital zircon U–Pb ages of the Huangqikou Formation of the Changcheng System in the Qianlishan Area, western North China Craton and their geological implications. *Acta Geol. Sichuan* **2021**, *41*, 33–39. (In Chinese with English Abstract)

67. Li, W.J. Geochronology and Geochemistry of the Helanshan-Qianlishan Granites in North China. Master's Dissertation, Sun Yat-sen University, Guangzhou, China, 2018; pp. 1–103, (In Chinese with English Abstract).

68. Lu, C.S.; Qian, J.H.; Yin, C.Q.; Gao, P.; Guo, M.J.; Zhang, W.F. Ultrahigh temperature metamorphism recorded in the Lüliang Complex, Trans-North China Orogen: P–T–t evolution and heating mechanism. *Precambrian Res.* **2022**, *383*, 106900. [CrossRef]

69. Wiedenbeck, M.; Allé, P.; Corfu, F.; Griffin, W.L.; Meier, M.; Oberli, F.; Quadt, A.; Roddick, J.C.; Spiegel, W. Three natural zircon standards for U-Th-Pb, Lu-Hf, trace element and REE analyses. *Geostand. Newsl.* **1995**, *19*, 1–23. [CrossRef]

70. Reed, W. *Certificate of Analysis: Standard Reference Materials 610 and 611*; National Institute of Standards and Technology: Gaithersburg, MD, USA, 1992.

71. Sláma, J.; Kosler, J.; Condon, D.; Crowley, J.L.; Gerdes, A.; Hanchar, J.M.; Horstwood, M.S.A.; Morris, G.A.; Nasdala, L.; Norberg, N.; et al. Plešovice zircon–A new natural reference material for U–Pb and Hf isotopic microanalysis. *Chem. Geol.* **2008**, *249*, 1–35. [CrossRef]

72. Liu, Y.S.; Hu, Z.C.; Gao, S.; Günther, D.; Xu, J.; Gao, C.G.; Chen, H.H. In situ analysis of major and trace elements of anhydrous minerals by LA-ICP-MS without applying an internal standard. *Chem. Geol.* **2008**, *257*, 34–43. [CrossRef]

73. Ludwig, K.R. *Isoplot/Ex Version 4.15: A Geochronological Toolkit for Microsoft Excel*; Berkeley Geochronology Center Special Publication: Berkeley, CA, USA, 2012; Volume 5, p. 75.

74. Rubatto, D. Zircon trace element geochemistry: Partitioning with garnet and the link between U–Pb ages and metamorphism. *Chem. Geol.* **2002**, *184*, 123–138. [CrossRef]

75. Ramsay, J.G. *Folding and Fracturing of Rocks*; McGraw-Hill: New York, NY, USA, 1967; p. 568.

76. Liu, S.J.; Dong, C.Y.; Xu, Z.Y.; Santosh, M.; Ma, M.Z.; Xie, H.Q.; Liu, D.Y.; Wan, Y.S. Palaeoproterozoic episodic magmatism and high-grade metamorphism in the North China Craton: Evidence from SHRIMP zircon dating of magmatic suites in the Daqingshan area. *Geol. J.* **2013**, *48*, 429–455. [CrossRef]

77. Li, W.J.; Yin, C.Q.; Long, X.P.; Zhang, J.; Xia, X.P.; Wang, L.J. Paleoproterozoic S-type granites from the Helanshan Complex in Inner Mongolia: Constraints on the provenance and the Paleoproterozoic evolution of the Khondalite Belt, North China Craton. *Precambrian Res.* **2017**, *299*, 195–209. [CrossRef]
78. Qiao, H.Z.; Yin, C.Q.; Xiao, W.J.; Zhang, J.; Qian, J.H.; Wu, S.J. Paleoproterozoic polyphase deformation in the Helanshan Complex: Structural and geochronological constraints on the tectonic evolution of the Khondalite Belt, North China Craton. *Precambrian Res.* **2022**, *368*, 106468. [CrossRef]
79. Gan, S.F.; Qian, X.L. A plate-tectonic model for the evolution of the Daqingshan granulite belt in Inner Mongolia, China. *Acta Geol. Sin.* **1996**, *70*, 298–308, (In Chinese with English Abstract).
80. Yu, H.F.; Sun, D.Y. Deformational-Metamorphic Evolution of Ductile Shear Zone in Daqingshan Area. *J. Chang. Univ. Earth Sci.* **1996**, *26*, 310–315, (In Chinese with English Abstract).
81. Liu, T.J.; Li, W.M.; Liu, Y.J.; Jin, W.; Shao, Y.L. Rheology of the anatectic mid-lower Crust in the Paleoproterozoic orogenic belt in Daqingshan. Inner Mongolia. *Acta Petrol. Sin.* **2020**, *55*, 459–486, (In Chinese with English Abstract).
82. Liu, T.J.; Li, W.M.; Liu, Y.J.; Jin, W.; Zhao, Y.L.; Iqbal, M.Z. Deformation characteristics of the high-grade metamorphic and anatectic rocks in the Daqingshan Paleoproterozoic orogenic belt, Inner Mongolia: A case study from the Shijiaqu-Xuehaigou area. *Precambrian Res.* **2022**, *374*, 106644. [CrossRef]
83. Xu, Z.Y.; Liu, Z.H.; Yang, Z.S. Structures of early metamorphic strata in the khondalite series in the Daqingshan-Wulashan area, Inner Mongolia: Results of the sub-horizontal bedding-parallel detachment deformation in the lower crust. *J. Stratigr.* **2005**, *29*, 423–432, (In Chinese with English Abstract).
84. Xu, Z.Y.; Liu, Z.H.; Yang, Z.S.; Wu, X.W.; Chen, X.F. Structure of metamorphic strata of the khondalite series in the Daqingshan-Wulashan area, central Inner Mongolia, China, and their geodynamic implications. *Geol. Bull. China* **2007**, *26*, 526–536. (In Chinese with English Abstract)
85. Liu, Z.H.; Pan, B.W.; Li, P.C.; Zhu, K.; Dong, X. Ductile Shear Zone in High-Grade Metamorphic Rocks and Its Rheomorphic Mechanism in the Daqing Mountain Area, Inner Mongolia. *Earth Sci.* **2017**, *42*, 2105–2116, (In Chinese with English Abstract).
86. Qiao, H.Z.; Yin, C.Q.; Zhang, J.; Qian, J.H.; Wu, S.J. New Discovery of ~1866 Ma high-temperature mylonite in the Helanshan Complex: Marking a late-stage ductile shearing in the Khondalite Belt, North China Craton. *Acta Geol. Sin. (Engl. Ed.)* **2021**, *95*, 1418–1419. [CrossRef]

Article

Timing and Tectonic Implications of the Development of the Orosirian Qianlishan Ductile Shear Zones in the Khondalite Belt, North China Craton

Hengzhong Qiao [1,2,*], Miao Liu [2] and Chencheng Dai [2]

[1] Department of Geology, Northwest University, Xi'an 710069, China
[2] College of Tourism and Geographical Science, Leshan Normal University, Leshan 614000, China
* Correspondence: qiaohzh@lsnu.edu.cn

Abstract: Orogen-parallel ductile shear zones are conspicuous structures in the Khondalite Belt, but the timing of shearing remains poorly understood. Here, we present field-based structural and zircon U-Pb geochronological studies on the newly discovered Qianlishan ductile shear zones (QDSZ) in the Khondalite Belt. Our results show that the nearly E-W-trending QDSZ are characterized by steeply S(SW)-dipping mylonitic foliations and mainly display a top-to-N(NE) sense of shearing. Two pre-kinematic intrusions yielded zircon crystallization ages of 2055 ± 17 Ma and 1947 ± 9 Ma, providing the maximum age limit for the QDSZ. Additionally, zircon overgrowth rims from three high-temperature mylonites gave metamorphic ages of 1902 ± 8 Ma, 1902 ± 26 Ma and 1884 ± 12 Ma, interpreted to record the timing of development of the QDSZ. Integrated with previous studies, we propose that the Qianlishan Complex suffered three phases of Orosirian deformation (D1–D3), of which the D3 deformation led to the development of the QDSZ. Deformation events D1, D2 and D3 are considered to have occurred at ca. 1.97–1.93 Ga, 1.93–1.90 Ga and 1.90–1.82 Ga, respectively. These events document that the Khondalite Belt underwent a protracted (>100 Myr) orogenic history in response to the collision between the Yinshan and Ordos blocks.

Keywords: shear zone; mylonite; zircon U-Pb age; Paleoproterozoic; Khondalite Belt; North China Craton

Citation: Qiao, H.; Liu, M.; Dai, C. Timing and Tectonic Implications of the Development of the Orosirian Qianlishan Ductile Shear Zones in the Khondalite Belt, North China Craton. *Minerals* **2024**, *14*, 561. https://doi.org/10.3390/min14060561

Academic Editor: Paris Xypolias

Received: 10 May 2024
Revised: 25 May 2024
Accepted: 27 May 2024
Published: 29 May 2024

1. Introduction

Shear zones are typically known as tabular high-strain zones that play a major role in accommodating tectonic strain, displacement and fluid migration in the lithosphere [1–6]. Such zones are important structures in orogeny and can occur at a variety of scales, which significantly influence the geometry and evolution of orogenic belts [4,7–9]. Mylonites are characteristic rock types of shear zones at middle to lower crustal levels, and they generally develop a range of ductile structures and fabrics, providing valuable information on the kinematic and tectonothermal processes during orogeny [10–13]. Therefore, the recognition and anatomy of ductile shear zones is a key requirement in geological investigations that attempt to unravel the deformation history and tectonic evolution of orogenic belts [4,14–17]. To achieve this goal, detailed field-based structural analysis integrated with geochronological studies on ductile shear zones is fundamentally necessary [8,12,13].

The Khondalite Belt has been widely regarded as an Orosirian continent-continent collisional orogen in the northwestern North China Craton (Figure 1) [18–23]. It extends nearly E-W for ~1000 km and is mainly composed of four high-grade metamorphic complexes, which from west to east are the Helanshan, Qianlishan, Wulashan-Daqingshan and Jining complexes [18,24]. Of particular interest is the widespread occurrence of NE to E-striking ductile shear zones in the Khondalite Belt, such as the Zongbieli, Xiashihao-Jiuguan and Xuwujia ductile shear zones of the Helanshan, Wulashan-Daqingshan and Jining complexes, respectively [24–28]. However, there is no evidence of orogen-parallel ductile shear zones in the Qianlishan Complex. Meanwhile, the formation and evolution of ductile shear zones in the Khondalite Belt have been poorly constrained, which inevitably hampers a better understanding of the polyphase deformation in the belt. Based on field investigations, we newly discovered nearly E-W-trending ductile shear zones in the Qianlishan Complex, named together as the Qianlishan ductile shear zones (QDSZ). In this study, we conducted field-based structural observations and LA-ICP-MS zircon U-Pb geochronology on the QDSZ. Combined with previous works, particularly the compilation of deformation-related geochronological data of the Khondalite Belt, these results shed new light on the deformation history and tectonic evolution of the Khondalite Belt.

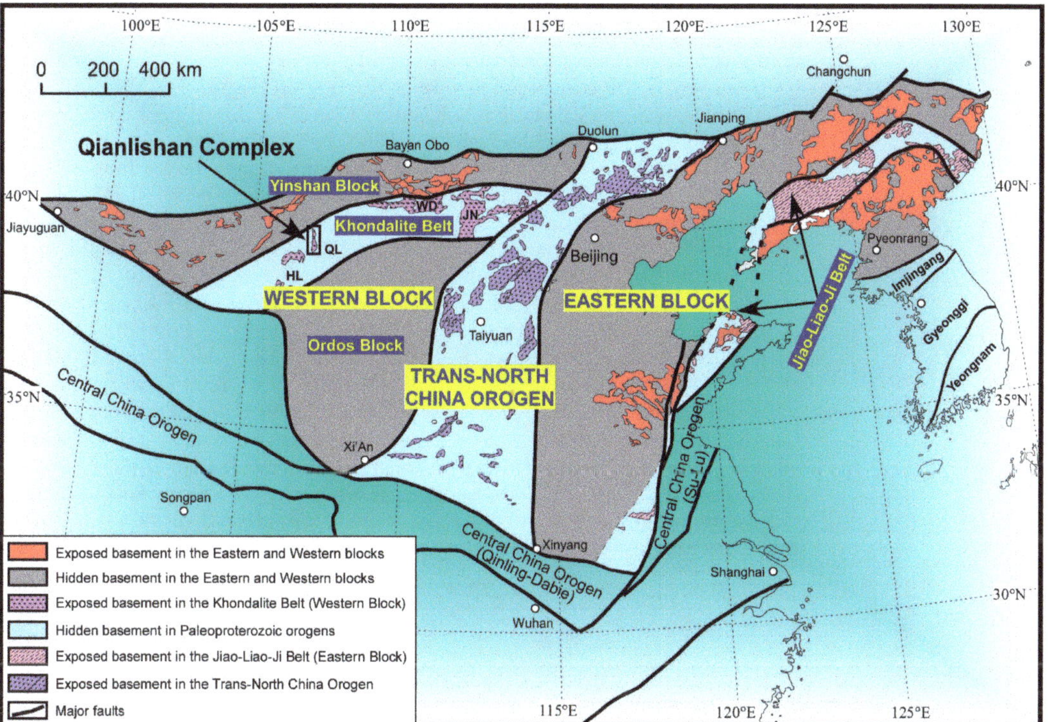

Figure 1. Tectonic subdivision of the Precambrian basement of the North China Craton (modified after [18]). Notes: HL, Helanshan Complex; QL, Qianlishan Complex; WD, Wulashan-Daqingshan Complex; JN, Jining Complex.

2. Geological Setting

The North China Craton, one of the oldest cratons in the world, is proposed to have formed from the assembly of several Archean to Paleoproterozoic continental blocks along orogenic belts (Figure 1) [18,19,29–31]. In the western North China Craton, the Yinshan Block in the north was considered to collide with the Ordos Block in the south at ~1.95 Ga, resulting in the formation of the nearly E-W-trending Khondalite Belt [18–23,28,32–41]. This belt is dominated by polydeformed high-grade metasedimentary rocks (i.e., the khondalites), mainly including garnet-sillimanite gneisses, felsic paragneisses, quartzites, marbles and calc-silicate rocks [18,24,42]. Spatially associated with the khondalites are S-type granites, charnockites, minor TTG gneisses, intermediate-mafic magmatic rocks and mafic granulites [18,24,42]. Numerous geochronological studies reveal that the protolith of the khondalites was deposited at ca. 2.00–1.95 Ga, and then suffered regional upper-amphibolite to granulite-facies metamorphism at ca. 1.95–1.80 Ga [20,32–38,40,43–57]. Of these, high-pressure pelitic granulites preserved similar clockwise P-T paths with post-peak near-isothermal decompression, and they indicate tectonic processes involving initial crustal thickening and subsequent rapid exhumation, related to the collisional orogeny between the Yinshan and Ordos blocks [37,50,52,53,58–63]. Noticeably, a series of orogen-parallel ductile shear zones commonly occurred in the Khondalite Belt, for example, the NE-SW to E-W-trending Zongbieli shear zones in the Helanshan Complex, the nearly E-W-trending Xiashihao-Jiuguan shear zone in the Wulashan-Daqingshan Complex and the NE(E)-SW(W)-trending Xuwujia shear zone in the Jining Complex [24–28,64,65].

The Qianlishan Complex is located in the western segment of the Khondalite Belt and is unconformably covered by Mesoproterozoic sedimentary rocks of the Changcheng-Jixian System (Figures 1 and 2) [18,21,24,65,66]. The complex predominantly comprises Paleoproterozoic granitoid plutons and the khondalites that have been known as the Qianlishan Group in this region (Figure 2) [20,21,24,39]. This group consists of the Chaganguole, Qianligou and Habuqigai formations, in which high-pressure pelitic granulites were found and characterized by a peak mineral assemblage composed of garnet + kyanite + K-feldspar + plagioclase + biotite + quartz, with corresponding P-T conditions at 11–15 kbar and 800–850 °C [37,60]. Available geochronological data of the Qianlishan Group demonstrate that U-Pb ages of detrital zircons dominantly range from 2.20 to 2.00 Ga, with a notable peak at ~2.03 Ga [20,37,67]. Metamorphic zircons yielded age groups of ~1.95 Ga and ~1.92 Ga, of which the former and latter were related to granulite-facies peak metamorphism and post-peak decompression processes, respectively [20,37,67]. Meanwhile, multiple magmatic pulses have been dated in the studied area at ca. 2.06 Ga, 1.95 Ga, 1.92 Ga and 1.88 Ga [20,21,39,68]. Additionally, previous structural investigations revealed two major phases of Paleoproterozoic deformation (D1–D2) in the Qianlishan Complex [21,24,39,69]. D1 mainly generated overturned to recumbent isoclinal folds (F1), transposition foliations/gneissosities (S1) and mineral lineations (L1) [21,69]. D2 produced tight to open doubly plunging upright folds (F2) [21,39]. Deformation ages of D1 and D2 were roughly constrained at 1976–1932 Ma and 1932–1899 Ma, respectively [21,39].

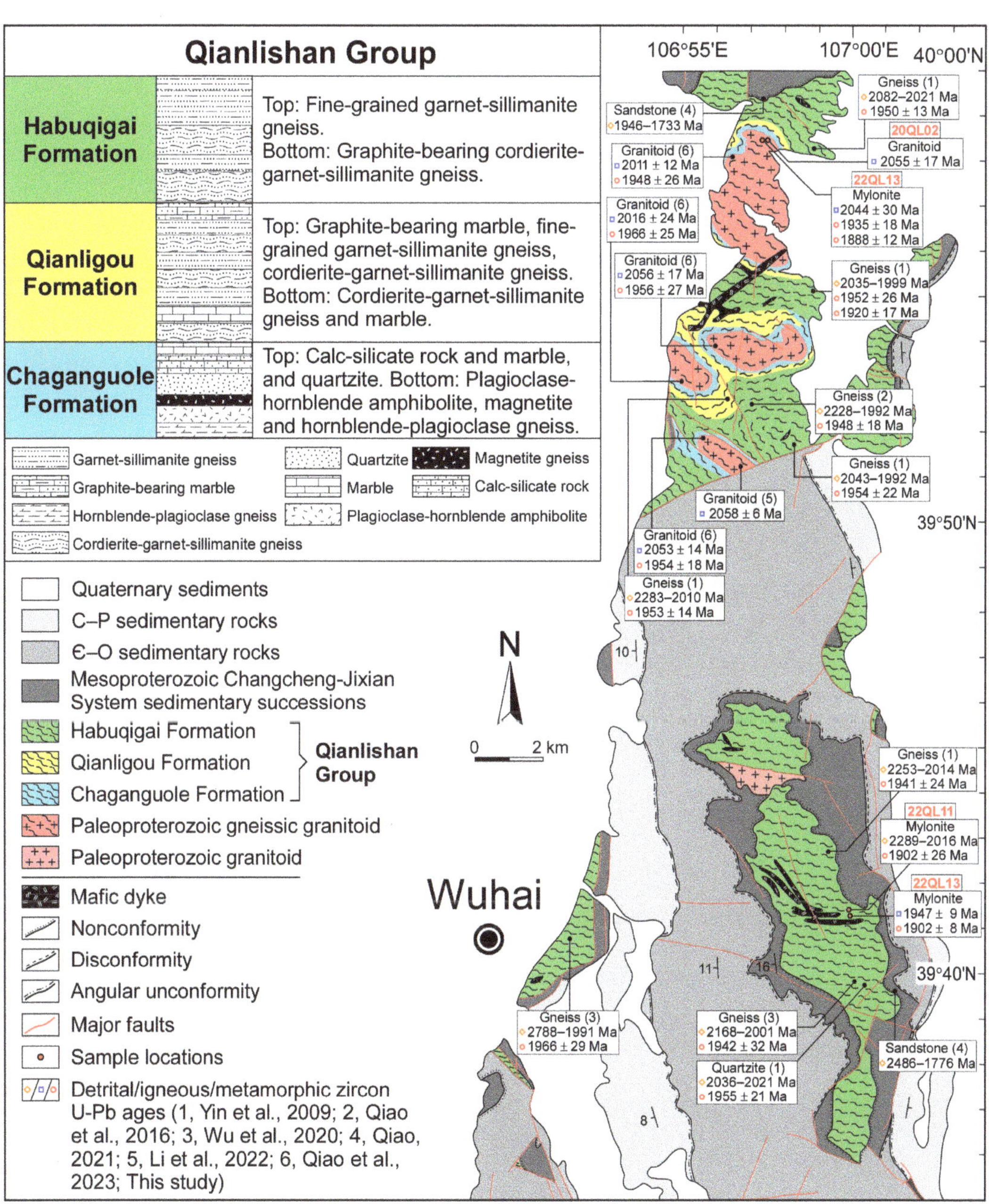

Figure 2. Simplified geological map of the Qianlishan Complex (modified after [20,21,39]). Representative geochronological data are from [20,37,39,66–68].

3. Samples and Methods

In this study, we conducted field-based (micro-)structural analysis to investigate the geometries, kinematics and temperature conditions of the QDSZ. Additionally, we collected four shear zone-related samples for zircon U-Pb geochronology, including three high-temperature mylonites (Samples 22QL03, 22QL13 and 22QL11) and a pre-kinematic granitoid pluton (Sample 20QL02). Zircons were extracted by magnetic and heavy liquid separation techniques. Subsequently, they were handpicked, mounted in epoxy resin, polished and photographed under reflected and transmitted light. The internal textures of these zircons were then studied through cathodoluminescence (CL) images. Zircon U-Pb analyses were performed by Laser Ablation Inductively Coupled Plasma Mass Spectrometry (LA-ICP-MS) at the Guangzhou Tuoyan Analytical Technology Co., Ltd., Guangzhou, China. Detailed analytical methods and procedures are similar to those in [70]. The spot size and frequency of the laser were set to 30 μm and 6 Hz, respectively. Zircon 91500 and glass NIST610 were used as external standards for U-Pb dating and trace element calibration, respectively. Zircon Plešovice was treated as an unknown sample to monitor the accuracy of acquired U-Pb data. The software ICPMSDataCal was used to conduct off-line inspection and integration of background and analytical signals, time-drift correction, and quantitative calibration for U-Pb dating and trace element analysis [71]. Concordia diagrams and age calculations were made using Isoplot/Ex_ver4.15 [72]. The uncertainties on weighted mean ages and intercept ages were quoted at the 95% confidence level (2σ). The zircon U-Pb data of this study are presented as Supplementary Material (Table S1).

4. Results

4.1. Structural Observations

The QDSZ mainly consist of a series of small narrow high-strain zones (e.g., Figures 3a,e and 4), but there is no map view of the large-scale shear zones observed in the Qianlishan Complex. These zones dominantly strike E-W or SEE-NWW and are characterized by mylonitized rocks with varying deformation intensities. For example, a 1–2 m wide E-W-trending ductile shear zone was remarkably developed within a granitoid pluton, which suffered intensive shear deformation to become granitic mylonite (Figure 3). Regionally, the mylonitic foliations (Sm) in the QDSZ dip mostly to the S(SW) with steep to sub-vertical angles (Figures 3 and 4). At some localities, moderately N(NE)-plunging stretching lineations (Lm) can be seen on the mylonitic foliation (Figure 4f). The appearance of kinematic indicators, including S-C fabrics, σ- and δ-type porphyroclasts and felsic aggregates (Figure 5), suggest a nearly top-to-N(NE) sense of shearing.

In addition, it is worth noting that quartz grains in felsic and granitic mylonites generally occur as ribbons with straight boundaries (e.g., Figures 3c,f and 4b,d,i), and they exhibit lobate to rectangular shapes in thin sections (Figures 3d and 4c), which implies grain boundary migration recrystallization (GBM) dominates. Furthermore, these ribbons are alternated by fine-grained recrystallized feldspar-rich layers (Figures 3c,d and 4c). In some cases, feldspar fish are locally observed (Figure 3d). The above-stated (micro-)structures indicate that the QDSZ developed under high-temperature deformation conditions, probably >650 °C [10,11,73]. This interpretation is also supported by the observation of a series of felsic segregates related to melting during the mylonitization, which commonly appear along the mylonitic foliation (e.g., Figures 4d,g and 5c).

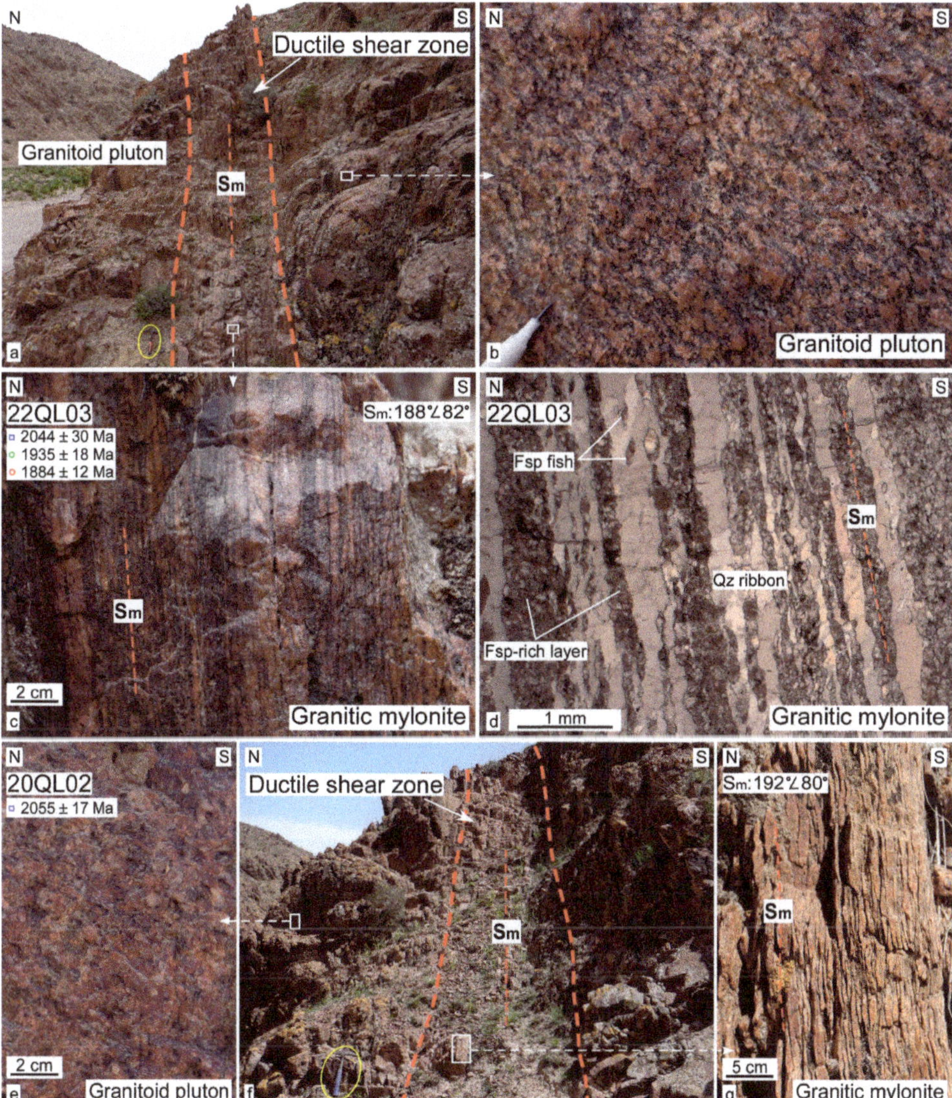

Figure 3. Representative photos showing shear zone-related structures and dated samples in the northern Qianlishan Complex. (**a,b**) E-W-trending ductile shear zone with a width of 1–2 m developed in a pre-kinematic granitoid pluton. (**c**) Within the shear zone, the granitoid was converted into granitic mylonite (Sample 22QL03). (**d**) Sample 22QL03 features strongly aligned quartz ribbons and fine-grained recrystallized feldspar-rich layers. Feldspar fish can be locally observed. (**e–g**) In another outcrop, a pre-kinematic granitoid pluton (Sample 20QL02) was affected by shear zone activity and developed mylonitic foliation. The blue square, and green and red circles indicate zircon U-Pb ages of magmatic cores, and inner and outer metamorphic overgrowth rims, respectively. Sm, mylonitic foliation; Fsp, feldspar; Qz, quartz.

Figure 4. Representative photos showing shear zone-related structures and dated samples in the southern Qianlishan Complex. (**a**) Pre-kinematic granitic dyke (~1936 Ma) that underwent the shear deformation was converted into granitic mylonite. (**b,c**) The granitic mylonite is characterized by the preferential orientation of feldspar-rich layers, rectangular polycrystalline quartz ribbons, and sillimanite and garnet porphyroclasts that define the mylonitic foliation (Sm). (**d**) Anatectic segregates commonly occur along the steeply S(SW)-dipping mylonitic foliation (Sm). (**e**) Pre-kinematic granitic dyke deformed by shear zone activity to form granitic mylonite (Sample 22QL13). (**f**) Moderately N(NE)-plunging stretching lineations (Lm) locally developed on the mylonitic foliation. (**g–i**) Metasedimentary rocks of the Habuqigai Formation mylonitized to a variable degree. Of these, quartz ribbons with straight boundaries can be observed in felsic mylonite (Sample 22QL11). Blue square, black triangle and red circle indicate zircon U-Pb ages of magmatic cores, inherited cores, and metamorphic overgrowth rims, respectively. Sil, sillimanite; Grt, garnet; Qz, quartz.

Figure 5. Kinematic indicators mainly showing a top-to-N(NE) sense of shearing in the QDSZ. (**a**) S-C fabrics and σ-type porphyroclasts. (**b**) σ-type felsic aggregate. (**c**) Sm-parallel anatectic segregates with asymmetric structures. (**d**) δ-type felsic aggregate. Grt, garnet.

4.2. Zircon U-Pb Geochronology

4.2.1. Pre-Kinematic Granitoid Pluton (Sample 20QL02)

Sample 20QL02 was collected from a granitoid pluton in the northern Qianlishan Complex (Figure 2; 39°58′31.32″ N, 106°57′43.21″ E). This granitoid was cut by an E-W-trending ductile shear zone (Figure 3e,f), indicating that it was formed prior to the shear zone activity. The sample mainly consists of fine-grained K-feldspar, plagioclase, quartz and hornblende, and is characterized by the typical granular texture (Figure 3e). Zircons separated from the sample are dominantly euhedral, prismatic and 250–450 μm in size. CL images reveal that these zircons have concentrically oscillatory-zoned cores with high luminescence (Figure 6a), interpreted to be of igneous origin. In some cases, they are surrounded by dark structureless rims that are considered to result from metamorphic processes, but these rims are too narrow to be analyzed (Figure 6a). A total of six spots with Th/U ratio of 0.64–0.85 were analyzed (discordance degree $\leq$ 6.03%; Table S1). They yielded a weighted mean ^{207}Pb/^{206}Pb age of 2055 ± 17 Ma (n = 6, MSWD = 2.8; Figure 7a). Consistently, these spots also defined an upper intercept age of 2047 ± 17 Ma (n = 6, MSWD = 6.7).

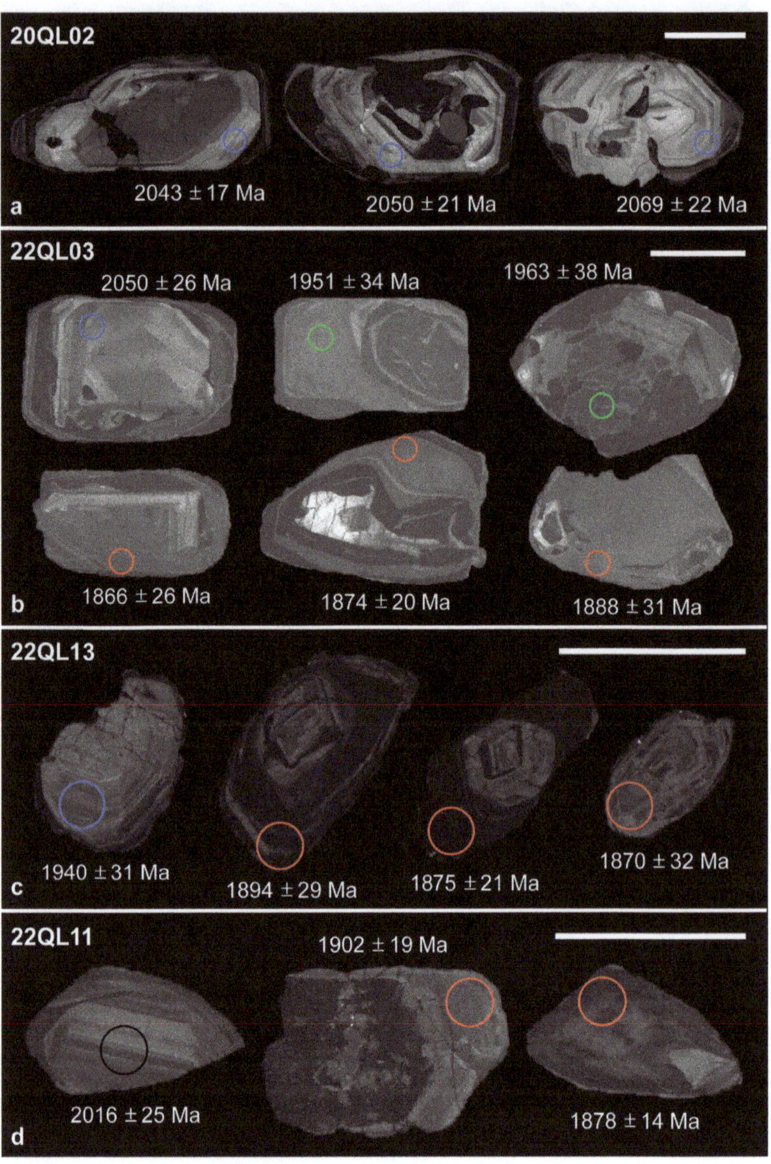

Figure 6. Representative CL images with their U-Pb ages of zircons from shear zone-related samples of the Qianlishan Complex. (**a**) Sample 20QL02. (**b**) Sample 22QL03. (**c**) Sample 22QL13. (**d**) Sample 22QL11. Circles indicate the analytical spots. Blue circle, magmatic zircon; black circle, inherited zircon; green circle, inner metamorphic rim; red circle, outer metamorphic rim. All scale bars are 100 μm.

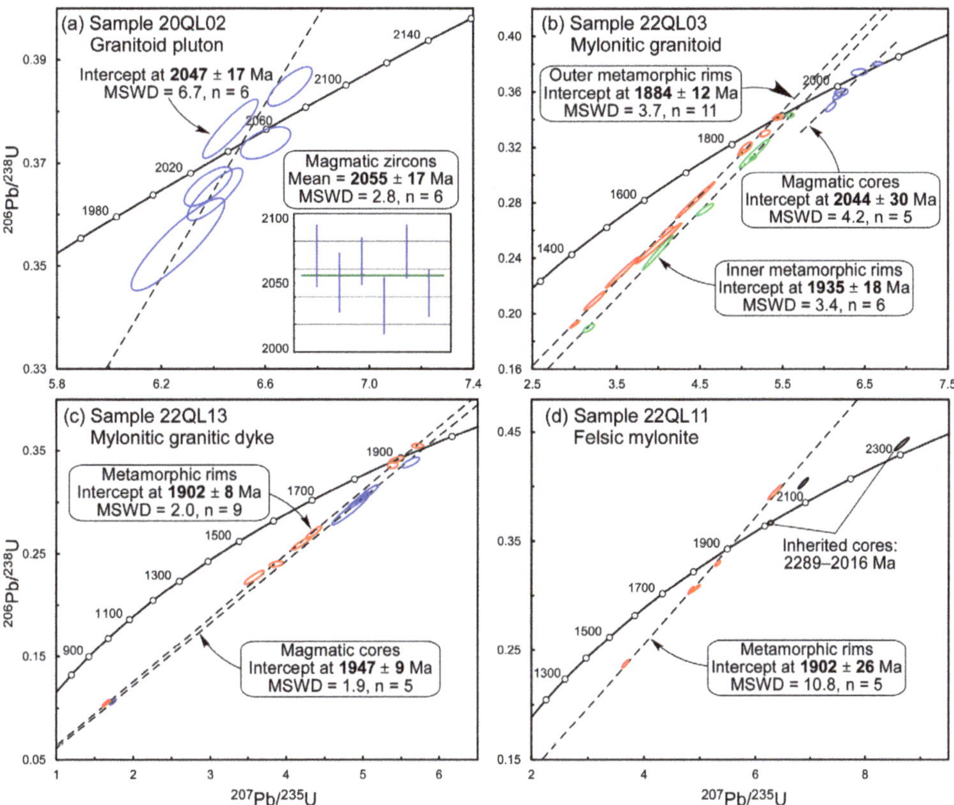

Figure 7. Concordia diagrams for zircon U-Pb data from shear zone-related samples of the Qianlishan Complex. (**a**) Sample 20QL02. (**b**) Sample 22QL03. (**c**) Sample 22QL13. (**d**) Sample 22QL11. Blue ellipse, magmatic zircon; black ellipse, inherited zircon; green ellipse, inner metamorphic rim; red ellipse, outer metamorphic rim. Uncertainties on the weighted mean age and the upper intercept ages were quoted at the 95% confidence level (2σ).

4.2.2. Mylonitic Granitoid Pluton (Sample 22QL03)

This sample was taken from a mylonitic granitoid adjacent to Sample 20QL02 (Figure 2; 39°58′31.35″ N, 106°57′41.53″ E). Within the shear zone, the granitoid underwent intensive ductile deformation to become a granitic mylonite (Sample 22QL03; Figure 3a–c). Sample 22QL03 is featured by sub-vertical mylonitic foliation (Sm), defined by the preferential orientation of fine-grained feldspar-rich layers and quartz ribbons (Figure 3c,d). Zircons extracted from this sample are euhedral to subhedral, prismatic or stubby and 200–350 μm in size. In CL images, they generally displayed complex textures, most of which possessed bright patchy or oscillatory-zoned cores (Figure 6b), indicative of magmatic origin. Remarkably, these zircon cores mostly retained double overgrowth rims, including the inner and outer rims (Figure 6b). The inner rims were commonly dark, nebulous or chaotic, and surrounded by the structureless outer rims (Figure 6b). Both the inner and outer rims were interpreted to be of metamorphic origin. A total of twenty-two spots were analyzed, of which five spots (discordance degree ≤ 6.10%, Table S1) corresponded to zircon cores and defined an upper intercept age of 2044 ± 30 Ma (n = 5, MSWD = 4.3; Figure 7b). Six spots on the inner metamorphic rims plot along a discordia line with an upper intercept age of 1935 ± 18 Ma (n = 6, MSWD = 3.4; Figure 7b). The remaining eleven spots corresponded to the outer metamorphic overgrowth rims, of which the two most concordant data (discordance degree $\leq$ 1.35%, Table S1) gave apparent ^{207}Pb/^{206}Pb ages of 1885 $\pm$ 29 Ma and 1866 $\pm$ 26 Ma. These eleven spots defined an upper intercept age of 1884 $\pm$ 12 Ma (n = 11, MSWD = 3.7; Figure 7b).

4.2.3. Mylonitic Granitic Dyke (Sample 22QL13)

Sample 22QL13 was collected from a mylonitic granitic dyke of the southern Qianlishan Complex (Figure 2; 39°41′13.44″ N, 107°00′01.50″ E). The sample was composed of strongly aligned grains of K-feldspar, plagioclase and quartz (Figure 4e). Most zircons from this sample were euhedral to subhedral, prismatic and 50–200 μm in size. CL images show that they are characterized by typical core-rim textures with bright and oscillatory-zoned cores (Figure 6c), evidently of magmatic origin. These cores are generally surrounded by dark, nebulous or structureless overgrowth rims, interpreted to be of metamorphic origin (Figure 6c). Fourteen spots were analyzed in this sample, and five spots on magmatic cores gave an upper intercept age of 1947 $\pm$ 9 Ma (n = 5, MSWD = 1.9; Figure 7c). The remaining nine spots were performed on metamorphic overgrowth rims, of which the most concordant spot yielded a ^{207}Pb/^{206}Pb age of 1899 $\pm$ 42 Ma (discordance degree = 1.84%, Figure 7c; Table S1). These nine spots defined an upper intercept age of 1902 $\pm$ 8 Ma (n = 9, MSWD = 2.0; Figure 7c).

4.2.4. Felsic Mylonite (Sample 22QL11)

Sample 22QL11 is a felsic mylonite of the southern Qianlishan Complex (Figure 2; 39°41′17.53″ N, 109°00′00.38″ E). Metasedimentary rocks of the Habuqigai Formation were subjected to ductile shear deformation and transformed into mylonites (Figure 4g,h). The mylonitic foliation (Sm) is characterized by the preferred alignment of feldspar-rich layers and quartz ribbons with straight boundaries (Figure 4i). Zircon grains from this sample are mainly subhedral to anhedral, stubby and 100–250 μm in size. CL images reveal that these zircons mostly possess patchy or oscillatory-zoned cores surrounded by nebulous overgrowth rims (Figure 6d). The cores and rims are regarded to be of magmatic and metamorphic origin, respectively. A total of eight spots were analyzed in this sample, of which three spots correspond to inherited cores and yielded concordant ^{207}Pb/^{206}Pb ages of 2289 $\pm$ 12 Ma, 2026 $\pm$ 12 Ma and 2016 $\pm$ 25 Ma (discordance degree $\leq$ 7.43%, Figure 7d; Table S1). The remaining five spots were analyzed on metamorphic overgrowth rims and defined an upper intercept age of 1902 $\pm$ 26 Ma (n = 5, MSWD = 10.8; Figure 7d).

5. Discussion

5.1. Timing of the Development of the Qianlishan Ductile Shear Zones (QDSZ)

Combined with field-based structural observations, geochronological data from shear zone-related intrusions and mylonites can help us to constrain the deformation age of the QDSZ. In the northern Qianlishan Complex, a small-scale ductile shear zone reworked a pre-kinematic granitoid (Figures 2 and 3). Magmatic zircons from this rock (Sample 20QL02, Figure 3e,f) yielded a weighted mean ^{207}Pb/^{206}Pb age of 2055 $\pm$ 17 Ma (Figures 6a and 7a). It is interpreted as the crystallization age of the granitoid and provides the maximum age limit for the shear zone (Figure 8a). Zircons from the granitic mylonite (Sample 22QL03, Figure 3a,c) within the shear zone mostly possess magmatic cores surrounded by both inner and outer metamorphic overgrowth rims (Figure 6b). The zircon cores gave a crystallization age of 2044 $\pm$ 30 Ma (Figure 7b), consistent with that obtained from Sample 20QL02. On the one hand, the inner rims yielded a metamorphic age of 1935 $\pm$ 18 Ma (Figure 7b). A similar metamorphic age of 1948 $\pm$ 26 Ma was obtained from a gneissic granitoid in another outcrop of the same pre-kinematic pluton (Figure 2) [20,21,39]. These metamorphic ages of 1948–1935 Ma are largely coherent with those of ~1950 Ma from high-pressure granulites of the Qianlishan Group (Figure 8a) [20,37,67], indicating that the granitoid also experienced regional metamorphism. On the other hand, the outer overgrowth rims gave a metamorphic age of 1884 $\pm$ 12 Ma (Figure 7b). This age was interpreted to most likely record the timing of the shear zone activity in the northern Qianlishan Complex.

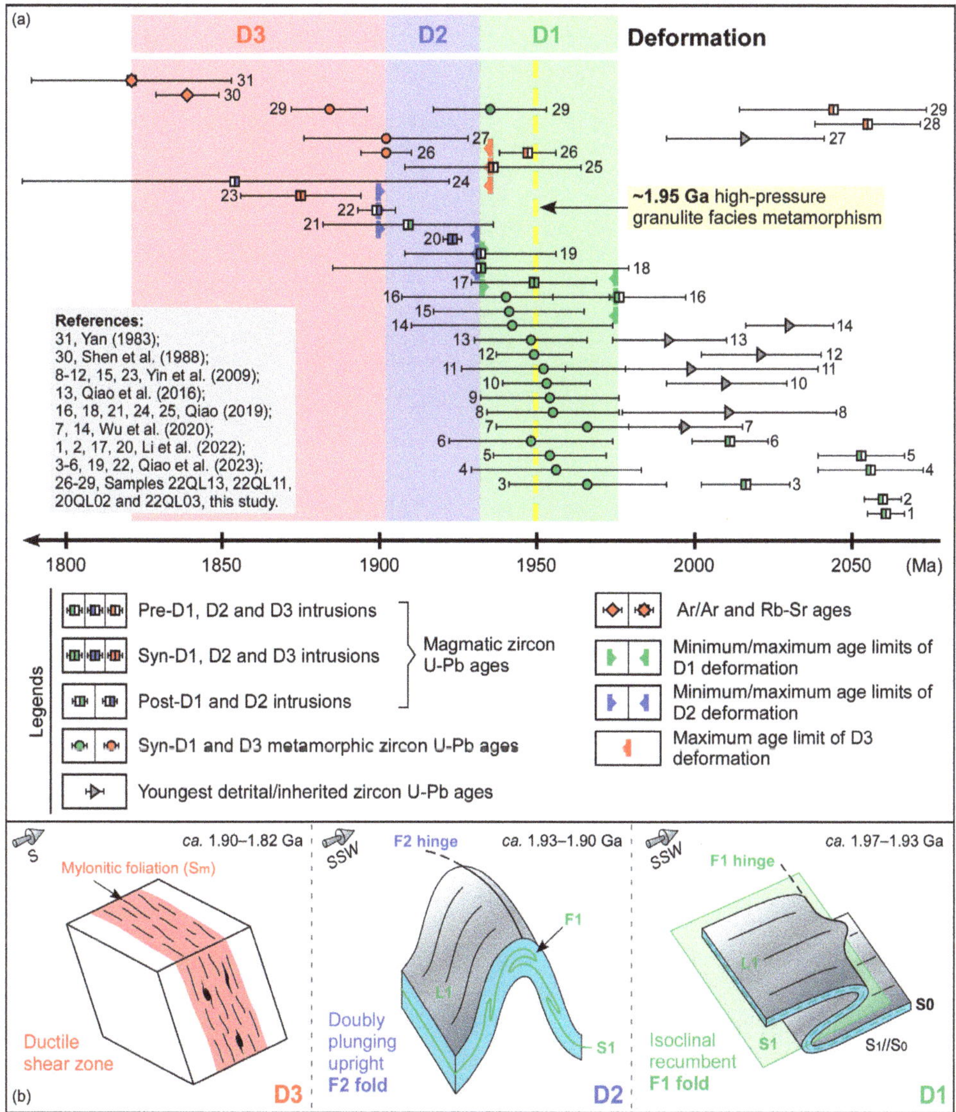

Figure 8. Schematic diagram showing the key geochronological data and polyphase deformation (D1–D3) in the Qianlishan Complex. (**a**) Summary of available geochronological data for the Qianlishan Complex. (**b**) The Qianlishan Complex suffered three phases of Orosirian deformation (D1–D3). Of these, the D3 deformation gave rise to the nearly E-W-trending Qianlishan ductile shear zones. It is here considered that the development of the QDSZ started at ca. 1902–1884 Ma and continued to ca. 1839–1821 Ma. Descriptions of the D1 and D2 structures can be found in [21,69]. Geochronological data are from [20,37,39,67 69,74,75] and this study. See the text for details.

In the southern Qianlishan Complex, a pre-kinematic granitic dyke was mylonitized (Sample 22QL13, Figure 4e). Zircons from this mylonite show typical core–rim textures (Figure 6c), of which the magmatic cores gave a $^{207}Pb/^{206}Pb$ age of 1947 ± 9 Ma (Figure 7c), regarded as the crystallization age of the dyke. Similarly, a mylonitized pre-kinematic granitic dyke was dated at 1936 ± 28 Ma (Figure 4a,b) [21,69]. These crystallization ages of 1947–1936 Ma indicate that the mylonitization must have occurred at some time after ~1936 Ma (Figure 8a). In this way, zircon overgrowth rims from Sample 22QL13 yielded a metamorphic age of 1902 ± 8 Ma (Figure 7c), considered to probably reflect the timing of the shear zone activity. This age is in good agreement with the dating results of Sample 22QL11 (Figure 4h,i), a felsic mylonite whose zircon overgrowth rims gave a metamorphic age of 1902 ± 26 Ma (Figures 6d and 7d). It is also worth noting that these ages of ~1902 Ma are comparable to that of 1884 ± 12 Ma obtained from the granitic mylonite (Sample 22QL03) of the northern Qianlishan Complex. Additionally, an Rb-Sr mineral isochron age of 1821 ± 32 Ma was obtained from a deformed metapelite of the southern Qianlishan Complex [74]. Shen et al. (1988) also reported a similar biotite ^{40}Ar/^{39}Ar plateau age of 1839 ± 10 Ma and interpreted that these ages of 1839–1821 Ma record the timing of a late-stage tectonothermal event after the granulite-facies peak metamorphism [75], probably related to the ductile shear zone activity. Therefore, we regard that the development of the QDSZ started at ca. 1902–1884 Ma and continued to ca. 1839–1821 Ma (Figure 8a).

5.2. Three Phases of Orosirian Deformation (D1–D3) in the Qianlishan Complex

Previous structural investigations show that the Qianlishan Complex underwent at least two major phases of deformation (D1–D2) in the late Paleoproterozoic (Figure 8b) [21,24,39,69]. The D1 deformation generated small-scale NWW-trending over-turned to recumbent isoclinal F1 folds, originally sub-horizontal penetrative S1 transposition foliations/gneissosities and NNE-SSW-oriented L1 mineral lineations [21]. Based on our previous geochronological results of pre-D1 and post-D1 leucocratic dykes (Table 1), D1 was regarded to roughly happen in the period between ~1976 Ma and ~1932 Ma (Figure 8) [21,39,69]. The extensive zircon U-Pb data also demonstrate that the Paleoproterozoic granitoids and the Qianlishan Group were together subjected to syn-D1 granulite-facies peak metamorphism at ~1950 Ma (Table 1; Figure 8) [20,21,37,39,67–69].

Table 1. Summary of available deformation-related geochronological data for the Khondalite Belt.

No. [1]	Sample	Rock Type and Structural Feature	Age (Ma)	Type [2]	n	MSWD	Interpretation	References
Qianlishan Complex								
1	-	Deformed metapelite	1821 ± 32	MIA	-	-	D3-related process	[74]
2	N80-48	Deformed metapelite	1839 ± 10	Bt/PA	-	-	D3-related process	[75]
3	22QL03	Granitic mylonite	1884 ± 12	M/UI	11	3.7	Syn-D3 metamorphism	This study
			1935 ± 18	M/UI	6	3.4	Syn-D1 metamorphism	
			2044 ± 30	I/UI	5	4.2	Pre-D3 intrusion	
4	20QL02	Granitoid cut by DSZ *	2055 ± 17	I/WM	6	2.8	Pre-D3 intrusion	This study
5	22QL13	Mylonitized granitic dyke	1902 ± 8	M/UI	9	2.0	Syn-D3 metamorphism	This study
			1947 ± 9	I/UI	5	1.9	Pre-D3 intrusion	
6	22QL11	Felsic mylonite	1902 ± 26	M/UI	5	10.8	Syn-D3 metamorphism	This study
			2289–2016	D/SG	3	-	Inheritance	
7	QL32-1	Mylonitized granitic dyke	1936 ± 28	I/UI	12	6.2	Pre-D3 intrusion	[21]
8	QL27-1	F2-cutting undeformed pegmatite dyke	1854 ± 68	I/UI	10	4.2	Post-D2 intrusion	[21]
			2118–1976	X/SG	10	-	Inheritance	
9	QL12	F2-cutting undeformed pegmatite dyke	1899 ± 6	I/UI	9	1.9	Post-D2 intrusion	[39]
10	QL14	S1-parallel and F2-folded leucocratic dyke	1932 ± 24	I/UI	18	6.5	Pre-D2 intrusion	[39]
11	QL08-2	S1-cutting leucogranitic dyke	1909 ± 27	I/UI	14	9.5	Post-D1 intrusion	[21]
			2089–1940	X/SG	9	-	Inheritance	
12	QL24-1	S1-cutting leucocratic dyke	1932 ± 47	I/UI	12	16	Post-D1 intrusion	[21]
			2082–1977	X/SG	6	-	Inheritance	
13	QL01-5	Isoclinally F1-folded leucogranitic dyke	1940 ± 33	M/UI	5	3.4	Syn-D1 metamorphism	[21]
			1976 ± 21	I/UI	7	6.1	Pre-D1 intrusion	
			2689–1997	X/SG	13	-	Inheritance	
14	QL19	S1-foliated granitoid	1948 ± 26	M/UI	12	2.2	Syn-D1 metamorphism	[39]
			2011 ± 12	I/WM	8	1.2	Pre-D1 intrusion	
15	QL20	S1-foliated granitoid	1954 ± 18	M/UI	6	1.17	Syn-D1 metamorphism	[39]
			2053 ± 14	I/WM	13	3.2	Pre-D1 intrusion	
16	QL05	S1-foliated granitoid	1956 ± 27	M/UI	7	5.2	Syn-D1 metamorphism	[39]
			2056 ± 17	I/WM	6	1.9	Pre-D1 intrusion	
17	QL03	S1-foliated granitoid	1966 ± 25	M/UI	17	8.0	Syn-D1 metamorphism	[39]
			2016 ± 24	I/WM	5	3.4	Pre-D1 intrusion	

Table 1. *Cont.*

No. [1]	Sample	Rock Type and Structural Feature	Age (Ma)	Type [2]	n	MSWD	Interpretation	References
Helanshan Complex								
18	HL07	Granitic mylonite	1866 ± 12	M/UI	33	1.6	Syn-D3 metamorphism	[65]
			1951 ± 5	I/WM	6	0.22	Pre-D3 intrusion	
19	HL14	S1-foliated granitic dyke	1867 ± 21	M/UI	13	0.48	Syn-D3 metamorphism	[28]
			1954 ± 13	I/UI	24	0.1	Syn-D1 intrusion	
			2049 ± 32	X/WM	3	0.02	Inheritance	
20	HL12	S1-cutting and F2-folded leucocratic dyke	1911 ± 43	I/UI	11	0.73	Syn-D2 intrusion	[28]
21	HL09	Isoclinally F1-folded leucocratic dyke	1949 ± 8	I/UI	15	0.45	Syn-D1 intrusion	[28]
22	HL31	Isoclinally F1-folded metasandstone	1951 ± 9	M/UI	22	0.03	Syn-D1 metamorphism	[28]
			2173–2018	D/SG	22	-	Inheritance	
23	HL13	S1-foliated granitoid	1954 ± 14	M/UI	14	0.06	Syn-D1 metamorphism	[28]
			2032 ± 34	I/WM	6	3.9	Pre-D1 intrusion	
24	HL18	S1-foliated leucocratic dyke	1954 ± 13	M/UI	21	0.08	Syn-D1 metamorphism	[28]
			2039 ± 17	I/WM	7	0.66	Pre-D1 intrusion	
Wulashan-Daqingshan Complex								
25	NOR24-2	Granitic mylonite	1814 ± 13	Bt/PA	-	-	D3 deformation	[27]
26	NOR8-3	Mylonitic felsic gneiss	1819 ± 14	Bt/WM	-	-	D3 deformation	[27]
27	NOR14-1	Mylonitic garnet-biotite gneiss	1885 ± 20	Bt/PA	-	-	D3 deformation	[27]
28	-	Granite cutting DSZ	1819 ± 3	I/SG	1	-	Post-D3 intrusion	[64]
29	NM0922	Undeformed coarse-grained syenogranite	1822 ± 17	I/WM	12	0.82	Post-D3 intrusion	[54]
30	NM0919	Deformed syenogranite with augen structure in DSZ	1853 ± 6	M/SG	1	-	Syn-D3 metamorphism	[54]
			1964 ± 4	I/WM	13	1.00	Pre-D3 intrusion	
31	NOR114-2	Sm-parallel granitic dyke in DSZ	1858 ± 21	I/WM	13	0.22	Syn-D3 intrusion	[27]
32	NM1112	Strongly Sm-foliated syenogranite vein in DSZ	1866 ± 23	M/UI	6	0.97	Syn-D3 metamorphism	[54]
			2112 ± 13	I/UI	6	0.55	Pre-D3 intrusion	
33	NM0915	Lm-lineated metagabbro dyke in DSZ	1884 ± 11	M/WM	5	1.3	Syn-D3 metamorphism	[47]

Table 1. *Cont.*

No. [1]	Sample	Rock Type and Structural Feature	Age (Ma)	Type [2]	n	MSWD	Interpretation	References
			1951 ± 9	I/WM	9	1.2	Pre-D3 intrusion	
34	NM1119	Strongly Sm-foliated charnockitic gneiss in the DSZ	1896 ± 17	M/WM	4	0.38	Syn-D3 metamorphism	[54]
			1987 ± 19	I/WM	5	0.80	Pre-D3 intrusion	
35	NM0814-2	Lm-lineated meta-gabbro dyke in DSZ	1906 ± 13	M/WM	6	1.0	Syn-D3 metamorphism	[47]
			1968 ± 8	I/WM	20	1.3	Pre-D3 intrusion	
36	NM1004	S1-cutting undeformed pegmatite vein	1841 ± 9	I/WM	14	1.5	Post-D1 intrusion	[76]
37	16BT13-1	S1-cutting granitic vein	1874 ± 17	I/WM	8	0.12	Post-D1 intrusion	[77]
38	NM0615	S1-foliated syenogranite	1952 ± 6	I/WM	4	1.07	Syn-D1 intrusion	[54]
39	NM0920	Metagabbro dyke with a weak S1 foliation	1960 ± 7	I/WM	8	1.2	Syn-D1 intrusion	[47]
40	NM0614-2	Strongly S1-foliated syenogranite	1939 ± 17	M/WM	8	0.27	Syn-D1 metamorphism	[54]
			2047 ± 25	I/SG	1	-	Pre-D1 intrusion	
Jining Complex								
41	D06JN023	Mylonitic metapelite in DSZ	1866 ± 22	M/WM	24	0.26	Syn-D3 metamorphism	[78]
42	D06JN034	Granite cut by DSZ	1957 ± 19	I/WM	30	0.78	Pre-D3 intrusion	[78]
43	13AZS14	Pegmatite dyke cutting S1 foliation	1918 ± 3	I/WM	19	1.1	Post-D1 intrusion	[56]
44	13AZS12	S1-foliated metaleucogranite	1950 ± 26	I/UI	15	2.5	Syn-D1 intrusion	[56]
45	13AZS11	S1-foliated metaleucogranite	1954 ± 18	I/UI	10	1.5	Syn-D1 intrusion	[56]

[1] Numbers also indicate the dated samples and references in Figure 9. [2] Bt, I, M, X and D indicate biotite, igneous, metamorphic, xenocrystic and detrital zircon, respectively. MIA, PA, UI, WM and SG denote mineral isochron age, plateau age, upper intercept age, weighted mean age and single grain age, respectively. Except for single grain age, the remaining ages are presented at the 2σ level. * Abbreviations: DSZ, ductile shear zone.

Subsequently, the D2 deformation mainly produced NW(W)-SE(E)-trending tight to open doubly plunging upright F2 folds that have variably deflected the previous D1 structures (Figure 8b) [21,69]. It is considered that D2 coevally developed with post-peak decompression processes [21,69]. The crystallization ages of pre-D2 and post-D2 leucocratic dykes indicate that D2 approximately took place at some time between ~1932 Ma and ~1899 Ma (Table 1; Figure 8) [21,39]. Notably, the superposition of F2 antiforms on S1 foliations gave rise to a series of gneiss domes, particularly in the northern Qianlishan Complex [21,39]. Qiao et al. (2023) proposed that these well-preserved domal structures are among the most representative Paleoproterozoic gneiss domes of the Khondalite Belt, which were less affected by post-D2 deformation [39].

In this study, nearly E-W-striking ductile shear zones were discovered in the Qianlishan Complex (e.g., Figures 3 and 4). As discussed above, available geochronological data show that they occurred between ~1902 Ma and ~1821 Ma (Table 1; Figures 6–8). Here, we attribute the QDSZ to the D3 deformation that affected the studied area. Within these high-strain zones, the khondalites and pre-kinematic intrusions were variably subjected to mylonitization and shear deformation (Figures 3–5). Our (micro-)structural observations reveal that the QDSZ were probably developed under high-temperature (T > 650 °C) deformation conditions (e.g., Figure 3c,d and Figure 4b,d,i). These shear zones are characterized by steep to sub-vertical S(SW)-dipping mylonitic foliations and mainly exhibit a top-to-N(NE) sense of shearing (Figures 3–5). In summary, the Qianlishan Complex records three phases of Orosirian deformation, namely D1 (1.97–1.93 Ga), D2 (1.93–1.90 Ga) and D3 (1.90–1.82 Ga) (Figure 8) [21,39,69], respectively.

5.3. Implications for the Deformation History of the Khondalite Belt

Integrated with previous studies, structural analysis and dating results of the Qianlishan Complex can provide important constraints on the geochronological framework of the deformation history of the Khondalite Belt. As shown in Table 1 and Figure 9, we compiled available deformation-related geochronological data from forty-five samples from the Khondalite Belt. Similar to the Qianlishan Complex, other high-grade complexes in the Khondalite Belt experienced three phases of deformation (D1–D3) in the late Paleoproterozoic [21,24,26–28,39]. As mentioned earlier, it is considered that D1 structures of the Qianlishan Complex were formed at ca. 1976–1932 Ma [21,39]. This is consistent with structural observations in the Helanshan Complex, where two syn-D1 leucocratic dykes that have been isoclinally F1-folded or S1-foliated were dated at 1954 ± 13 Ma and 1949 ± 8 Ma [28]. Similar ~1950 Ma syn-D1 leucocratic dykes were also reported in the Wulashan-Daqingshan and Jining complexes [47,54,56]. Additionally, the D1 deformation is inferred to be simultaneously accompanied by high-pressure granulite-facies peak metamorphism [21,22,28,69]. Regionally, high-pressure granulites of the Helanshan, Qianlishan, Wulashan-Daqingshan and Jining complexes yielded metamorphic zircon ages of 1963–1946 Ma, 1966–1941 Ma, 1965–1943 Ma and 1957–1945 Ma, respectively [20–22,33,35,37,47,54,65,78]. Subsequently, the timing of the D2 deformation in the Qianlishan Complex was constrained at ca. 1932–1899 Ma [21,39,69]. This interpretation is further supported by the crystallization age of a 1911 ± 43 Ma syn-D2 leucocratic dyke that truncated S1 foliations and was deflected by F2 folds in the Helanshan Complex [28].

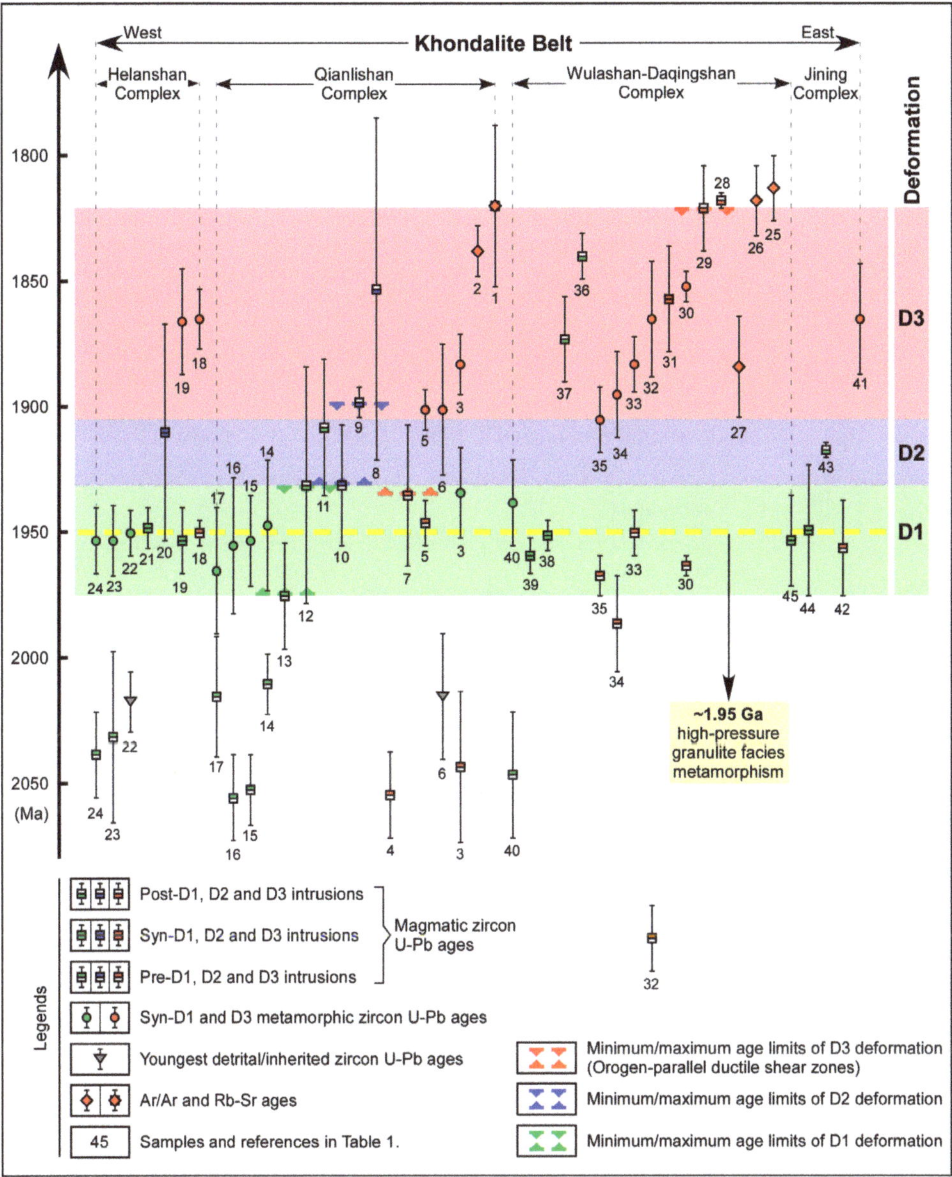

Figure 9. Geochronological framework of three-phase (D1–D3) Orosirian deformation in the Khondalite Belt. Deformation-related geochronological data were compiled from forty-five samples, indicating that the D1, D2 and D3 deformation events occurred at 1.97–1.93 Ga, 1.93–1.90 Ga and 1.90–1.82 Ga, respectively. Structural features, age interpretations and references of dated samples are listed in Table 1. See the text for details.

Later, a series of orogen-parallel ductile shear zones appeared throughout the Khondalite Belt [24,26–28,64,65,77]. Of these, the D3 deformation led to the development of ductile shear zones in the Qianlishan Complex at ca. 1902–1821 Ma (Figure 8). Comparably, five mylonites from the Wulashan-Daqingshan Complex were dated and their zircon overgrowth rims yielded metamorphic ages ranging from 1906 ± 13 Ma to 1853 ± 6 Ma (Table 1; Figure 9) [47,54]. These ages are coherent with our inference that the widespread shear zone activity of the Khodnalite Belt probably started at ~1900 Ma. Similar metamorphic overgrowth rims with ages of 1866 ± 12 Ma and 1866 ± 22 Ma were obtained in high-temperature mylonites from the Helanshan and Jining complexes [65,78]. Remarkably, Gong et al. (2014) reported three biotite ^{40}Ar/^{39}Ar ages of 1885 ± 20 Ma, 1819 ± 14 Ma and 1814 ± 13 Ma on mylonites from the Wulashan-Daqingshan Complex and speculated that they corresponded to the ductile shearing event [27]. Notably, a biotite ^{40}Ar/^{39}Ar age of 1839 ± 10 Ma was also obtained in the Qianlishan Complex [75] and interpreted to be related to the D3 tectonothermal event. Additionally, the E-W-trending ductile shear zones of the Wulashan-Daqingshan Complex were cut by two post-kinematic undeformed granitic intrusions with crystallization ages of 1822 ± 17 Ma and 1819 ± 14 Ma [54,64], providing the minimum age limit for the shear deformation. Thus, the D3-related orogen-parallel shear zones of the Khondalite Belt are interpreted to have approximately developed in the period of 1.90–1.82 Ga (Figure 9). Considering all the data together, we propose that the Khondalite Belt experienced a prolonged orogenic history (>100 Myr) and three phases of Orosirian deformation (D1–D3) at ca. 1.97–1.82 Ga, related to the NNE-SSW-directed collision between the Yinshan and Ordos blocks [18,19,21–23].

6. Conclusions

In this study, we report for the first time the nearly E-W-trending ductile shear zones in the Qianlishan Complex of the Khondalite Belt. They are characterized by steep to sub-vertical S(SW)-dipping mylonitic foliations and mainly display a top-to-N(NE) sense of shearing. Microstructures indicate that these shear zones probably developed under high-temperature (T > 650 °C) deformation conditions. Geochronological data reveal that the development of the Qianlishan ductile shear zones approximately started at ca. 1902–1884 Ma and continued to ca. 1839–1821 Ma. The Qianlishan Complex suffered three phases of Orosirian deformation (D1–D3), of which the D3 deformation gave rise to the aforementioned orogen-parallel ductile shear zones. Deformation events D1, D2 and D3 are regarded to have occurred at ca. 1.97–1.93 Ga, 1.93–1.90 Ga and 1.90–1.82 Ga, respectively. The polyphase deformation (D1–D3) document that the Khondalite Belt underwent a protracted (>100 Myr) orogenic history in response to the collision between the Yinshan and Ordos blocks.

Supplementary Materials: The following supporting information can be downloaded at: https://www.mdpi.com/article/10.3390/min14060561/s1, Table S1: LA-ICP-MS zircon U-Pb data of shear zone-related dating samples in the Qianlishan Complex, Khondalite Belt.

Author Contributions: Conceptualization, H.Q.; methodology, H.Q., M.L. and C.D.; investigation, H.Q., M.L. and C.D.; data curation, H.Q.; writing—original draft preparation, H.Q., M.L. and C.D.; writing—review and editing, H.Q.; supervision, H.Q.; funding acquisition, H.Q. All authors have read and agreed to the published version of the manuscript.

Funding: This research was funded by the National Natural Science Foundation of China (Grant No. 42002221), China Postdoctoral Science Foundation (Grant No. 2022M712569), Science and Technology Department of Sichuan Province (Grant No. MZGC20230103) and Leshan Normal University (Grant No. KYCXTD2023-2, KYPY2023-0006, RC202009).

Data Availability Statement: The original contributions presented in the study are included in the article/supplementary material, further inquiries can be directed to the corresponding author.

Acknowledgments: The editors and reviewers are thanked for their helpful and constructive comments on the manuscript.

Conflicts of Interest: The authors declare no conflicts of interest.

References

1. Ramsay, J.G.; Graham, R.H. Strain variation in shear belts. *Can. J. Earth Sci.* **1970**, *7*, 786–813. [CrossRef]
2. Sibson, R.H. Fault rocks and fault mechanisms. *J. Geol. Soc.* **1977**, *133*, 191–213. [CrossRef]
3. Ramsay, J. Shear zone geometry: A review. *J. Struct. Geol.* **1980**, *2*, 83–99. [CrossRef]
4. Fossen, H.; Cavalcante, G.C.G. Shear zones—A review. *Earth-Sci. Rev.* **2017**, *171*, 434–455. [CrossRef]
5. Zhang, H.; Hou, G.; Zhang, B.; Tian, W. Kinematics, temperature and geochronology of the Qingyi ductile shear zone: Tectonic implications for late Neoarchean microblock amalgamation in the Western Shandong Province, North China craton. *J. Struct. Geol.* **2022**, *161*, 104645. [CrossRef]
6. Li, W.; Cao, S.; Dong, Y.; Zhan, L.; Tao, L. Crustal anatexis and initiation of the continental-scale Chongshan strike-slip shear zone on the southeastern Tibetan Plateau. *Tectonics* **2023**, *42*, e2023TC007864. [CrossRef]
7. Cao, S.; Neubauer, F. Deep crustal expressions of exhumed strike-slip fault systems: Shear zone initiation on rheological boundaries. *Earth-Sci. Rev.* **2016**, *162*, 155–176. [CrossRef]
8. Oriolo, S.; Wemmer, K.; Oyhantçabal, P.; Fossen, H.; Schulz, B.; Siegesmund, S. Geochronology of shear zones—A review. *Earth-Sci. Rev.* **2018**, *185*, 665–683. [CrossRef]
9. Qiu, E.; Zhang, Y.; Larson, K.P.; Li, B. Dating strike-slip ductile shear through combined zircon-, titanite-and apatite U–Pb geochronology along the southern Tan-Lu Fault zone, East China. *Tectonics* **2023**, *42*, e2022TC007734. [CrossRef]
10. Stipp, M.; StuÈnitz, H.; Heilbronner, R.; Schmid, S.M. The eastern Tonale fault: A 'natural laboratory' for crystal plastic deformation of quartz over a temperature range from 250 to 700 °C. *J. Struct. Geol.* **2002**, *24*, 1861–1884. [CrossRef]
11. Passchier, C.W.; Trouw, R. *Microtectonics*; Springer: Berlin/Heidelberg, Germany, 2005.
12. Ribeiro, B.V.; Kirkland, C.L.; Kelsey, D.E.; Reddy, S.M.; Hartnady, M.I.; Faleiros, F.M.; Rankenburg, K.; Liebmann, J.; Korhonen, F.J.; Clark, C. Time-strain evolution of shear zones from petrographically constrained Rb–Sr muscovite analysis. *Earth Planet. Sci. Lett.* **2023**, *602*, 117969. [CrossRef]
13. Sun, S.; Dong, Y. High temperature ductile deformation, lithological and geochemical differentiation along the Shagou shear zone, Qinling Orogen, China. *J. Struct. Geol.* **2023**, *167*, 104791. [CrossRef]
14. van der Pluijm, V.A.; Mezger, K.; Cosca, M.A.; Essene, E.J. Determining the significance of high-grade shear zones by using temperature-time paths, with examples from the Grenville orogen. *Geology* **1994**, *22*, 743–746. [CrossRef]
15. Reiners, P.W.; Ehlers, T.A.; Zeitler, P.K. Past, present, and future of thermochronology. *Rev. Mineral. Geochem.* **2005**, *58*, 1–18. [CrossRef]
16. Law, R.D. Deformation thermometry based on quartz c-axis fabrics and recrystallization microstructures: A review. *J. Struct. Geol.* **2014**, *66*, 129–161. [CrossRef]
17. Simonetti, M.; Carosi, R.; Montomoli, C.; Law, R.D.; Cottle, J.M. Unravelling the development of regional-scale shear zones by a multidisciplinary approach: The case study of the Ferrière-Mollières Shear Zone (Argentera Massif, Western Alps). *J. Struct. Geol.* **2021**, *149*, 104399. [CrossRef]
18. Zhao, G.; Sun, M.; Wilde, S.A.; Li, S. Late Archean to Paleoproterozoic evolution of the North China Craton: Key issues revisited. *Precambrian Res.* **2005**, *136*, 177–202. [CrossRef]
19. Zhao, G.; Cawood, P.A.; Li, S.; Wilde, S.A.; Sun, M.; Zhang, J.; He, Y.; Yin, C. Amalgamation of the North China Craton: Key issues and discussion. *Precambrian Res.* **2012**, *222*, 55–76. [CrossRef]
20. Yin, C.; Zhao, G.; Sun, M.; Xia, X.; Wei, C.; Zhou, X.; Leung, W. LA-ICP-MS U–Pb zircon ages of the Qianlishan Complex: Constrains on the evolution of the Khondalite Belt in the Western Block of the North China Craton. *Precambrian Res.* **2009**, *174*, 78–94. [CrossRef]
21. Yin, C.; Qiao, H.; Lin, S.; Li, C.; Zhang, J.; Qian, J.; Wu, S. Deformation history of the Qianlishan complex, Khondalite belt, North China: Structures, ages and tectonic implications. *J. Struct. Geol.* **2020**, *141*, 104176. [CrossRef]
22. Yin, C.; Zhao, G.; Xiao, W.; Lin, S.; Gao, R.; Zhang, J.; Qian, J.; Gao, P.; Qiao, H.; Li, W. Paleoproterozoic accretion and assembly of the Western Block of North China: A new model. *Earth-Sci. Rev.* **2023**, *241*, 104448. [CrossRef]
23. Guo, J.; Peng, P.; Chen, Y.; Jiao, S.; Windley, B.F. UHT sapphirine granulite metamorphism at 1.93–1.92 Ga caused by gabbronorite intrusions: Implications for tectonic evolution of the northern margin of the North China Craton. *Precambrian Res.* **2012**, *222*, 124–142. [CrossRef]
24. Lu, L.; Xu, X.; Liu, F. *Early Precambrian Khondalites in North China: Changchun*; Changchun Publishing House: Changchun, China, 1996.
25. Guo, J.; Zhai, M. Mylonite of granulite facies in Xuwujia, Nei Monggol. *Chin. J. Geol.* **1992**, *27*, 190–192. (In Chinese with English Abstract)
26. Gan, S.; Qian, X. A plate-tectonic model for the evolution of the Daqingshan granulite belt in Inner Mongolia, China. *Acta Geol. Sin.* **1996**, *70*, 298–308. (In Chinese with English Abstract) [CrossRef]
27. Gong, W.; Hu, J.; Wu, S.; Chen, H.; Qu, H.; Li, Z.; Liu, Y.; Wang, L. Possible southwestward extrusion of the Ordos Block in the Late Paleoproterozoic: Constraints from kinematic and geochronologic analysis of peripheral ductile shear zones. *Precambrian Res.* **2014**, *255*, 716–733. [CrossRef]
28. Qiao, H.; Yin, C.; Xiao, W.; Zhang, J.; Qian, J.; Wu, S. Paleoproterozoic polyphase deformation in the Helanshan Complex: Structural and geochronological constraints on the tectonic evolution of the Khondalite Belt, North China Craton. *Precambrian Res.* **2022**, *368*, 106468. [CrossRef]

29. Zhao, G.; Zhai, M. Lithotectonic elements of Precambrian basement in the North China Craton: Review and tectonic implications. *Gondwana Res.* **2013**, *23*, 1207–1240. [CrossRef]

30. Kusky, T.M.; Polat, A.; Windley, B.F.; Burke, K.C.; Dewey, J.F.; Kidd, W.S.F.; Maruyama, S.; Wang, J.; Deng, H.; Wang, Z.; et al. Insights into the tectonic evolution of the North China Craton through comparative tectonic analysis: A record of outward growth of Precambrian continents. *Earth-Sci. Rev.* **2016**, *162*, 387–432. [CrossRef]

31. Wei, C.; Zhai, M.; Wang, B. Four phases of Orosirian metamorphism in the north North China Craton (NNCC): Insights into the regional tectonic framework and evolution. *Earth-Sci. Rev.* **2023**, *241*, 104449. [CrossRef]

32. Yin, C.; Zhao, G.; Guo, J.; Sun, M.; Xia, X.; Zhou, X.; Liu, C. U–Pb and Hf isotopic study of zircons of the Helanshan Complex: Constrains on the evolution of the Khondalite Belt in the Western Block of the North China Craton. *Lithos* **2011**, *122*, 25–38. [CrossRef]

33. Liu, P.; Liu, F.; Liu, C.; Liu, J.; Wang, F.; Xiao, L.; Cai, J.; Shi, J. Multiple mafic magmatic and high-grade metamorphic events revealed by zircons from meta-mafic rocks in the Daqingshan–Wulashan Complex of the Khondalite Belt, North China Craton. *Precambrian Res.* **2014**, *246*, 334–357. [CrossRef]

34. Liu, P.; Liu, F.; Cai, J.; Liu, C.; Liu, J.; Wang, F.; Xiao, L.; Shi, J. Spatial distribution, P–T–t paths, and tectonic significance of high-pressure mafic granulites from the Daqingshan–Wulashan Complex in the Khondalite Belt, North China Craton. *Precambrian Res.* **2017**, *303*, 687–708. [CrossRef]

35. Cai, J.; Liu, F.; Liu, P. Paleoproterozoic multistage metamorphic events in Jining metapelitic rocks from the Khondalite Belt in the North China Craton: Evidence from petrology, phase equilibria modelling and U–Pb geochronology. *J. Asian Earth Sci.* **2017**, *138*, 515–534. [CrossRef]

36. Jiao, S.; Fitzsimons, I.C.; Zi, J.; Evans, N.J.; Mcdonald, B.J.; Guo, J. Texturally controlled U–Th–Pb monazite geochronology reveals Paleoproterozoic UHT metamorphic evolution in the Khondalite belt, North China craton. *J. Petrol.* **2020**, *61*, egaa023. [CrossRef]

37. Wu, S.; Yin, C.; Davis, D.W.; Zhang, J.; Qian, J.; Qiao, H.; Xiao, Y.; Liu, J. Metamorphic evolution of high-pressure felsic and pelitic granulites from the Qianlishan Complex and tectonic implications for the Khondalite Belt, North China Craton. *Geol. Soc. Am. Bull.* **2020**, *132*, 2253–2266. [CrossRef]

38. Jiang, X.; Yu, S.; Liu, Y.; Li, S.; Lv, P.; Peng, Y.; Gao, X.; Ji, W.; Li, C.; Xie, W. Episodic metamorphism and anatexis within the Khondalite Belt, North China Craton: Constraint from Late-Paleoproterozoic fluid-fluxed melting of the Daqingshan Complex. *Precambrian Res.* **2022**, *369*, 106504. [CrossRef]

39. Qiao, H.; Deng, P.; Li, J. Geochronological Constraints on the Origin of the Paleoproterozoic Qianlishan Gneiss Domes in the Khondalite Belt of the North China Craton and Their Tectonic implications. *Minerals* **2023**, *13*, 1361. [CrossRef]

40. Shi, Q.; Xu, Z.; Santosh, M.; Ding, D.; Zhao, Z.; Li, C.; Gao, X.; Ma, W.; Xu, Y.; Li, H.; et al. Mantle Magmatism, Metamorphism and Anatexis: Evidence from Geochemistry and Zircon U–Pb-Hf Isotopes of Paleoproterozoic S-type Granites, Khondalite Belt of the North China Craton. *Int. Geol. Rev.* **2023**, *65*, 943–968. [CrossRef]

41. Zhu, W.; Tian, W.; Wang, B.; Zhang, Y.; Wei, C. Paleoproterozoic Crust–Mantle Interaction in the Khondalite Belt, North China Craton: Constraints from Geochronology, Elements, and Hf-O-Sr-Nd Isotopes of the Layered Complex in the Jining Terrane. *Minerals* **2023**, *13*, 462. [CrossRef]

42. Zhai, M. Khondalite revisited-record of special geological processes on Earth. *Acta Petrol. Sin.* **2022**, *96*, 2967–2997. (In Chinese with English Abstract) [CrossRef]

43. Xia, X.; Sun, M.; Zhao, G.; Luo, Y. LA-ICP-MS U–Pb geochronology of detrital zircons from the Jining Complex, North China Craton and its tectonic significance. *Precambrian Res.* **2006**, *144*, 199–212. [CrossRef]

44. Xia, X.; Sun, M.; Zhao, G.; Wu, F.; Xu, P.; Zhang, J.; Luo, Y. U–Pb and Hf isotopic study of detrital zircons from the Wulashan khondalites: Constraints on the evolution of the Ordos Terrane, Western Block of the North China Craton. *Earth Planet. Sci. Lett.* **2006**, *241*, 581–593. [CrossRef]

45. Wan, Y.; Song, B.; Liu, D.; Wilde, S.A.; Wu, J.; Shi, Y.; Yin, X.; Zhou, H. SHRIMP U–Pb zircon geochronology of Palaeoproterozoic metasedimentary rocks in the North China Craton: Evidence for a major Late Palaeoproterozoic tectonothermal event. *Precambrian Res.* **2006**, *149*, 249–271. [CrossRef]

46. Wan, Y.; Liu, D.; Dong, C.; Xu, Z.; Wang, Z.; Wilde, S.A.; Yang, Y.; Liu, Z.; Zhou, H. The Precambrian Khondalite Belt in the Daqingshan area, North China Craton: Evidence for multiple metamorphic events in the Palaeoproterozoic era. *Geol. Soc. London, Spec. Publ.* **2009**, *323*, 73–97. [CrossRef]

47. Wan, Y.; Xu, Z.; Dong, C.; Nutman, A.; Ma, M.; Xie, H.; Liu, S.; Liu, D.; Wang, H.; Cu, H. Episodic Paleoproterozoic (~2.45, ~1.95 and ~1.85 Ga) mafic magmatism and associated high temperature metamorphism in the Daqingshan area, North China Craton: SHRIMP zircon U–Pb dating and whole-rock geochemistry. *Precambrian Res.* **2013**, *224*, 71–93. [CrossRef]

48. Dong, C.; Liu, D.; Li, J.; Wang, Y.; Zhou, H.; Li, C.; Yang, Y.; Xie, L. Palaeoproterozoic Khondalite Belt in the western North China Craton: New evidence from SHRIMP dating and Hf isotope composition of zircons from metamorphic rocks in the Bayan Ul-Helan Mountains area. *Chin. Sci. Bull.* **2007**, *52*, 2984–2994. (In Chinese with English Abstract) [CrossRef]

49. Zhou, X.; Geng, Y. Metamorphic age of the khondalites in the Helanshan region: Constraints on the evolution of the Western block in the North China Craton. *Acta Petrol. Sin.* **2009**, *25*, 1843–1852. (In Chinese with English Abstract)

50. Jiao, S.; Guo, J.; Harley, S.L.; Peng, P. Geochronology and trace element geochemistry of zircon, monazite and garnet from the garnetite and/or associated other high-grade rocks: Implications for Palaeoproterozoic tectonothermal evolution of the Khondalite Belt, North China Craton. *Precambrian Res.* **2013**, *237*, 78–100. [CrossRef]

51. Jiao, S.; Guo, J.; Wang, L.; Peng, P. Short-lived high-temperature prograde and retrograde metamorphism in Shaerqin sapphirine-bearing metapelites from the Daqingshan terrane, North China Craton. *Precambrian Res.* **2015**, *269*, 31–57. [CrossRef]
52. Jiao, S.; Fitzsimons, I.C.; Guo, J. Paleoproterozoic UHT metamorphism in the Daqingshan Terrane, North China Craton: New constraints from phase equilibria modeling and SIMS U–Pb zircon dating. *Precambrian Res.* **2017**, *303*, 208–227. [CrossRef]
53. Santosh, M.; Liu, D.; Shi, Y.; Liu, S. Paleoproterozoic accretionary orogenesis in the North China Craton: A SHRIMP zircon study. *Precambrian Res.* **2013**, *227*, 29–54. [CrossRef]
54. Liu, S.; Dong, C.; Xu, Z.; Santosh, M.; Ma, M.; Xie, H.; Liu, D.; Wang, Y. Palaeoproterozoic episodic magmatism and high-grade metamorphism in the North China Craton: Evidence from SHRIMP zircon dating of magmatic suites in the Daqingshan area. *Geol. J.* **2013**, *48*, 429–455. [CrossRef]
55. He, X.; Yin, C.; Long, X.; Qian, J.; Wang, L.; Qiao, H. Archean to Paleoproterozoic continental crust growth in the Western Block of North China: Constraints from zircon Hf isotopic and whole-rock Nd isotopic data. *Precambrian Res.* **2017**, *303*, 105–116. [CrossRef]
56. Wang, L.; Guo, J.; Yin, C.; Peng, P. Petrogenesis of ca. 1.95 Ga meta-leucogranites from the Jining Complex in the Khondalite Belt, North China Craton: Water-fluxed melting of metasedimentary rocks. *Precambrian Res.* **2017**, *303*, 355–371. [CrossRef]
57. Gou, L.; Zi, J.; Dong, Y.; Liu, X.; Li, Z.; Xu, X.; Zhang, C.; Liu, L.; Long, X.; Zhao, Y. Timing of two separate granulite-facies metamorphic events in the Helanshan complex, North China Craton: Constraints from monazite and zircon U–Pb dating of pelitic granulites. *Lithos* **2019**, *350*, 105216. [CrossRef]
58. Zhou, X.; Zhao, G.; Geng, Y. Helanshan high-pressure pelitic granulites: Petrological evidence for collision event in the Western Block of the North China Craton. *Acta Petrol. Sin.* **2010**, *26*, 2113–2121. (In Chinese with English Abstract) [CrossRef]
59. Cai, J.; Liu, F.; Liu, P.; Liu, C.; Wang, F.; Shi, J. Metamorphic P–T path and tectonic implications of pelitic granulites from the Daqingshan Complex of the Khondalite Belt, North China Craton. *Precambrian Res.* **2014**, *241*, 161–184. [CrossRef]
60. Yin, C.; Zhao, G.; Wei, C.; Sun, M.; Guo, J.; Zhou, X. Metamorphism and partial melting of high-pressure pelitic granulites from the Qianlishan Complex: Constraints on the tectonic evolution of the Khondalite Belt in the North China Craton. *Precambrian Res.* **2014**, *242*, 172–186. [CrossRef]
61. Yin, C.; Zhao, G.; Sun, M. High-pressure pelitic granulites from the Helanshan Complex in the Khondalite Belt, North China Craton: Metamorphic P-T path and tectonic implications. *Am. J. Sci.* **2015**, *315*, 846–879. [CrossRef]
62. Xu, X.; Gou, L.; Long, X.; Dong, Y.; Liu, X.; Zi, J.; Li, Z.; Zhang, C.; Liu, L.; Zhao, J. Phase equilibrium modelling and SHRIMP zircon U–Pb dating of medium-pressure pelitic granulites in the Helanshan complex of the Khondalite Belt, North China Craton, and their tectonic implications. *Precambrian Res.* **2018**, *314*, 62–75. [CrossRef]
63. Xu, X.; Gou, L.; Dong, Y.; Zhang, C.; Long, X.; Zhao, Y.; Zhou, F.; Tian, Z. Metamorphic P–T evolution of the pelitic granulites from the Helanshan in the North China Craton revealed by phase equilibrium modeling and garnet trace elements: Implications for Paleoproterozoic collisional orogenesis. *Precambrian Res.* **2013**, *394*, 107093. [CrossRef]
64. Wang, H.; Yuan, G.; Xing, H.; Wang, J. One the tectonic implication for Xiashihao-Jiuguan ductile shear zone in Guyang-Wuchuan area, Inner Mongolia. *Prog. Precambrian Res.* **1999**, *22*, 1–20. (In Chinese with English Abstract)
65. Qiao, H.; Yin, C.; Zhang, J.; Qian, J.; Wu, S. New Discovery of ~1866 Ma high-temperature mylonite in the Helanshan Complex: Marking a late-stage Ductile Shearing in the Khondalite Belt, North China Craton. *Acta Geol. Sin. Engl. Ed.* **2021**, *95*, 1418–1419. [CrossRef]
66. Qiao, H.Z. Detrital zircon U–Pb ages of the Huangqikou Formation of the Changcheng System in the Qianlishan Area, western North China Craton and their geological implications. *Acta Geol. Sichuan* **2021**, *41*, 33–39. (In Chinese with English Abstract)
67. Qiao, H.; Yin, C.; Li, Q.; He, X.; Qian, J.; Li, W. Application of the revised Ti-in-zircon thermometer and SIMS zircon U-Pb dating of high-pressure pelitic granulites from the Qianlishan-Helanshan Complex of the Khondalite Belt, North China Craton. *Precambrian Res.* **2016**, *276*, 1–13. [CrossRef]
68. Li, W.; Yin, C.; Lin, S.; Li, W.; Gao, P.; Zhang, J.; Qian, J.; Qiao, H. Paleoproterozoic tectonic evolution from subduction to collision of the Khondalite Belt in North China: Evidence from multiple magmatism in the Qianlishan Complex. *Precambrian Res.* **2022**, *368*, 106471. [CrossRef]
69. Qiao, H. Structural and Geochronological Studies of the Qianlishan-Helanshan Complex in North China. Ph. D. Thesis, Sun Yat-sen University, Guangzhou, China, 2019. (In Chinese with English Abstract)
70. Lu, C.; Qian, J.; Yin, C.; Gao, P.; Guo, M.; Zhang, W. Ultrahigh temperature metamorphism recorded in the Lüliang Complex, Trans-North China Orogen: P–T–t evolution and heating mechanism. *Precambrian Res.* **2022**, *383*, 106900. [CrossRef]
71. Liu, Y.; Hu, Z.; Gao, S.; Günther, D.; Xu, J.; Gao, C.; Chen, H. In situ analysis of major and trace elements of anhydrous minerals by LA-ICP-MS without applying an internal standard. *Chem. Geol.* **2008**, *257*, 34–43. [CrossRef]
72. Ludwig, K.R. *Isoplot/Ex Version 4.15: A Geochronological Toolkit for Microsoft Excel*; Berkeley Geochronology Center Special Publication: Berkeley, CA, USA, 2012; Volume 5, p. 75.
73. Hippertt, J.; Rocha, A.; Lana, C.; Egydio-Silva, M.; Takeshita, T. Quartz plastic segregation and ribbon development in high-grade striped gneisses. *J. Struct. Geol.* **2001**, *23*, 67–80. [CrossRef]
74. Yan, Y. A study on P-T condition and mineralogy of the metamorphic complex of the Qianlishan Group, Nei Mongol. In *Petrology Research (Second Collection)*; Institute of Geology, Chinese Academy of Science, Ed.; Geological Publishing House: Beijing, China, 1983. (In Chinese with English Abstract)

75. Shen, Q.; Fu, Y.; Zhang, M. ^{40}Ar/^{39}Ar dating on biotite from the Qianlishan Group in Bohai Bay, Nei Monggol. *Bull. Inst. Geol. Chin. Acad. Geol. Sci.* **1988**, *18*, 68–74. (In Chinese with English Abstract)
76. Ma, M.; Wan, Y.; Xu, Z.; Liu, S.; Xie, H.; Dong, C.; Liu, D. Late Paleoproterozoic K-feldspar pegmatite veins in Daqingshan area, North China Craton: SHRIMP age and Hf composition of zircons. *Geol. Bull. China* **2012**, *31*, 825–833. (In Chinese with English Abstract)
77. Liu, T.; Li, W.; Liu, Y.; Jin, W.; Zhao, Y.; Iqbal, M.Z. Deformation characteristics of the high-grade metamorphic and anatectic rocks in the Daqingshan Paleoproterozoic orogenic belt, Inner Mongolia: A case study from the Shijiaqu-Xuehaigou area. *Precambrian Res.* **2022**, *374*, 106644. [CrossRef]
78. Li, X.; Yang, Z.; Zhao, G.; Grapes, R.; Guo, J. Geochronology of khondalite-series rocks of the Jining Complex: Confirmation of depositional age and tectonometamorphic evolution of the North China craton. *Int. Geol. Rev.* **2011**, *53*, 1194–1211. [CrossRef]

Article

Geochronology and Geological Implications of Paleoproterozoic Post-Collisional Monzogranitic Dykes in the Ne Jiao-Liao-Ji Belt, North China Craton

Yan Zhao [1,2,3], Junchao Lyu [1,*], Xu Han [4], Shoufa Lin [2], Peng Zhang [1,3], Xueming Yang [5] and Cong Chen [1,3]

[1] Shenyang Centre, China Geological Survey, Shenyang 110034, China; zhaoyan@mail.cgs.gov.cn (Y.Z.)
[2] Department of Earth and Environmental Sciences, University of Waterloo, Waterloo, ON N2L 3G1, Canada
[3] Key Laboratory of Deep Mineral Resources Exploration and Evaluation, Department of Natural Resource of Liaoning Province, Shenyang 110032, China
[4] School of Resources and Civil Engineering, Liaoning Institute of Science and Technology, Benxi 117004, China
[5] Manitoba Geological Survey, Winnipeg, MB R3C 0V8, Canada
* Correspondence: lvjunchao@mail.cgs.gov.cn

Abstract: Hardly any previous studies have focused on the granitic dykes which intrude into the Paleoproterozoic Liaohe Group in the Liaodong Peninsula, northeast of the North China Craton. In situ zircon U-Pb dating, Lu-Hf isotopic and geochemical analyses on three representative monzogranite dykes were taken in this study. These dykes have relatively high content of SiO_2 (72.20%–74.78%) and K_2O (2.83%–6.37%), and have characteristics of high-K calc-alkaline to shoshonite series. Two dyke samples have I-type granite features and have high Sr/Y ratios and positive Eu anomalies, showing an adakitic feature. Another dyke has a high ratio of Ga/Al, but has a low Zr saturation temperature, which differs from the typical A-type granite. Zircon grains from these three dykes have typical magmatic zoning in CL images and yield consistent U-Pb ages of ~1859–1852 Ma, which are interpreted as the crystallization ages of these dykes. Hf isotopic analyses yield mainly negative $\varepsilon_{Hf}(t)$ values and T_{DM2} ages of 2782–2430 Ma, similar to those of the 2.2–2.1 Ga granitoids and meta-sedimentary rocks (the Liaohe Group), indicating these monzogranitic dykes may have been sourced from melting of Paleoproterozoic granitoids and meta-sedimentary rocks. The monzogranitic dykes were generated under a post-collisional geological setting after the Jiao-Liao-Ji orogeny process.

Keywords: monzogranitic dykes; U-Pb zircon dating; Lu-Hf isotopes; post-collisional granites; Jiao-Liao-Ji Belt; North China Craton

Citation: Zhao, Y.; Lyu, J.; Han, X.; Lin, S.; Zhang, P.; Yang, X.; Chen, C. Geochronology and Geological Implications of Paleoproterozoic Post-Collisional Monzogranitic Dykes in the Ne Jiao-Ji Belt, North China Craton. *Minerals* **2023**, *13*, 928. https://doi.org/10.3390/min13070928

Academic Editor: Aleksei V. Travin

Received: 10 May 2023
Revised: 27 June 2023
Accepted: 4 July 2023
Published: 11 July 2023

1. Introduction

As one of the oldest cratons in the world, the North China Craton (NCC) is considered to be composed of the Eastern Block, the Western Block and Trans-North China Orogen [1,2]. Three major Paleoproterozoic orogenic belts have been recognized within the NCC [3], of which the Jiao-Liao-Ji Belt (JLJB) in the Eastern Block extends more than 1000 km ([4]; Figure 1). Great attention had been paid to the debate on the early Paleoproterozoic tectonic evolution model of the JLJB, and up to four tectonic models have been proposed as follows: an intra-continental rift opening and closing model [5,6], an arc–continent collision model [7–11], a rifting ocean followed by late subduction and collision model [6,12,13] and a back-arc basin (or retro-arc foreland basin) opening and closure model [14–17].

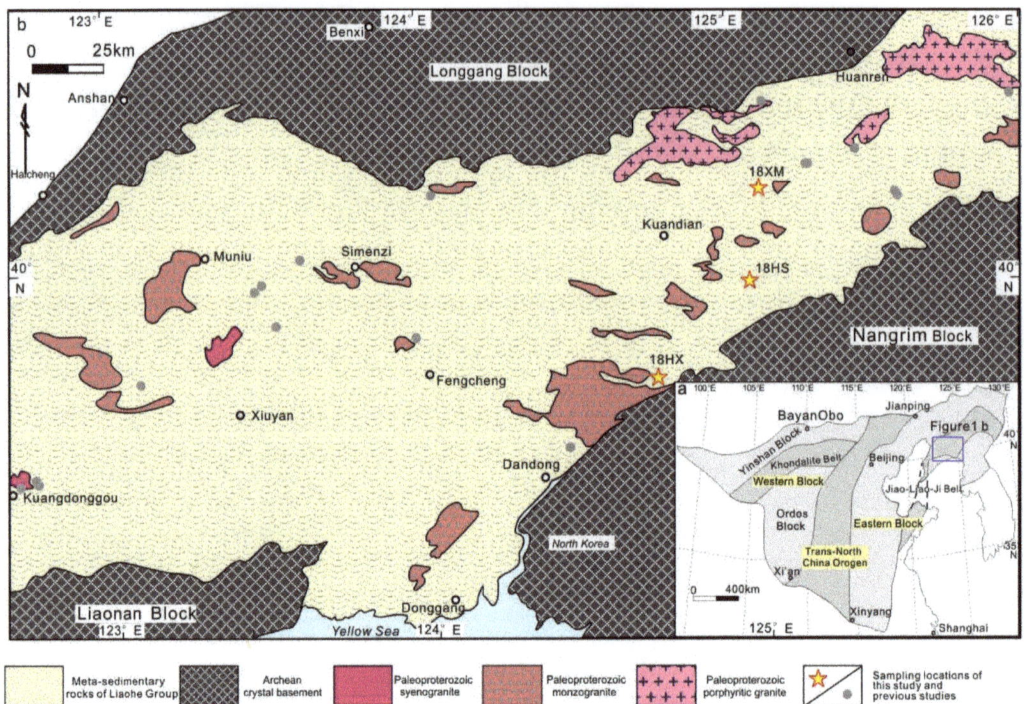

Figure 1. Simplified geological map of the Liaodong Peninsula (**b**) and part of the Jiao-Liao-Ji Belt in the North China Craton (**a**). (Modified after [18,19]). Locations of the monzogranitic dykes of this study are indicated.

Compared to the controversy regarding the early stage of its tectonic evolution, studies on the late Paleoproterozoic JLJB orogenic processes are not as well developed. Previous research has focused on the porphyritic granites [20], syenite [21,22], granitic leucosomes in granulites [23], metamorphic rocks [24] and pegmatite veins [25] from ~1880 Ma to 1860 Ma. The granitoids may have been distributed under a geological setting related to the JLJB complex and long-term evolvement of the orogeny process. Although some granitoids are interpreted as originating under a post-collisional setting after the orogeny process, hardly any robust field evidence with corresponding geochronological evidence has been provided.

The Paleoproterozoic granitoid dykes, which were newly discovered during recent geological mapping fieldwork in the Liaodong Peninsula in Northeast China, enable us to better understand the post-collision process of the JLJB. Here, we report results of geochemical, zircon U-Pb ages, and Hf isotope studies of the three dykes in the Kuandian and Dandong areas (Figure 1), aiming to constrain the age and petrogenesis of post-collisional granites and to better understand the tectonic evolution process of the JLJB.

2. Geological Settings

The JLJB in the Eastern Block of the NCC is located between the Archean Longgang and the Liaonan-Nangrim blocks and extends into the Jiaodong Peninsula to the southwest (Figure 1a; [1,26,27]). Voluminous sedimentary rocks, granitoids, volcanic rocks and metamorphic rocks were involved in the intense orogenic process in ~1.9 Ga and experienced related deformation [28–33]. The Liaohe Group in the Liaoning Province is correlated with the Laoling and Ji'an groups in the Jilin Province and the Fenzishan and Jingshan groups in the Shandong Province. The Liaohe Group consists of meta-sedimentary rocks and some meta-volcanic rocks and is subdivided into the Langzishan, Lieryu, Gaojiayu, Dashiqiao, and Gaixian formations [4,5,34]. Previous researchers summarized the widely distributed igneous rocks data and distinguished five magmatic episodes [20,21] as follows: ca. 2190–2160 Ma A2-type granites with minor basaltic dykes and tuffs, ca. 2160–2110 Ma tholeiitic rocks, ca. 2110–2080 Ma mafic rocks and aluminous A2-type granites, ca. 2010–1885 Ma adakitic granites related to regional metamorphism and ca. 1875–1850 Ma post-collisional granites.

Post-collisional granites are represented by the biotite-bearing and garnet-bearing porphyritic granites in the Kuandian and Huanren areas [20], syenite in the Kuangdonggou area [22], and quartz diorite in the Qinghe area [24]. These undeformed granitoids intruded into the deformed Paleoproterozoic meta-sedimentary rocks and early gneissic granitoids. In addition, hardly any reliable post-collisional contemporary mafic intrusion or dykes are reported at this period within the JLJB.

3. Sample Materials and Analytical Methods

3.1. Sample Materials

The granitic dykes from the Kuandian and Dandong areas, eastern Liaodong Peninsula, had a width of one to two meters and a length of tens of meters. Unlike the leucosomes in granulites and gneiss, which had a width of a few centimeters and paralleled the foliation of granulites as [23] reported, these granitic dykes are observed to intrude into the Paleoproterozoic Gaixian and Gaojiayu formations either cutting through or following the foliation of gneiss (Figure 2).

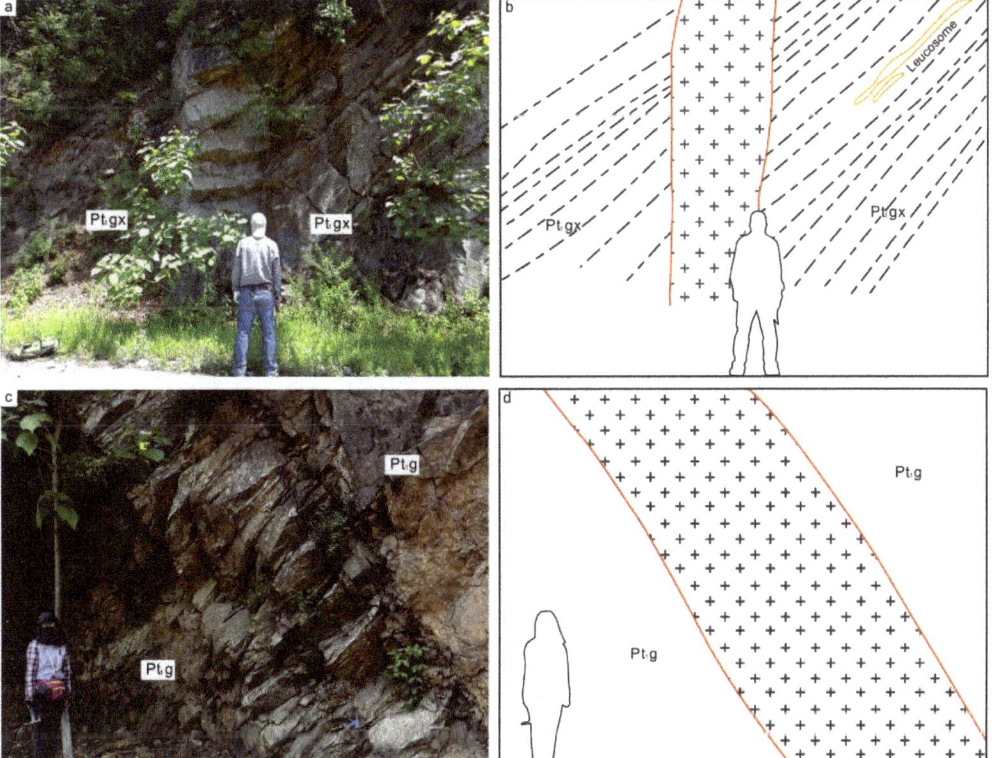

Figure 2. *Cont.*

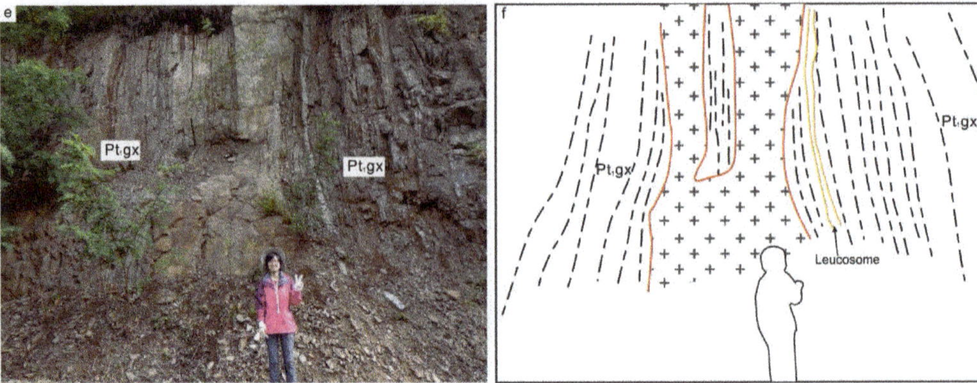

Figure 2. Field photographs and sketches of the studied monzogranitic dykes in the Kuandian and Dandong areas within the Jiao-Liao-Ji Belt: (**a,b**) sample 18XM, (**c,d**) sample 18HX, (**e,f**) sample 18HS ((**a**): a monzogranitic dyke intrudes into the Paleoproterozoic Gaixian Formation gneiss and cuts the foliation of the gneiss near Kuandian County; (**c**): a monzogranitic dyke intrudes into the Paleoproterozoic Gaojiayu Formation granulite in the Hushan area; (**e**): a granitic dyke intrudes into the Gaixian Formation schist and gneiss near the Hongshi area) Pt_1g: Paleoproterozoic Gaojiayu Formation granulite; Pt_1gx: Paleoproterozoic Gaixian Formation schist and gneiss.

Sample 18XM (40°53′11″ N, 125°8′40″ E) was collected from a 1 m wide monzogranite dyke that intrudes into the Gaixian Formation schist and gneiss (Figure 2a). Sample 18HX (40°22′41″ N, 124°43′40″ E) was collected from a 2 m wide dyke that intrudes into the Gaojiayu Formation granulite near Hushan Town (Figure 2c). Sample 18HS is a coarse-grained granite (40°41′00″ N, 125°9′01″ E) collected from a 1.5 m wide dyke that intrudes into the Gaixian Formation schist and gneiss near Hongshi Town (Figure 2e).

3.2. Thin Section Petrography

The thin sections of the studied granitoids were prepared for optical petrography at the Shenyang Institute of Geology and Mineral Resources, Shenyang, China. Small pieces of rock were cut from the field samples with a diamond blade, and then mounted on a petrography carrier glass (~27 mm × 47 mm). The mounted sections were polished by using a range of ever-finer abrasive powders down to a thickness of 50 μm. Based on the polished thin sections, the microstructure, texture and mineral modal contents in vol% were estimated by point counting using an optical petrography microscope in both transmitted and polarized light.

3.3. Major and Trace Elements

After removal of altered surfaces, fresh whole-rock samples were crushed and ground to 200 mesh size in an agate mill. Chemical analyses were conducted at the Shenyang Institute of Geology and Mineral Resources, Shenyang, China. X-ray fluorescence (XRF, PANALYTICAL, Almelo, Holland) (AXIOS-Minerals) using fused glass disks and inductively coupled plasma mass spectrometry (ICP–MS, Agilent company, Santa Clara, CA, USA) (Agilent 7500a with a shield torch) were used to measure major and trace element compositions, respectively. The detailed sample-digesting procedure is as follows: Sample powder (200 mesh) was placed in an oven at 105 °C and dried for 12 h, then ~1.0 g dried sample was accurately weighted and placed in the ceramic crucible and then heated in a muffle furnace at 1000 °C for 2 h. After cooling to 400 °C, this sample was placed in the drying vessel and weighted again in order to calculate the loss on ignition (LOI). Sample powder (0.6 g) was mixed with 6.0 g cosolvent ($Li_2B_4O_7$:$LiBO_2$:LiF = 9:2:1) and 0.3 g oxidant (NH_4NO_3) in a Pt crucible, which was placed in the furnace at 1150 °C for 14 min. Then, this melting sample was quenched with air for 1 min to produce flat discs on the fire brick for the XRF analyses. Fe_2O_3 and FeO content were measured directly, while the FeO^T was calculated (Supplementary Table S1). The detection limit of XRF differs from 0.001% to 1% following GB/T 14506.1-2010 standard. For trace element analysis, (1) 50 mg sample powder was accurately weighed and placed in a Teflon bomb; (2) 1 mL HNO_3 and 1 mL HF were slowly added into the Teflon bomb; (3) Teflon bomb was placed in a stainless steel pressure jacket and heated to 190 °C in an oven for >24 h; (4) after cooling, the Teflon bomb was opened and placed on a hotplate at 140 °C and evaporated to incipient dryness, and then 1 mL HNO_3 was added and evaporated to dryness again; (5) 1 mL of HNO_3, 1 mL of MQ water and 1 mL internal standard solution of 1 ppm In were added, and the Teflon bomb was resealed and placed in the oven at 190 °C for >12 h; (6) the final solution was transferred to a polyethylene bottle and diluted to 100 g by the addition of 2% HNO_3. Precision and accuracy were better than 5% for major elements and 10% for trace elements based on repeated analyses of the USGS standards BHVO-1, BCR-2, and AGV-1 [35]. The detection limit of ICP-MS differed from 0.003 to 1×10^{-6} following GB/T 14506.30-2010 standard.

3.4. Zircon LA-ICP-MS U-Pb Dating

Zircon grains from the granitoid samples were extracted using heavy-liquid and magnetic separation, and purified by hand-picking under a binocular microscope. Zircons were mounted and polished before cathodoluminescent (CL) images were taken to examine the internal structure and potential inclusions of individual grains. U-Pb dating and trace element analysis of zircon were simultaneously conducted by LA-ICP-MS (Agilent company, Santa Clara, CA, USA) (Agilent 7700e) at the Wuhan Sample Solution Analytical Technology Co., Ltd., Wuhan, China. Laser sampling was performed using a GeolasPro laser ablation system that consists of a COMPexPro 102 ArF excimer laser (wavelength of 193 nm and maximum energy of 200 mJ) and a MicroLas optical system. Helium was applied as a carrier gas. Argon was used as the make-up gas and was mixed with the carrier gas via a T-connector before entering the ICP system. The laser beam diameter was 35 μm and the repetition rate was 5 Hz. Each spot analysis consisted of a ~5 s background measurement and a 45 s sample measurement. The $^{207}Pb/^{206}Pb$, $^{206}Pb/^{238}U$, $^{207}Pb/^{235}U$ and $^{208}Pb/^{232}Th$ values were corrected for instrumental isotopic and elemental fractionation effects using zircon standard 91500. An Excel-based software, ICPMSDataCal (ICPMSdatacal Excel2016, Redmond, WA, USA), was used to perform off-line selection and integration of the background. It also analyzed signals, time–drift correction and quantitative calibration for trace element analysis and U-Pb dating [36]. Concordia diagrams and weighted mean calculations were made using Isoplot/Ex_ver3 (Ludwig, Berkeley, CA, USA) [37].

3.5. Zircon Lu-Hf Ratio Analyses

The in situ Lu–Hf isotope analyses ($n = 42$) were also conducted by MC-ICP-MS at the Wuhan Sample Solution Analytical Technology Co., Ltd. (Wuhan, China). A stationary spot used a beam diameter of ~55 μm. Helium was used to transport the ablated sample aerosol mixed with argon from the laser-ablation cell to the MC-ICP-MS torch by a mixing chamber. $^{176}Lu/^{175}Lu = 0.02658$ and $^{176}Yb/^{173}Yb = 0.796218$ ratios were determined to correct for the isobaric interferences of ^{176}Lu and ^{176}Yb on ^{176}Hf [38]. The $^{176}Hf/^{177}Hf$ and $^{176}Lu/^{177}Hf$ ratios of the 91500 standard zircon were 0.282270 ± 0.000023 (2σ, $n = 15$) and 0.00028, similar to the commonly accepted $^{176}Hf/^{177}Hf$ ratio of 0.282284 ± 0.000003 (1σ) measured using the solution method [39]. Zircon international standard GJ-1 was used as the reference standard, with a weighted mean $^{176}Hf/^{177}Hf$ ratio of 0.282006 ± 32 (2SD, $n = 24$).

4. Results

4.1. Petrography

The 18XM sample has a fine- to medium-grained texture and a grey to light-brown color (Figure 3a,b). Minerals in the monzogranite include quartz (~30 vol%), plagioclase

(~30 vol%), orthoclase (~35 vol%) and biotite (<5 vol%). The 18HX sample has a character-istic medium-grained texture and a grey color (Figure 3c). Quartz (~30 vol%), plagioclase (~33 vol%), orthoclase (~25 vol%), biotite (~10 vol%) and minor muscovite (~2 vol%) comprise the minerals in this monzogranite. The 18HS sample has a characteristic coarse-grained texture and a grey to brown color (Figure 3d). Plagioclase (~40 vol%), orthoclase (~35 vol%), quartz (~23 vol%) and minor biotite (~2 vol%) comprise this monzogranite.

Figure 3. Photomicrographs of the studied monzogranitic dykes in the Kuandian and Dandong areas (images are without analysator). (**a,b**) Sample 18XM, (**c**) sample 18HX, (**d**) sample 18HS (Bt: biotite; Or: orthoclase; Pl: plagioclase; Qz: quartz).

4.2. Major and Trace Elements

The 18XM samples have a relatively high content of SiO_2 (72.81–74.78 wt%) and K_2O (5.52–5.70 wt%), and plot into the metaluminous and peraluminous field (Figure 4b). The 18HS and 18HX granitic dyke samples also have a relatively high content of SiO_2 (72.20–74.08 wt%) and K_2O (2.83–6.37 wt%) and plot into the fields of high-K calc-alkaline to shoshonite series. These two samples have medium Al_2O_3 contents, and in the A/NK ver-sus A/CNK diagram they are mainly metaluminous to weakly peraluminous (Figure 4b). Major and trace element content of the three analyzed granitoid dykes are given in Supplementary Table S1 and Figure 5. The 18HS and 18HX samples exhibit similar chondrite-normalized rare earth element (REE) patterns with marked positive Eu anomalies (Eu/Eu* = 2.16–4.70). The 18XM samples are also enriched in light REE but with strong negative Eu anomalies (Eu/Eu* = 0.31–0.53).

4.3. Zircon U-Pb Geochronology

CL images of representative zircons from the three samples are listed in Supplementary Table S2 and Figure 6. Zircon grains have similar euhedral to subhedral shape and are about 70 to 200 µm long. Th/U ratios of the testing grains range from 0.11 to 0.86, indicating an igneous origin. Fifteen spots from sample 18XM and thirty spots from sample 18HS are analyzed, yielding intercept U-Pb age of 1859 ± 33 Ma (MSWD = 2.1) and 1852 ± 10 Ma (MSWD = 1.02), respectively. Twenty-five spots from sample 18HX yield weighted mean ^{207}Pb/^{206}Pb ages of 1856 ± 12 Ma (MSWD = 1.3) (Figure 6).

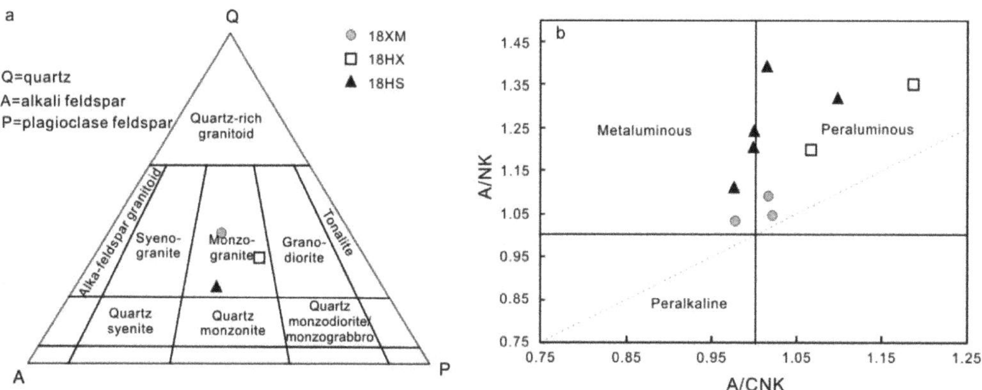

Figure 4. Discrimination diagrams for the monzogranitic dykes in the Jiao-Liao-Ji Belt. (**a**) Quartz-alkali feldspar-plagioclase feldspar classification diagram (after [40]); (**b**) A/CNK versus A/NK after Maniar and Piccoli [41].

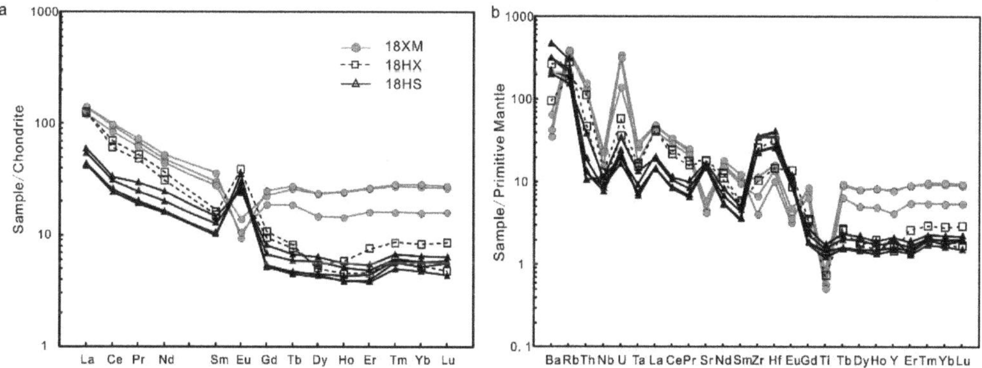

Figure 5. CI chondrite-normalized REE (**a**) and primitive mantle-normalized trace elements distribution (**b**) patterns for the monzogranitic dykes in the Jiao-Liao-Ji Belt (CI chondrite and primitive mantle values are from [42]).

4.4. Zircon Lu-Hf Isotopic Characteristics

Zircon analytical spots with concordant U-Pb ages are analyzed for their Lu-Hf isotope contents. Results are listed in Supplementary Table S3 and Figure 7. Sample 18HX yielded negative $\varepsilon_{Hf}(t)$ values, ranging between −4.0 and −0.9. The initial ^{176}Hf/^{177}Hf ratios ranged from 0.281482 to 0.281607, with calculated T_{DM1} ages of 2447–2302 Ma and T_{DM2} ages of 2782–2566 Ma. Sample 18XM also yielded negative $\varepsilon_{Hf}(t)$ values, ranging between −2.8 and −6.8. The initial ^{176}Hf/^{177}Hf ratios ranged from 0.281484 to 0.281540, with calculated T_{DM1} ages of 2377–2452 Ma and T_{DM2} ages of 2696–2826 Ma. Similarly, initial ^{176}Hf/^{177}Hf ratios of zircons from 18HS were from 0.281573 to 0.281669, with calculated T_{DM1} ages of 2338–2206 Ma and T_{DM2} ages of 2640–2430 Ma. The $\varepsilon_{Hf}(t)$ values for 18HS ranged from −2.5 and +1.2.

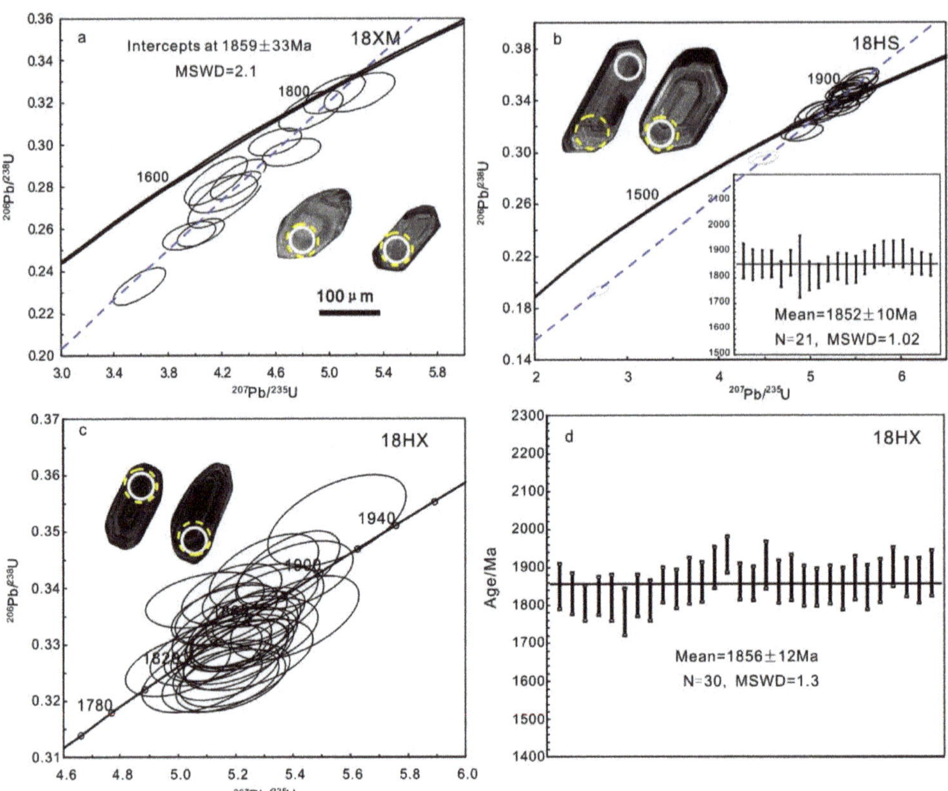

Figure 6. Concordia diagrams for U-Pb analyses of zircons from monzogranitic dykes in the Jiao-Liao-Ji Belt, the 35-μm analytical spots of U-Pb (white ring) and 50 μm analytical spots of Hf (yellow ring) analysis of represented zircon grains are also shown. Respective error range (ellipse and bar) of 2SD is shown. (**a**) Sample 18XM; (**b**) sample 18HS; (**c**,**d**) sample 18HX.

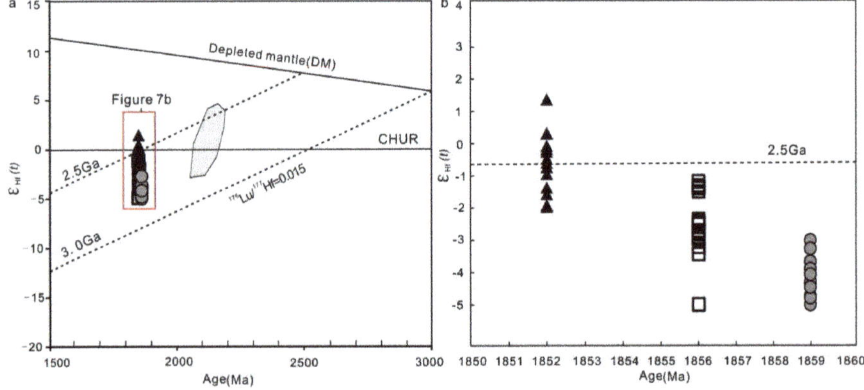

Figure 7. Zircon [207]Pb/[206]Pb ages versus $\varepsilon_{Hf}(t)$ values of the monzogranitic dykes in the Jiao-Liao-Ji Belt (**a**,**b**). Lu-Hf evolution line for depleted mantle is constrained by a present-day [176]Hf/[177]Hf ratio of 0.28325 [43] and [176]Lu/[177]Hf ratio of 0.0384 [38]. The gray polygon is the Paleoproterozoic Hf isotopic data from [20]. Legend is similar with that in Figure 5.

5. Discussion

5.1. U-Pb Ages of the Monzogranitic Dykes and the Granitoids Geochronology within the JLJB

Three monzogranitic dykes were observed cutting through foliation of the deformed Liaohe Group meta-sedimentary rocks during the fieldwork (Figure 2). Zircon grains from these dykes exhibit typical oscillatory zoning in CL images, with most Th/U ratios greater than 0.2, indicating a magmatic origin [44]. These zircon grains yield consistent U-Pb ages of ~1859 to 1852 Ma (Supplementary Table S2, Figure 6), showing that these monzogranitic dykes crystallized ~1860–1850 Ma.

Previous researchers have summarized igneous rocks and their relations with the tectonic evolution process of the JLJB, and established different evolution stages [4,18,19]. Here, we also list Late Paleoproterozoic granitoids types and available ages in Liaoning and Jilin provinces in Table 1. Taking these studies into consideration, three stages of granitoids could be recognized as follows. An early magmatic event of ~2200 to 2140 Ma, indicated by A-type and I-type granites, which are also known as the Liaoji Granites [26,45–47]. A middle period consisting of ~1890 to 1860 Ma biotite-bearing and garnet-bearing porphyritic granites, pegmatite and adakitic granites. They were believed to be accompanied by peak metamorphism in the JLJB [16,25,33] or by Paleoproterozoic oceanic slab subduction [48]. A post-collisional magmatic event at ~1855 to 1840 Ma, is represented by the monzogranite dykes in this study, syenite in the Kuangdonggou area, and sporadically pegmatite dykes [21,25].

Table 1. Geological characteristics and chronological results of Late Paleoproterozoic granites in the Liao-Ji region.

Intrusion Location	Location/GPS	Lithology	Methods	Testing Spots	Age/Ma	Interpretation	Reference
Sanjiazi	40°44'22″ N 123°15'32″ E	Pegmatite	LA ICP-MS	30	1814 ± 20 Ma	Crystallization age	[25]
Sanjiazi	40°51'11″ N 123°19'27″ E	Pegmatite	LA ICP-MS	26	1869 ± 22 Ma	Crystallization age	[25]
Sanjiazi	40°42'06″ N 123°13'30″ E	Pegmatite	LA ICP-MS	37	1873 ± 17 Ma	Crystallization age	[25]
Yangmugan	40°38'16″ N 125°07'20″ E	Pegmatite	LA ICP-MS	41	1842 ± 7 Ma	Crystallization age	[25]
Yangmugan	40°38'01″ N 125°07'43″ E	Pegmatite	LA ICP-MS	30	1866 ± 13 Ma	Crystallization age	[25]
Xiaomiao Dyke	40°53'11″ N 125°08'40″ E	Monzogranite	LA ICP-MS	15	1859 ± 33 Ma	Crystallization age	This paper
Hongshi Dyke	40°22'41″ N, 124°43'40″ E	Monzogranite	LA ICP-MS	30	1852 ± 10 Ma	Crystallization age	This paper
Hushan Dyke	40°41'00″ N, 125°9'01″ E	Monzogranite	LA ICP-MS	29	1856 ± 12 Ma	Crystallization age	This paper
Dadingzi	Eastern Qingchegnzi Town, Dandong City	Monzogranite	SIMS	12	1869 ± 16 Ma	Crystallization age	[49]
Nantaizi	40°31'42″ N, 122°48'37″ E	Monzogranite	LA ICP-MS	31	1851 ± 11 Ma	Crystallization age	[50]
Housongshu	40°28'12″ N, 122°56'12″ E	Trondhjemite	LA ICP-MS	16	1892 ± 16 Ma	Crystallization age	[50]
Taipingshao	10 km north of Taipingshao Town Dandong City	Granite	LA ICP-MS	28	1892 ± 38 Ma	Crystallization age	[51]
Taipingshao	12 km north of Taipingshao Town Dandong City	Granite	LA ICP-MS	27	1859 ± 36 Ma	Crystallization age	[51]
Kuangdonggou	Kuangdonggou Town Gaizhou City	Syenite	LA ICP-MS	19	1879 ± 17 Ma	Crystallization age	[22]
Kuangdonggou	Kuangdonggou Town Gaizhou City	Syenite	LA ICP-MS	17	1874 ± 18 Ma	Crystallization age	[22]
Kuangdonggou	5 km west of Kuangdonggou Town Gaizhou City	Diorite	LA ICP-MS	12	1870 ± 18 Ma	Crystallization age	[22]
Sizhanggunzi	Sizhanggunzi village Gaizhou City	Granodiorite	LA ICP-MS	19	1871.2 ± 9.3 Ma	Crystallization age	[52]
Qinghe	Qianjin Village Qinghe Town Ji'an City	Quartz diorite	SIMS	12	1877 ± 15 Ma	Crystallization age	[21]

Table 1. *Cont.*

Intrusion Location	Location/GPS	Lithology	Methods	Testing Spots	Age/Ma	Interpretation	Reference
Shuangcha	Shuangcha Village Ji'an City	Garnet-bearing porphyritic granite	LA ICP-MS	30	1890 ± 21 Ma	Crystallization age	[53]
Wuleishan	Northwest of Sanjiazi Town Anshan City	Porphyritic granite	SIMS	12	1830.5 ± 5.9 Ma	Crystallization age	[49]
Shuangcha	Huangweizi Village Ji'an City	porphyritic biotite mozogranite	SIMS	17	1872 ± 9 Ma	Crystallization age	[21]
Shuangcha	Zhongxing Village Ji'an City	Rapakivi granite	SIMS	18	1867 ± 13 Ma	Crystallization age	[21]
Lujiapuzi	Pulepu Town Fushun City	Rapakivi granite	SIMS	19	1847 ± 40 Ma	Crystallization age	[21]
13th Gou	13th Gou Baishan City Jilin Province	Garnet-bearing porphyritic granite	LA ICP-MS	20 11	1868 ± 9 Ma 1848 ± 13 Ma	Crystallization age, Metamorphic age	[20]
12th Gou	12th Gou Baishan City Jilin Province	Garnet-bearing porphyritic granite	LA ICP-MS	42 9	1872 ± 6 Ma 1851 ± 14 Ma	Crystallization age, Metamorphic age	[20]
Huadian	Huadian Town Ji'an City Jilin Province	Garnet-bearing porphyritic granite	LA ICP-MS	26 13	1871 ± 7 Ma 1850 ± 12 Ma	Crystallization age, Metamorphic age	[20]
Huadian	Huadian Town Ji'an City Jilin Province	Garnet-bearing porphyritic granite	LA ICP-MS	26 10	1866 ± 2 Ma 1850 ± 4 Ma	Crystallization age, Metamorphic age	[20]
Huadian	Taishang Town Ji'an City Jilin Province	Garnet-bearing porphyritic granite	LA ICP-MS	35 11	1869 ± 2 Ma 1850 ± 4 Ma	Crystallization age, Metamorphic age	[20]
Laoheishan	Laoheishan Village Dandong City Liaoning Province	Garnet-bearing porphyritic granite	LA ICP-MS	25 11	1872 ± 8 Ma 1851 ± 12 Ma	Crystallization age, Metamorphic age	[20]
Jiguanshan	Jiguanshan Town Dandong City Liaoning Province	Garnet-bearing porphyritic granite	LA ICP-MS	31 13	1870 ± 7 Ma 1850 ± 11 Ma	Crystallization age, Metamorphic age	[20]
Laoheishan	Laoheishan Village Dandong City Liaoning Province	Garnet-bearing porphyritic granite	LA ICP-MS	27 11	1868 ± 3 Ma 1846 ± 5 Ma	Crystallization age, Metamorphic age	[20]
Jiguanshan	Jiguanshan Town Dandong City Liaoning Province	Garnet-bearing porphyritic granite	LA ICP-MS	25 12	1870 ± 3 Ma 1842 ± 4 Ma	Crystallization age, Metamorphic age	[20]
12th Gou	12th Gou Baishan City Jilin Province	Biotite-bearing porphyritic granite	LA ICP-MS	34 8	1868 ± 6 Ma 1849 ± 13 Ma	Crystallization age, Metamorphic age	[20]
11th Gou	11th Gou Baishan City Jilin Province	Biotite-bearing porphyritic granite	LA ICP-MS	20 14	1872 ± 7 Ma 1849 ± 9 Ma	Crystallization age, Metamorphic age	[20]
Qinghe	Qinghe Town Ji'an City Jilin Province	Biotite-bearing porphyritic granite	LA ICP-MS	24 19	1865 ± 7 Ma 1849 ± 9 Ma	Crystallization age, Metamorphic age	[20]
Sipingxiang	Sipingxiang Fushun City Liaoning Province	Biotite-bearing porphyritic granite	LA ICP-MS	28 10	1872 ± 7 Ma 1850 ± 13 Ma	Crystallization age, Metamorphic age	[20]
Taipingshao	West of Taipingshao Town Dandong City Liaoning Province	Biotite-bearing porphyritic granite	LA ICP-MS	17 11	1867 ± 10 Ma 1842 ± 12 Ma	Crystallization age, Metamorphic age	[20]
Yahe	East of Yahe Town Dandong City Liaoning Province	Biotite-bearing porphyritic granite	LA ICP-MS	33 16	1865 ± 6 Ma 1849 ± 8 Ma	Crystallization age, Metamorphic age	[20]
Gulouzi	North of Gulouzi Town Dandong City Liaoning Province	Biotite-bearing porphyritic granite	LA ICP-MS	18 14	1864 ± 8 Ma 1844 ± 9 Ma	Crystallization age, Metamorphic age	[20]
11th Gou	11th Gou Baishan City Jilin Province	Biotite-bearing porphyritic granite	LA ICP-MS	28 11	1864 ± 2 Ma 1846 ± 4 Ma	Crystallization age, Metamorphic age	[20]

Table 1. *Cont.*

Intrusion Location	Location/GPS	Lithology	Methods	Testing Spots	Age/Ma	Interpretation	Reference
Qinghe	Qinghe Town Ji'an City Jilin Province	Biotite-bearing porphyritic granite	LA ICP-MS	40 27	1867 ± 2 Ma 1847 ± 3 Ma	Crystallization age, Metamorphic age	[20]
Taipingshao	West of Taipingshao Town Dandong City Liaoning Province	Biotite-bearing porphyritic granite	LA ICP-MS	38 14	1869 ± 3 Ma 1849 ± 3 Ma	Crystallization age, Metamorphic age	[20]
11th Gou	11th Gou Baishan City Jilin Province	Flesh-red porphyritic granite	LA ICP-MS	30 24	1868 ± 8 Ma 1849 ± 8 Ma	Crystallization age, Metamorphic age	[20]
Taipingshao	West of Taipingshao Town Dandong City Liaoning Province	Flesh-red porphyritic granite	LA ICP-MS	39 8	1866 ± 6 Ma 1846 ± 13 Ma	Crystallization age, Metamorphic age	[20]

5.2. Genesis of the Monzogranitic Dykes

The monzogranite dykes in this study have a similar high SiO_2 content and alkalis and have low content of MgO, FeO^T and CaO (Supplementary Table S1). In the primitive mantle-normalized trace element diagram (Figure 5b), the 18HS and 18HX samples are enriched in large ion lithosphere elements (LILEs) with positive Eu (Figure 5a) and Ba, U anomalies and depleted in some high field-strength elements (HFSEs). The 18HS and 18HX samples also exhibit high Sr/Y ratio, indicating an adakitic granite feature (Figure 8a). The 18XM samples show relatively high content of U and HREE, but exhibit similar overall trace elements to those in 18HS and 18HX. The 18 XM samples have A-type granite features as their high ratios of Ga/Al, while the 18 HS and 18 HX samples show I-type granite characteristics (Figure 9). Further Zr saturation calculation results reveal that 18XM samples have the lowest temperature of 688 °C to 770 °C (Supplementary Table S1), rather than the high temperature of typical A-type granites [54–57]. Considering the distribution of the studied monzogranitic dykes within the JLJB, they should have been generated under a thickened crust tectonic setting.

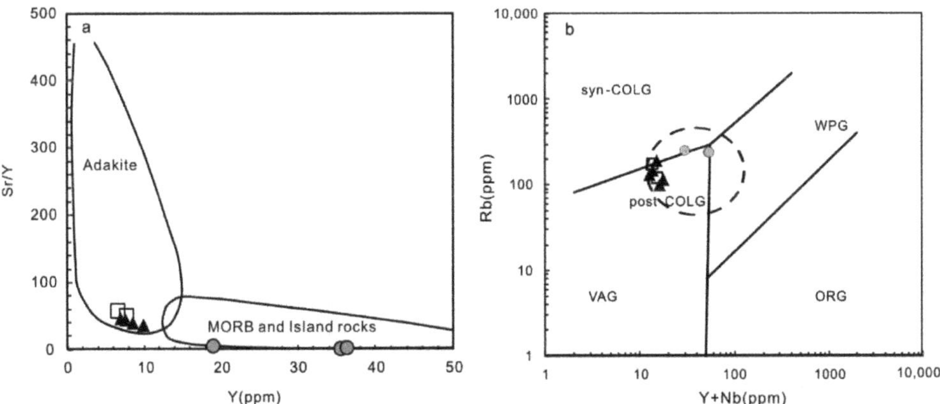

Figure 8. Sr/Y vs. Y diagram for the monzogranitic dykes in the Jiao-Liao-Ji Belt (**a**, after [58] and Rb-(Y + Nb) diagrams (**b**, after [59,60]). The circle in Figure 8a represents a post-collisional geodynamic setting of Pearce et al. [60] (MORB: mid ocean ridge basalt; ORG: orogenic granite; syn-COLG: syn-collisional granite; COLG: collisional granitoids; OA: oceanic arc; VAG: volcanic arc granite; WPG: within-plate granite). Legend is similar to that in Figure 5.

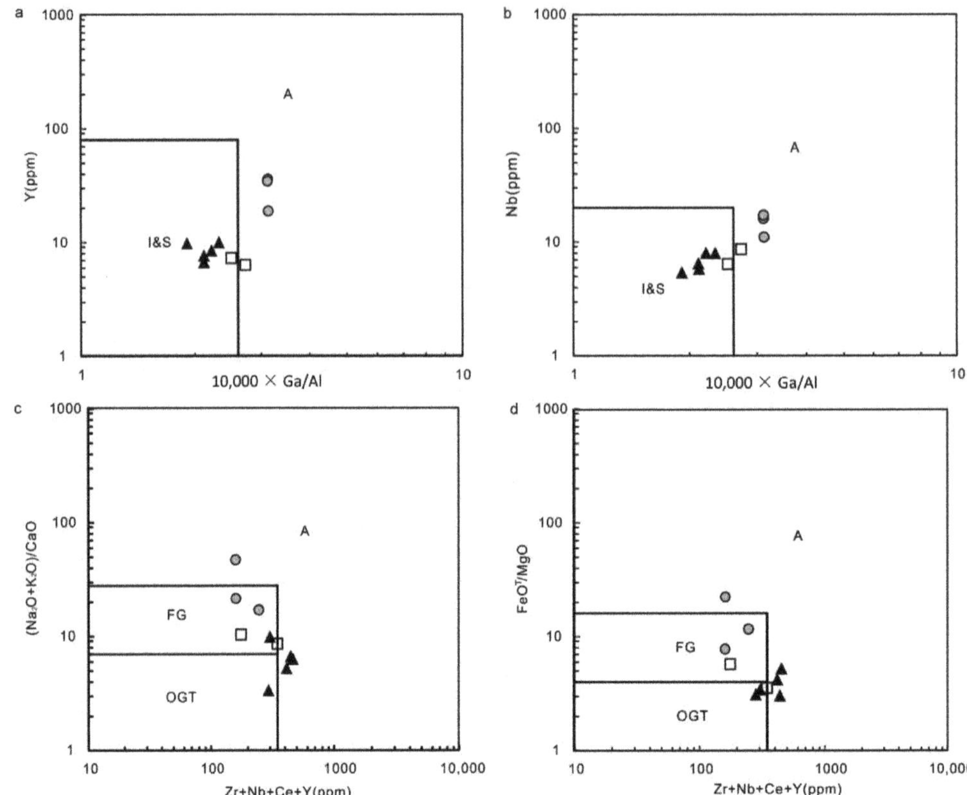

Figure 9. Y (**a**), Nb (**b**) versus 10,000 × Ga/Al plots and (Na$_2$O + K$_2$O)/CaO (**c**), FeO^T/MgO (**d**) versus Zr + Nb + Ce + Y plots (after [54]) A, I and S = A-, I- and S-type granites, respectively; FG = fractionated felsic granites; OGT = unfractionated M-, I- and S-type granites, legend is similar to that in Figure 5.

Hf isotope analysis also illustrated similar conclusions. Hf isotope data show T_{DM2} ages of 2640–2430 Ma, and reveal that most of the negative $\varepsilon_{Hf}(t)$ values plot near the 2.5 Ga crustal evolution line with a few positive values (Supplementary Table S3; Figure 7). These results are similar to those of the 2.2–2.1 Ga granitoids by previous studies [20,28,30,51]. These granitic dykes have low Mg$^{\#}$ and variable LILE/HFSE ratios (Supplementary Table S1), suggesting that they may have been derived from low-degree partial melting of a thickened crust, rather than subduction of oceanic crust. This magmatic event should have been triggered by delamination-induced extension at the late stage of the orogenic process of the JLJB.

5.3. Implications to the Evolution Model of the JLJB

Although there are matters of debate on the early evolution model of the JLJB, there is a consensus understanding on the orogenic process in Late Paleoproterozoic (e.g., [14,61–64]). The Longgang block and the Liaonan-Nangrim block convergence began at ~2.0 or 1.95 Ga [16,26,65], and widespread metamorphism and deformation are also believed to have taken place at ~1.9 Ga [19,66–69] or ~1.88–1.85 Ga [70,71].

Termination time of the JLJB orogeny was only roughly defined by corresponding granitoids geochronological studies. Garnet-bearing and biotite-bearing porphyritic granites (~1870–1865 Ma), rapakivi granites and alkaline syenite (~1872–1850 Ma) were reported within the JLJB as products of anorogenic magmatism by previous studies and interpreted as representing the termination of the JLJB orogeny [20,22,72,73]. These geochronological results also met previous metamorphic study results in the JLJB, as ~1.95–1.85 Ga HP and MP granulites in the Jiaobei Block from the Jiaodong Peninsula, the South Liaohe Group in Liaoning Province, and the Ji'an Group in Jilin Province were reported [23,68,69,74,75].

Monzogranitic dykes in this study provide robust constraint on post-collisional granitoids in the JLJB orogeny process. Three granitic dykes observed in this study exhibit manifested post-collisional granite field features by cutting through foliation of the Liaohe Group meta-sedimentary rocks which deformed at ~1.9 Ga by previous studies [23,31,63] (Figure 2). These granitic dykes have transitional features of I- to A-type granites, probably distributing under a post-collisional tectonic setting as [76] illustrated. These monzogranite dykes show features of post-collisional granites in the model proposed by Pearce et al. [59,60] (Figure 8b). Considering these features of the granitoid dykes, the whole-rock geochemistry and Hf isotopic results, we believe these dykes represent exhumation and extension in the final collisional of the orogeny process. The termination collisional time of the JLJB orogen was constrained to be not later than 1860 Ma by these monzogranite dykes.

6. Conclusions

Based on a study of zircon U-Pb geochronology, Hf isotopes and whole-rock geochemistry of the monzogranitic dykes, the following conclusions can be drawn.

(1) The monzogranite dykes cut through the foliation of the Liaohe Group meta-sedimentary rocks which deformed at ~1900 Ma. Zircons from the monzogranite dyke have euhedral to subhedral shape and high ratio of Th/U, and yield a consistent zircon U-Pb age of ~1859–1852 Ma.
(2) Two dyke samples have I-type granite and adakitic granite features; the other dyke has a characteristic of A-type granite. Hf isotope data show T_{DM2} ages of 2640–2430 Ma and reveal that most of the $\varepsilon_{Hf}(t)$ values plot near the 2.5 Ga crustal evolution line. Such geochemistry and Hf isotope studies indicate they may have been generated under a delamination of thickened crust tectonic setting.
(3) The termination collisional time of the JLJB orogen was constrained to be not later than 1860 Ma by these monzogranitic dykes.

Supplementary Materials: The following supporting information can be downloaded at: https://www.mdpi.com/article/10.3390/min13070928/s1, Table S1: Major(wt%) and trace ($\times 10^{-6}$) element data for representative samples of the Paleoproterozoic monzogranitic dykes from the Kuandian and Dandong area within the Jiao-Liao-Ji Belt, (TZr(°C): Zircon saturations temperatures by [77]) and (δEu/Eu* = Eu N/(Sm N $\times$ Gd N)1/2, N denotes the chondrite normalization (Sun and McDonough, 1989) [42]). Table S2: LA-ICP-MS zircon U-Pb analyses for representative samples of Paleoproterozoic monzograntic dykes from the Kuandian and Dandong area within the Jiao-Liao-Ji Belt, Table S3: Lu-Hf isotopic compositions of the in situ zircons of Paleoproterozoic monzogranitic dykes in the Kuandian and Dandong area within the Jiao-Liao-Ji Belt.

Author Contributions: Conceptualization, Y.Z. and X.H.; methodology, P.Z. and J.L.; investigation, Y.Z. and C.C.; writing—original draft preparation, Y.Z.; writing—review and editing, Y.Z., X.H., S.L. and X.Y. All authors have read and agreed to the published version of the manuscript.

Funding: This research was funded by the National Science Foundation of China (Grant No. 42072104; 41502093), Geological Survey Project of China (Grant No. DD20230329; DD20230102), Scientific Research Foundation of Education Department of Liaoning Province (Grant No. LJKQZ20222473, LJKZ1075) and the China Scholarship Council.

Data Availability Statement: Not applicable.

References

1. Zhao, G.C.; Sun, M.; Wilde, S.A.; Li, S.Z. Late Archean to Paleoproterozoic evolution of the North China Craton: Key issues revisited. *Precambrian Res.* **2005**, *136*, 177–202. [CrossRef]
2. Wilde, S.A.; Zhao, G.C. Archean to Paleoproterozoic evolution of the North China Craton. *J. Asian Earth Sci.* **2005**, *24*, 519–522. [CrossRef]
3. Zhai, M.G.; Santosh, M. Metallogeny of the North China Craton: Link with secular changes in the evolving Earth. *Gondwana Res.* **2013**, *24*, 275–297. [CrossRef]
4. Liu, F.L.; Liu, P.H.; Wang, F.; Liu, C.H.; Cai, J. Progresses and overviews of voluminous meta-sedimentary series within the Paleoproterozoic Jiao-Liao-Ji orogenic/mobile belt, North China Craton. *Acta Petrol. Sin.* **2015**, *31*, 2816–2846, (In Chinese with English abstract).
5. Li, S.Z.; Zhao, G.C.; Sun, M.; Han, Z.Z.; Hao, D.F.; Luo, Y.; Xia, X.P. Deformation history of the Paleoproterzoic Liaohe Group in the Eastern Block of the North China Craton. *J. Asian Earth Sci.* **2005**, *24*, 659–674. [CrossRef]
6. Zhao, G.C.; Cawood, P.A.; Li, S.Z.; Wilde, S.A.; Sun, M.; Zhang, J.; He, Y.H.; Yin, C.Q. Amalgamation of the North China Craton: Key issues and discussion. *Precambrian Res.* **2012**, *222–223*, 55–76. [CrossRef]
7. Zhang, Q.S.; Yang, Z.S.; Liu, L.D. *Early Crust and Mineral Deposits of Liaodong Peninsula*; Geological Publishing House: Beijing, China, 1988; pp. 218–450, (In Chinese with English abstract).
8. Sun, M.; Armstrong, R.L.; Lambert, R.S.; Jiang, C.C.; Wu, J.H. Petrochemistry and Sr, Pb and Nd isotopic geochemistry of Paleoproterozoic Kuandian Complex, the eastern Liaoning province, China. *Precambrian Res.* **1993**, *62*, 171–190.
9. Bai, J. *The Pre-Cambrian Geology and Pb–Zn Mineralization in the Northern Margin of North China Platform*; Geological Publishing House: Beijing, China, 1993; pp. 47–89. (In Chinese)
10. Li, Z.; Chen, B.; Wei, C.J.; Wang, C.X.; Han, W. Provenance and tectonic setting of the Paleoproterozoic metasedimentary rocks from the Liaohe Group, Jiao–Liao–Ji Belt, North China Craton: Insights from detrital zircon U-Pb geochronology, whole–rock Sm–Nd isotopes, and geochemistry. *J. Asian Earth Sci.* **2015**, *111*, 711–732. [CrossRef]
11. Li, Z.; Chen, B.; Yan, X.L. The Liaohe Group: An insight into the Paleoproterozoic tectonic evolution of the Jiao-Liao-Ji Belt, North China Craton. *Precambrian Res.* **2019**, *326*, 174–195. [CrossRef]
12. Li, Z.; Wei, C.J.; Chen, B.; Fu, B.; Gong, M.Y. Late Neoarchean reworking of the Mesoarchean crustal remnant in northern Liaoning, North China Craton: A U-Pb-Hf-O-Nd perspective. *Gondwana Res.* **2020**, *80*, 350–369. [CrossRef]
13. Zhu, K.; Liu, Z.H.; Xu, Z.Y.; Wang, X.A.; Cui, W.L.; Hao, Y.J. Petrogenesis and tectonic implications of two types of Liaoji granitoid in the Jiao-Liao-Ji Belt, North China Craton. *Precambrian Res.* **2019**, *331*, 1–19. [CrossRef]
14. Li, S.Z.; Zhao, G.C.; Santosh, M.; Liu, X.; Dai, L.; Suo, Y.H.; Tam, P.K.; Song, M.; Wang, P.C. Paleoproterozoic structural evolution of the southern segment of the Jiao–Liao–Ji belt, North China Craton. *Precambrian Res.* **2012**, *200–203*, 59–73. [CrossRef]
15. Zhao, G.C.; Zhai, M.G. Lithotectonic elements of Precambrian basement in the North China Craton: Review and tectonic implications. *Gondwana Res.* **2013**, *23*, 1207–1240. [CrossRef]
16. Meng, E.; Liu, F.L.; Liu, P.H.; Liu, C.H.; Yang, H.; Wang, F.; Shi, J.R.; Cai, J. Petrogenesis and tectonic significance of Paleoproterozoic meta-mafic rocks from central Liaodong Peninsula, northeast China: Evidence from zircon U-Pb dating and in situ Lu-Hf isotopes, and whole-rock geochemistry. *Precambrian Res.* **2014**, *247*, 92–109. [CrossRef]
17. Xu, W.; Liu, F.L.; Tian, Z.H.; Liu, L.S.; Ji, L.; Dong, Y.S. Source and petrogenesis of Paleoproterozoic meta-mafic rocks intruding into the North Liaohe Group: Implications for back-arc extension prior to the formation of the Jiao-Liao-Ji Belt, North China Craton. *Precambrian Res.* **2018**, *207*, 66–81. [CrossRef]
18. Xu, W.; Liu, F.L. Geochronological and geochemical insights into the tectonic evolution of the Paleoproterozoic Jiao-Liao-Ji Belt, Sino-Korean Craton. *Earth Sci. Rev.* **2019**, *193*, 162–198. [CrossRef]
19. Yang, C.W.; Liu, J.L.; Yang, H.X.; Zhang, C.; Feng, J.; Lu, T.J.; Sun, Y.Q.; Zhang, J. Tectonics of the Paleoproterozoic Jiao-Liao-Ji orogenic belt in the Liaodong peninsula, North China Craton: A review. *J. Asian Earth Sci.* **2019**, *176*, 141–156. [CrossRef]
20. Liu, F.L.; Liu, C.H.; Itano, K.; Iizuka, T.; Cai, J.; Wang, F. Geochemistry, U-Pb dating, and Lu-Hf isotopes of zircon and monazite of porphyritic granites within the Jiao-Liao-Ji orogenic belt: Implications for petrogenesis and tectonic setting. *Precambrian Res.* **2017**, *300*, 78–106. [CrossRef]
21. Lu, X.P.; Wu, F.Y.; Guo, J.H.; Yin, C.J. Late Paleoproterozoic granitic magmatism and crustal evolution in the Tonghua region, northeast China. *Acta Petrol. Sin.* **2005**, *21*, 721–736, (In Chinese with English abstract).
22. Yang, J.H.; Wu, F.Y.; Xie, L.W.; Liu, X.M. Petrogenesis and tectonic implications of Kuangdonggou syenites in the Liaodong peninsula, east North China Craton: Constraints from in-situ zircon U-Pb ages and Hf isotopes. *Acta Petrol. Sin.* **2007**, *23*, 263–276, (In Chinese with English abstract).
23. Liu, F.L.; Liu, L.S.; Cai, J.; Liu, P.H.; Wang, F.; Liu, C.H.; Liu, J.H. A widespread Paleoproterozoic partial melting event within the Jiao-Liao-Ji Belt, North China Craton: Zircon U-Pb dating of granitic leucosomes within politic granulites and its tectonic implications. *Precambrian Res.* **2019**, *326*, 155–173. [CrossRef]
24. Lu, X.P.; Wu, F.Y.; Zhang, Y.B.; Zhao, C.B.; Guo, C.L. Emplacement age and tectonic setting of the Paleoproterozoic Liaoji Granites in Tonghua area, southern Jilin Province. *Acta Petrol. Sin.* **2004**, *20*, 381–392, (In Chinese with English abstract).
25. Yang, H.; Wang, H.; Liu, J.H. Zircon U-Pb dating and its geological significance of granitic pegmatites from the Kuandian and Sanjiazi area in eastern Liaodong Province. *Acta Petrol. Sin.* **2017**, *33*, 2675–2688, (In Chinese with English abstract).

26. Li, S.Z.; Zhao, G.C. SHRIMP U-Pb zircon geochronology of the Liaoji granitoids: Constraints on the evolution of the Paleoproterozoic Jiao-Liao-Ji belt in the Eastern Block of the North China Craton. *Precambrian Res.* **2007**, *158*, 1–16. [CrossRef]
27. Wu, F.Y.; Li, Q.L.; Yang, J.H.; Kim, J.N.; Han, R.H. Crustal growth and evolution of the Rangnim Massif, northern Korean Peninsula. *Acta Petrol. Sin.* **2016**, *32*, 2933–3294, (In Chinese with English abstract).
28. Li, Z.; Chen, B.; Wang, J.L. Geochronological framework and geodynamic implications of mafic magmatism in the Liaodong Peninsula and adjacent regions, North China Craton. *Acta Geol. Sin.* **2016**, *90*, 138–153, (In Chinese with English abstract). [CrossRef]
29. Li, Z.; Chen, B.; Wei, C.J. Is the Paleoproterozoic Jiao-Liao-Ji Belt (North China Craton) a rift? *Int. J. Earth Sci.* **2017**, *106*, 355–375. [CrossRef]
30. Li, Z.; Li, J.; Chen, B. Early Precambrian tectonic-thermal events: Coupled U-Pb-Hf of detrital zircons from Jiao-Liao-Ji Belt, North China Craton. *Arab. J. Geosci.* **2018**, *11*, 424–440. [CrossRef]
31. Li, Z.; Meng, E.; Wang, C.Y.; Li, Y.G. Early Precambrian tectono-thermal events in Southern Jilin Province, China: Implications for the evolution of Neoarchean to Paleoproterozoic crust in the northeastern North China Craton. *Miner. Petrol.* **2019**, *113*, 185–205. [CrossRef]
32. Xiao, C.H.; Liu, X.C.; Zhao, Y.; Zhang, Q.; Chen, Z.L.; Liu, J.M.; Wei, C.S.; Yao, X.F.; Wang, W.; Xie, L.Q. Structural controls and Re-Os dating of molybdenite of the Wulong gold deposit, NE China. *Earth Sci.* **2020**, *45*, 3982–3997, (In Chinese with English abstract).
33. Zhao, Y.; Lin, S.F.; Zhang, P.; Yang, X.M.; Gu, Y.C.; Bi, Z.W.; Kou, L.L. Geochronology and Geochemical Characteristics of Paleoproterozoic Syn-orogenic Granitoids and Constraints on the Geological Evolution of the Jiao-Liao-Ji Orogenic Belt, North China Craton. *Precambrian Res.* **2021**, *365*, 106386. [CrossRef]
34. Chen, J.S.; Xing, D.H.; Liu, M.; Li, B.; Yang, H.; Tian, D.X.; Yang, F.; Wang, Y. Zircon U-Pb chronology and geological significance of felsic volcanic rocks in the Liaohe Group from the Liaoyang area, Liaoning Province. *Acta Petrol. Sin.* **2017**, *33*, 2792–2810, (In Chinese with English abstract).
35. Rudnick, R.L.; Gao, S. Composition of the continental crust. In *The Crust, Treatise in Geochemistry*; Rudnick, R.L., Ed.; Elsevier-Pergamon: Oxford, UK, 2003; Volume 3, pp. 1–64.
36. Liu, Y.S.; Hu, Z.C.; Gao, S.; Guünther, D.; Xu, J.; Gao, C.G.; Chen, H.H. In situ analysis of major and trace elements of anhydrous minerals by LA-ICP-MS without applying an internal standard. *Chem. Geol.* **2008**, *257*, 34–43. [CrossRef]
37. Ludwig, K.R. *Isoplot/Ex, Version 3: A Geochronological Toolkit for Microsoft Excel*; Berkeley Geochronology Centre: Berkeley, CA, USA, 2003.
38. Griffin, W.L.; Pearson, N.J.; Belousova, E.; Jackson, S.E.; Van Achterbergh, E.; O'Reilly, S.Y.; Shee, S.R. The Hf–isotope composition of cratonic mantle: LAM-MC-ICPMS analysis of zircon megacrysts in kimberlites. *Geochim. Cosmochim. Acta* **2000**, *64*, 133–147. [CrossRef]
39. Woodhead, J.D.; Hergt, J.M. A preliminary appraisal of seven natural zircon reference materials for in situ Hf-isotope analysis. *Geost Geoanalytical Res.* **2005**, *29*, 183–195. [CrossRef]
40. Streckeisen, A.L. Classification of the common igneous rocks by means of their chemical composition: A provisional attempt. *Neues Jahrb. Fur Mineral.* **1976**, *1*, 1–15.
41. Maniar, P.D.; Piccoli, P.M. Tectonic discrimination of granitoids. *Geol. Soc. Am. Bull.* **1989**, *101*, 635–643. [CrossRef]
42. Sun, S.S.; McDonough, W.F. Chemical and isotopic systematics of oceanic basalts; implications for mantle composition and processes. *Geol. Soc. Lond. Spec. Publ.* **1989**, *42*, 313–345. [CrossRef]
43. Nowell, G.M.; Kempton, P.D.; Noble, S.R.; Fitton, J.G.; Saunders, A.D.; Mahoney, J.J.; Taylor, R.N. High precision Hf isotope measurements of MORB and OIB by thermal ionisation mass spectrometry: Insights into the depleted mantle. *Chem. Geol.* **1998**, *149*, 211–233. [CrossRef]
44. Hoskin, P.O.; Schaltegger, U. The composition of zircon and igneous and metamorphic petrogenesis. *Rev. Miner. Geochem.* **2003**, *53*, 27–62. [CrossRef]
45. Song, Y.H.; Yang, F.C.; Yan, G.L.; Wei, M.H.; Shi, S.S. SHRIMP U-Pb ages and Hf isotopic compositions of Paleoproterozoic granites from the Eastern part of Liaoning Province and their tectonic significance. *Acta Geol. Sin.* **2016**, *90*, 2620–2636, (In Chinese with English abstract).
46. Ren, Y.W.; Wang, H.C.; Kang, J.L.; Chu, H.; Tian, H. Paleoproterozoic magmatic events in the Hupiyu area in Yingkou, Liaoning Province and their geological significance. *Acta Geol. Sin.* **2017**, *91*, 2456–2472, (In Chinese with English abstract).
47. Zhao, Y.; Zhang, P.; Li, Y.; Li, D.T.; Chen, J.; Bi, Z.W. Geochemistry of two types of Paleoproterozoic granites, and zircon U-Pb dating, and Lu-Hf isotopic characteristics in the Kuandian area within the Jiao-Liao-Ji belt: Implications for regional tectonic setting. *Geol. J.* **2020**, *55*, 7564–7580. [CrossRef]
48. Zhang, P.; Zhao, Y.; Yang, H.Z.; Kou, L.L. Discrimination and its significance of the Sizhanggunzi granite in Kuangdonggou area, eastern Liaoning Province. *J. Northeast. Univ. (Nat. Sci.)* **2016**, *37*, 1349–1352, (In Chinese with English abstract).
49. Yang, M.C.; Chen, B.; Yan, C. Petrological, geochronological, geochemical and Sr-Nd-Hf isotopic constraints on the petrogenesis of the Shuangcha Paleoproterozoic metaporphyritic granite in the southern Jilin Province: Tectonic implications. *Acta Petrol. Sin.* **2015**, *31*, 1573–1588, (In Chinese with English abstract).

50. Wang, X.P.; Peng, P.; Wang, C.; Yang, S.Y. Petrogenesis of the 2115 Ma Haicheng mafic sills from the Eastern North China Craton: Implications for an intra-continental rifting. *Gondwana Res.* **2016**, *39*, 347–364. [CrossRef]

51. Bi, J.H.; Ge, W.C.; Xing, D.H.; Yang, H.; Dong, Y.; Tian, D.X.; Chen, H.J. Paleoproterozoic meta-rhyolite and meta-dacite of the Liaohe Group, Jiao-Liao-Ji Belt, North China Craton: Petrogenesis and implications for tectonic setting. *Precambrian Res.* **2018**, *314*, 306–324. [CrossRef]

52. Zhao, Y.; Kou, L.L.; Zhang, P.; Bi, Z.W.; Li, D.T.; Chen, C. Characteristics of Geochemistry and Hf isotope from meta-gabbro in the Longchang area, eastern of North China Craton: Implications on the evolution of the Jiao-Liao-Ji Paleoproterozoic orogeny belt. *Earth Sci.* **2019**, *44*, 3333–3345, (In Chinese with English abstract).

53. Wang, X.P.; Peng, P.; Wang, C.; Yang, S.Y. Nature of three episodes of Paleoproterozoic magmatism (2180 Ma, 2115 Ma and 1890 Ma) in the Liaoji belt, North China with implications for tectonic evolution. *Precambrian Res.* **2017**, *298*, 252–267. [CrossRef]

54. Drummond, M.S.; Defant, M.J. A model for trondhjemite-tonalitedacite genesis and crustal growth via slab melting: Archean to modern comparisons. *J. Geophys. Res. Solid Earth* **1990**, *95*, 21503–21521. [CrossRef]

55. Whalen, J.B.; Currie, K.L.; Chappell, B.W. A-type granite: Geochemical characteristics, discrimination and petrogenesis. *Contrib. Miner. Petrol.* **1987**, *95*, 407–419. [CrossRef]

56. Pearce, J.; Harris, N.B.W.; Tindle, A.G. Trace element discrimination diagrams for the tectonic interpretation of granitic rocks. *J. Petrol.* **1984**, *25*, 956–983. [CrossRef]

57. Pearce, J. Sources and settings of granitic rocks. *Epis. J. Int. Geosci.* **1996**, *19*, 120–125. [CrossRef]

58. Ma, Y.F.; Liu, Y.J.; Qin, T.; Sun, W.; Zang, Y.Q. Carboniferous granites in the Jalaid Banner area, middle Great Xing'an Range, NE China: Petrogenesis, tectonic background and orogeny accretionary implications. *Acta Petrol. Sin.* **2018**, *34*, 2931–2955.

59. Ma, Y.F.; Liu, Y.J.; Qin, T.; Sun, W.; Zang, Y.Q.; Zhang, Y.J. Late Devonian to early Carboniferous magmatism in the western Songliao–Xilinhot block, NE China: Implications for eastward subduction of the Nenjiang oceanic lithosphere. *Geol. J.* **2020**, *55*, 2208–2231. [CrossRef]

60. Ma, Y.F.; Liu, Y.J.; Wang, Y.; Qin, T.; Chen, H.J.; Sun, W.; Zang, Y.Q. Late Carboniferous mafic to felsic intrusive rocks in the central Great Xing'an Range, NE China: Petrogenesis and tectonic implications. *Int. J. Earth Sci.* **2020**, *109*, 761–783. [CrossRef]

61. Zhai, M.G.; Santosh, M. The early Precambrian odyssey of North China Craton: A synoptic overview. *Gondwana Res.* **2011**, *20*, 6–25. [CrossRef]

62. Liu, P.H.; Liu, F.L.; Yang, H.; Wang, F.; Liu, J.H. Protolith ages and timing of peak and retrograde metamorphism of the high pressure granulites in the Shandong Peninsula, eastern North China Craton. *Geosci. Front.* **2012**, *3*, 923–943. [CrossRef]

63. Peng, P.; Wang, X.P.; Windley, B.F.; Guo, J.H.; Zhai, M.G.; Li, Y. Spatial distribution of 1950–1800 Ma metamorphic events in the North China Craton: Implications for tectonic subdivision of the craton. *Lithos* **2014**, *202*, 250–266. [CrossRef]

64. Wu, M.L.; Lin, S.F.; Wan, Y.S.; Gao, J.F. Crustal evolution of the Eastern Block in the North China Craton: Constraints from zircon U–Pb geochronology and Lu–Hf isotopes of the Northern Liaoning Complex. *Precambrian Res.* **2016**, *275*, 35–47. [CrossRef]

65. Zhou, X.W.; Zhao, G.C.; Wei, C.J.; Geng, Y.S.; Sun, M. Metamorphic evolution and Th-U-Pb zircon and monazite geochronology of high-pressure politic granulites in the Jiaobei massif of the North China Craton. *Am. J. Sci.* **2008**, *308*, 328–350. [CrossRef]

66. Luo, Y.; Sun, M.; Zhao, G.C.; Li, S.Z.; Xu, P.; Ye, K.; Xia, X.P. LA-ICP-MS U-Pb zircon ages of the Liaohe Group in the Eastern Block of the North China Craton: Constraints on the evolution of the Jiao-Liao-Ji Belt. *Precambrian Res.* **2004**, *134*, 349–371. [CrossRef]

67. Liou, J.G.; Tsujimori, T.; Chu, W.; Zhang, R.Y.; Wooden, J.L. Protolith and metamorphic ages of the Haiyangsuo Complex, eastern China: A non-UHP exotic tectonic slab in the Sulu ultrahigh-pressure terrane. *Miner. Petrol.* **2006**, *88*, 207–226. [CrossRef]

68. Tam, P.Y.; Zhao, G.C.; Liu, F.L.; Zhou, X.W.; Sun, M.; Li, S.Z. Timing of metamorphism in the Paleoproterozoic Jiao-Liao-Ji Belt: New SHRIMP U–Pb zircon dating of granulites, gneisses and marbles of the Jiaobei massif in the North China Craton. *Gondwana Res.* **2011**, *19*, 150–162. [CrossRef]

69. Liu, J.H.; Liu, F.L.; Ding, Z.J.; Liu, C.H.; Yang, H.; Liu, P.H.; Wang, F.; Meng, E. The growth, reworking and metamorphism of early Precambrian crust in the Jiaobei terrane, the North China Craton: Constraints from U-Th-Pb and Lu–Hf isotopic systematics, and REE concentrations of zircon from Archean granitoid gneisses. *Precambrian Res.* **2013**, *224*, 287–303. [CrossRef]

70. Wan, Y.S.; Song, B.; Liu, D.Y.; Wilde, S.A.; Wu, J.S.; Shi, Y.R.; Yin, X.Y.; Zhou, H.Y. SHRIMP U-Pb zircon geochronology of Paleoproterozoic metasedimentary rocks in the North China Craton: Evidence for a major Late Paleoproterozoic tectonothermal event. *Precambrian Res.* **2006**, *149*, 249–271. [CrossRef]

71. Meng, E.; Wang, C.Y.; Li, Y.G.; Li, Z.; Yang, H.; Cai, J.; Ji, L.; Ji, M.Q. Zircon U-Pb-Hf isotopic and whole–rock geochemical studies of Paleoproterozoic metasedimentary rocks in the northern segment of the Jiao-Liao-Ji Belt, China: Implications for provenance and regional tectonic evolution. *Precambrian Res.* **2017**, *298*, 472–489. [CrossRef]

72. Lu, X.P.; Wu, F.Y.; Guo, J.H.; Wilde, S.A.; Yang, J.H.; Liu, X.M.; Zhang, X.O. Zircon U-Pb geochronological constraints on the Paleoproterozoic crustal evolution of the Eastern Block in the North China Craton. *Precambrian Res.* **2006**, *146*, 138–164. [CrossRef]

73. Liu, J.; Zhang, J.; Liu, Z.H.; Yin, C.Q.; Zhao, C.; Li, Z.; Yang, Z.J.; Dou, S.Y. Geochemical and geochronological study on the Paleoproterozoic rock assemblage of the Xiuyan region: New constraints on an integrated rift-and collision tectonic process involving the evolution of the Jiao-Liao-Ji Belt, North China Craton. *Precambrian Res.* **2018**, *310*, 179–197. [CrossRef]

74. Liu, J.; Zhang, J.; Liu, Z.H.; Yin, C.Q.; Xu, Z.Y.; Cheng, C.Q.; Zhao, C.; Wang, X. Late Paleoproterozoic crustal thickening of the Jiao-Liao-Ji belt, North China Craton: Insights from ca. 1.95–1.88 Ga syn-collisional adakitic granites. *Precambrian Res.* **2021**, *355*, 106120. [CrossRef]

75. Yu, H.C.; Liu, J.; He, Z.H.; Liu, Z.H.; Cheng, C.Q.; Hao, Y.J.; Zhao, C.; Zhang, H.Q.; Dong, Y.C. Geochronology and Zircon Hf isotope of the Paleoproterozoic Gaixian Formation in the Southeastern Liaodong Peninsula: Implication for the tectonic evolution of the Jiao-Liao-Ji Belt. *Minerals* **2022**, *2022*, 792. [CrossRef]
76. Wang, Q.; Mc Dermott, F.; Xu, J.F.; Bellon, H.; Zhu, Y.T. Cenozoic K-rich adakitic volcanic rocks in the Hohxil area, northern T ibet:Lower-crustal m elting in an intracontinental setting. *Geology* **2005**, *33*, 465–468. [CrossRef]
77. Miller, F.H.; McDowell, S.M.; Mapes, R.W. Hot and cold granites? Implications of zircon saturation temperatures and preservation of inheritance. *Geology* **2003**, *31*, 529–532. [CrossRef]

 *minerals*

Article

Petrogenesis of the Early Paleoproterozoic Felsic Metavolcanic Rocks from the Liaodong Peninsula, NE China: Implications for the Tectonic Evolution of the Jiao-Liao-Ji Belt, North China Craton

Changquan Cheng [1], Jin Liu [2,*], Jian Zhang [1,3], Hongxiang Zhang [2], Ying Chen [1], Xiao Wang [1], Zhongshui Li [2] and Hongchao Yu [2]

[1] School of Earth Sciences and Engineering, Sun Yat-sen University, Guangzhou 510275, China
[2] College of Earth Sciences, Jilin University, Changchun 130061, China
[3] Department of Earth Sciences, The University of Hong Kong, Pokfulam Road, Hong Kong, China
* Correspondence: liujin@jlu.edu.cn

Abstract: The early Paleoproterozoic (ca. 2.2–2.1 Ga) tectonic evolution of the Jiao–Liao–Ji belt (JLJB) is a continuous hot topic and remains highly controversial. Two main tectonic regimes have been proposed for the JLJB, namely arc-related setting and intra-continental rift. Abundant ca. 2.2–2.1 Ga volcanic rocks were formed in the JLJB, especially in the Liaodong Peninsula. These ca. 2.2–2.1 Ga volcanic rocks therefore could host critical information for the evolution of the JLJB. In this study, we report a suit of ca. 2.2–2.1 Ga felsic metavolcanic rocks in the Liaodong Peninsula of the JLJB to provide new insights into the above issue. Zircon U-Pb dating reveals that the felsic metavolcanic rocks were erupted at 2185–2167 Ma. They have variable εHf(t) values (-0.70 to $+9.69$), high SiO_2 (66.30–75.30 wt.%) and relatively low TiO_2 (0.03–0.78 wt.%), tFe_2O_3 (0.55–5.03 wt.%), MgO (0.17–8.76 wt.%), Cr (9.16–67.30 ppm), Co (2.01–7.00 ppm) and Ni (3.90–25.70 ppm) contents with enrichments in light rare earth element (REE) and large ion lithophile element (LILE), and depletions in heavy REE and high field strength element (HFSE). Geochemical and isotopic results indicate that the felsic metavolcanic rocks were sourced from partial melting of ancient Archean TTG rocks and juvenile lower crustal materials. Combined with coeval A-type granites, bimodal volcanic rocks and the absence of typical arc magmatism, the most likely tectonic regime at ca. 2.2–2.1 Ga for the JLJB is an intra-continental rift.

Keywords: felsic metavolcanic rocks; geochemistry; intra-continental rift; Jiao-Liao-Ji Belt; North China Craton

Citation: Cheng, C.; Liu, J.; Zhang, J.; Zhang, H.; Chen, Y.; Wang, X.; Li, Z.; Yu, H. Petrogenesis of the Early Paleoproterozoic Felsic Metavolcanic Rocks from the Liaodong Peninsula, NE China: Implications for the Tectonic Evolution of the Jiao-Liao-Ji Belt, North China Craton. *Minerals* **2023**, *13*, 1168. https://doi.org/ 10.3390/min13091168

Academic Editor: Jaroslav Dostal

Received: 13 August 2023
Revised: 30 August 2023
Accepted: 1 September 2023
Published: 3 September 2023

1. Introduction

The North China Craton (NCC) experienced continuous evolution processes of continental crust formation, subduction, collision and cratonization during the Archean to Paleoproterozoic [1–9]. It can be generally divided into the Western Block (WB), the Eastern Block (EB) and the intervening North-South-trending Trans-North China Orogen (TNCO) [3,10,11] (Figure 1a). The WB can be further subdivided into two microblocks (i.e., Yinshan and Ordos blocks), and they are separated by the Paleoproterozoic East-West-trending Khondalite Belt [1,12]. The EB is formed via the amalgamation of the Longgang and Nangrim blocks along the northeast-southwest-trending Jiao-Liao-Ji Belt (JLJB) during the Paleoproterozoic [3,13–17].

The JLJB underwent polyphase structural deformation, magmatism, metamorphism and sedimentation during the Paleoproterozoic orogenesis [18–24] (Figure 1b). Although numerous geochronological, geochemical, metamorphic, and structural studies have been carried out in the JLJB [13,19,22,24–32], its tectonic evolution during the Paleoproterozoic remains highly controversial [21,22,24,32,33]. Currently, there are three main tectonic models: (1) the first model involves oceanic subduction and arc/continent-continent collision between the Longgang and Nangrim blocks [27,30,34–39]; (2) the second model suggests the opening and closure of an intra-continental rift for the tectonic evolution of the JLJB [13,14,25,40–42]; (3) the third model proposes that the JLJB experienced early intra-continental rift and subsequent oceanic subduction and collision [3,17,21–23]. In recent years, increasingly studies suggest an extension tectonic regime for the JLJB during the early Paleoproterozoic (ca. 2.2–2.1 Ga) [14,21,22,31,32,38,43–46]. However, there is still a key dispute whether such an extension is a consequence of back-arc basin or intra-continental rift [22,24,32,38,43,44,46]. Therefore, the ca. 2.2–2.1 Ga magmatism can serve as a significant object of study to provide crucial constrains for the above dispute.

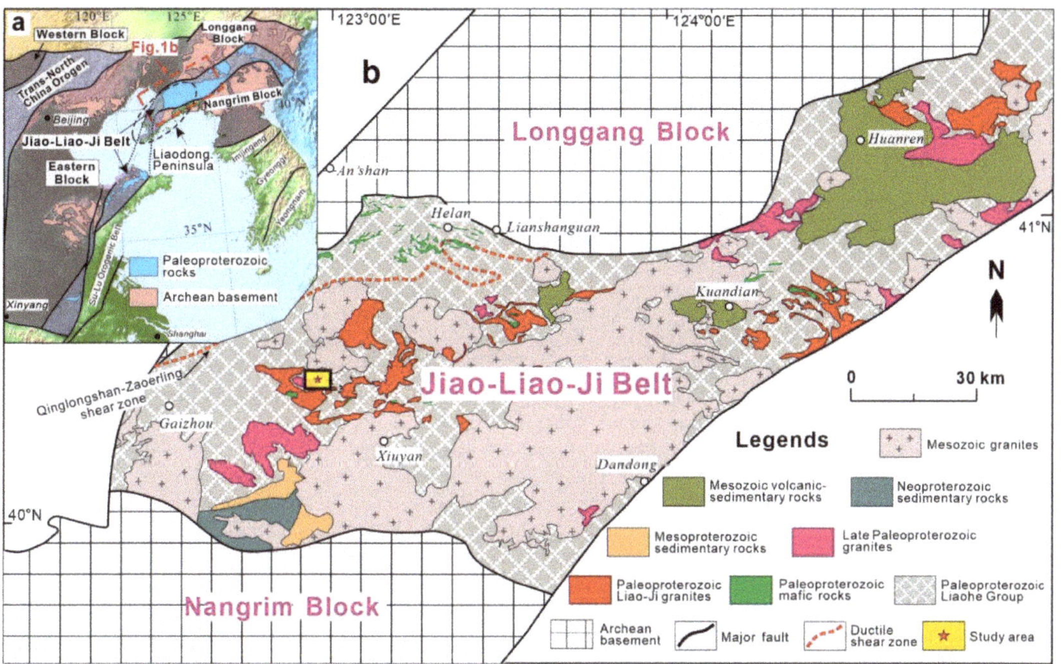

Figure 1. (**a**) Simplified map showing tectonic units of the eastern North China Craton [1,3]; (**b**) Geological map of the Liaodong Peninsula showing main lithological units [13].

Most recently, we implemented detailed geological investigations and identified a suit of felsic metavolcanic rocks in the Houxianyu area (Figure 2), where there is one of the best exposures in the northern Liaodong Peninsula, NE China. In this contribution, new petrology, zircon U-Pb geochronology, bull-rock major and trace elements and zircon Lu-Hf isotopic compositions are reported for the newly identified felsic metavolcanic rocks in the Liaodong Peninsula, with the aim of deciphering their timing of formation, petrogenesis, and tectonic setting, and thus further providing new constrains for the tectonic evolution of the JLJB during the Paleoproterozoic.

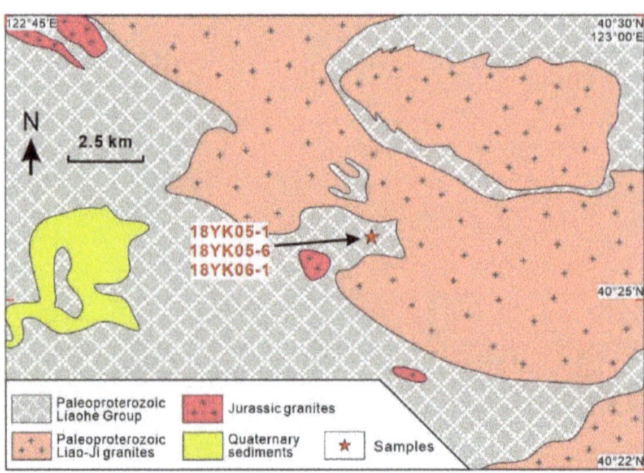

Figure 2. Simplified geological map of the Houxianyu area showing the sampling location.

2. Geological Background

The Paleoproterozoic JLJB is a ca. 1200 km long orogenic belt in the eastern part of the NCC, and geographically it traverses the northern Korean Peninsula, southern Jilin Province, Liaodong and Jiaodong peninsulas, and Anhui Province from northeast to southwest [3,47] (Figure 1a). It is tectonically bounded by the two Archean microblocks (i.e., the Longgang Block to the northwest and the Nangrim Block to the southeastern) [3] (Figure 1a). The Longgang Block mainly consists of abundant Neoarchean tonalite-trondhjemite-granodiorite (TTG) gneisses and supracrustal rocks (e.g., pelitic rock, amphibolite and granulite) and a small volume of Eoarchean to Mesoarchean granitoids [7,48–54]. The Nangrim Block, located in southern Liaoning Province and in North Korea, is composed of Neoarchean granitoid and supracrustal rocks [55–58]. The JLJB mainly consists of Paleoproterozoic granitoids, metamafic intrusions and voluminous metavolcanic-metasedimentary successions with variable metamorphic grades ranging from greenschist to amphibolite facies and locally granulite facies [13,15,18,21–23,25,29,31,32,38,45,59].

As a major terrane of the JLJB, the Liaodong Peninsula is mainly composed of various Paleoproterozoic granitic and mafic intrusive rocks and the Paleoproterozoic Liaohe Group [13,22,23,30,31,44,46,59–61] (Figure 1b). The Paleoproterozoic granitoids are widespread in the Liaodong Peninsula, including ca. 2.20–2.16 Ga Liaoji gneissic granitoids (mainly A-type granites) and ca. 1.90–1.85 Ga post-collision associated granitoids (e.g., undeformed syenite and porphyritic monzogranite) [21–23,28,45,61]. The Paleoproterozoic mafic intrusive rocks are mainly distributed in the northern Liaodong Peninsula (e.g., Helan, Longchang, Jidongyu areas), and a small number of mafic intrusions are also distributed in the Xiuyan and Kuandian areas [31,33,36,38,44,46]. They contain greenschist, plagioclase amphibolite, meta-gabbro and meta-diabase, and have emplacement ages of ca. 2.15–2.10 Ga. The ca. 2.20–2.10 Ga felsic and mafic intrusive rock assemblages were generally deformed and show obvious gneissic structures [21,22,31,38,46]. The Paleoproterozoic Liaohe Group, from bottom to top, can be further subdivided into five units, including the Langzishan, Li'eryu, Gaojiayu, Dashiqiao and Gaixian formations [13,18]. The Liaohe Group is a suit of volcanic-sedimentary rocks and mainly comprises schist, marble, biotite gneiss, fine-grained felsic gneiss, meta-clastic rock, meta-rhyolite and meta-basalt [18]. Traditionally, the Liaohe Group, on the basis of metamorphic and lithological features, is subdivided into the North Liaohe and South Liaohe groups [13].

During the Late Paleoproterozoic, the JLJB underwent widespread and intense high-grade metamorphism [15–17,26,29,62]. More and more mafic and pelitic granulites, characterized by clockwise P-T paths, have been identified in the Liaodong and Jiaodong peninsulas [15–17,20,26,29,62,63]. Such high-grade metamorphism is interpreted as the result of the ca. 1.90 Ga collision event between the Nangrim and Longgang blocks, which marks the final assembly of the EB in the eastern NCC [3]. After the Late Paleoproterozoic orogenic event, the crystalline basement of the eastern NCC was overlain by Meso- to Neoproterozoic sedimentary rocks and experienced intense magmatism and structural deformation during the Mesozoic [64].

3. Petrography

A suit of felsic metavolcanic rocks of this study were newly identified in the Houxianyu area of the northern Liaodong Peninsula (Figure 2). These felsic rocks belong to metamorphosed volcano-sedimentary rock units of the Li'eryu Formation, Liaohe Group. The Paleoproterozoic and Mesozoic granitoids are in tectonic or intrusive contacts with these volcanic sequences. The felsic metavolcanic rocks of this study mainly consist of fine-grained meta-dacite and fine-grained meta-rhyolite, with layered or massive structures (Figure 3). The sample lithologies and locations and the methods applied to each rock sample are summarized in Table 1.

Figure 3. Representative field photographs and microphotographs of the felsic metavolcanic rocks. (**a–c**) Fine-grained meta-dacites (Sample 18YK05-9); (**d–f**) Fine-grained meta-rhyolites (Sample 18YK05-6). Abbreviations: Pl, plagioclase; Qtz, quartz; Kfs, K-feldspar; Bt, biotite.

Table 1. The simplified list of the felsic metavolcanic rock samples in this study.

Sample	Lithology	GPS	Methods
18YK05-1	Meta-dacite		Zircon U-Pb dating and Lu-Hf isotopic analyses
18YK05-2	Meta-dacite	40°25′56.7″ N; 122°54′32.8″ E	Whole-rock geochemistry
18YK05-3	Meta-dacite		Whole-rock geochemistry
18YK05-6	Meta-rhyolite		Zircon U-Pb dating and Lu-Hf isotopic analyses
18YK05-7	Meta-dacite	40°25′56.5″ N; 122°54′32.5″ E	Whole-rock geochemistry
18YK05-8	Meta-rhyolite		Whole-rock geochemistry
18YK05-9	Meta-dacite		Whole-rock geochemistry
18YK06-1	Meta-rhyolite		Zircon U-Pb dating and Lu-Hf isotopic analyses
18YK06-2	Meta-rhyolite	40°25′57.9″ N; 122°54′34.1″ E	Whole-rock geochemistry
18YK06-4	Meta-dacite		Whole-rock geochemistry

The fine-grained meta-dacites are grey to white in color and occur as deformed layers (Figure 3a). They generally display weakly gneissic structures and fine-grained crystalloblastic textures, with a typical mineral assemblage of plagioclase (~37 vol.%), quartz (~35 vol.%), biotite (~18 vol.%), K-feldspar (~5 vol.%) and minor accessory minerals (~5 vol.%; e.g., tourmaline, zircon, apatite and some opaque minerals) (Figure 3b,c). The majority of these minerals are subhedral to anhedral, with grain sizes of 200–600 μm (Figure 3c).

The fresh surfaces of the fine-grained meta-rhyolites are light in color, and the rocks generally occur as blocks in field outcrops (Figure 3d). They are granular and show fine-grained crystalloblastic textures (Figure 3e). Compared with the fine-grained meta-dacites, the dominant mineral assemblage of the fine-grained meta-rhyolites contains fewer dark-colored minerals, and it is composed of plagioclase (~45 vol.%), quartz (~35 vol.%), K-feldspar (~12 vol.%), and a few other minerals (~8 vol.%; e.g., biotite, tourmaline, zircon and apatite) (Figure 3e,f). Most of these minerals are subhedral to anhedral, with grain sizes ranging from 200 to 1000 μm (Figure 3f).

4. Analytical Methods

4.1. Zircon U-Pb Dating

Three representative felsic metavolcanic rock samples were collected for the zircon U-Th-Pb isotope and trace element analyses (Table 1). Zircon grains were separated from crushed samples using combined techniques of magnetic separation and standard density methods. Typical and clear zircon grains without cracks were handpicked through the binocular microscope, and then fixed on an epoxy disk. After being ground and polished, the zircon grains were imaged by cathodoluminescence (CL) and reflected/transmitted light. Zircon U-Th-Pb isotopic compositions, as well as trace element concentrations, were measured using the instruments of laser ablation (LA) inductively coupled plasma (ICP)-mass spectrometry (MS) at the Yanduzhongshi Geological Analysis Laboratories, Beijing, China. A total of 100 spots on zircon grains were analyzed, with a beam size of 30 μm in diameter. The data acquisition durations for each analysis spot included 20–30 s blank background acquisition and a sample measuring time of 50 s. Standard zircon 91,500 and NIST610 were utilized as external standards for the U-Th-Pb isotopes and trace element compositions of dated zircon grains. Quantitative calibrations for acquired original isotopic and trace element data were carried out using the ICPMSDataCal software [65]. Sample weighted mean or intercept age calculations and Concordia diagrams were performed through the Isoplot program [66].

4.2. Bulk-Rock Geochemistry

A total of seven felsic metavolcanic rock samples were selected for bulk-rock major and trace element analyses (Table 1). Prior to geochemical analysis, fresh rock samples were crushed and powdered to <200 mesh in a completely cleaned agate mill. Bulk-rock major element concentrations were measured using a Leeman Prodigy ICP-optical emission spectroscopy (OES) system at the State Key Laboratory of Geological Processes and Mineral Resources, China University of Geosciences, Beijing, China. Loss on ignition (LOI) values were determined through placing 1 g of bulk-rock powders in the muffle furnace for two hours at a temperature condition of 1000 °C, and then cooling and reweighing residual samples. The acquired results yielded analytical uncertainties better than 2% for most of the major elements. Bulk-rock trace and rare earth element concentration analyses were performed through an Agilent 7500a quadrupole ICP-MS instrument system. The acquired results yielded analytical accuracies generally better than 5% for most of the trace elements.

4.3. Zircon Lu-Hf Isotopic Analyses

The above three felsic metavolcanic rock samples, which had been dated by LA-ICP-MS, were used for in situ zircon Hf isotopic composition analyses at the Yanduzhongshi Geological Analysis Laboratories, Beijing, China. All analyses were conducted using the LA-MC-ICP-MS instrument system (a Neptune multi-collector (MC)-ICP-MS instrument equipped with a NewWave UP213 LA system). A total of 45 representative zircon Lu-Hf test spots were analyzed, similar or close to U-Pb dating domains. The beam size for each spot was ~40 μm in diameter. The test conditions included 16 J/cm² laser energy density and an 8 Hz laser repetition rate. More specific and detailed analytical procedures, instrument conditions and experimental processes were described by Wu et al. (2006) [67].

5. Analytical Results

5.1. Zircon U-Pb Geochronology

Zircon U-Pb isotopic data, CL images and Concordia diagrams for three felsic metavolcanic rock samples (18YK05-1, 18YK05-6 and 18YK06-1) are presented in Supplementary Table S1, Figures 4 and 5, respectively.

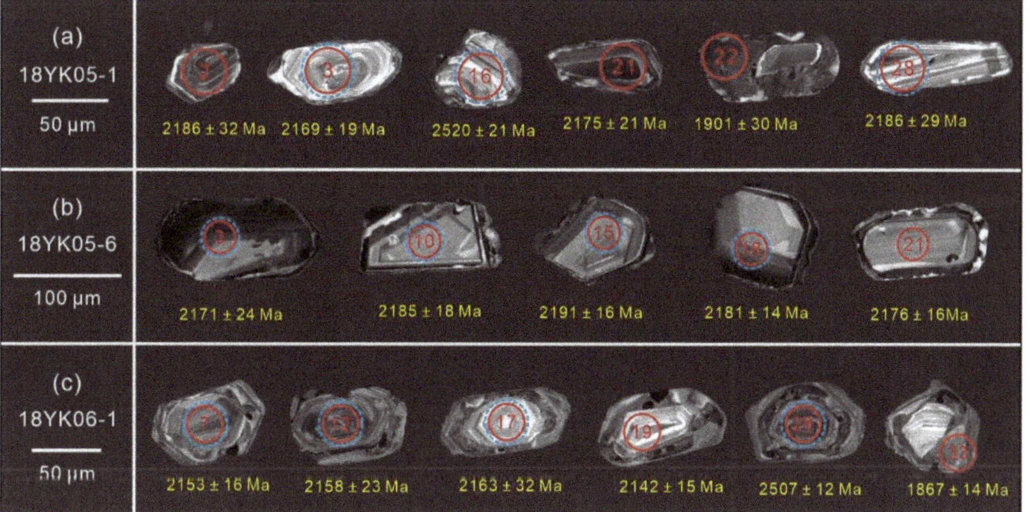

Figure 4. Representative zircon cathodoluminescence (CL) images of the felsic metavolcanic rocks. (**a**) Sample 18YK05-1; (**b**) Sample 18YK05-6; (**c**) Sample 18YK06-1. The blue dashed circle and internal red circle represent the positions of Hf isotopic and U-Pb age analyses, respectively.

As shown in the CL images (Figure 4a), zircon crystals from Sample 18YK05-1 are generally subhedral to anhedral and grey to white in color, with grain lengths of approximately 50–100 μm and length/width ratios of 1.2–2.5. Most of the zircon grains exhibit clear and well-developed oscillatory zoning, or characteristic core-rim structures (Figure 4a). Combined with their relatively high Th/U ratios of 0.47–1.18 (Supplementary Table S1), these zircon grains and cores are similar to those of felsic igneous rocks, indicating a magmatic origin. Nineteen U-Pb analysis spots on the above zircon domains yield apparent ^{207}Pb/^{206}Pb dates between 2188 ± 17 Ma and 2163 ± 24 Ma, with a weighted mean ^{207}Pb/^{206}Pb age of 2174 ± 10 Ma (n = 19, MSWD = 0.12), and all of them have concordances better than 96% and fall on or near the Concordia line (Figure 5a). Therefore, the age of 2174 ± 10 Ma represents the magma eruption age of Sample 18YK05-1. Six analyses on zircon core domains gave old apparent ^{207}Pb/^{206}Pb dates of 2529–2479 Ma, suggesting that these zircon cores could be inherited zircons derived from the Neoarchean basement rocks of the Longgang Block (Supplementary Table S1 and Figure 5a). In addition, two analyses on zircon grains or rims without obvious internal structures gave ^{207}Pb/^{206}Pb dates 1901 ± 30 Ma and 1898 ± 30 Ma, consistent with the Late Paleoproterozoic regional metamorphic event in the JLJB.

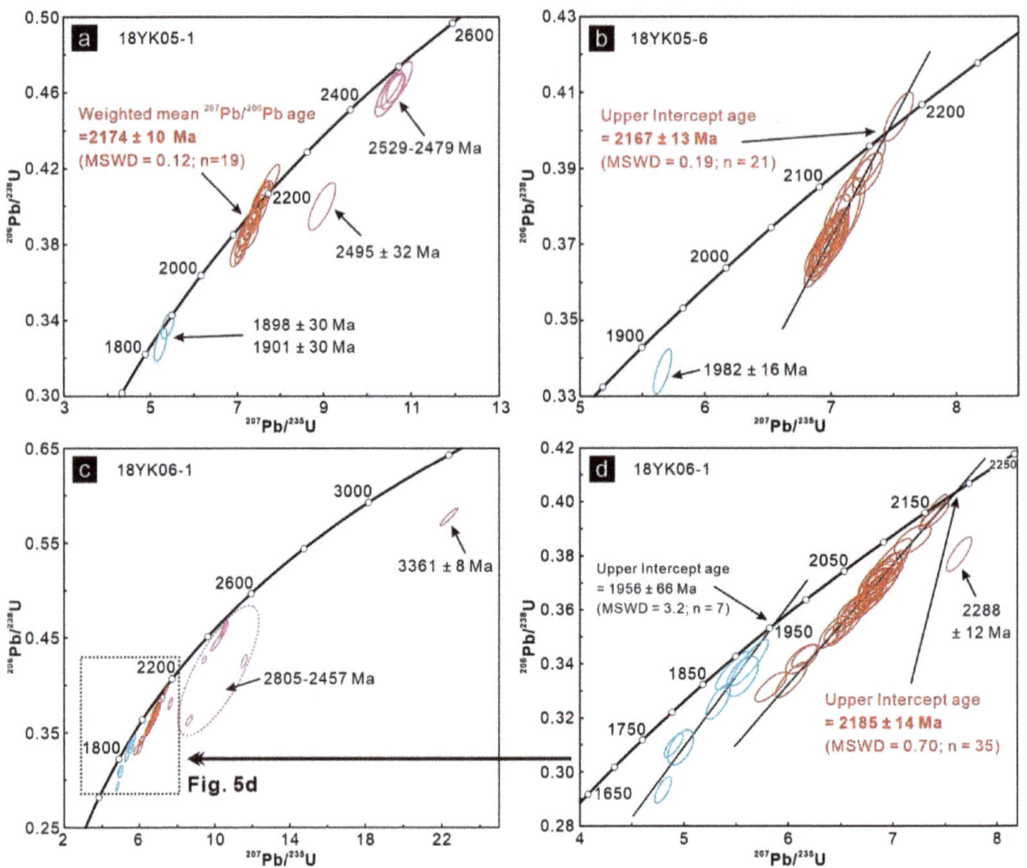

Figure 5. Zircon U-Pb Concordia diagrams of the felsic metavolcanic rocks. (**a**) Sample 18YK05-1; (**b**) Sample 18YK05-6; (**c**) Sample 18YK06-1; (**d**) Partial enlarged view of Figure 5c (Sample 18YK06-1).

Zircon grains from Sample 18YK05-6 are euhedral to subhedral crystals and black to grey in color in the CL images (Figure 4b), with grain lengths ranging from 75 to 150 μm and aspect ratios of 1.1–2.0. The majority of zircon crystals have well-developed oscillatory growth zoning, and some of them display clear core-rim textures (Figure 4b). In addition, they show relatively high and variable Th/U ratios varying from 0.13 to 0.85 (Supplementary Table S1). All of these features indicate that these zircon grains or cores have a magmatic origin. Twenty-one U-Pb analysis spots with concordances better than 95% yield apparent ^{207}Pb/^{206}Pb dates between 2196 ± 21 Ma and 2163 ± 14 Ma (Supplementary Table S1). They define a discordant line and give an upper intercept age of 2167 ± 13 Ma (n = 21, MSWD = 0.19), which is interpreted as the magma eruption age of Sample 18YK05-6 (Figure 5b).

Zircon crystals from Sample 18YK06-1 are irregularly shaped, grey to white in color and subhedral to anhedral in CL images (Figure 4c), with grain lengths of approximately 70–100 μm and length/width ratios ranging from 1.0 to 2.0. Most of them exhibit obvious core-rim structures with well-developed oscillatory zoned cores surrounded by structureless and blurred rims (Figure 4c). In addition, some zircon grains with no obvious core-rim structures are completely structureless or show clear oscillatory growth zoning throughout them. Thirty-five U-Pb analyses on zircon cores or domains with distinct oscillatory growth zoning, with high Th/U ratios of 0.30–1.08 (mostly >0.5), yield apparent ^{207}Pb/^{206}Pb dates between 2174 ± 12 Ma and 2065 ± 18 Ma (Supplementary Table S1). These analyses define a discordant line and give an upper intercept age of 2185 ± 14 Ma (n = 35, MSWD = 0.70), representing the magma eruption age of Sample 18YK06-1 (Figure 5c,d). Eight analyses yield old apparent $^{207}Pb/^{206}Pb$ dates of 3361–2288 Ma (Supplementary Table S1 and Figure 5c), which indicates that these zircon grains could be inherited in origin and derived from ancient surrounding rocks (e.g., Archean TTG rocks). In addition, seven analyses on zircon rims or domains without distinct internal structures have relatively low Th/U ratios of 0.06–0.14 (Supplementary Table S1). They yield young apparent $^{207}Pb/^{206}Pb$ dates between 1956 ± 21 Ma and 1867 ± 14 Ma, with an upper intercept age of 1956 ± 66 Ma (n = 7, MSWD = 3.2) (Supplementary Table S1 and Figure 5c,d), which represents the metamorphic age of Sample 18YK06-1.

In summary, these felsic metavolcanic rocks were formed during the early Paleoproterozoic (2185–2167 Ma) and underwent the late Paleoproterozoic regional metamorphic event (1956–1898 Ma).

5.2. Whole-Rock Geochemistry

The whole-rock major and trace element data and calculated parameters for the seven felsic metavolcanic rock samples are presented in Table 2. These samples are acid rocks with high SiO_2 (66.30–75.30 wt.%; average 69.51 wt.%) concentrations. They also contain relatively low TiO_2 (0.03–0.78 wt.%; average 0.45 wt.%), total Fe_2O_3 (tFe_2O_3; 0.55–5.03 wt.%; average 2.85 wt.%) and MnO (0.01–0.04 wt.%; average 0.02 wt.%) contents and show a large range of Al_2O_3 (10.57–17.23 wt.%), Na_2O (0.99–9.01 wt.%) and K_2O (0.04–7.15 wt.%). As shown in the protolith discrimination diagrams (Figure 6a,b) these felsic rock samples exhibit low TiO_2, Ni contents and Zr/TiO_2 ratios and mainly fall into the field of volcanic rocks rather than sedimentary rocks. In addition, most zircon grains from these samples are subhedral with the characteristics of magmatic oscillatory growth zonings, differing from detrital zircon grains from sedimentary rocks with relatively rounded shapes (Figure 4). Previous field and petrological studies also identified a suit of metavolcanic rocks (meta-rhyolites and meta-dacites) in the Li'eryu Formation, Liaohe Group [21,60,68,69]. In the geochemical classification diagrams (Figure 6c,d), these samples also mainly plot in the fields of rhyolite and dacite. Therefore, the felsic rocks in this study are a suit of metavolcanic rocks and mainly comprise meta-rhyolites and meta-dacites.

Table 2. Bulk-rock major (wt.%) and trace (ppm) element data of the felsic metavolcanic rocks in the Liaodong Peninsula of the Jiao-Liao-Ji Belt.

Sample	18YK05-2	18YK05-3	18YK05-7	18YK05-8	18YK05-9	18YK06-2	18YK06-4
SiO_2	69.95	67.24	66.43	74.31	66.54	75.30	66.77
TiO_2	0.61	0.54	0.11	0.03	0.57	0.78	0.50
Al_2O_3	12.57	11.67	17.23	14.96	14.04	10.57	15.33
tFe_2O_3	4.59	4.75	0.68	0.55	2.05	5.03	2.33
MnO	0.03	0.04	0.01	0.02	0.02	0.02	0.02
MgO	6.32	8.76	1.17	0.17	4.76	4.12	1.64
CaO	1.64	2.72	2.67	1.99	2.40	0.12	2.08
Na_2O	0.99	1.00	9.01	5.70	1.04	1.37	2.96
K_2O	0.04	0.18	0.61	1.18	7.15	1.02	6.95
P_2O_5	0.14	0.14	0.07	0.02	0.10	0.18	0.15
LOI%	2.87	2.49	1.58	0.57	0.53	0.36	0.51
Total	99.75	99.52	99.57	99.50	99.19	98.88	99.23
La	7.10	18.68	18.75	11.10	32.84	10.43	35.88
Ce	17.17	43.00	38.72	24.56	81.06	25.78	61.40
Pr	2.23	4.95	4.40	3.03	9.46	3.26	5.08
Nd	10.18	18.76	15.75	11.91	35.78	13.58	16.66
Sm	2.65	3.27	2.72	2.90	6.37	3.23	2.60
Eu	0.48	0.49	0.45	0.58	1.00	0.70	0.44

Table 2. *Cont.*

Sample	18YK05-2	18YK05-3	18YK05-7	18YK05-8	18YK05-9	18YK06-2	18YK06-4
Gd	2.88	2.92	1.95	2.93	4.93	3.15	2.48
Tb	0.45	0.42	0.24	0.49	0.68	0.51	0.34
Dy	2.92	2.64	1.25	3.17	4.03	3.28	2.13
Ho	0.63	0.54	0.23	0.62	0.79	0.68	0.47
Er	1.81	1.56	0.63	1.75	2.23	1.98	1.48
Tm	0.28	0.25	0.10	0.28	0.32	0.29	0.24
Yb	1.89	1.66	0.64	1.75	1.96	1.80	1.73
Lu	0.30	0.28	0.10	0.24	0.29	0.28	0.31
Y	16.30	14.28	7.84	20.40	18.83	16.27	15.13
tREE	50.97	99.44	85.95	65.30	181.74	68.95	131.25
LREE/HREE	3.57	8.68	15.68	4.82	10.93	4.76	13.28
$(La/Yb)_N$	2.70	8.06	20.86	4.55	12.00	4.14	14.88
δEu	0.53	0.49	0.60	0.61	0.54	0.67	0.53
δCe	1.06	1.10	1.05	1.04	1.13	1.08	1.12
Li	13.26	17.28	11.06	10.73	63.88	64.08	25.38
Sc	10.51	11.45	1.50	2.33	5.25	13.48	12.58
V	97.42	140.20	20.92	21.50	45.20	90.29	25.11
Cr	54.30	45.62	10.54	9.16	56.36	67.30	61.46
Co	5.13	7.00	3.51	3.28	4.40	6.22	2.01
Ni	21.06	19.25	4.00	3.90	22.96	25.70	21.86
Cu	0.75	0.92	9.39	12.08	4.33	1.78	4.35
Zn	53.02	81.30	34.72	35.22	37.94	16.17	32.63
Ga	15.45	16.33	18.66	16.75	18.36	17.61	24.03
Rb	1.05	9.99	15.23	17.71	148.42	149.30	75.74
Sr	187.64	148.84	1112.80	963.60	122.30	108.26	239.40
Zr	216.60	222.20	132.28	134.42	241.20	191.52	251.82
Nb	1.45	2.40	1.81	3.64	15.68	11.62	16.22
Cs	0.15	1.12	0.54	0.42	15.65	11.40	5.47
Ba	3.04	6.78	197.78	197.28	89.52	758.80	107.74
Hf	5.04	5.24	3.00	3.08	5.58	4.78	6.06
Ta	0.10	0.17	0.12	0.25	1.07	0.69	1.01
Pb	4.79	4.34	8.03	10.65	4.05	2.03	7.30
Th	11.20	11.24	4.55	2.14	13.57	7.36	14.82
U	0.60	0.71	0.69	0.68	0.91	0.73	1.65

The felsic metavolcanic rocks contain relatively low Cr (9.16–67.30 ppm; average 43.53 ppm), Co (2.01–7.00 ppm; average 4.51 ppm), and Ni (3.90–25.70 ppm; average 16.96 ppm) (Table 2). Their right-inclined chondrite-normalized REE patterns show prominent enrichments in light REEs (LREEs) and slighter enrichments in heavy REEs (HREEs), with total REE (tREE) contents of 50.97–181.74 ppm, high $(La/Yb)_N$ ratios of 2.70–20.86, and obviously negative Eu anomalies (δEu = 0.49–0.67; average 0.57) (Table 2 and Figure 7a). In the spider diagram of primitive mantle- (PM-) normalized trace elements (Figure 7b), most felsic metavolcanic rock samples are relatively enriched in large ion lithophile elements (LILEs; e.g., Rb, Ba and K), Th and U, and show negative anomalies of high field strength elements (HFSEs; e.g., Nb, Ta, Ti and P) and Sr.

5.3. Zircon Lu-Hf Isotopic Compositions

The zircon Lu-Hf isotopic results for the three dated samples (18YK05-1, 18YK05-6 and 18YK06-1) are presented in Figure 8 and Table 3. Thirty-four Lu-Hf analyses on magmatic zircon grains (2185–2167 Ma) from the three felsic metavolcanic rock samples yield $^{176}Lu/^{177}Hf$ ratios ranging from 0.000326 to 0.002066 and $^{176}Hf/^{177}Hf$ ratios ranging from 0.281430 to 0.281683 (Table 3). They display variable $\varepsilon Hf(t)$ values varying from −0.70 to +9.69 (Table 3) and fall into the field between the depleted mantle (DM) evolutionary line and chondrite uniform reservoir (CHUR) line in the $\varepsilon Hf(t)$ versus age diagram (Figure 8). The calculated two-stage Hf model ages (T_{DM2}) for the felsic metavolcanic rocks range from

2903 to 2152 Ma (Table 3). In addition, eleven analyses on inherited zircon grains, with ancient apparent ^{207}Pb/^{206}Pb ages of 3361–2288 Ma, have ^{176}Lu/^{177}Hf ratios of 0.000809 to 0.001462, ^{176}Hf/^{177}Hf ratios of 0.281010 to 0.281539, and a large range of εHf(t) values (−1.85 to +14.99) (Table 3).

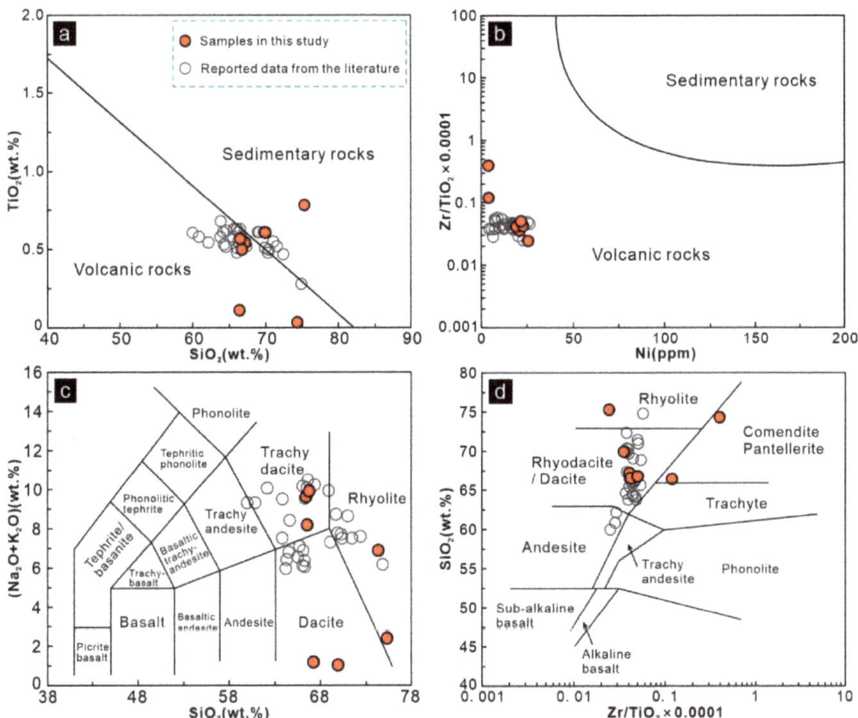

Figure 6. Geochemical classification diagrams of the felsic metavolcanic rocks. (**a**) TiO$_2$ versus SiO$_2$ diagram [70]; (**b**) Zr/TiO$_2$ × 0.0001 versus Ni diagram [71]; (**c**) (Na$_2$O + K$_2$O) versus SiO$_2$ diagram [72]; (**d**) SiO$_2$ versus Zr/TiO$_2$ × 0.0001 diagram [73]. Reported data of the felsic metavolcanic rocks are from [21,60,68,69].

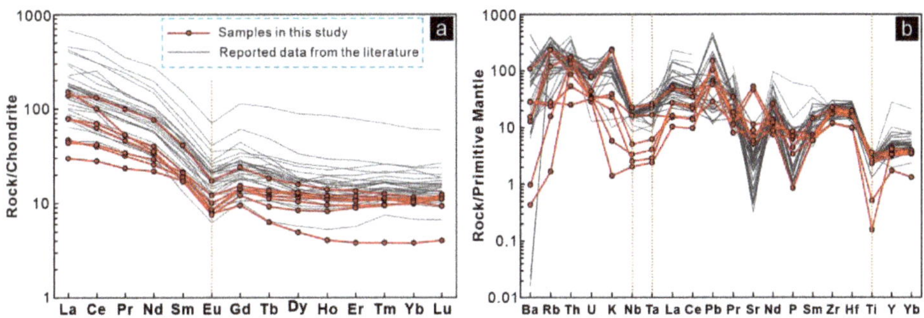

Figure 7. (**a**) Chondrite-normalized REE pattern diagram; (**b**) Primitive mantle-normalized spider diagram. Reported data of the felsic metavolcanic rocks are from [21,60,68,69]. The normalized values are from [74].

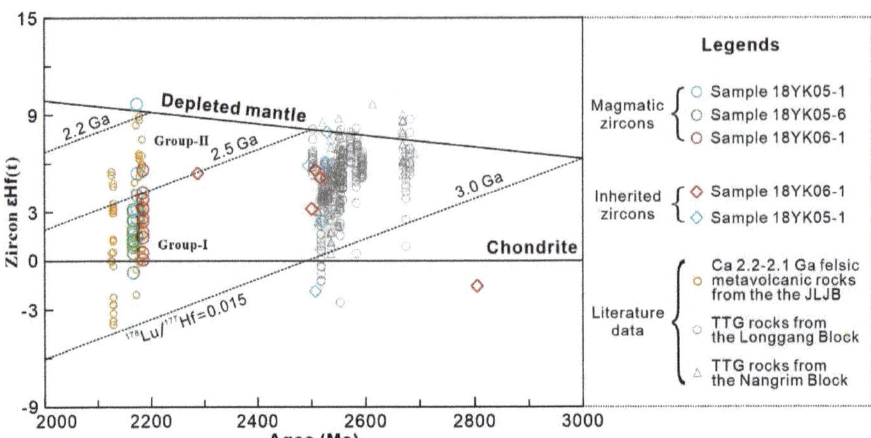

Figure 8. The εHf(t) versus age diagram of the felsic metavolcanic rocks. Reported Hf isotopic data of the felsic metavolcanic rocks are from [60,68]. Reported Hf isotopic data of the TTG rocks in the Longgang Block are from [52,53,75,76]. Reported Hf isotopic data of the TTG rocks in the Nangrim Block are from [57].

Table 3. Zircon Hf isotopic data of the felsic metavolcanic rocks in the Liaodong Peninsula of the Jiao-Liao-Ji Belt.

Sample	t (Ma)	^{176}Lu/^{177}Hf	2σ	^{176}Hf/^{177}Hf	2σ	εHf(t)	T_{DM1} (Ma)	T_{DM2} (Ma)
Sample 18YK05-1								
18YK05-1-01	2527	0.000914	0.000065	0.281383	0.000027	6.08	2603	2649
18YK05-1-03	2174	0.000326	0.000006	0.281683	0.000028	9.69	2160	2152
18YK05-1-07	2174	0.001437	0.000045	0.281609	0.000020	5.43	2326	2415
18YK05-1-10	2174	0.000838	0.000013	0.281546	0.000018	4.05	2377	2500
18YK05-1-13	2507	0.000872	0.000015	0.281171	0.000024	−1.85	2888	3118
18YK05-1-14	2495	0.001190	0.000016	0.281411	0.000021	5.89	2584	2636
18YK05-1-16	2520	0.000809	0.000005	0.281281	0.000018	2.45	2735	2865
18YK05-1-17	2174	0.001382	0.000019	0.281496	0.000015	1.50	2479	2657
18YK05-1-24	2174	0.000809	0.000010	0.281521	0.000015	3.22	2409	2551
18YK05-1-26	2174	0.000820	0.000011	0.281503	0.000015	2.57	2434	2591
18YK05-1-27	2529	0.001021	0.000007	0.281440	0.000036	7.95	2533	2536
18YK05-1-28	2174	0.002066	0.000054	0.281572	0.000033	3.19	2418	2553
Sample 18YK05-6								
18YK05-6-01	2167	0.000842	0.000014	0.281487	0.000017	1.80	2457	2720
18YK05-6-02	2167	0.001164	0.000021	0.281430	0.000018	−0.70	2556	2903
18YK05-6-03	2167	0.001750	0.000002	0.281544	0.000015	2.52	2436	2669
18YK05-6-04	2167	0.001029	0.000032	0.281529	0.000016	3.04	2411	2630
18YK05-6-05	2167	0.001356	0.000029	0.281471	0.000018	0.49	2513	2817
18YK05-6-06	2167	0.001740	0.000023	0.281529	0.000018	1.99	2457	2709
18YK05-6-09	2167	0.001099	0.000008	0.281476	0.000016	1.03	2489	2777
18YK05-6-10	2167	0.001206	0.000035	0.281484	0.000018	1.17	2485	2767
18YK05-6-13	2167	0.000927	0.000029	0.281529	0.000019	3.17	2405	2620
18YK05-6-15	2167	0.001490	0.000013	0.281498	0.000019	1.26	2484	2761
18YK05-6-16	2167	0.001339	0.000026	0.281481	0.000018	0.87	2497	2789
18YK05-6-17	2167	0.000959	0.000016	0.281477	0.000014	1.30	2478	2757
18YK05-6-18	2167	0.000960	0.000013	0.281487	0.000016	1.63	2465	2733
18YK05-6-19	2167	0.000727	0.000015	0.281501	0.000017	2.47	2431	2671

Table 3. *Cont.*

Sample	t (Ma)	^{176}Lu/^{177}Hf	2σ	^{176}Hf/^{177}Hf	2σ	εHf(t)	T_{DM1} (Ma)	T_{DM2} (Ma)
				Sample 18YK06-1				
18YK06-1-04	2185	0.001448	0.000052	0.281558	0.000016	3.84	2398	2521
18YK06-1-05	2185	0.001308	0.000018	0.281447	0.000018	0.09	2543	2752
18YK06-1-06	2185	0.001060	0.000012	0.281509	0.000017	2.67	2441	2593
18YK06-1-07	2185	0.001071	0.000007	0.281592	0.000018	5.62	2327	2412
18YK06-1-10	2185	0.001233	0.000007	0.281548	0.000017	3.79	2399	2525
18YK06-1-11	2185	0.001376	0.000013	0.281536	0.000016	3.17	2424	2563
18YK06-1-12	2185	0.000958	0.000007	0.281479	0.000015	1.77	2475	2649
18YK06-1-14	2501	0.000898	0.000005	0.281318	0.000015	3.21	2691	2805
18YK06-1-17	2185	0.001109	0.000008	0.281477	0.000017	1.47	2488	2668
18YK06-1-26	2805	0.001231	0.000045	0.281010	0.000017	−1.55	3135	3329
18YK06-1-28	2507	0.000990	0.000009	0.281387	0.000023	5.63	2603	2661
18YK06-1-32	3361	0.001162	0.000008	0.281118	0.000027	14.99	2982	2758
18YK06-1-34	2288	0.001462	0.000005	0.281539	0.000018	5.42	2425	2504
18YK06-1-36	2185	0.001016	0.000016	0.281550	0.000016	4.21	2381	2499
18YK06-1-37	2185	0.001163	0.000008	0.281530	0.000016	3.27	2418	2557
18YK06-1-39	2185	0.001575	0.000026	0.281524	0.000025	2.43	2454	2608
18YK06-1-40	2517	0.001081	0.000014	0.281370	0.000018	5.12	2632	2700
18YK06-1-45	2185	0.001735	0.000051	0.281476	0.000016	0.51	2531	2726
18YK06-1-46	2185	0.001551	0.000048	0.281545	0.000017	3.22	2423	2560

The present-day ^{176}Hf/^{177}Hf and ^{176}Lu/^{177}Hf ratios are 0.282772 and 0.0332 for the chondrite, and 0.28325 and 0.0384 for depleted mantle. The ^{176}Lu/^{177}Hf ratio of the average continental crust is 0.015. The decay constant of ^{176}Lu is 6.54×10^{-12} a^{-1}.

6. Discussion

6.1. Early Paleoproterozoic (ca. 2.2–2.1 Ga) Magmatism in the Liaodong Peninsula of the JLJB

The early Paleoproterozoic (concentrated at ca. 2.2–2.1 Ga) is one of the most developed periods of magmatic activity in the JLJB [24]. A large number of ca. 2.2–2.1 Ga igneous rocks are widely distributed in the Liaodong Peninsula of the JLJB [21,22,31,33,38,44,45,60,61,68] (Figure 1b). They are mainly composed of three lithological units, including meta-volcanic rock, granitic intrusion, and mafic intrusion. Characteristically, the meta-volcanic rocks and granitic intrusions were mainly formed at ca. 2205–2150 Ma and ca. 2185–2150 Ma, respectively, but the mafic intrusions were emplaced relatively late (concentrated at ca. 2160–2100 Ma) (Figure 9). The meta-volcanic sequences are mainly outcropped in the North and South Liaohe Groups and characterized by bimodal volcanic rocks [21,40,77,78]. They display various lithologies, mainly composed of meta-rhyolite, meta-dacite, fine-grained gneiss, amphibolite, and meta-basalt. The granitic intrusions, also known as Liaoji granites, are generally outcropped as gneissic plutons in the eastern Liaoning and southern Jilin areas [22] (Figure 1b). The Liaoji granites with typical A-type granite affinity and bimodal volcanic sequences are widely regarded as the products of intra-continental rift [21,22,79,80]. Voluminous meta-mafic intrusions are distributed in the northern part of the Liaodong Peninsula (i.e., Haicheng and Helan areas), and minor intrusions are distributed in the Xiuyan and Kuandian areas [31,38,44,46,79] (Figure 1b). These mafic intrusions, comprising meta-gabbro, meta-diabase, and amphibolite, were thought to show close genetic association with extensional tectonic regimes (e.g., intra-continental rift and back-arc basin) [31,38,44,46,79].

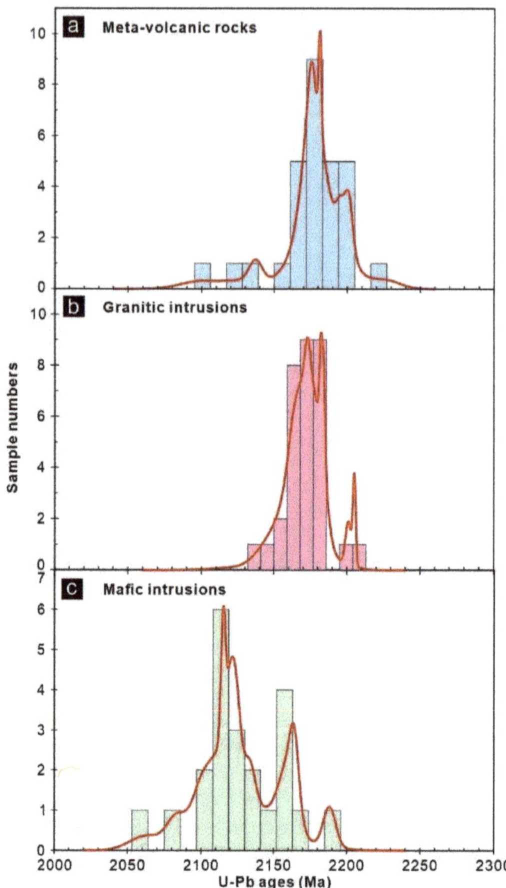

Figure 9. U-Pb age histograms of the early Paleoproterozoic magmatic rocks in the Liaodong Peninsula of the JLJB. Data of the metavolcanic rocks are from [21,27,39,60,68,69,81–84] and this study. Data of the granitic intrusions are from [21,22,45,68,77,80,81,84–89]. Data of the mafic intrusions are from [31,33,36,38,44,46].

6.2. Petrogenesis

The early Paleoproterozoic felsic metavolcanic rocks in this study underwent low-grade metamorphism as a result of a late Paleoproterozoic tectothermal event. Therefore, it is necessary to evaluate the metamorphism effects on their bulk-rock major and trace element data. All of the samples contain lower loss on ignition (LOI) contents (0.36–2.87 wt.%; average 1.27 wt.%) (Table 2), suggesting that the effects of low-grade metamorphism on geochemical compositions could be negligible. Such an explanation is also supported by the lack of obvious Ce anomalies (δCe = 1.04–1.13; average 1.08) in all samples (Table 2 and Figure 7a). Therefore, the geochemical data of felsic metavolcanic rocks in this study can be used to effectively discuss and trace their petrogenesis and tectonic implications.

The felsic metavolcanic rock samples show relatively variable tREE contents (50.97–181.74 ppm), $(La/Yb)_N$ ratios (2.70–20.86), and negative Eu anomalies (δEu = 0.49–0.67) (Table 2), which implies that they could experience a fractional crystallization process. The negative correlation between SiO_2 and Al_2O_3 suggests the felsic metavolcanic rocks experienced the plagioclase fractionation (Figure 10a), which is also consistent with their negative Eu anomalies in the REE pattern diagram (Figure 7a). These samples exhibit a negative evolutionary trend between Er and Dy (Figure 10b), which suggests the hornblende fractionation [90]. Such an explanation is supported by the decrease in V contents with decreasing Cr concentrations [91] (Figure 10c). However, the lack of an obvious evolutionary trend between Th and V implies that the biotite fractionation process is negligible [90] (Figure 10d). In addition, their relatively variable tFe_2O_3 (0.55–5.03 wt.%) and TiO_2 (0.03–0.78 wt.%) abundances are indicative of Fe-Ti oxide (e.g., magnetite and titanite) accumulation. Therefore, the felsic metavolcanic rocks in this study experienced a certain degree of fractional crystallization.

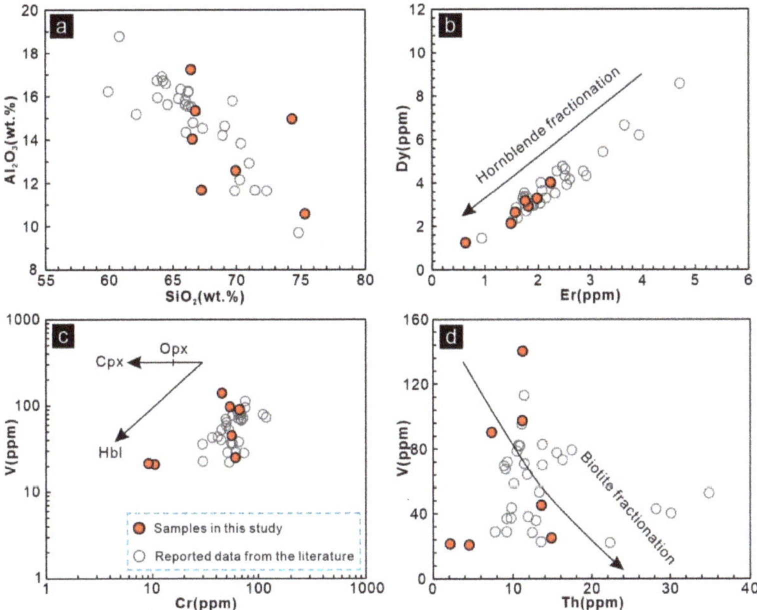

Figure 10. (**a**) Al_2O_3 versus SiO_2 diagram; (**b**) Dy versus Er diagram; (**c**) V versus Cr diagram [91]; (**d**) V versus Th diagram. Reported data of the felsic metavolcanic rocks are from [21,60,68,69].

The felsic metavolcanic rocks lack the mafic enclaves, which indicates that these samples may not have been contaminated by mantle-derived materials. Such an interpretation is also consistent with their low TiO_2 (0.03–0.78 wt.%), tFe_2O_3 (0.55–5.03 wt.%), Cr (9.16–67.30 ppm), Co (2.01–7.00 ppm), and Ni (3.90–25.70 ppm) (Table 2). The petrogenesis of felsic magmatism generally includes two main mechanisms: (1) fractional crystallization of mantle-derived magmas [91–95] and (2) partial melting of crustal materials [60,68,96–99]. As discussed above, although the primary magma of felsic metavolcanic rocks in this study inevitably experienced a certain degree of fractional crystallization during the later stage of magma eruption to the surface, their petrogenesis mechanism could not be the fractional crystallization of mantle-derived magmas. Firstly, the differentiation process of basaltic magmas will produce a complete set of magmatic evolutionary sequences, including basaltic, andesitic and rhyolitic magmatisms. However, the early Paleoproterozoic meta-volcanic sequences in the JLJB are characterized by bimodal volcanic rocks [21,40,77,78]. Secondly, multiple positive correlations between La contents and La/Sm, La/Yb, La/Hf ratios strongly suggest that the petrogenesis of the felsic metavolcanic rocks is mainly controlled by the partial melting process rather than fractional crystallization (Figure 11a–c), which is also supported by the positive correlation between Ce contents and Ce/Zr ratios (Figure 11d). These rock samples contain relatively high SiO_2 (66.30–75.30 wt.%) and low tFe_2O_3 (0.55–5.03 wt.%), MgO (0.17–8.76 wt.%), Cr (9.16–67.30 ppm), Co (2.01–7.00 ppm) and Ni (3.90–25.70 ppm) contents (Table 2), suggesting that they were mainly derived from partial melting of crustal rocks rather than mantle peridotite. Thirdly, felsic magmatism generated by the fractional crystallization of mantle-derived magmas generally exhibits depleted and homogeneous zircon Hf compositions. Nevertheless, the zircon samples from the felsic metavolcanic rocks in the Liaodong Peninsula of the JLJB exhibit a large range of εHf(t) values, and some samples even fall below the CHUR line and show enriched zircon Hf compositions (Figure 8). Therefore, the fractional crystallization mechanism may not play a major role in geochemical compositions. In summary, the petrogenesis of the felsic metavolcanic rocks in this study are most likely the mechanism (2): partial melting of pre-existing crustal materials.

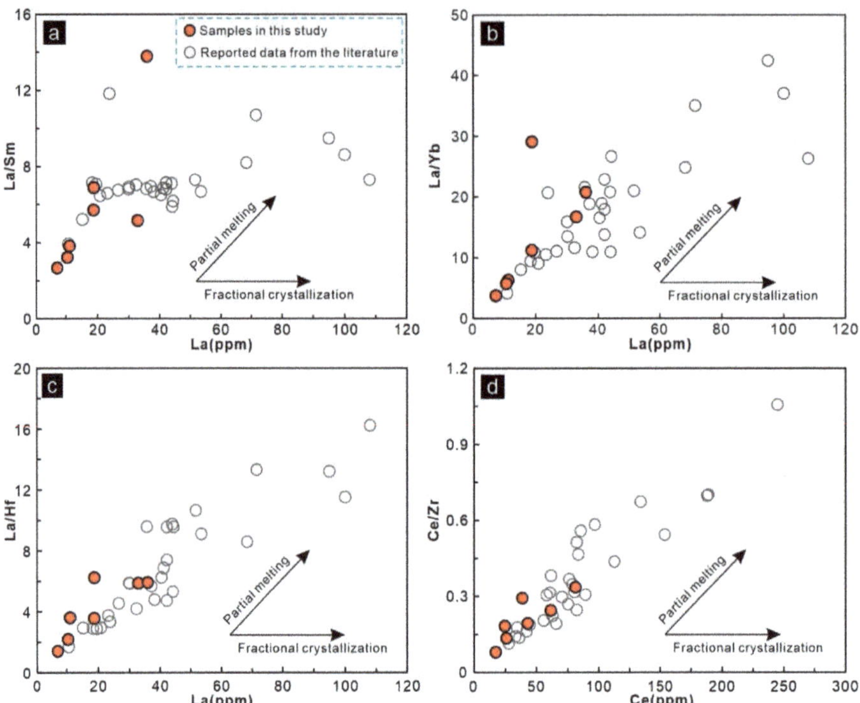

Figure 11. (**a**) La/Sm versus La diagram; (**b**) La/Yb versus La diagram; (**c**) La/Hf versus La diagram; (**d**) Ce/Zr versus Ce diagram. Reported data of the felsic metavolcanic rocks are from [21,60,68,69].

Zircon CL images and U-Pb geochronological results reveal that the felsic metavolcanic rocks of this study contain many inherited magmatic zircon grains with ancient apparent ^{207}Pb/^{206}Pb ages ranging from 3361 ± 8 Ma to 2288 ± 12 Ma (concentrated at the Neoarchean) (Supplementary Table S1 and Figure 4a,c). Therefore, the Neoarchean TTG rocks, accounting for the majority of continental crust rocks in the Archean Longgang and Nangrim blocks [52,53,57], are potential protoliths for the felsic metavolcanic rocks. In this study, we collected zircon Hf isotopic data from reported early Paleoproterozoic felsic metavolcanic rocks in the Liaodong Peninsula and Neoarchean TTG rocks in the Longgang and Nangrim blocks. Zircon grains from the early Paleoproterozoic felsic metavolcanic rocks exhibit heterogeneous zircon Hf isotopic compositions with a broad range of εHf(t) values (approximately −4 to +9) (Figure 8). They can be divided into two groups based on their different εHf(t) values and T_{DM2} ages. Zircons of the Group-I show similar ancient T_{DM2} ages of ca. 3.0–2.5 Ga and low εHf(t) values (about −4 to +4), which indicates that the felsic metavolcanic rocks could be derived from partial melting of ancient Archean

TTG rocks. Such an explanation is also supported by their T_{DM2} ages and evolutionary trends being similar to those of Neoarchean TTG rocks (Figure 8). Differently, zircons of the Group-II are close to the DM evolutionary line with higher $\varepsilon Hf(t)$ values (about +4 to +9) and younger T_{DM2} ages (ca. 2.5–2.2 Ga), which suggests that juvenile mafic materials in the lower crust could be another source for the felsic metavolcanic rocks (Figure 8). In summary, the early Paleoproterozoic felsic metavolcanic rocks in the Liaodong Peninsula of the JLJB were most likely derived from partial melting of Archean TTG rocks and juvenile crustal materials.

6.3. Tectonic Implications

The early Paleoproterozoic tectonic evolution of the JLJB has long been controversial [13,14,24,32,33]. Therefore, contemporaneous magmatic rocks can provide valuable insights into this issue. There is a view that ca. 2.2–2.1 Ga metavolcanic sequences, mafic and granitic intrusions with arc-type geochemical characteristics (enrichment in LILE and depletion in HFSE) were considered to be associated with magmatic arc-related environments [27,33,38,46,60,61,68]. However, many studies revealed that the JLJB went through an intra-continental rift system during the early Paleoproterozoic. Firstly, the Longgang and Nangrim blocks on both sides of the JLJB have similar Archean basement rocks geochemically and geochronologically [3,40], which indicates that the two blocks previously belonged to a unified continent. Secondly, the existence of regional large-scale mafic dykes generally represents a crustal extension tectonic regime. In the whole JLJB, abundant ca. 2.2–2.1 Ga mafic intrusions were identified in southern Jilin Province and on the Liaodong Peninsula [31,38,44,46,79]. Thirdly, the metavolcanic sequences in the JLJB are mainly composed of bimodal volcanic rocks [21,40,77,78], which indicates that the JLJB experienced an intra-continental rift rather than magmatic arc during the early Paleoproterozoic. Fourthly, a large number of Liaoji granites with an affinity of A-type granite also suggest an intra-continental rift setting [21,22,43,45,80]. Finally, the absence of coeval typical arc magmatism in the JLJB could be in disagreement with an arc-related setting [3,18,21,22]. Therefore, these lines of evidence indicate that the JLJB most likely went through an intra-continental rift during the early Paleoproterozoic. In this study, the ca. 2185–2167 Ma felsic metavolcanic rocks are closely related to A-type granite and metavolcanic rocks in the spatial and temporal distribution. It can be inferred that they were also generated in an intra-continental rift setting.

Liu et al. (2020) [22] identified a suit of synchronous A-type and adakitic granites, which were formed in an intra-continental rift trigged by lithospheric delamination at ca. 2.2 Ga. Generally, the lithospheric delamination could result in the decompression melting and upwelling of asthenosphere mantle and generate the mafic volcanic and intrusive rocks in the JLJB. Meanwhile, the upwelling of hyperthermal mantle-derived magmas will further induce the appearance of initial rift and partial melting of crustal rocks and produce the numerous A-type Liaoji granites and felsic volcanic rocks of this study. Such a delamination-rift model can effectively explain the special early Paleoproterozoic lithological assemblages in the JLJB.

A further extension of continental rift would have resulted in an initial ocean basin and separated unified Eastern Block into two continental blocks (i.e., the Longgang and Nangrim blocks). A subsequent regional high-grade tectothermal event at ca. 1.95–1.85 Ga resulted in the collision between the Longgang and Nangrim blocks [15–17,26,29,63]. Correspondingly, the felsic metavolcanic rocks in this study also recorded metamorphic ages of 1956–1898 Ma (Figure 5). Notably, increasing studies from metamorphic geology, especially the discovery of a high-pressure mafic/pelitic granulite with a clockwise P-T-t path, strongly revealed that the JLJB experienced oceanic subduction before the final collision [16,17,26,63]. Therefore, the most likely tectonic scenario is that the JLJB went through an intra-continental rift system during the early Paleoproterozoic (ca. 2.2–2.1 Ga), followed by a transition phase from rift to oceanic subduction, and later a continent–continent collision at ca. 1.95–1.85 Ga [3,21,22,31].

7. Conclusions

1. The felsic metavolcanic rocks in the Liaodong Peninsula were erupted at 2185–2167 Ma and underwent a late metamorphic event at 1956–1898 Ma.
2. Geochemical and isotopic results suggest that the felsic metavolcanic rocks were derived from partial melting of Archean TTG rocks and Paleoproterozoic juvenile lower crustal materials.
3. The JLJB most likely experienced an intra-continental rift at ca. 2.2–2.1 Ga.

Supplementary Materials: The following supporting information can be downloaded at: https://www.mdpi.com/article/10.3390/min13091168/s1, Table S1: Zircon U-Pb data for the felsic metavolcanic rocks in the Liaodong Peninsula of the Jiao-Liao-Ji Belt.

Author Contributions: Conceptualization, C.C., J.L. and J.Z.; methodology, C.C., J.L. and H.Z.; software, C.C.; validation, C.C.; formal analysis, C.C., H.Z. and Y.C.; investigation, C.C., J.L., H.Z., Y.C., X.W., Z.L. and H.Y.; data curation, C.C.; writing—original draft preparation, C.C.; writing—review and editing, J.L. and J.Z.; supervision, J.L. and J.Z.; project administration, J.L. and J.Z.; funding acquisition, J.L. and J.Z. All authors have read and agreed to the published version of the manuscript.

Funding: This research was funded by the National Natural Science Foundation of China (Grant Nos. 42025204 and 42172212).

Data Availability Statement: Not applicable.

Acknowledgments: We sincerely thank the Academic Editor and three anonymous reviewers for their constructive and helpful comments. We also thank the Managing Editor and Assistant Editor very much for their editorial work.

Conflicts of Interest: The authors declare no conflict of interest.

References

1. Zhao, G.C.; Sun, M.; Wilde, S.A.; Li, S.Z. Late Archean to Paleoproterozoic evolution of the North China Craton: Key issues revisited. *Precambrian Res.* **2005**, *136*, 177–202. [CrossRef]
2. Zhao, G.C.; Wilde, S.A.; Guo, J.H.; Cawood, P.A.; Sun, M.; Li, X.P. Single zircon grains record two Paleoproterozoic collisional events in the North China Craton. *Precambrian Res.* **2010**, *177*, 266–276. [CrossRef]
3. Zhao, G.C.; Cawood, P.A.; Li, S.Z.; Wilde, S.A.; Sun, M.; Zhang, J.; He, Y.H.; Yin, C.Q. Amalgamation of the North China Craton: Key issues and discussion. *Precambrian Res.* **2012**, *222–223*, 55–76. [CrossRef]
4. Santosh, M.; Wilde, S.A.; Li, J.H. Timing of Paleoproterozoic ultrahigh-temperature metamorphism in the North China Craton: Evidence from shrimp U-Pb zircon geochronology. *Precambrian Res.* **2007**, *159*, 178–196. [CrossRef]
5. Zhai, M.G.; Santosh, M. The early Precambrian odyssey of North China Craton: A synoptic overview. *Gondwana Res.* **2011**, *20*, 6–25. [CrossRef]
6. Zhai, M.G.; Santosh, M. Metallogeny of the North China Craton: Link with secular changes in the evolving Earth. *Gondwana Res.* **2013**, *24*, 275–297. [CrossRef]
7. Wan, Y.S.; Liu, D.Y.; Nutman, A.P.; Zhou, H.Y.; Dong, C.Y.; Yin, X.Y.; Ma, M.Z. Multiple 3.8-3.1 Ga tectono-magmatic events in a newly discovered area. *J. Asian Earth Sci.* **2012**, *54–55*, 18–30. [CrossRef]
8. Wan, Y.S.; Xie, S.W.; Yang, C.H.; Kröner, A.; Ma, M.Z.; Dong, C.Y.; Du, L.L.; Xie, H.Q.; Liu, D.Y. Early Neoarchean (~2.7 Ga) tectono-thermal events in the North China Craton: A synthesis. *Precambrian Res.* **2014**, *247*, 45–63. [CrossRef]
9. Kusky, T.M.; Polat, A.; Windley, B.F.; Burke, K.C.; Dewey, J.F.; Kidd, W.S.F.; Maruyama, S.; Wang, J.P.; Deng, H.; Wang, Z.S.; et al. Insights into the tectonic evolution of the North China Craton through comparative tectonic analysis: A record of outward growth of Precambrian continents. *Earth-Sci. Rev.* **2016**, *162*, 387–432. [CrossRef]
10. Zhang, J.; Zhao, G.C.; Li, S.Z.; Sun, M.; Chan, L.S.; Shen, W.L.; Liu, S.W. Structural pattern of the Wutai Complex and its constraints on the tectonic framework of the Trans-North China Orogen. *Precambrian Res.* **2012**, *222–223*, 212–229. [CrossRef]
11. Zhang, J.; Zhao, G.C.; Shen, W.L.; Li, S.Z.; Sun, M. Aeromagnetic study of the Hengshan–Wutai–Fuping region: Unraveling a crustal profile of the Paleoproterozoic Trans-North China Orogen. *Tectonophysics* **2015**, *662*, 208–218. [CrossRef]
12. Yin, C.Q.; Zhao, G.C.; Wei, C.J.; Sun, M.; Guo, J.H.; Zhou, X.W. Metamorphism and partial melting of high-pressure pelitic granulites from the Qianlishan Complex: Constraints on the tectonic evolution of the Khondalite Belt in the North China Craton. *Precambrian Res.* **2014**, *242*, 172–186. [CrossRef]
13. Li, S.Z.; Zhao, G.C.; Sun, M.; Han, Z.Z.; Hao, D.F.; Luo, Y.; Xia, X.P. Deformation history of the Paleoproterozoic Liaohe Group in the Eastern Block of the North China Craton. *J. Asian Earth Sci.* **2005**, *24*, 659–674. [CrossRef]
14. Li, S.Z.; Zhao, G.C.; Santosh, M.; Liu, X.; Dai, L.M.; Suo, Y.H.; Tam, P.Y.; Song, M.C.; Wang, P.C. Paleoproterozoic structural evolution of the southern segment of the Jiao-Liao-Ji belt, North China Craton. *Precambrian Res.* **2012**, *200–203*, 59–73. [CrossRef]

15. Tam, P.Y.; Zhao, G.C.; Sun, M.; Li, S.Z.; Iizuka, Y.; Ma, G.S.K.; Yin, C.Q.; He, Y.H.; Wu, M.L. Metamorphic P-T path and tectonic implications of medium-pressure pelitic granulites from the Jiaobei massif in the Jiao-Liao-Ji Belt, North China Craton. *Precambrian Res.* **2012**, *220–221*, 177–191. [CrossRef]

16. Tam, P.Y.; Zhao, G.C.; Zhou, X.W.; Sun, M.; Guo, J.H.; Yin, C.Q.; Wu, M.L.; He, Y.H. Metamorphic P-T path and implications of high-pressure pelitic granulites from the Jiaobei massif in the Jiao-Liao-Ji Belt, North China Craton. *Gondwana Res.* **2012**, *22*, 104–117. [CrossRef]

17. Tam, P.Y.; Zhao, G.C.; Zhou, X.W.; Guo, J.H.; Sun, M.; Li, S.Z.; Yin, C.Q.; Wu, M.L. Petrology and metamorphic P-T path of high-pressure mafic granulites from the Jiaobei massif in the Jiao-Liao-Ji Belt, North China Craton. *Lithos* **2012**, *155*, 94–109. [CrossRef]

18. Liu, F.L.; Liu, P.H.; Wang, F.; Liu, C.H.; Cai, J. Progresses and overviews of voluminous meta-sedimentary series within the Paleoproterozoic Jiao-Liao-Ji orogenic/mobile belt, North China Craton. *Acta Petrol. Sin.* **2015**, *31*, 2816–2846. (In Chinese with English Abstract)

19. Tian, Z.H.; Liu, F.L.; Windley, B.F.; Liu, P.H.; Wang, F.; Liu, C.H.; Wang, W.; Cai, J.; Xiao, W.J. Polyphase structural deformation of low- to medium-grade metamorphic rocks of the Liaohe Group in the Jiao-Liao-Ji Orogenic Belt, North China Craton: Correlations with tectonic evolution. *Precambrian Res.* **2017**, *303*, 641–659. [CrossRef]

20. Zou, Y.; Zhai, M.G.; Santosh, M.; Zhou, L.G.; Zhao, L.; Lu, J.S.; Shan, H.X. High-pressure pelitic granulites from the Jiao-Liao-Ji Belt, North China Craton: A complete P-T path and its tectonic implications. *J. Asian Earth Sci.* **2017**, *134*, 103–121. [CrossRef]

21. Liu, J.; Zhang, J.; Liu, Z.H.; Yin, C.Q.; Zhao, C.; Li, Z.; Yang, Z.J.; Dou, S.Y. Geochemical and geochronological study on the Paleoproterozoic rock assemblage of the Xiuyan region: New constraints on an integrated rift-and- collision tectonic process involving the evolution of the Jiao-Liao-Ji Belt, North China Craton. *Precambrian Res.* **2018**, *310*, 179–197. [CrossRef]

22. Liu, J.; Zhang, J.; Yin, C.Q.; Cheng, C.Q.; Liu, X.G.; Zhao, C.; Chen, Y.; Wang, X. Synchronous A-type and adakitic granitic magmatism at ca. 2.2 Ga in the Jiao–Liao–Ji belt, North China Craton: Implications for rifting triggered by lithospheric delamination. *Precambrian Res.* **2020**, *342*, 105629. [CrossRef]

23. Liu, J.; Zhang, J.; Liu, Z.H.; Yin, C.Q.; Xu, Z.Y.; Cheng, C.Q.; Zhao, C.; Wang, X. Late Paleoproterozoic crustal thickening of the Jiao–Liao–Ji belt, North China Craton: Insights from ca. 1.95–1.88 Ga syn-collisional adakitic granites. *Precambrian Res.* **2021**, *355*, 106120. [CrossRef]

24. Xu, W.; Liu, F.L. Geochronological and geochemical insights into the tectonic evolution of the Paleoproterozoic Jiao-Liao-Ji belt. Sino-Korean Craton. *Earth-Sci. Rev.* **2019**, *193*, 162–198. [CrossRef]

25. Luo, Y.; Sun, M.; Zhao, G.C.; Li, S.Z.; Xu, P.; Ye, K.; Xia, X.P. LA–ICP–MS U-Pb zircon ages of the Liaohe Group in the Eastern Block of the North China Craton: Constraints on the evolution of the Jiao–Liao–Ji Belt. *Precambrian Res.* **2004**, *134*, 349–371. [CrossRef]

26. Zhou, X.W.; Zhao, G.C.; Wei, C.J.; Geng, Y.S.; Sun, M. Metamorphic evolution and Th-U-Pb zircon and monazite geochronology of high-pressure pelitic granulites in the Jiaobei massif of the North China Craton. *Am. J. Sci.* **2008**, *308*, 328–350. [CrossRef]

27. Li, Z.; Chen, B. Geochronology and geochemistry of the Paleoproterozoic meta-basalts from the Jiao-Liao-Ji Belt, North China Craton: Implications for petrogenesis and tectonic setting. *Precambrian Res.* **2014**, *255*, 653–667. [CrossRef]

28. Liu, F.L.; Liu, C.H.; Itano, K.; Iizuka, T.; Cai, J.; Wang, F. Geochemistry, U-Pb dating, and Lu-Hf isotopes of zircon and monazite of porphyritic granites within the Jiao-Liao-Ji orogenic belt: Implications for petrogenesis and tectonic setting. *Precambrian Res.* **2017**, *300*, 78–106. [CrossRef]

29. Liu, P.H.; Liu, F.L.; Tian, Z.H.; Cai, J.; Ji, L.; Wang, F. Petrological and geochronological evidence for Paleoproterozoic granulite-facies metamorphism of the South Liaohe Group in the Jiao-Liao-Ji Belt, North China Craton. *Precambrian Res.* **2019**, *327*, 121–143. [CrossRef]

30. Wang, F.; Liu, F.L.; Liu, P.H.; Cai, J.; Schertl, H.P.; Ji, L.; Liu, L.S.; Tian, Z.H. In situ zircon U-Pb dating and whole rock geochemistry of metasedimentary rocks from South Liaohe Group, Jiao-Liao-Ji orogenic belt: Constraints on the depositional and metamorphic ages, and implications for tectonic setting. *Precambrian Res.* **2017**, *303*, 764–780. [CrossRef]

31. Cheng, C.Q.; Zhang, J.; Liu, J.; Zhao, C.; Yin, C.Q.; Qian, J.H.; Gao, P.; Liu, X.G.; Chen, Y. Geochemistry and petrogenesis of ca. 2.1 Ga meta-mafic rocks in the central Jiao–Liao–Ji Belt, North China Craton: A consequence of intracontinental rifting or subduction? *Precambrian Res.* **2022**, *370*, 106553. [CrossRef]

32. Liu, C.H.; Liu, F.L.; Zhao, G.C.; Tian, Z.H.; Cai, J.; Zhu, J.J.; Sun, X. Using multiple methods to better constrain the depositional age of a meta-sedimentary succession: An example of the Laoling Group from the northeastern North China Craton. *Precambrian Res.* **2023**, *392*, 107058. [CrossRef]

33. Meng, E.; Liu, F.L.; Liu, P.H.; Liu, C.H.; Yang, H.; Wang, F.; Shi, J.R.; Cai, J. Petrogenesis and tectonic significance of Paleoprotero-zoic meta-mafic rocks from central Liaodong Peninsula, northeast China: Evidence from zircon U-Pb dating and in situ Lu–Hf isotopes, and whole-rock geochemistry. *Precambrian Res.* **2014**, *247*, 92–109. [CrossRef]

34. Bai, J. *The Precambrian Geology and Pb-Zn Mineralization in the Northern Margin of North China Platform*; Geological Publishing House: Beijing, China, 1993.

35. Wang, H.C.; Ren, Y.W.; Lu, S.N.; Kang, J.L.; Chu, H.; Yu, H.B.; Zhang, C.J. Stratigraphic units and tectonic setting of the Paleoproterozoic Liao-Ji orogen. *Acta Geosci. Sin.* **2015**, *36*, 583–598. (In Chinese with English Abstract)

36. Yuan, L.L.; Zhang, X.H.; Xue, F.H.; Han, C.M.; Chen, H.H.; Zhai, M.G. Two episodes of Paleoproterozoic mafic intrusions from Liaoning province, North China Craton: Petrogenesis and tectonic implications. *Precambrian Res.* **2015**, *264*, 119–139. [CrossRef]

37. Meng, E.; Wang, C.Y.; Li, Y.G.; Li, Z.; Yang, H.; Cai, J.; Ji, L.; Jin, M.Q. Zircon U-Pb-Hf isotopic and whole-rock geochemical studies of Paleoproterozoic metasedimentary rocks in the northern segment of the Jiao-Liao-Ji Belt, China: Implications for provenance and regional tectonic evolution. *Precambrian Res.* **2017**, *298*, 472–489. [CrossRef]
38. Xu, W.; Liu, F.L.; Tian, Z.H.; Liu, L.S.; Ji, L.; Dong, Y.S. Source and petrogenesis of Paleoproterozoic meta-mafic rocks intruding into the North Liaohe Group: Implications for back-arc extension prior to the formation of the Jiao-Liao-Ji Belt, North China Craton. *Precambrian Res.* **2018**, *307*, 66–81. [CrossRef]
39. Li, Z.; Chen, B.; Yan, X.L. The Liaohe Group: An insight into the Paleoproterozoic tectonic evolution of the Jiao-Liao-Ji Belt, North China Craton. *Precambrian Res.* **2019**, *326*, 174–195. [CrossRef]
40. Zhang, Q.S.; Yang, Z.S. *Early Crust and Mineral Deposits of Liaodong Peninsula*; Geological Publishing House: Beijing, China, 1988.
41. Luo, Y.; Sun, M.; Zhao, G.C.; Li, S.Z.; Ayers, J.C.; Xia, X.P.; Zhang, J.H. A comparison of U-Pb and Hf isotopic compositions of detrital zircons from the North and South Liaohe Groups: Constraints on the evolution of the Jiao-Liao-Ji Belt, North China Craton. *Precambrian Res.* **2008**, *163*, 279–306. [CrossRef]
42. Li, S.Z.; Zhao, G.C.; Sun, M.; Han, Z.Z.; Zhao, G.T.; Hao, D.F. Are the South and North Liaohe Groups different exotic terranes? Nd isotope constraints. *Gondwana Res.* **2006**, *9*, 198–208. [CrossRef]
43. Lan, T.G.; Fan, H.R.; Yang, K.F.; Cai, Y.C.; Wen, B.J.; Zhang, W. Geochronology, mineralogy and geochemistry of alkali-feldspar granite and albite granite association from the Changyi area of Jiao-Liao-Ji Belt: Implications for Paleoproterozoic rifting of eastern North China Craton. *Precambrian Res.* **2015**, *266*, 86–107. [CrossRef]
44. Wang, X.P.; Peng, P.; Wang, C.; Yang, S.Y. Petrogenesis of the 2115 Ma Haicheng mafic sills from the eastern North China Craton: Implications for an intra-continental rifting. *Gondwana Res.* **2016**, *39*, 347–364. [CrossRef]
45. Wang, X.P.; Peng, P.; Wang, C.; Yang, S.Y.; Söderlund, U.; Su, X.D. Nature of three episodes of Paleoproterozoic magmatism (2180 Ma, 2115 Ma and 1890 Ma) in the Liaoji Belt, North China with implications for tectonic evolution. *Precambrian Res.* **2017**, *298*, 252–267. [CrossRef]
46. Xu, W.; Liu, F.L.; Santosh, M.; Liu, P.H.; Tian, Z.H.; Dong, Y.S. Constraints of mafic rocks on a Paleoproterozoic back-arc in the Jiao-Liao-Ji Belt, North China Craton. *J. Asian Earth Sci.* **2018**, *166*, 195–209. [CrossRef]
47. Liu, C.H.; Zhao, G.C.; Liu, F.L.; Cai, J. The southwestern extension of the JiaoLiao-Ji belt in the North China Craton: Geochronological and geochemical evidence from the Wuhe Group in the Bengbu area. *Lithos* **2018**, *304–307*, 258–279. [CrossRef]
48. Liu, D.Y.; Wilde, S.A.; Wan, Y.S.; Wu, J.S.; Zhou, H.Y.; Dong, C.Y.; Yin, X.Y. New U-Pb and Hf isotopic data confirm Anshan as the oldest preserved segment of the North China Craton. *Am. J. Sci.* **2008**, *308*, 200–231. [CrossRef]
49. Liu, J.; Liu, Z.H.; Zhao, C.; Wang, C.J.; Peng, Y.B.; Zhang, H. Petrogenesis and zircon LA-ICP-MS U-Pb dating of newly discovered Mesoarchean gneisses on the northern margin of the North China Craton. *Int. Geol. Rev.* **2017**, *59*, 1575–1589. [CrossRef]
50. Wan, Y.S.; Ma, M.Z.; Dong, C.Y.; Xie, H.Q.; Xie, S.W.; Ren, P.; Liu, D.Y. Widespread late Neoarchean reworking of Meso- to Paleoarchean continental crust in the Anshan-Benxi area, North China Craton, as documented by U-Pb-Nd-Hf-O isotopes. *Am. J. Sci.* **2015**, *315*, 620–670. [CrossRef]
51. Wang, W.; Liu, S.W.; Santosh, M.; Wang, G.H.; Bai, X.; Guo, R.R. Neoarchean intra-oceanic arc system in the Western Liaoning Province: Implications for the Early Precambrian crust-mantle geodynamic evolution of the Eastern Block of the North China Craton. *Earth-Sci. Rev.* **2015**, *150*, 329–364. [CrossRef]
52. Wang, W.; Liu, S.W.; Cawood, P.A.; Bai, X.; Guo, R.R.; Guo, B.R.; Wang, K. Late Neoarchean subduction-related crustal growth in the Northern Liaoning region of the North China Craton: Evidence from ~2.55 to 2.50 Ga granitoid gneisses. *Precambrian Res.* **2016**, *281*, 200–223. [CrossRef]
53. Wang, W.; Cawood, P.A.; Liu, S.W.; Guo, R.R.; Bai, X.; Wang, K. Cyclic formation and stabilization of Archean lithosphere by accretionary orogenesis: Constraints from TTG and potassic granitoids, North China Craton. *Tectonics* **2017**, *36*, 1724–1742. [CrossRef]
54. Li, Z.S.; Shan, X.L.; Liu, J.; Zhang, J.; Liu, Z.H.; Cheng, C.Q.; Wang, Z.G.; Zhao, C.; Yu, H.C. Late Neoarchean TTG and monzogranite in the northeastern North China Craton: Implications for partial melting of a thickened lower crust. *Gondwana Res.* **2023**, *115*, 201–223. [CrossRef]
55. Zhao, G.C.; Cao, L.; Wilde, S.A.; Sun, M.; Choe, W.J.; Li, S.Z. Implications based on the first SHRIMP U-Pb zircon dating on Precambrian granitoid rocks in North Korea. *Earth Planet. Sci. Lett.* **2006**, *251*, 365–379. [CrossRef]
56. Zhao, L.; Zhai, M.G.; Nutman, A.P.; Oh, C.W.; Bennett, V.C.; Zhang, Y.B. Archean basement components and metamorphic overprints of the Rangnim Massif in the northern part of the Korean Peninsula and tectonic implications for the Sino-Korean Craton. *Precambrian Res.* **2020**, *344*, 105735. [CrossRef]
57. Wang, M.J.; Liu, S.W.; Fu, J.H.; Wang, K.; Guo, R.R.; Guo, B.R. Neoarchean DTTG gneisses in southern Liaoning Province and their constraints on crustal growth and the nature of the Liao-Ji Belt in the Eastern Block. *Precambrian Res.* **2017**, *303*, 183–207. [CrossRef]
58. Zhai, M.G.; Zhang, X.H.; Zhang, Y.B.; Wu, F.Y.; Peng, P.; Li, Q.L.; Li, Z.; Guo, J.H.; Li, T.S.; Zhao, L.; et al. The geology of North Korea: An overview. *Earth-Sci. Rev.* **2019**, *194*, 57–96. [CrossRef]
59. Yu, H.; Liu, J.; He, Z.; Liu, Z.; Cheng, C.; Hao, Y.; Zhao, C.; Zhang, H.; Dong, Y. Geochronology and Zircon Hf Isotope of the Paleoproterozoic Gaixian Formation in the Southeastern Liaodong Peninsula: Implication for the Tectonic Evolution of the Jiao-Liao-Ji Belt. *Minerals* **2022**, *12*, 792. [CrossRef]

60. Bi, J.H.; Ge, W.C.; Xing, D.H.; Yang, H.; Dong, Y.; Tian, D.X.; Chen, H.J. Palaeoproterozoic meta-rhyolite and meta-dacite of the Liaohe Group, Jiao-Liao-Ji Belt, North China Craton: Petrogenesis and implications for tectonic setting. *Precambrian Res.* **2018**, *314*, 306–324. [CrossRef]

61. Zhu, K.; Liu, Z.H.; Xu, Z.Y.; Wang, X.A.; Cui, W.L.; Hao, Y.J. Petrogenesis and tectonic implications of two types of Liaoji granitoid in the Jiao–Liao–Ji Belt, North China Craton. *Precambrian Res.* **2019**, *331*, 105369. [CrossRef]

62. Cai, J.; Liu, F.L.; Liu, P.H.; Wang, F.; Meng, E.; Wang, W.; Yang, H.; Ji, L.; Liu, L.S. Discovery of granulite-facies metamorphic rocks in the Ji'an area, northeastern Jiao–Liao–Ji Belt, North China Craton: Metamorphic P-T evolution and geological implications. *Precambrian Res.* **2017**, *303*, 626–640. [CrossRef]

63. Cai, J.; Liu, F.L.; Liu, C.H. A unique Paleoproterozoic HP–UHT metamorphic event recorded by the Bengbu mafic granulites in the southwestern Jiao–Liao–Ji Belt, North China. Craton. *Gondwana Res.* **2020**, *80*, 244–274. [CrossRef]

64. Zhang, S.H.; Zhao, Y.; Davis, G.A.; Ye, H.; Wu, F. Temporal and spatial variations of Mesozoic magmatism and deformation in the North China Craton: Implications for lithospheric thinning and decratonization. *Earth-Sci. Rev.* **2014**, *162*, 387–432. [CrossRef]

65. Liu, Y.S.; Gao, S.; Hu, Z.C.; Gao, C.G.; Zong, K.Q.; Wang, D.B. Continental and oceanic crust recycling-induced melt-peridotite interactions in the Trans-North China Orogen: U-Pb dating, Hf isotopes and trace elements in zircons from mantle xenoliths. *J. Petrol.* **2010**, *51*, 537–571. [CrossRef]

66. Ludwig, K.R. *ISOPLOT 3.00: A Geochronological Toolkit for Microsoft Excel*; Berkeley Geochronology Center: Berkeley, CA, USA, 2003.

67. Wu, F.Y.; Yang, Y.H.; Xie, L.W.; Yang, J.H.; Xu, P. Hf isotopic compositions of the standard zircons and baddeleyites used in U-Pb geochronology. *Chem. Geol.* **2006**, *234*, 105–126. [CrossRef]

68. Dong, Y.; Bi, J.H.; Xing, D.H.; Ge, W.C.; Yang, H.; Hao, Y.J.; Ji, Z.; Jing, Y. Geochronology and geochemistry of Liaohe Group and Liaoji granitoid in the Jiao-Liao-Ji Belt, North China Craton: Implications for petrogenesis and tectonic evolution. *Precambrian Res.* **2019**, *332*, 105399. [CrossRef]

69. Chen, J.S.; Xing, D.H.; Liu, M.; Li, B.; Yang, H.; Tian, D.X.; Yang, F.; Wang, Y. Zircon U-Pb chronology and geological significance of felsic volcanic rocks in the Liaohe Group from the Liaoyang area, Liaoning Province. *Acta Petrol. Sin.* **2017**, *33*, 2792–2810. (In Chinese with English Abstract)

70. Tarney, J. Geochemistry of Archaean high grade gneisses with implications as to the origin and evolution of the Precambrian curst. In *The Early History of the Earth*; Windley, B.F., Ed.; John Wiley: Hoboken, NJ, USA, 1976.

71. Winchester, J.A.; Park, R.G.; Holland, J.G. The geochemistry of Lewisian semipelitic schists from the Gairloch District, Wester Ross. *Scott. J. Geol.* **1980**, *16*, 165–179. [CrossRef]

72. Middlemost, E.A.K. Naming materials in the magma/igneous rock system. *Earth-Sci. Rev.* **1994**, *37*, 215–224. [CrossRef]

73. Winchester, J.A.; Floyd, P.A. Geochemical magma type discrimination: Application to altered and metamorphosed basic igneous rocks. *Earth Planet. Sci. Lett.* **1977**, *28*, 459–469. [CrossRef]

74. Sun, S.S.; McDonough, W.F. Chemical and isotopic systematics of oceanic basalts: Implications for mantle composition and processes. *Geol. Soc. Lond. Spec. Publ.* **1989**, *42*, 313–345. [CrossRef]

75. Wang, M.J.; Liu, S.W.; Wang, W.; Wang, K.; Yan, M.; Guo, B.R.; Bai, X.; Guo, R.R. Petrogenesis and tectonic implications of the Neoarchean North Liaoning tonalitic-trondhjemitic gneisses of the North China Craton, North China. *J. Asian Earth Sci.* **2016**, *131*, 12–39. [CrossRef]

76. Cheng, C.Q.; Liu, J.; Zhang, J.; Chen, Y.; Yin, C.Q.; Liu, X.G.; Qian, J.H.; Gao, P.; Wang, X. Petrogenesis of newly identified Neoarchean granitoids in the Qingyuan of NE China: Implications on crustal growth and reworking of the North China Craton. *J. Asian Earth Sci.* **2022**, *236*, 105333. [CrossRef]

77. Sun, M.; Armstrong, R.L.; Lambert, R.S.; Jiang, C.C.; Wu, J.H. Petrochemistry and Sr, Pb and Nd isotopic geochemistry of Paleoproterozoic Kuandian Complex, the eastern Liaoning Province, China. *Precambrian Res.* **1993**, *62*, 171–190.

78. Peng, Q.M.; Palmer, M.R. The Paleoproterozoic boron deposits in eastern Liaoning, China: A metamorphosed evaporate. *Precambrian Res.* **1995**, *72*, 185–197. [CrossRef]

79. Liu, J.H.; Wang, X.J.; Chen, H. Intracontinental extension and geodynamic evolution of the Paleoproterozoic Jiao-Liao-Ji Belt, North China Craton: Insights from coeval A-type granitic and mafic magmatism in eastern Liaoning Province. *Geol. Soc. Am. Bull.* **2021**, *133*, 1765–1792. [CrossRef]

80. Li, S.Z.; Zhao, G.C. SHRIMP U-Pb zircon geochronology of the Liaoji granitoids: Constraints on the evolution of the Paleoproterozoic Jiao-Liao-Ji belt in the Eastern Block of the North China Craton. *Precambrian Res.* **2007**, *158*, 1–16. [CrossRef]

81. Wan, Y.S.; Song, B.; Liu, D.Y.; Wilde, S.A.; Wu, J.S.; Shi, Y.R.; Yin, X.Y.; Zhou, H.Y. SHRIMP U-Pb zircon geochronology of Paleoproterozoic metasedimentary rocks in the North China Craton: Evidence for a major Late Paleoproterozoic tectonothermal event. *Precambrian Res.* **2006**, *149*, 249–271. [CrossRef]

82. Zhang, Y.F.; Liu, J.D.; Xiao, R.G.; Wang, S.Z.; Wang, J.; Bao, D.J. The Hyalotourmalites of Houxianyu Borate Deposit in Eastern Liaoning: Zircon Features and SHRIMP Dating. Earth Sci. *J. China Univ. Geosci.* **2010**, *35*, 985–999. (In Chinese with English Abstract)

83. Li, Z.; Chen, B.; Liu, J.W.; Zhang, L.; Yang, C. Zircon U-Pb ages and their implications for the South Liaohe Group in the Liaodong Peninsula, Northeast China. *Acta Petrol. Sin.* **2015**, *31*, 1589–1605. (In Chinese with English Abstract)

84. Chen, B.; Li, Z.; Wang, J.; Zhang, L.; Yan, X. Liaodong peninsula ~2.2 Ga magmatic event and its geological significance. *J. Jilin Univ.* **2016**, *46*, 303–320. (In Chinese with English Abstract)

85. Lu, X.; Wu, F.; Lin, J.; Sun, D.; Zhang, Y.; Guo, C. Geochronological successions of the early Precambrian granitic magmatism in southern Liaodong Peninsula and its constraints on tectonic evolution of the north China craton. *Chin. J. Geol.* **2004**, *39*, 123–138. (In Chinese with English Abstract)

86. Yang, M.C.; Chen, B.; Yan, C. Petrogenesis of Paleoproterozoic gneissic granites from Jiao-Liao-Ji Belt of North China Craton and their tectonic implications. *J. Earth Sci. Environ.* **2015**, *37*, 31–51. (In Chinese with English Abstract)

87. Song, Y.H.; Yang, F.C.; Yan, G.L.; Wei, M.H.; Shi, S.S. SHRIMP U-Pb age and Hf isotopic compositions of Paleoproterozoic granites from the eastern part of Liaoning Province and their tectonic significance. *Acta Geol. Sin.* **2016**, *90*, 2620–2636. (In Chinese with English Abstract)

88. Wang, P.S.; Dong, Y.S.; Li, F.Q.; Gao, B.S.; Gan, Y.C.; Chen, M.S.; Xu, W. Paleoproterozoic granitic magmatism and geological significance in Huanghuadian area, eastern Liaoning Province. *Acta Petrol. Sin.* **2017**, *33*, 2708–2724. (In Chinese with English Abstract)

89. Wang, X.J.; Liu, J.H.; Ji, L. Zircon U-Pb chronology, geochemistry and their petrogenesis of Paleoproterozoic monzogranitic gneisses in Kuandian area, eastern Liaoning Province, Jiao-Liao-Ji Belt, North China Craton. *Acta Petrol. Sin.* **2017**, *33*, 2689–2707. (In Chinese with English Abstract)

90. Zhou, Y.Y.; Sun, Q.Y.; Zhao, T.P.; Zhai, M.G. Petrogenesis of the early Paleoproterozoic low-δ^{18}O potassic granites in the southern NCC and its possible implications for no confluence of glaciations and magmatic shutdown at ca. 2.3 Ga. *Precambrian Res.* **2021**, *361*, 106258. [CrossRef]

91. Yang, J.H.; Chung, S.L.; Wilde, S.A.; Wu, F.Y.; Chu, M.F.; Lo, C.H.; Fan, H.R. Petrogenesis of post-orogenic syenites in the Sulu Orogenic Belt, East China: Geochronological, geochemical and Nd–Sr isotopic evidence. *Chem. Geol.* **2005**, *214*, 99–125. [CrossRef]

92. Foland, K.A.; Allen, J.C. Magma sources for Mesozoic anorogenic granites of the White Mountain magma series, New England, USA. *Contrib. Miner. Petrol.* **1991**, *109*, 195–211. [CrossRef]

93. Turner, S.P.; Foden, J.D.; Morrison, R.S. Derivation of some A-type magmas by fractionation of basaltic magma; an example from the Padthaway Ridge, South Australia. *Lithos* **1992**, *28*, 151–179. [CrossRef]

94. Sisson, T.W.; Ratajeski, K.; Hankins, W.B.; Glazner, A.F. Voluminous granitic magmas from common basaltic sources. *Contrib. Miner. Petrol.* **2005**, *148*, 635–661. [CrossRef]

95. Eliwa, H.A.; El-Bialy, M.Z.; Murata, M. Ediacaran post-collisional volcanism in the Arabian-Nubian Shield: The high-K calc-alkaline Dokhan Volcanics of Gabal Samr El-Qaa (592 ± 5 Ma), North Eastern Desert, Egypt. *Precambrian Res.* **2014**, *246*, 180–207. [CrossRef]

96. Beard, J.S.; Lofgren, G.E. Dehydration melting and water-saturated melting of basaltic and andesitic greenstones and amphibolites at 1, 3, and 6.9 kb. *J. Petrol.* **1991**, *32*, 365–401. [CrossRef]

97. Nakajima, K.; Arima, M. Melting experiments on hyrous low-K tholeiite: Implications for the genesis of tonalitic crust in the Izu-Bonin-Mariana arc. *Isl. Arc* **1998**, *7*, 359–373. [CrossRef]

98. Frost, C.D.; Frost, B.R.; Chamberlain, K.R.; Edwards, B. Petrogenesis of the 1.43 Ga Sherman batholith, SE, Wyoming, USA: A reduced, rapakivi-type anorogenic granite. *J. Petrol.* **1999**, *40*, 1771–1802. [CrossRef]

99. Maurice, A.E.; Gharib, M.E.; Wilde, S.A.; Ali, K.A.; Osman, M.S.M. Subduction to post-collisional volcanism in the Northern Arabian-Nubian Shield: Genesis of Cryogenian/Ediacaran intermediate-felsic magmas and the lifespan of a Neoproterozoic mature island arc. *Precambrian Res.* **2021**, *358*, 106148. [CrossRef]

Article

Detrital Zircon LA-ICP-MS U-Pb Ages of the North Liaohe Group from the Lianshanguan Area, NE China: Implications for the Tectonic Evolution of the Paleoproterozoic Jiao-Liao-Ji Belt

Jinhui Gao [1], Weimin Li [1,2,*], Yongjiang Liu [3,4], Yingli Zhao [1,5], Tongjun Liu [1] and Quanbo Wen [1]

[1] College of Earth Sciences, Jilin University, Changchun 130061, China; gaojh21@mails.jlu.edu.cn (J.G.)
[2] Key Laboratory of Mineral Resources Evaluation in Northeast Asia, Ministry of Natural Resources, Changchun 130026, China
[3] MOE Key Lab of Submarine Geoscience and Prospecting Techniques, Institute for Advanced Ocean Study, College of Marine Geosciences, Ocean University of China, Qingdao 266100, China
[4] Laboratory for Marine Mineral Resources, Qingdao National Laboratory for Marine Science and Technology, Qingdao 266100, China
[5] Research Center of Paleontology & Stratigraphy, Jilin University, Changchun 130026, China
* Correspondence: weiminli@jlu.edu.cn

Citation: Gao, J.; Li, W.; Liu, Y.; Zhao, Y.; Liu, T.; Wen, Q. Detrital Zircon LA-ICP-MS U-Pb Ages of the North Liaohe Group from the Lianshanguan Area, NE China: Implications for the Tectonic Evolution of the Paleoproterozoic Jiao-Liao-Ji Belt. *Minerals* **2023**, *13*, 708. https://doi.org/10.3390/min13050708

Academic Editor: José Francisco Santos

Received: 24 April 2023
Revised: 18 May 2023
Accepted: 18 May 2023
Published: 22 May 2023

Abstract: The Liaohe Group, which is a significant lithostratigraphic unit within the Paleoproterozoic Jiao-Liao-Ji Belt situated between the Longgang and Liaonan-Nangrim blocks, comprises the Langzishan, Li'eryu, Gaojiayu, Dashiqiao, and Gaixian formations, which are characterized mainly by a clastic-rich sequence with an interlayered bimodal-volcanic sequence, carbonate-rich sequence, and (meta-)pelite-rich sequence. Currently, the tectonic background and evolution of the Liaohe Group remain contentious. Based on the study of detrital zircon geochronology and the zircon trace element characteristics in the Langzishan and Li'eryu formations in the North Liaohe Group in the Lianshanguan area, NE China, this paper reveals the formations' provenances, depositional ages, and relationships with Paleoproterozoic granitoids (the Liao-Ji granites). The present results, in conjunction with previous studies, indicate that the depositional age of the Langzishan Formation is 2136 Ma and that of the Li'eryu Formation is 1974 Ma. The provenances of the Langzishan Formation and the Li'eryu Formation are mainly characterized by Neoarchean-to-early-Paleoproterozoic basement rocks (~2.6–2.4 Ga) and the Liao-Ji granites (~2.2–2.0 Ga), respectively. Moreover, the coeval mafic and metasedimentary rocks of the Liaohe Group exhibit characteristics of an extensional environment, which is represented by the tectonic setting of a back-arc basin. Notably, the Upper Langzishan Formation records a prominent shift in sedimentary environment from a passive continental margin to an active continental margin. In terms of the tectonic evolution of the North Liaohe Group and the Jiao-Liao-Ji Belt, our proposed model suggests that the Archean basement rocks in the northern part of the continental block, along with a limited contribution from the Paleoproterozoic Liao-Ji granites, served as the primary sources for the Langzishan Formation. Subsequently, the rapid deposition of the Li'eryu Formation was influenced by intense magmatism and subsequent erosion of the subduction-related magmatic arc (the Liao-Ji granites) within a back-arc basin environment. Lastly, the deposition of clastic materials from the Longgang blocks and the Liao-Ji granites resulted in the formation of the Gaojiayu, Dashiqiao, and Gaixian formations.

Keywords: detrital zircon U-Pb ages; North Liaohe Group; provenance; Jiao-Liao-Ji Belt; North China Craton

1. Introduction

The North China Craton (NCC) is one of the oldest cratons in the world and is bounded by the Late Paleozoic–Early Mesozoic Central Asian Orogenic Belt to the north, the Early Paleozoic–Mesozoic Qilian-Qingling-Dabie orogen to the southwest, and the

Mesozoic Sulu Orogen to the east (Figure 1a) [1–4]. Generally, the NCC is considered to have been formed by the Paleoproterozoic amalgamation (~1.85 Ga) of two distinct Archean-to-Paleoproterozoic blocks named the Eastern and Western blocks separated by the Trans-North China Orogen or Central Orogenic Belt [5–13]. Additionally, the Eastern and Western blocks are viewed as assemblies of several micro-blocks that collided to form the critical Paleoproterozoic orogenic belts, such as the Khondalite Belt (KB) and Jiao-Liao-Ji Belt (JLJB). However, the evolutionary histories of these micro-blocks are still highly controversial [5,10,14–16]. Therefore, recognizing the Paleoproterozoic orogenic belts is a significant achievement in understanding the history of the NCC.

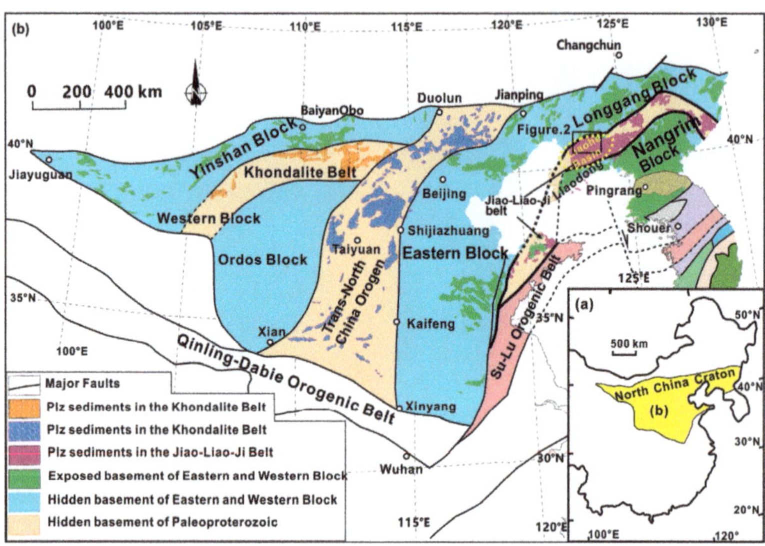

Figure 1. (**a**) Outline Map of Central and Eastern China. (**b**) Tectonic division of the North China Craton. (modified after Zhao et al., 2005 [10]).

Among the Paleoproterozoic orogenic belts in the NCC, the JLJB separates the Eastern Block into the Longgang Block (LB) in the northwest and the Liaonan-Nangrim Block (LNB) in the southeast. As previously reported, this belt records a long and extremely complex history of magmatism, multi-metamorphic evolution, tectonic deformation, and crustal overgrowth and reworking, but its origin and tectonic evolution remain controversial despite a large number of recent studies on the belt [17–27]. Possible explanations for the origin and evolution of the JLJB include: (a) the opening and closing of an intracontinental rift [20,28–32], (b) arc-continent collision [33–37], and (c) a combination of the two previous scenarios, with intracontinental rifting progressing to the formation of a new ocean basin or back-arc basin then being subsequently closed by subduction [13,27,38–40].

Zircon is a common accessory mineral in most of the detrital sedimentary rocks (especially sandstones) due to its stable crystal structure, and because of the high closure temperature of the U-Th-Pb isotope system, it remains stable from low to high metamorphism conditions and weathering transport. Therefore, detrital zircon U-Pb geochronology has been widely used to define the maximum sedimentary age of sedimentary strata [41] and trace the provenance of the sediments [42], significantly contributing to the understanding of early crustal tectono-thermal evolution [43].

The detrital zircons collected from meta-sedimentary rocks in the North and South Liaohe groups of the JLJB are significantly important and commonly studied. Previous research has determined the ages of clastic and metamorphic zircons in sedimentary rocks of the Liaohe Group, yielding age peaks of 2200–2000 Ma and ~2500 Ma and determining the sedimentary age of the metasedimentary rocks to be 2.0–1.9 Ga [44–46]. Although previous studies have examined the detrital zircons in these sedimentary rocks, discrepancies still exist in determining the sedimentary age of the Liaohe Group [36,44,45,47,48].

By conducting U-Pb dating of these zircons (15 samples with 17–74 valid data for each sample), this study aims to determine the provenances, maximum depositional ages, and depositional environments of the North Liaohe Group in the Langzishan area, NE China. Furthermore, by comparing the similarities and differences between the North and South Liaohe groups, this study will provide a more detailed understanding of the tectonic setting and evolution of the Liaohe Group, as well as of the JLJB.

2. Geological Setting

The JLJB is a Paleoproterozoic orogenic belt located in the northeastern part of the Eastern Block of the NCC. It extends from southern Jilin Province to eastern Shandong Province, tectonically separating the Precambrian Eastern Block into the LNB and the LB (Figure 1) [9,10,20]. The belt is primarily composed of greenschist to granulite facies-metamorphic sedimentary rocks and widespread granitoids, along with a small number of mafic intrusions and volcanic successions (Figure 1) [29,49].

2.1. Paleoproterozoic (Meta-)Volcanic-Sedimentary Sequences

According to the distribution of the Paleoproterozoic volcanic-sedimentary sequence, the JLJB can be divided into two sub-belts: the northern sub-belt, comprising the Laoling Group in southern Jilin, the North Liaohe Group in eastern Liaoning, and the Fenzishan Group, and the southern sub-belt, including the Ji'an, South Liaohe, and Jingshan groups and possibly the Macheonryeong Group in North Korea [10,12,27,28]. These successions are transitional, starting from a basal clastic-rich sequence and lower bimodal-volcanic sequence, going through a middle carbonate-rich sequence, and ending with an upper pelite-rich sequence [50].

In the Liaodong peninsula, the Liaohe Group comprises a thick sequence of metamorphosed volcanic-sedimentary rocks. In the northern part of the Liaohe Group, the lithology mainly consists of clastic and carbonate rocks, while in the southern part, it is predominantly composed of a volcanic-clastic-carbonate succession. These two groups are separated by faults defined by the locations of the Gaixian-Ximucheng-Taziling-Jiangcaodianzi-Aiyang areas [28,32,51].

Traditionally, the Liaohe Group has been divided into five formations from bottom to top: the Langzishan, Li'eryu, Gaojiayu, Dashiqiao, and Gaixian formations. It is noteworthy that the lowermost Langzishan Fm. is only distributed in the North Liaohe Group, while the Gaixian Fm. is mostly found in the South Liaohe Group [28,52].

2.2. Paleoproterozoic Magmatism

The Paleoproterozoic granites are widely exposed in the JLJB and are mainly divided into two sequences in the Liaodong Peninsula: the Liao-Ji granites and rapakivis [53]. The Liao-Ji granites are mostly distributed in the southern part, with zircon U-Pb ages ranging from 1.93 Ga to 2.30 Ga. Based on their main mineral assemblages (quartz, plagioclase, perthite, biotite, and, in some cases, amphibole, magnetite, or tourmaline) and geochemical affinities, the Liao-Ji granites are classified into magnetite monzonitic granites (I-type Liao-Ji granite) [38,39,54–56] and amphibole monzonitic granites (A-type Liao-Ji granite) [20,57,58]. According to Liu et al. [59], the Liao-Ji granites resulted from the partial melting of the Archean basement rocks, and their gneissic structures have been attributed to syn-emplacement deformation. However, Qu et al. [60] and Chen et al. [61] argued that the foliation of the Liao-Ji granites resulted from post-emplacement deformation, and that the granites belonged to migmatites derived from both crustal and mantle sources.

The rapakivi granites are mostly distributed in the eastern part of the JLJD, with zircon U-Pb ages ranging from 1.8 Ga to 1.9 Ga. They are generally composed of quartz, microcline, garnet, and biotite, which are their major mineral assemblages [17,60,62,63]. The genesis of the rapakivi granites in the JLJB has been interpreted differently. For instance, Li et al. [38]

suggested that they resulted from the detachment of the orogenic belt or the upwelling of mantle magma, while Chen et al. [61] proposed that they were post-orogenic granites, a mixture of magmas from both crustal and mantle sources. Yang et al. [54] believed that the rapakivi granites in the Shuangcha area were formed in 1890 ± 21 Ma, marking the end of the orogenic processes of the JLJB.

2.3. Geology of the Study Area

The study area is situated in the Lianshanguan region of eastern Liaoning Province (Figure 2), which is characterized by the widespread exposure of the Archean basement and Paleoproterozoic strata, including the North Liaohe Group, a minor amount of Neoarchean supracrustal rocks (Anshan Group in Anshan-Benxi Area Liaoning), Paleoproterozoic igneous rocks, and Phanerozoic strata (Figures 2 and 3) [64–66].

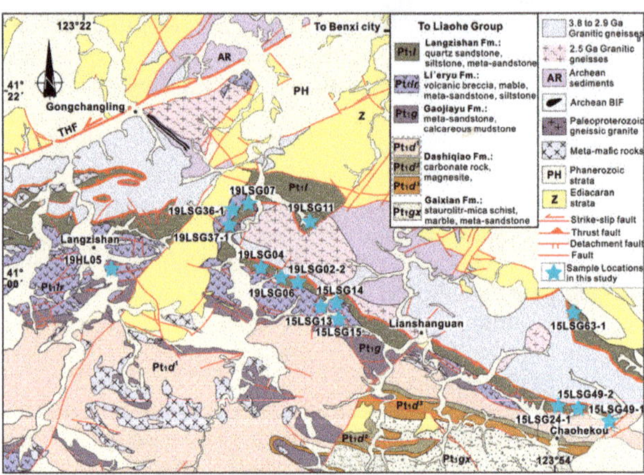

Figure 2. Stratigraphic structural map of the Lianshanguan area, NE China, showing the detailed sampling locations. (Modified from Wen et al., 2021 [67]).

			Sample	Lithology
the Li' eryu Fm. (Pt₁/r, ~2.2–1.9 Ga)			19LSG13	Siltstone
			15LSG36–1	Meta-quartz sandstone
			19LSG15	
				Meta-sandstone
			19HL05	
				Mable
				Volacnic breccia
the Langzishan Fm. (Pt₁/r, ~2.5–2.2 Ga)	Upper			Slate
				Phyllite
			19LSG11	Siltstone
			19LSG14	Garnet mica schist
	Lower		19LSG06	Siltstone
			15LSG63–1	Meta-sandstone
			15LSG49–2	Feldspathic quartz sandstone
			15LSG24–1	
			19LSG04	Quartz sandstone
			19LSG02–2	
			15LSG49–1	Quartz sandstone
			15LSG37–1	(pebbly) Quartz sandstone
			19LSG07	
NCC' basement (~3.8–2.5 Ga)				Archean granites

Figure 3. Stratigraphic histogram of sampling sites.

The Archean basement in the Lianshanguan area, mainly consisting of granitic gneisses and migmatites, forms a short-axis, anticline-like dome that stretches approximately 40 km in the NW direction and has a width of 5–10 km [68]. The outer edge of the dome is overlain by the North Liaohe Group and Sinian strata, and an unconformity separates it from the overlying Langzishan Fm., resulting in the formation of a typical weathering paleocrust, particularly in the Sandaoling area. In addition, numerous quartz veins have been developed in the granitic basement and truncated by the unconformity surface (Figure 4a). The granitic conglomerate sandstone—located above—and a medium-sized sandy conglomerate are present. A large-scale ductile shear zone has been developed on the southern edge of the dome (Figure 4b–d). Based on kinematic indicators including asymmetric folds, rotated boudins, and offset markers, the ductile zone shows a dextral shearing sense. The Paleoproterozoic igneous rocks here consist of both deformed felsic granites and mafic rocks with ages ranging from approximately 2.2 to 1.9 Ga [24,51,69,70].

Figure 4. Unconformity contact relationship between the Langzishan Fm. and Lianshanguan granite (basement). Red arrows represent the shearing directions. (**a**) Sketch of unconformity interface. (**b**) Fold deformation of the Langzishan Fm. (**c**) Mica schist and cataclastic quartz schist. (**d**) Dextral sheared schist and granitic vein of the Langzishan Fm.

The Langzishan Fm. (Pt_1l), the lowermost unit of the North Liaohe Group, primarily comprises garnet-bearing kyanite-cordierite-mica schists, mica-quartz schists, two mica schists, and quartzites with coarse sandstone and conglomerate at the base (Figure 3). Calcite marbles, tremolite-dolomite marbles, phosphates, and uranium deposits are also present within the formation. The formation is unconformably in contact with the Neoarchean basement of the LB. The Li'eryu Fm. (Pt_1lr) is characterized by boron-bearing (meta-)volcanic rocks and clastic sedimentary series, including meta-rhyolitic and mafic volcanics, sandstones, and marbles at the base, and meta-sandstones and tremolite (dolomite) marbles above. The presence of numerous volcanic rocks with arc-affinity suggests intense magmatic activity near subduction zones and magmatic arcs [71]. The Gaojiayu Fm. (Pt_1g) is defined by carbonaceous clastic and clay rocks with minor volcanics [72,73]. It consists mainly of garnet-bearing mica schists, biotite leptynites, and diopside-dolomite marbles with minor rhyolitic and mafic volcanics at the base. The Dashiqiao Fm. (Pt_1d) is represented by carbonate rocks with minor clastic rocks, comprising dolomitic or calcic marbles in its lower and upper sections and mica schists in its middle section. The Gaixian Fm.

(Pt$_1$*gx*) is a meta-pelitic sequence primarily composed of staurolite-garnet-mica schists, sillimanite-mica schists, mica leptynites, and phyllites with thin marble layers.

3. Petrographic Description

The (meta-)sedimentary rocks in the Langzishan and Li'eryu formations from the North Liaohe Group were chosen for detrital zircon LA-ICP-MS U-Pb dating, and these included (pebbly) quartz sandstones (19LSG07, 15LSG37-1, 15LSG49-1, 19LSG02-2, and 19LSG04), feldspathic quartz sandstones (15LSG24-1 and 15LSG49-2), meta-sandstone (15LSG63-1), siltstones (19LSG06 and 19LSG11), and garnet mica schist (19LSG14) from the Langzishan Fm. Furthermore, the meta-sandstones (19HL05-2 and 19LSG15), siltstone (19LSG13), and quartz sandstone (15LSG36-01) from the Li'eryu Fm. were also included (Figures 2 and 3). The petrographic features of representative samples are outlined below (Only representative surface samples have been selected for detailed description, and all sample descriptions are available within the Supplementary Materials).

3.1. Representative Samples from the Langzishan Fm

Samples 19LSG07 and 15LSG37-1 were obtained from the granitic conglomerate weathering paleocrust located at the contact zone between the Langzishan Fm. and the Lianshanguan granitic rock mass (basement). These pebbly sandstones exhibit a massive structure and are composed of well-sorted, subangular clastic particles. Specifically, the samples contain ~10% quartz, ~35% microcline feldspar, ~25% plagioclase feldspar, and ~25% lithic fragments, along with a small amount of chlorite and amphibole. Siliceous cementation is observed, with both inlaid cementation and particle support being evident (Figure 5a,b).

Sample 19LSG02-2 is a quartz sandstone, with the clastic particles mainly consisting of quartz (93%), alkali feldspar (5%), and a small quantity of impurities (2%). Alkali feldspar is mostly microcline with gridiron twining (Figure 5c).

Sample 19LSG04 is a terrigenous quartz sandstone, primarily composed of quartz (~90%), with small amounts of alkaline feldspar (8%), dark minerals (2%), and micas (Figure 5d).

Sample 19LSG06 is a calcareous siltstone consisting mainly of quartz (~90%), with minor feldspar (8%) and lithic fragments (2%). A thin calcite vein with severe alteration is present, and the cement is calcareous, shows contact cementation, and is particle-supported (Figure 5e).

Lastly, sample 19LSG14 is a garnet-mica schist with a porphyroblastic texture, and the minerals in it mainly consist of biotite (~30%), quartz (45%), and feldspar (20%) with a small quantity of garnet (~5%). The grain size of the garnet phenocryst is up to 2.0 cm in diameter (Figure 5f).

3.2. Representative Samples from the Li'eryu Fm

Sample 19LSG15 is a meta-sandstone, characterized by a mineral assemblage of quartz (43%), alkaline feldspar (25%), calcite (15%), biotite (10%), muscovite (5%), and a small proportion of dark minerals (2%). The rock's cement is calcareous, with contact cementation and particle support being observed (Figure 5g).

Sample 19LSG13 is a siltstone, exhibiting a silty structure. The clastic particles are mainly composed of quartz (75%), feldspar (18%), and muscovite (5%), with a small proportion of dark minerals (2%). The rock is well-sorted, with contact cementation and particle support evident (Figure 5h).

Figure 5. Representative thin-section micrographs of the Langzishan formation and the Li'leyu formation of the North Liaohe Group. Image (**a**) is from the sample 19LSG07 (quartz sandstone), (**b**) is from the sample 15LSG37-1 (quartz sandstone), (**c**) is from the sample 19LSG02-2 (quartz sandstone), (**d**) is from the sample 19LSG04 (quartz sandstone), (**e**) is from the sample 19LSG06 (siltstone), (**f**) is from the sample 19LSG14 (garnet mica schist), (**g**) is from the sample 19LSG15 (meta-sandstone), and (**h**) is from the sample 19LSG13 (siltstone). The abbreviations for the minerals are as follows: Qtz—quartz, Mc—microcline, Pl—plagioclase, Ms—muscovite, Chl—chlorite, Cal—calcite, Grt—garnet, and Bt—biotite.

4. Zircon U-Pb LA-ICP-MS Dating

4.1. Analytical Techniques

Zircon grains were separated using the conventional magnetic separation technique and hand picking methods at Yuneng Mineral Separation Company in Hebei Province. The separated zircon particles were fixed with epoxy resin and polished until the cores were exposed, and then coated with carbon. Prior to the determination of the sample, the sample surface was cleaned with 3% HNO3 to remove the C-coating. Transmitted light, reflected light, and cathodoluminescent (CL) image acquisitions (Figure 6) were conducted at the Beijing Gaonianlinghang Corporation. LA-ICP-MS U-Pb zircon dating was performed at the Key Laboratory of Mineral Resources Evaluation in Northeast Asia,

Ministry of Land and Resources, Jilin University, Changchun, China. CL image phase formation was obtained using the Momo CL3+ cathodoluminescent device produced by the company Gatan, which is based in the UK. Zircon dating was conducted using the latest generation Agilent 7900 ICP-MS with a shield torch from the company Agilent. The laser ablation system comprised the ComPex102 Excimerlaser (working material ArF, wavelength 193 nm) from Lambdaphysik company in Germany and an optical system from the company Microlas. Helium was used as the carrier gas for the ablation material, and the spot beam diameter was 32 μm. The frequency was 10 Hz, the laser energy was 90 mJ, the gas background acquisition time of each analysis point was 20 s, and the signal acquisition time was 40 s. Zircon 91500 was used as an external standard for age calibration, and NIST SRM610 silicate glass was applied for instrument optimization. The test procedures and lead correction methods have previously been specified in the literature [74]. In order to minimize the effects of common lead loss in ancient zircon (dated to >1000 Ma), $^{207}Pb/^{206}Pb$ age was used, and $^{206}Pb/^{238}U$ age was used for Zircon dated to <1000 Ma [75]. The error quoted for isotope ratios and ages (i.e., standard error) was 1σ. The GLITTER (ver. 4.4, Macquarie University, Balaclava Rd, Macquarie Park NSW 2109) and Isoplot (Ver. 4.15) [76] programs were used for data processing, age calculations, and concordia plots. The analytical data can be found in Supplementary Table S1.

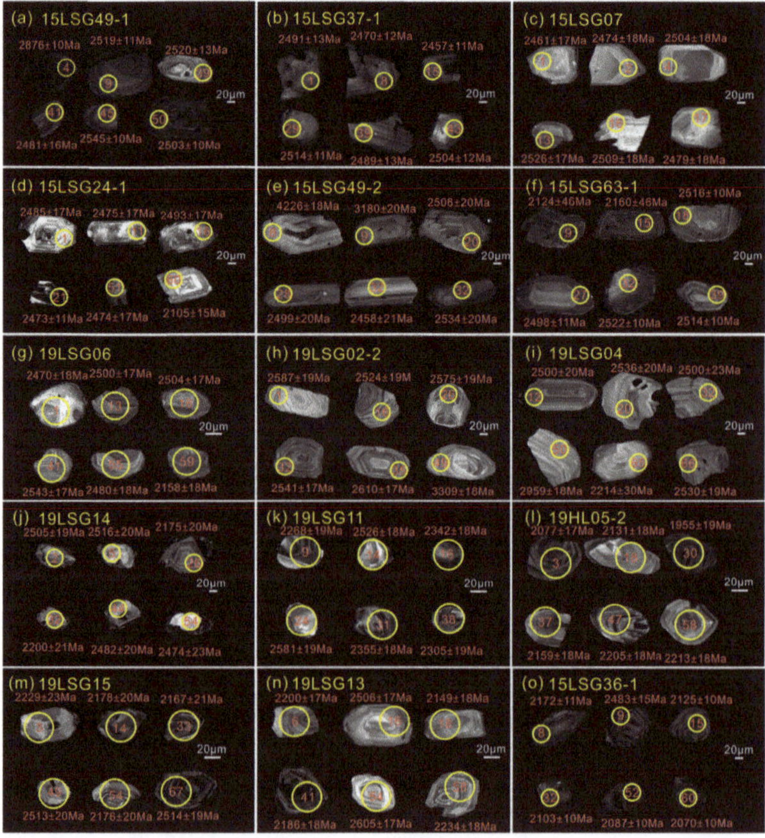

Figure 6. (**a–c**) Representative CL images of zircon from sedimentary rocks in Lower Langzishan Fm., (**d–k**) representative CL images of zircon from sedimentary rocks in Upper Langzishan Fm., (**l–o**) representative CL images of zircon from sedimentary rocks in Li'eryu Fm. Yellow circles indicate the experimental positions.

4.2. Results of the Detrital Zircon U-Pb LA-ICP-MS Ages

Due to the large number of samples, many of those displaying similar features and data in the main text are only briefly grouped and described. For detailed descriptions of individual samples, please refer to the Supplementary Materials.

In the Lower Langzishan Fm., a total of 520 zircons were analyzed (Supplementary Table S1), and these showed oscillatory zonings with Th/U ratios ranging from 0.17 to 4.17, indicating that they are typical magmatic zircons. Zircon grains from the quartz sandstone (15LSG49-1) in the bottom layer of the Langzishan Fm. were analyzed. Out of the 60 zircons analyzed, 50 yielded the ^{207}Pb/^{206}Pb age of 2433–2736 Ma, with an age peak at 2538 ± 13 Ma (Figures 6a and 7a). Zircon grains from pebbly quartz sandstones (15LSG37-1 and 19LSG07) from the Langzishan Fm. showed concordant data, with the ^{207}Pb/^{206}Pb ages ranging from 2190 to 3031 Ma and having an age peak at 2508 ± 17 Ma (Figure 6b,c and Figure 7b,c). Zircon grains from feldspathic quartz sandstones (15LSG24-1 and 15LSG49-2) had ^{207}Pb/^{206}Pb ages ranging from 2105 to 2564 Ma and 2368 to 2678 Ma, with age peaks at 2465 ± 16 Ma and 2525 ± 14 Ma, respectively (Figure 6d,e and Figure 7d,e). Zircon grains from metamorphic sandstone (15LSG63-1) showed the ^{207}Pb/^{206}Pb age range of 2457–2713 Ma, with a weighted average age of 2488 ± 13 Ma (Figures 6f and 7f). The siltstone (19LSG06) sample showed ^{207}Pb/^{206}Pb ages ranging from 2378 to 2622 Ma, with an age peak at 2516 ± 8.2 Ma (Figures 6g and 7g). Zircon grains from quartz sandstone (19LSG02-2 and 19LSG04) had the ^{207}Pb/^{206}Pb ages ranging from 2446 to 2747 Ma, with an age peak at 2563 ± 14 Ma (Figure 6h,i and Figure 7h,i).

A garnet mica schist (sample 19LSG14) and a siltstone (sample 19LSG11) were collected from the Upper Langzishan Fm., and the zircon grains from the schist showed ^{207}Pb/^{206}Pb ages ranging from 1945 to 2671 Ma, with a main age peak at 2505 ± 19 Ma and a secondary peak at 2185 ± 36 Ma (Figures 6j and 7j). Zircon grains from siltstone showed ^{207}Pb/^{206}Pb ages ranging from 2117 to 2581 Ma, with an age peak at 2341 ± 53 Ma (Figures 6k and 7k).

In the Li'eryu Fm., the analysis of zircon grains extracted from different samples of metamorphic sandstone (19IIL05, 19LSG15, 15LSG36-1) and siltstone (19LSG13) was conducted. The zircon grains were found to have short columnar or rounded morphology, with aspect ratios less than 2:1, and exhibited oscillatory zoning with Th/U ratios ranging from 0.11 to 1.20, indicating that they were typical magmatic zircons. The ^{207}Pb/^{206}Pb age of these zircon grains ranged from 1846 to 2270 Ma with age peaks at different times depending on the samples. The analysis also revealed older ages for some of the samples, ranging from 2479 to 2605 Ma (Figures 6i–o and 7i–o).

In summary, eleven samples from the Langzishan Fm. and four samples from the Li'eryu Fm. were systematically studied by using the detrital zircon LA-IC-MS U-Pb dating. A total of 709 effective data from a total of 894 analyses were chosen (90% < concordance < 110%) for further U-Pb age probability plots (Figure 8) and a discussion of the geological implications. Among them, the U-Pb age ranges of 433 zircons in the (pebbly) quartz sandstones, and the feldspathic quartz sandstones from the Lower Langzishan Fm., were 2105–4226 Ma, with the main age peak being at 2462 Ma. Besides, there were only 14 data distributed in the period 2105–2265 Ma (peak at ~2195 Ma) and 6 older data distributed in the period 2938–3309 Ma, barring one sample dated to 4226 Ma (Figure 8a). Going upward in the stratigraphic sequence, 75 valid data on the garnet mica schist in the Upper Langzishan Fm. were distributed from 1945 Ma to 2671 Ma, with the main peak being at 2458 Ma and a remarkable secondary peak being at 2179 Ma (Figure 8b). Moreover, 203 effective data (1846–2605 Ma) in the Li'eryu Fm. showed the main age peak being at 2160 Ma (Figure 8c). Only 1 younger age—1846 Ma (Th/U ratio 0.08)—was present, significantly indicating the range of metamorphic ages within the formation. Besides this, very few zircon grains discovered in the Li'eryu Fm. yielded U-Pb ages of 2506–2501 Ma ($n = 3$).

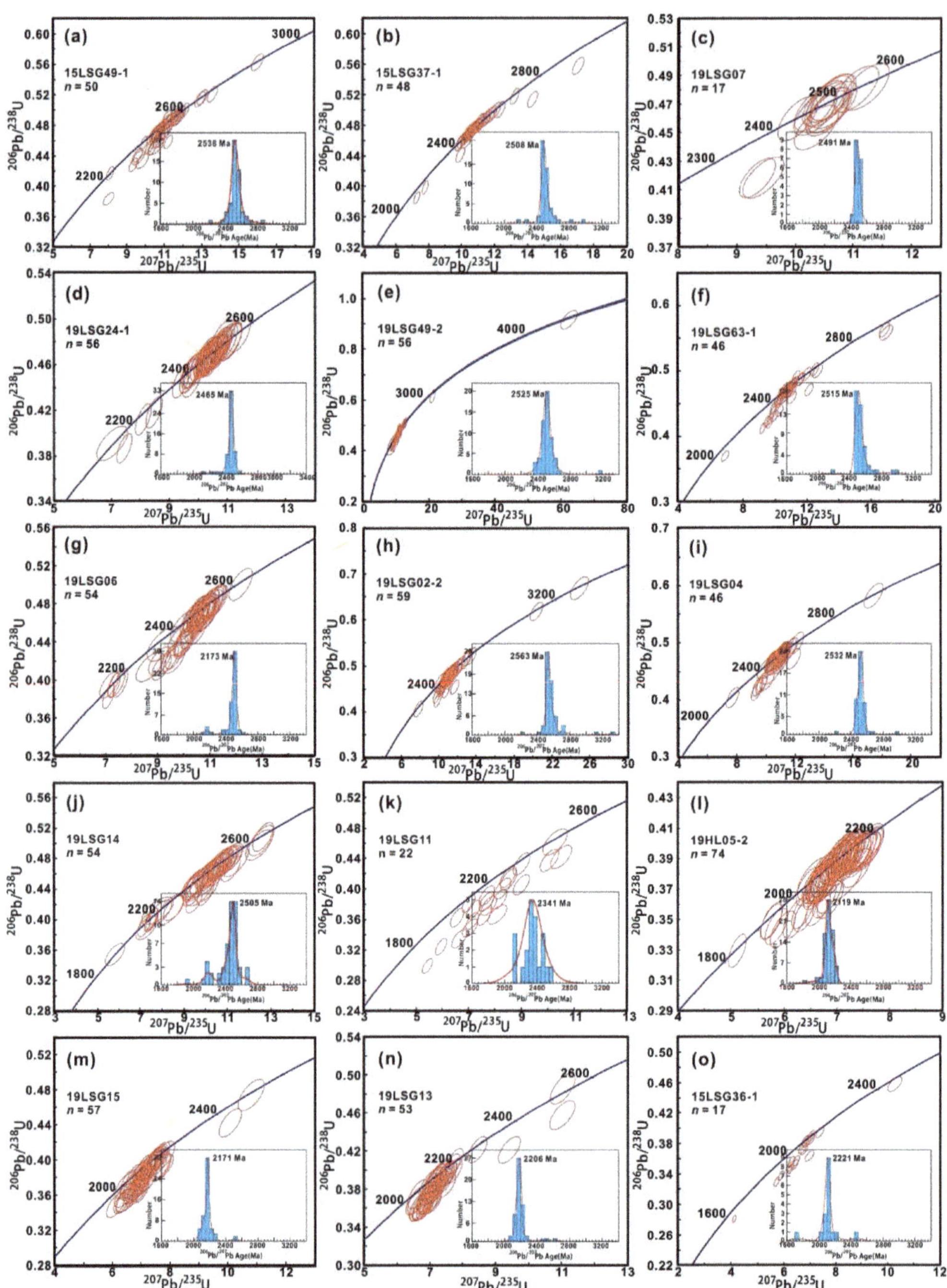

Figure 7. 206Pb/ 235U vs. 207Pb/ 238U diagrams and U-Pb age histograms (sub-figure) for the U-Pb detrital zircon analyses of sedimentary rocks from the Lower Langzishan Fm. (**a–c**), the Upper Langzishan Fm. (**d–k**) and the Li'eryu Fm. (**l–o**) of the North Liaohe Group.

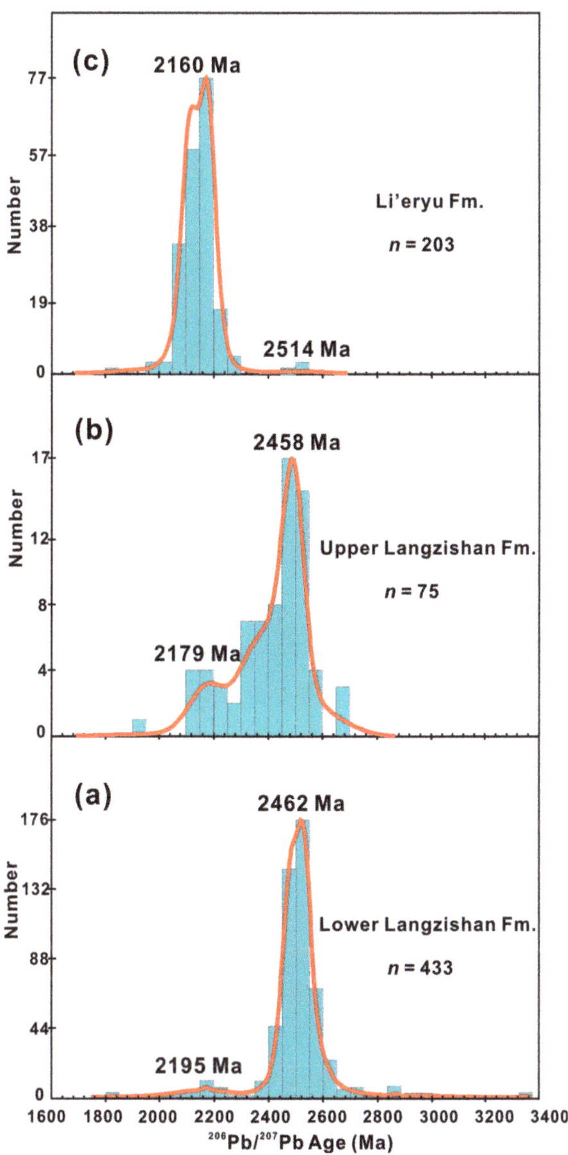

Figure 8. (**a**) U-Pb ages histograms of detrital zircons from metasedimentary rocks within the Lower Langzishan Fm., (**b**) U-Pb ages histograms of detrital zircons from metasedimentary rocks within the Upper Langzishan Fm., (**c**) U-Pb ages histograms of detrital zircons from metasedimentary rocks within the Li'eryu Fm.

5. Discussion

5.1. Provenance Analysis

The Langzishan and Li'eryu formations are characterized by two distinct age peaks, with ages of approximately 2.5 Ga and 2.2 Ga, respectively. Granitoids dated to ~2.5 Ga are widely developed, and represent the ages of the crystallized basement of the NCC [28,35,44,77–85]. The proportion of zircons aged ~2.2 Ga increases significantly in the upper layer of the Langzishan and Li'eryu formations, a finding that is consistent with what has been noted in a large number of 2.1–2.2 Ga magmatic rocks (the Liao-Ji granites) developed in the JLJB [20,24,25,35,37–39,48,56,86–89]. This result is in agreement with the results obtained by Liu et al. [90] from the dating of aluminum-rich metamorphic sedimentary rocks in the Liaohe Group (the main peak age is 2.05–2.15 Ga, and the secondary peak age is 2.45–2.55 Ga). The provenance of the Li'eryu Fm. has primarily been derived from the Liao-Ji granites, as determined through the analysis of peak ages and Lu-Hf isotopes, in a manner consistent with that of previous studies [44,83]. Although some detrital zircon U-Pb ages fall within the 2.8–3.3 Ga range, much like the ages of Eoarchean and middle Archean zircons found in the Anshan-Benxi area of NE China [91–93], no clastic zircons dating earlier than 3300 Ma have been identified. This suggests that unlike what happened in the Anshan-Benxi area, Eoarchean granites were not exhumed and deposited in the Langzishan Fm. in the Lianshanguan area.

Significantly, a few detrital zircons in the Langzishan Fm. have ages ranging from 2650 Ma to 2800 Ma, with a peak age of 2680 Ma (Figure 7b–d). Regional magmatism that is consistent with these ages is rare in the area and adjacent areas. This indicates two possibilities. The first is that the magmatic rocks of approximately 2680 Ma in the Langzishan Fm. and its adjacent areas have undergone weathering and removal during later tectonic thermal events. Secondly, recent discoveries have shown that some Neoarchean granites in the Anshan-Benxi area contain circa 2.7 Ga-inherited zircons and a small number of clastic zircons with an age of 2680 Ma, possibly originating from inherited zircons in the Neoarchean granites in the Benxi area [64]. Furthermore, only one older age—of approximately 4.2 Ga—is present, a finding which may be consistent with Hadean zircons documented in the eastern NCC [94,95], although the geological significance of these zircons is still unclear.

The detrital zircons provide a means of examining the sources of sediments, as their trace element compositions can be analyzed (Figure 9) [96]. The detrital zircons analyzed in the study showed a positive correlation between U and Y concentrations, suggesting derivation from acidic or intermediate acidic igneous rocks such as granite, syenite, and pegmatite. A few zircons from throughout the stratigraphy fall within the field of mafic rocks (Figure 9a,b). The Eu/Eu* distribution range (0.1–1) is more concentrated than that of Ce/Ce* (1–100) in the zircons from the study area, suggesting derivation from granites, nepheline-syenite pegmatites, and pegmatites. The Y concentrations and Yb/Sm display a linear distribution (Figure 9c), suggesting derivation from intermediate-silicic igneous plutons, such as granite, syenite pegmatites, and larvikites, with some zircons derived from mafic rocks. The positive correlation between the Nb and Ta concentrations of these zircons (Figure 9d) suggests derivation from granites, syenites, and mafic rocks, while a few zircons may have originated from carbonatites. Overall, Figure 9 suggests a common derivation for these zircons from intermediate-silicic igneous plutons. The felsic Archean basement of the NCC and the Paleoproterozoic Liao-Ji granites in the JLJB are likely provenances for the Langzishan Fm. and Li'eryu Fm., respectively.

Zircon trace element compositions can provide valuable insights into magmatic, metamorphic, and crustal processes and settings [96–98]. We used U/Yb versus Hf and Y diagrams to constrain their tectonic environments. The detrital zircons from both the Langzishan Fm. and the Li'eryu Fm. display similar characteristics when continental and oceanic crustal fields are plotted in, suggesting that they were derived from continental or mixed sources (Figure 10a,b). Additionally, some concordant detrital zircons have been plotted in the arc-related/orogenic fields in the Th/U-versus-Nb/Hf and Th/Nb-versus-Hf/Th diagrams, while a few zircons have been plotted in the within-plate/anorogenic field (Figure 10c,d).

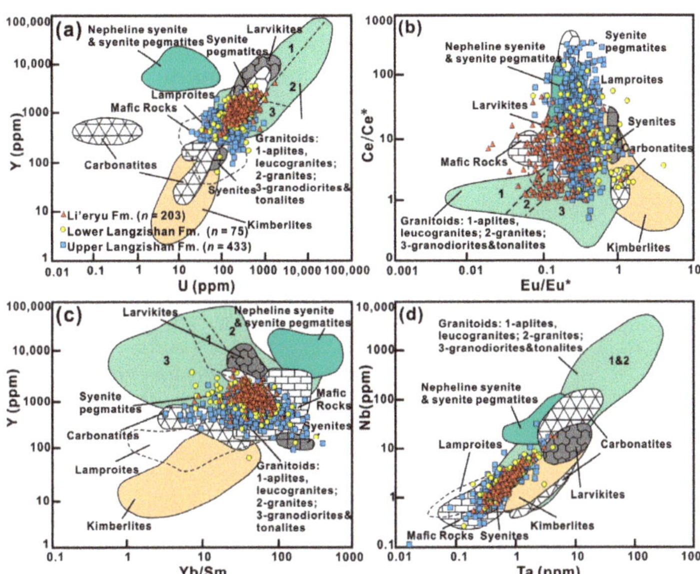

Figure 9. Zircon Y versus U (**a**), Ce/Ce* versus Eu/Eu* (**b**), Y versus Yb/Sm (**c**), and Nb versus Ta (**d**), with the images showing different compositions of zircons derived from different rock types (Belousova et al., 2002 [96]).

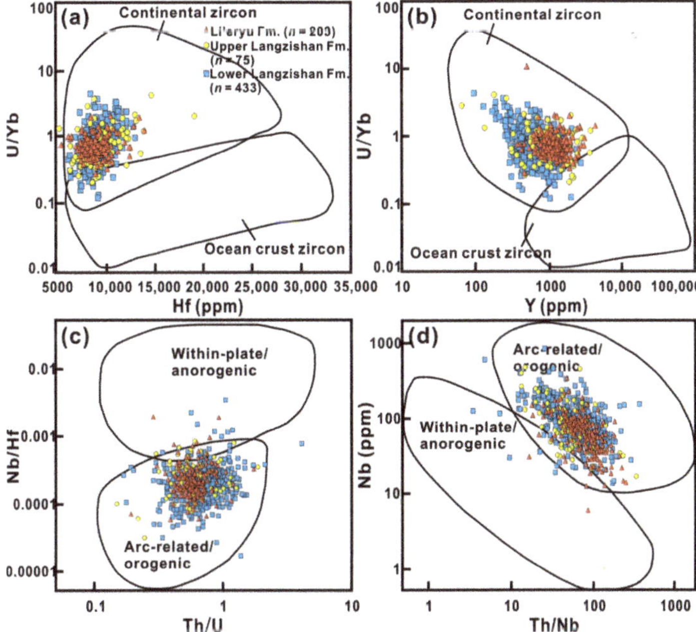

Figure 10. (**a**,**b**) Zircon Hf and Y versus U/Yb (Grimes et al., 2007 [98]), (**c**) Th/U versus Nb/Hf (Yang et al., 2012 [99]), and (**d**) Th/Nb versus Hf/Th (Yang et al., 2012 [99]).

In summary, it was found in this study that the detrital zircons in the study area were mainly derived from intermediate-silicic igneous rocks and formed in a tectonic setting of continental or mixed-continental-and-oceanic fields with arc-related/orogenic characteristics.

5.2. Depositional Age of the Langzishan and Li'eryu Formations

Due to the absence of volcanic rocks or pyroclastic components in the Langzishan Fm., its depositional age has long been a matter of debate. Some scholars argue that it formed between 2.05–1.93 Ga [12,44,48,84], while Xu et al. [27] propose an earlier depositional age of prior to 2.17 Ga. Based on zircon geochronology of aluminum-rich schist gneiss in the South and North Liaohe formations, Liu et al. [44] obtained an important age of ca. 1.95 Ga for the oldest metamorphic zircons in the Langzishan Fm. For the Li'eryu Fm., Wang et al. [88] reported a maximum depositional age from the South Liaohe Group based on the representative youngest individual zircon U-Pb ages of 2050 Ma. Furthermore, Xu et al. [27] compared the age distributions of detrital zircons and concluded that the deposition time of the Li'eryu Fm. was between 2.17–2.10 Ga. However, Chen et al. [100] concluded a geochronological study on acid volcanic rocks in the Liaoyang area and believed that there were three periods of magmatic activity in the Li'eryu Fm., i.e., 2190–2180 Ma, 2110–2100 Ma, and 1970–1960 Ma. Therefore, the previous deposition time of 2.05–2.1 Ga may represent the second period of Li'eryu Fm. magmatic activity, while the latest phase of magmatic activity should have been 1.95 Ga.

Over the past two decades, three methods have been commonly used to determine the depositional ages of sediments based on the youngest age populations in the detrital zircons. These methods include (1) using the age of the youngest detrital zircon as an older limit on deposition [101], (2) determining the peak ages of the youngest detrital zircons to represent the maximum age of the sedimentary rocks [101], and (3) calculating the weighted average ages of the three youngest detrital zircons [102]. For this study, we considered the primary detrital zircon data using the youngest age as the basis for examining the depositional age. Using Kernel Density Estimation (KDE) [103], the minimum crystallization age was found to be around 2113 ± 23 Ma ($n = 3$), indicating that the maximum depositional age of the Langzishan Fm. could not be earlier than 2136 Ma within 95% confidence. For the Li'eryu Fm., 203 valid data on detrital zircons were examined, and two ages of 1955 ± 19 Ma in primary magmatic zircons were identified in sample 19HL05. It was therefore concluded that there was a probability that the deposition time of the Li'eryu Fm. could not be earlier than 1974 Ma within 95% confidence. This finding is consistent with the results of previous studies based on the ages of interlayered volcanic rocks (~1.95 Ga) in the Li'eryu Fm. [100].

In summary, previous studies suggest an upper limit depositional age of ~1.95 Ga for the Liaohe Group, and based on the formation time of Liao-Ji granites (2.1–2.2 Ga), which are considered the primary sedimentary sources of the Liaohe Group, we propose a depositional age of 2.20–1.95 Ga for the Paleoproterozoic Liaohe Group. We also propose depositional time limits of ~2100 Ma for the Langzishan Fm. and ~1950 Ma for the Li'eryu Fm. based on our own experimental data. However, the sedimentary time limit of other strata has not been established in this study due to the lack of our own experimental data. Nonetheless, according to Wang et al. [88], the maximum depositional ages for the Gaojiayu, Dashiqiao, and Gaixian formations from the South Liaohe Group are 2069 Ma, 2043 Ma, and 1915 Ma, respectively, based on the representative youngest individual zircon U-Pb ages.

5.3. Sedimentary Similarities and Differences between the North and South Liaohe Groups

Luo et al. [44] proposed that the North and South Liaohe Groups formed simultaneously due to their similar provenances, based on a comparison of the U-Pb and Hf isotopic compositions of detrital zircons. In this study we combined our own findings with previous research to reveal differences in the sedimentary sequence and age distribution of detrital zircons in each formation of the Liaohe Group (Figure 11). For example, the South Liaohe Group has limited records on the Langzishan Fm., and there are virtually no records of the

Gaixian Fm. in the North Liaohe Group [19–21,40]. The Langzishan Fm. is only present in the North Liaohe Group, which could indicate two possibilities: it may not have been discovered in the South Liaohe Group, or it may have only been deposited in the northern margin of the LB during the Paleoproterozoic era [28,52]. Considering all the previous reports, we propose that the differences among the strata in the North and South Liaohe Groups could provide insights and evidence for the tectonic and evolutionary model of the Liaohe Group. This study collected a total of 5380 effective detrital zircon data, including 1232 data from the Langzishan Fm., 1159 data from the Li'eryu Fm., 1107 data from the Gaojiayu Fm., 642 data from the Dashiqiao Fm., and 1240 data from the Gaixian Fm. See Figure 11 for details on the geochronological data, which were grouped according to their exact locations for each formation.

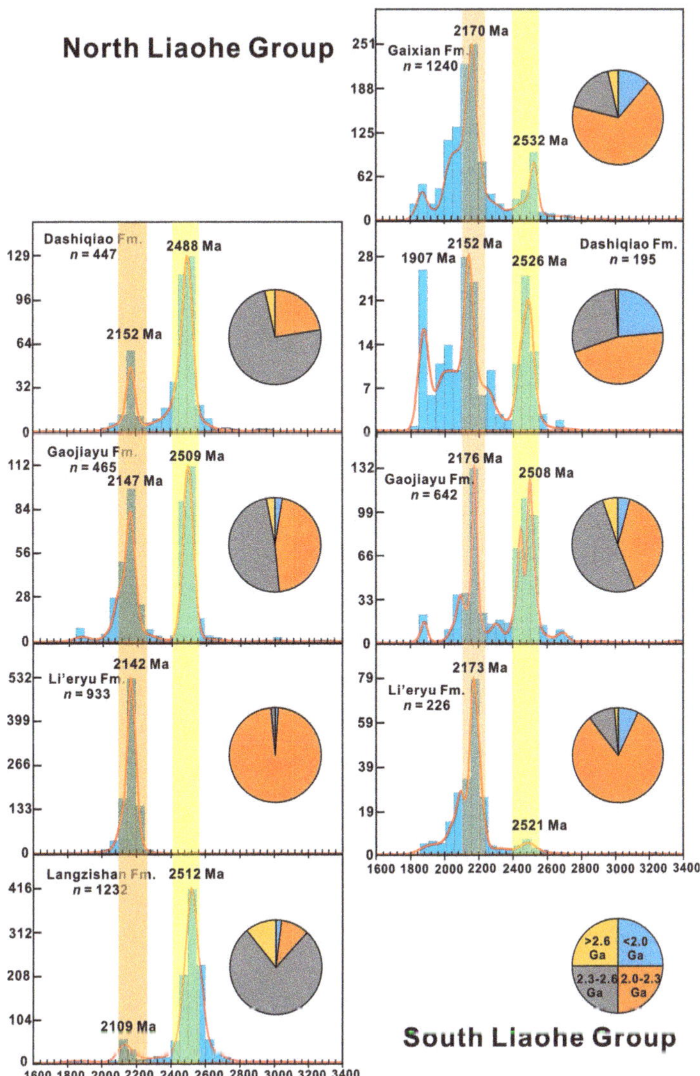

Figure 11. Collected U-Pb age distributions for the detrital zircons from (meta-)sedimentary rocks within the Liaohe Group (part of the data was collected from [28,35,44,52,83,104–113]).

Compared to what has been noted in other formations in the Liaohe Group, only the lower layers of the Langzishan Fm. contain detrital zircon ages from the Eoarchean to middle Archean eras, with a much higher proportion of Neoarchean (~2.5 Ga) ages than in other formations (Figure 11). Additionally, the Li'eryu Fm. has a relatively unimodal peak age (at ~2150 Ma) barring a few Archean ages (~2.52 Ga) that appear in the southern part. The detrital zircon ages show bimodal peaks at ~2.49–2.53 Ga and ~2.15–2.18 Ga for the Gaojiayu, Dashiqiao, and Gaixian formations. However, the Dashiqiao and Gaixian formations from the South Liaohe Group both have a younger secondary peak of ~1.91 Ga, which likely represents the metamorphic ages of these formations [114–116].

Figure 11 serves as a critical transitional record revealing that detrital zircon from the basement of the NCC vanishes in the Li'eryu Fm. (~2.5 Ga) but reappears in the upper strata. One possible explanation for the single peak age (~2.18–2.15 Ga) in the Li'eryu Fm. is based on the deposition of a substantial number of magmatic zircons (~2.2 Ga) from arc-related magmas (the Liao-Ji granites) [20,24,37,48,95,117,118]. During this period, intense magmatic activity accompanied the expansion and rapid deposition of a back-arc basin [37,48,95,116], possibly leading to the formation of a single age of provenance in the Li'eryu Fm. However, magmatic activity weakened during the deposition of the Gaojiayu, Dashiqiao, and Gaixian formations, indicating that the back-arc basin may have started to shrink. Consequently, the detrital zircon age distribution of these formations exhibits two peak ages originating from the Archean NCC basement (~2.5 Ga) and Paleoproterozoic Liao-Ji granites (~2.2 Ga).

It is worth noting that the detrital zircon ages from the Li'eryu Fm. of the South Liaohe Group exhibit two age peaks, i.e., the main peak of ~2.17 Ga and a secondary peak of ~2.52 Ga. In contrast, the age distribution from the Li'eryu Fm. in the North Liaohe Group has only a single peak of ~2.14 Ga (Figure 9). This significant difference in the age spectra suggests that the clastic materials in the Li'eryu Fm. have not been derived from the source area in the LB to the north, which lacks the records of the ~2.5 Ga Archean basement. This implies that the provenance of the Li'eryu Fm. may have been from the south. Moreover, the unimodal age spectra in the Li'eryu Fm. suggest a single provenance of ~2.1–2.2 Ga arc-affinity magmatites that were rapidly uplifted to the surface, likely forming a geomorphic feature that was high in the south and low in the north. The small amounts of clastics of ~2.5 Ga in age may be remnants of the ancient NCC's basement that are preserved in the magmatic arcs to the south. This finding, which further suggests that the Li'eryu Fm. might have undergone rapid deposition, is consistent with the hypothesis proposed by Wang et al. [104].

Furthermore, the correlation between the clastic zircon crystallization age (CA) and the sedimentary age of the host rock (DA) can constrain the tectonic environments of sedimentary rocks, based on differences in zircon formation and storage capacity in various tectonic settings [119]. Data from this study and from previous investigations of the Langzishan and Li'eryu formations were incorporated into the ^{206}Pb/^{207}Pb age map based on the accumulation ratio, which yielded depositional ages of 2136 Ma and 1974 Ma for the Langzishan and Li'eryu formations, respectively. As shown in Figure 12, the Lower Langzishan Fm. was represented by eight samples displaying CA-CD > 150 Ma at 5% of the zircon populations, indicating their deposition in a passive continental margin under extensional tectonic settings [119]. The Upper Langzishan Fm. was characterized by two samples showing CA-DA < 150 Ma at 5% of the zircon populations, and most of the data fell within the convergent basin. When combined with the age distribution characteristics of detrital zircons in the lower and upper layers of the Langzishan Fm., the results suggest that the sedimentary process of the Langzishan Fm. represents the transformation from a passive to an active continental margin. In contrast, all the samples from the Li'eryu Fm. showed CA-DA < 150 Ma at 5% of the zircon populations, and the main samples fell within the convergent basin, indicating that they were deposited on an active continental margin. This result is consistent with previous research results [27,109,113].

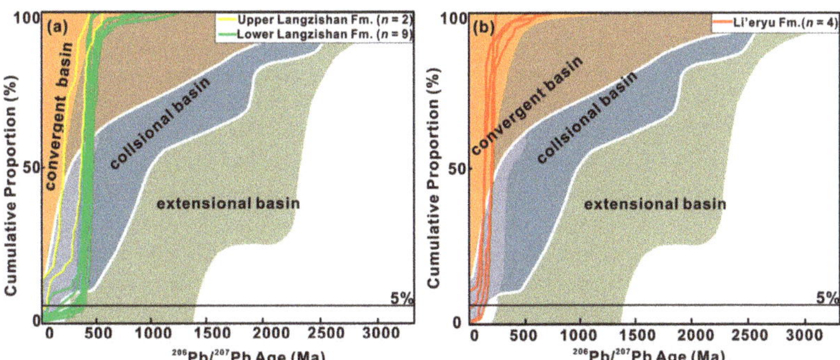

Figure 12. Detrital zircon ages vs. cumulative proportion diagrams for original basin settings of the Langzishan Fm. (**a**) and the Li'eryu Fm. (**b**) (Cawood et al., 2012 [119]).

5.4. Geological Significance and Sedimentary Background

As mentioned earlier, the tectonic evolution of the JLJB has been the subject of much debate, with three models proposed: (1) intra-continental rift opening and closure, (2) arc-continent collision, and (3) back-arc basin expansion and subsequent convergence [13,20,27–29,31–40,56]. Recently, models (2) and (3) have gained more support [27,30,36,37,40] despite ongoing controversy. The active continental margin tectonic model, first proposed by Bai et al. [33], posits that the LB and LNB represent two exotic blocks while the JLJB is an intervening island arc or back-arc basin. This viewpoint has been supported by various pieces of evidence, including: (1) studies of Paleoproterozoic Liao-Ji granites using geochronology and geochemistry, which revealed that the I- and A2-type ~2.2–2.0 Ga granites constituted a calc-alkaline series likely formed in the extensional tectonic setting of a continental arc [52,54,55,120] and/or back-arc basin [27,38,90,106,121–123]; (2) the reinterpretation of previously regarded (meta-)mafic rocks, which were believed to be examples of bimodal magmatism, together with the meta-volcanic rocks and Liao-Ji granites, that were developed in a rift environment as typical island arc basalts formed in a back-arc tectonic setting [15,24,68,124]; and (3) newly obtained detrital zircon U-Pb ages, combined with previous studies [27,104], showing unimodal age peaks of ~2.5 Ga for the Langzishan Fm. and ~2.15 Ga for the Li'eryu Fm., and bimodal age peaks at ~2.51 Ga and ~2.17 Ga for the Gaojiayu and Dashiqiao formations, respectively. Notably, these varying zircon age spectra are difficult to interpret in the context of intra-continental rift opening, particularly for the deposition of the Li'eryu Fm. In addition, the age patterns (ages vs. cumulative proportion; Figure 12) suggest that the depositional tectonic environment was a passive continental margin for the Langzishan Fm. and an active continental margin, interpreted as a back-arc basin environment [104], for the Li'eryu Fm.

In summary, based on a combination of previous findings and our own new results, the present tectonic scenario is more consistent with initial back-arc extension along an active continental margin followed by arc-continent collision for the Paleoproterozoic JLJB. We propose a rough evolutionary model (Figure 13) that encapsulates the following statements. (a) During the period of ca. > 2.2 Ga, the southeastern margin of the LB was a passive continental margin, especially during the depositional period for the Lower Langzishan Fm. The basement rocks of the LB to the north were the primary sources for the deposition of the Lower Langzishan Fm. (Figure 13a). (b) Subsequently, in ~2.2 Ga, oceanic subduction began beneath the southeastern margin of the LB, leading to the formation of arc-affinity magmatic rocks (the Liao-Ji granites). As a result, the depositional environment transformed into an active continental margin, and a small amount of subduction-related monzogranite gneiss (~2.2 Ga Liao-Ji granite) likely contributed to the clastics in the Upper Langzishan Fm. (Figure 13b). (c) With the continuous subduction of the oceanic crust, it is possible that an increasing subduction rate or angle of the oceanic plate led to vertical slab roll-back and the contemporaneous extension of the active continental margin [125]. In around 2.2–2.0 Ga, a back-arc basin was initially formed. Accompanied by lithospheric extension and mantle upwelling, the partial melting of ancient crustal material, together with minor contributions of juvenile mafic rocks, generated widespread A2-type granitoids in the JLJB [27,54,55,90,120–123]. During this time, a large volume of volcanic clastics (boron-bearing), and debris derived from the magmatic arc, were rapidly deposited, forming the Li'eryu Fm. in around 1.95 Ga (Figure 13c). Importantly, the single peak of the detrital zircon age spectrum probably indicates intensive subduction activity during the deposition of the Li'eryu Fm. After that, the Gaojiayu, Dashiqiao, and Gaixian formations were extensively deposited in the back-arc basin. The provenance from the LB gradually increased, supplying ancient (~2.5 Ga) basement materials together with clastics from the magmatic arc (~2.2–2.0 Ga Liao-Ji granites), possibly because the subduction became relatively slow or the back-arc basin began to shrink. Finally, in around 1.87–1.85 Ga [10,20,27,90], the back-arc basin was closed and formed the JLJB.

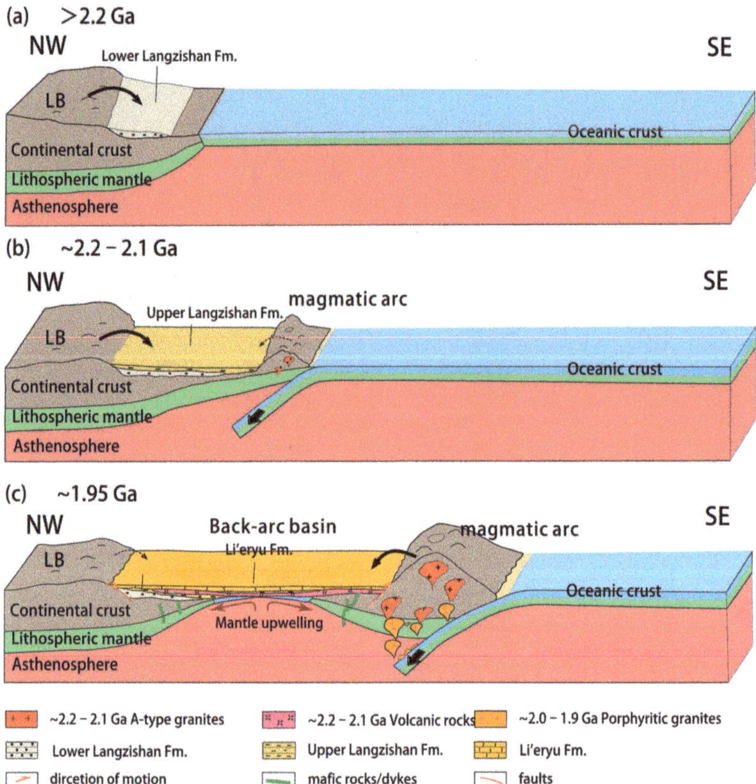

Figure 13. Schematic diagram of structural evolution of Liaohe Group. For the provenances of the Langzishan and Li'eryu formations, the ~3.3–2.5 Ga clastic materials in the sedimentary rocks of the lower Langzishan Fm. came from the erosion of the LB's basement on the NW side (**a**), minor amounts of ~2.2–2.1 Ga clastic materials of the upper Langzishan Fm. were derived from the erosion of a small amount of ~2.2–2.1 Ga granite on the SE-direction (**b**), and a large amount of 2.2–2.1 Ga clastic materials in the Li'eryu Fm. came mainly from the erosion of ~2.2–2.0 Ga Liao-Ji granites in the magmatic arc on the SE side (**c**).

6. Conclusions

Our study on zircon U-Pb geochronology and the trace element characteristics of sedimentary rocks from the Langzishan and Li'eryu formations of the North Liaohe Group of the JLJB of NCC allows us to add key constraints to the evolution of the Paleoproterozoic tectonics of the NCC. These include the following:

1. The depositional ages of the Langzishan Fm. and Li'eryu Fm. of the North Liaohe Group are 2136 Ma and 1976 Ma, respectively. The main provenances for these two formations are the Neoarchean basement rocks of the NCC (~2.6–2.4 Ga) and the Liao-ji granites (~2.2–2.0 Ga), respectively.
2. The sedimentary characteristics and geochronological evidence suggest that the depositional environment changed from a passive continental margin to an active continental margin during the deposition of the Upper Langzishan Fm.
3. The North and South Liaohe Groups exhibit similar sedimentary characteristics; however, some differences exist between the two, possibly due to variations in the local tectonic settings during deposition.
4. The North Liaohe Group was primarily supplied with clastic materials from the Archean basement rocks of the LB and a small amount of Paleoproterozoic Liao-Ji granites, which deposited the Langzishan Fm. The Li'eryu Fm. was rapidly deposited, likely due to intense magmatism and the erosion of the subduction-related magmatic arc (the Liao-Ji granites) in a back-arc basin environment. Finally, clastic materials from both the LB and Liao-Ji granites accumulated to form the Gaojiayu, Dashiqiao, and Gaixian formations.

Supplementary Materials: The following supporting information can be downloaded at: https://www.mdpi.com/article/10.3390/min13050708/s1, Figure S1: Supplementary thin-section micrographs of the Langzishan Fm. and the Li'eryu Fm. in the North; Table S1: Results of LA-ICP-MS U-Pb zircon dating for the Langzishan and the Li'eryu formatios from the North Liaohe Group in the JLJB.

Author Contributions: J.G.: Methodology, Formal analysis, Writing—Original Draft; W.L. (Corresponding author): Supervision, Conceptualization, Writing—Review & Editing; Y.L.: Supervision, Conceptualization; Y.Z.: Data Curation, Formal analysis; T.L.: Investigation; Q.W.: Data Curation. All authors have read and agreed to the published version of the manuscript.

Funding: This research was funded by the National Key R&D Program of China (Grant No. 2016YFC0666108-02) and National Natural Science Foundation of China (grant nos. 41872215 and 42172228).

Data Availability Statement: Not Applicable.

Acknowledgments: We are grateful to Yujie Hao, from the Key Laboratory of Mineral Resources Evaluation in Northeast Asia, Ministry of Land and Resources, Jilin University, for his help on zircon LA-ICP-MS U-Pb dating.

Conflicts of Interest: The authors declare no conflict of interest. The funders had no role in the design of the study, in the collection, analyses, or interpretation of data, in the writing of the manuscript, or in the decision to publish the results.

References

1. Sengör, A.M.C.; Natal'in, B.A.; Burtman, U.S. Evolution of the Altaid tectonic collage and Paleozoic crustal growth in Eurasia. *Nature* **1993**, *364*, 209–304. [CrossRef]
2. Xu, J.W.; Zhu, G. Tectonic models of the Tan-Lu fault zone, eastern China. *Geol. Rev.* **1994**, *36*, 771–784.
3. Windley, B.F.; Alexeiev, D.; Xiao, W.; Kröner, A.; Badarch, G. Tectonic models for accretion of the Central Asian Orogenic Belt. *J. Geol. Soc. Lond.* **2007**, *164*, 31–47. [CrossRef]
4. Xiao, W.; Li, S.Z.; Santosh, M.; Jahn, B.-M. Orogenic Belts in Central Asia: Correlations and connections. *J. Asian Earth Sci.* **2012**, *49*, 1–6. [CrossRef]
5. Santosh, M.; Liu, D.; Shi, Y.; Liu, S.J. Paleoproterozoic accretionary orogenesis in the North China Craton: A SHRIMP zircon study. *Precambrian Res.* **2012**, *227*, 29–54. [CrossRef]
6. Wilde, S.A.; Zhao, G.C.; Sun, M. Development of the North China Craton during the Late Archaean and its final amalgamation at 1.8Ga; some speculations on its position within a global Palaeoproterozoic Supercontinent. *Gondwana Res.* **2002**, *5*, 85–94. [CrossRef]
7. Wilde, S.A.; Cawood, P.A.; Wang, K.Y.; Nemchin, A.A. Granitoid evolution in the Late Archean Wutai Complex. North China Craton. *J. Asian Earth Sci.* **2005**, *24*, 597–613. [CrossRef]

8. Zhao, G.C.; Wilde, S.A.; Cawood, P.A.; Lu, L.Z. Thermal evolution of Archean basement rocks from the eastern part of the North China Craton and its bearing on tectonic setting. *Int. Geol. Rev.* **1998**, *40*, 706–721. [CrossRef]

9. Zhao, G.C.; Wilde, S.A.; Cawood, P.A.; Sun, M. Archean blocks and their boundaries in the North China Craton: Lithological, geochemical, structural and P-T path constraints and tectonic evolution. *Precambrian Res.* **2001**, *107*, 45–73. [CrossRef]

10. Zhao, G.C.; Sun, M.; Wilde, S.A.; Li, S. Late Archean to Paleoproterozoic evolution of the North China Craton: Key issues revisited. *Precambrian Res.* **2005**, *136*, 177–202. [CrossRef]

11. Zhao, G.C.; Cawood, P.A.; Li, S.Z.; Wilde, S.A.; Sun, M.; Zhang, J.; He, Y.H.; Yin, C.Q. Amalgamation of the North China Craton: Key issues and discussion. *Precambrian Res.* **2012**, *222–223*, 55–76. [CrossRef]

12. Zhao, G.C.; Cawood, P.A. Precambrian Geology of China. *Precambrian Res.* **2012**, *222–223*, 13–54. [CrossRef]

13. Zhao, G.C.; Zhai, M. Lithotectonic elements of Precambrian basement in the North China Craton: Review and tectonic implications. *Gondwana Res.* **2013**, *23*, 1207–1240. [CrossRef]

14. Kusky, T.; Li, J.; Santosh, M. The Paleoproterozoic North Hebei Orogen: North China craton's collisional suture with the Columbia supercontinent. *Gondwana Res.* **2007**, *12*, 4–28. [CrossRef]

15. Xu, W.; Liu, F.; Santosh, M.; Liu, P.; Tian, Z.; Dong, Y. Constraints of mafic rocks on a Paleoproterozoic back-arc in the Jiao-Liao-Ji belt, North China craton. *J. Asian Earth Sci.* **2018**, *166*, 195–209. [CrossRef]

16. Zhai, M.G.; Santosh, M. The early Precambrian odyssey of the North China Craton: A synoptic overview. *Gondwana Res.* **2011**, *20*, 6–25. [CrossRef]

17. Zhao, Y.; Zhang, P.; Li, Y.; Li, D.T.; Chen, J.; Bi, W.Z. Geochemistry of two types of Palaeoproterozoic granites, and zircon U–Pb dating, and Lu–Hf isotopic characteristics in the Kuandian area within the Jiao-Liao-Ji Belt: Implications for regional tectonic setting. *Acta Geol. Sin.* **2020**, *55*, 7564–7580. [CrossRef]

18. Kim, J.; Cho, M. Low-pressure metamorphism and leucogranite magmatism, north-eastern Yeongnam Massif. Korea: Implication for Paleopreoterozoic crustal evolution. *Precambrian Res.* **2003**, *122*, 235–251. [CrossRef]

19. Lee, B.C.; Oh, C.W.; Cho, D.L.; Yi, K. Paleoproterozoic (2.0–1.97 Ga) subduction related magmatism on the north–central margin of the Yeongnam Massif, Korean Peninsula, and its tectonic implications for reconstruction of the Columbia supercontinent. *Gondwana Res.* **2019**, *72*, 34–53. [CrossRef]

20. Li, S.Z.; Zhao, G.C. SHRIMP U-Pb zircon geochronology of the Liao-ji granitoids: 4, 397-403. Constraints on the evolution of the Paleoproterozoic Jiao-Liao-Ji belt in the Eastern block of the North China Craton. *Precambrian Res.* **2007**, *158*, 1–16. [CrossRef]

21. Liu, D.Y.; Nutman, A.P.; Compston, W.; Wu, J.S.; Shen, Q.H. Remnants of 3800 crust in the Chinese part of the Sino-Korean craton. *Geology* **1992**, *20*, 339–342. [CrossRef]

22. Liu, Y.; Li, S. Paleoproterozoic granite in Haicheng-Dashiqiao-Jidong area, Eastern Liaoning. *Liaoning Geol.* **1996**, *1*, 10–18.

23. Hu, G.W. The basic structural characteristics of the early Proterozoic Liaohe Group. *Bull. Tianjin Inst. Geol. Miner. Resour.* **1992**, *26*, 179–188, (In Chinese with English Abstract).

24. Meng, E.; Liu, F.L.; Liu, P.H.; Liu, C.H.; Yang, H.; Wang, F.; Shi, J.R.; Cai, J. Petrogenesis and tectonic significance of Paleoproterozoic meta-mafic rocks from central Liaodong Peninsula, northeast China: Evidence from zircon U-Pb dating and in situ Lu-Hf isotopes, and whole-rock geochemistry. *Precambrian Res.* **2014**, *247*, 92–109. [CrossRef]

25. Sun, M.; Armstrong, R.L.; Lambert, R.S.; Jiang, C.C.; Wu, J.H. Petrochemistry and Sr, Pb and Nd isotopic geochemistry of Palaeoproterozoic Kuandian Complex, the eastern Liaoning Province, China. *Precambrian Res.* **1993**, *62*, 171–190.

26. Wang, C.W.; Liu, Y.J.; Li, D.T. New evidences on the correlation of Liaohe Lithogroup between the southern and the northern regions in eastern Liaoning Province. *J. Chang. Univ. Sci. Technol.* **1997**, *1*, 17–24, (In Chinese with English Abstract).

27. Xu, W.; Liu, F.L. Geochronological and geochemical insights into the tectonic evolution of the Paleoproterozoic Jiao-Liao-Ji Belt, Sino-Korean Craton. *Earth Sci. Rev.* **2019**, *193*, 162–198. [CrossRef]

28. Li, S.Z.; Zhao, G.C.; Sun, M.; Han, Z.Z.; Hao, D.F.; Luo, Y.; Xia, X.P. Deformation history of the Paleoproterozoic Liaohe Group in the Eastern Block of the North China Craton. *J. Asian Earth Sci.* **2005**, *24*, 659–674. [CrossRef]

29. Li, S.Z.; Zhao, G.C.; Sun, M.; Han, Z.Z.; Zhao, G.T.; Hao, D.F. Are the South and North Liaohe Groups different exotic terranes Sm–Nd isotope constraints on the Jiao-Liao-Ji orogen. *Gondwana Res.* **2006**, *9*, 198–208. [CrossRef]

30. Li, S.Z.; Zhao, G.C.; Santosh, M.; Liu, X.; Dai, L.; Suo, Y.H.; Tam, P.K.; Song, M.; Wang, P.C. Paleoproterozoic structural evolution of the southern segment of the Jiao-Liao-Ji belt, North China Craton. *Precambr. Res.* **2012**, *200–203*, 59–73. [CrossRef]

31. Peng, Q.M.; Palmer, M.R. The Paleoproterozoic boron deposits in eastern Liaoning, China—a metamorphosed evaporite. *Precambrian Res.* **1995**, *72*, 185–197. [CrossRef]

32. Zhang, Q.; Yang, Z. *Early Crust and Mineral Deposits of Liaodong Peninsula*; Geological Publishing House: Beijing, China, 1988; (In Chinese with English Abstract).

33. Bai, J. *The Precambrian Geology and Pb-Zn Mineralization in the Northern Margin of North China Platform*; Geological Publishing House: Beijing, China, 1993; pp. 47–89, (In Chinese with English Abstract).

34. Faure, M.; Lin, W.; Monie, P.; Bruguier, O. Palaeoproterozoic arc magmatism and collision in Liaodong Peninsula (north-east China). *Terra Nova* **2004**, *16*, 75–80. [CrossRef]

35. Lu, X.P.; Wu, F.Y.; Guo, J.H.; Wilde, S.A.; Yang, J.H.; Liu, X.M.; Zhang, X.O. Zircon U–Pb geochronological constraints on the Palaeoproterozoic crustal evolution of the Eastern block in the North China Craton. *Precambrian Res.* **2006**, *146*, 138–164. [CrossRef]

36. Li, Z.; Chen, B. Geochronology and geochemistry of the Paleoproterozoic metabasalts from the Jiao-Liao-Ji Belt, North China Craton: Implications for petrogenesis and tectonic setting. *Precambrian Res.* **2014**, *255*, 653–667. [CrossRef]
37. Yuan, L.L.; Zhang, X.H.; Xue, F.H.; Han, C.M.; Chen, H.H.; Zhai, M.G. Two episodes of Paleoproterozoic mafic intrusions from Liaoning province, North China Craton: Petrogenesis and tectonic implications. *Precambrian Res.* **2015**, *264*, 119–139. [CrossRef]
38. Li, C.; Li, Z.; Yang, C. Palaeoproterozoic granitic magmatism in the northern segment of the Jiao-Liao-Ji Belt: Implications for orogenesis along the Eastern Block of the North China Craton. *Int. Geol. Rev.* **2017**, *60*, 217–241. [CrossRef]
39. Li, C.; Chen, B.; Li, Z.; Yang, C. Petrologic and geochemical characteristics of Paleoproterozoic monzogranitic gneisses from Xiuyan-Kuandian area in Liaodong Peninsula and their tectonic implications. *Acta Petrol. Sin.* **2017**, *33*, 963–977, (In Chinese with English Abstract).
40. Zhang, W.; Liu, F.L.; Cai, J.; Liu, C.H.; Liu, J.H.; Liu, P.H.; Liu, L.S.; Wang, F.; Yang, H. Geo-chemistry, zircon U-Pb dating and tectonic implications of the Palaeoproterozoic Ji'an and Laoling groups, northeastern Jiao-Liao-Ji Belt, North China Craton. *Precambrian Res.* **2018**, *314*, 264–287. [CrossRef]
41. Wan, Y.S.; Zhang, Q.D.; Song, T.R. SHRIMP ages of detrital zircons from the Changcheng System in the Ming Tombe area, Beijing: Constraints on the protolith nature and maximum depositional age of the Mesoproterozoic cover of the North China Craton. *Chin. Sci. Bull.* **2003**, *48*, 2500–2506.
42. Sircombe, K.N.; Freeman, M.J. Provenance of detrital zircons on the Western Australia coastline: Implications for the geologic history of the Perth Basin and denudation of the Yilgam Craton. *Geology* **1999**, *27*, 879–882. [CrossRef]
43. Wilde, S.A.; Valley, J.W.; Peck, W.H.; Graham, C.M. Evidence from detrital zircons for the existence of continental crust and oceans on the Earth 4.4 Gyr ago. *Nature* **2001**, *409*, 175–178. [CrossRef]
44. Luo, Y.; Sun, M.; Zhao, G.C.; Li, S.Z.; Ayers, J.C.; Xia, X.P.; Zhang, J.H. A comparison of U-Pb and Hf isotopic compositions of detrital zircons from the North and South Liaohe Groups: Constraints on the evolution of the Jiao-Liao-Ji Belt, North China Craton. *Precambrian Res.* **2008**, *163*, 279–306. [CrossRef]
45. Li, Z.; Chen, B.; Wei, C.J.; Wang, C.X.; Han, W. Provenance and tectonic setting of the Paleoproterozoic metasedimentary rocks from the Liaohe Group, Jiao-Liao-Ji Belt, North China Craton: Insights from detrital zircon U-Pb geochronology, whole-rock Sm-Nd isotopes, and geochemistry. *J. Asian Earth Sci.* **2015**, *111*, 711–732. [CrossRef]
46. Wang, X.J.; Liu, J.H.; Ji, L. Zircon U-Pb chronology, geochemistry and their petrogenesis of Paleoproterozoic monzogranitic gneisses in Kuandian area, eastern Liaoning Province, Jiao-Liao-Ji Belt, North China Craton. *Acta Petrol. Sin.* **2017**, *33*, 2689–2707.
47. Li, Z.; Chen, B.; Liu, J.W.; Zhang, L.; Yang, C. Zircon U-Pb ages and their implications for the South Liaohe Group inthe Liaodong Peninsula, Northeast China. *Acta Petrol. Sin.* **2015**, *31*, 1589–1605, (In Chinese with English Abstract).
48. Liu, F.L.; Liu, P.H.; Wang, F.; Liu, C.H.; Cai, J. Progresses and overviews of voluminous meta-sedimentary series within the Paleoproterozoic Jiao-Liao-Ji orogenic/mobile belt, North China Craton. *Acta Petrol. Sin.* **2015**, *31*, 2816–2846, (In Chinese with English Abstract).
49. Lu, X.P.; Wu, F.Y.; Lin, J.Q.; Sun, D.Y.; Zhang, Y.B.; Guo, C.L. Geochronological successions of the early Precambrian granitic magmatism in southern Liaoning Peninsula and its constraints on tectonic evolution of the North China Craton. *Chin. J. Geol.* **2004**, *39*, 123–139, (In Chinese with English Abstract).
50. Li, S.Z.; Yang, Z.S.; Liu, Y.J. Paleoproterozoic tectonic framework of the eastern North China Craton. *J. Changchun Univ. Sci. Tech.* **1995**, *25*, 14–21, (In Chinese with English Abstract).
51. Yang, Z.J.; Wang, W.; Zhao, Y.; Zhou, Y.H.; Zhang, J.; Sun, S.L.; Liu, C.C. Geochemistry and zircon U- Pb- Hf isotopes of Paleoproterozoic granitic rocks in Wangjiapuzi area, eastern Liaoning Province, and their geological significance. *Geol. Bull. Chin.* **2019**, *38*, 603–618, (In Chinese with English Abstract).
52. Dong, Y.; Bi, J.H.; Xing, D.H.; Ge, W.C.; Yang, H.; Hao, Y.J.; Ji, Z.; Jing, Y. Geochronology and geochemistry of Liaohe Group and Liaoji granitoid in the Jiao-Liao-Ji Belt, North China Craton: Implications for petrogenesis and tectonic evolution. *Precambrian Res.* **2019**, *332*, 105399. [CrossRef]
53. Li, S.Z.; Liu, Y.J.; Yang, Z.S.; Ma, R. Continental Dynamics and Regional Metamorphism of the Liaohe Group. *Geol. Rev.* **2001**, *47*, 9–18, (In Chinese with English Abstract).
54. Yang, M.C.; Chen, B.; Yan, C. Petrogenesis of Paleoproterozoic gneissic granites from Jiao-Liao-Ji Belt of North China Craton and their tectonic implications. *J. Earth Sci. Environ.* **2015**, *37*, 31–51, (In Chinese with English Abstract).
55. Chen, B.; Li, Z.; Wang, J.L.; Zhang, L.; Yan, X.L. Liaodong Peninsula ~2.2 Ga magmatic event and its geological significance. *J. Jilin Univ. (Earth Sci. Ed.)* **2016**, *46*, 303–320, (In Chinese with English Abstract).
56. Li, Z.; Chen, B.; Wei, C.J. Is the Paleoproterozoic Jiao-Liao-Ji Belt (North China Craton) a rift ntemational. *J. Earth Sci.* **2017**, *106*, 355–375.
57. Lan, T.G.; Fan, H.R.; Yang, K.F.; Cai, Y.C.; Wen, B.J.; Zhang, W. Geochronology, mineralogy and geochemistry of alkali-feldspar granite and albite granite association from the Changyi area of Jiao-Liao-Ji belt: Implications for Paleoproterozoic rifting of eastern north china craton. *Precambrian Res.* **2015**, *266*, 86–107. [CrossRef]
58. Teng, D.W.; Wang, Y.K.; Hao, X.J.; Liu, Z.H.; Zhu, K. Petrogenesis of Liao-ji Granites in Yongdian area of Liaoning and their constraints on tectonic evolution of Liao-Ji Mobile Belt. *Glob. Geol.* **2017**, *36*, 1105–1115, (In Chinese with English Abstract).
59. Liu, Y.; Li, S.; Bai, L. Structural evolution of ductile decollement zone occurred between Archaean gneiss complexes and Proterozoic Liaohe group in Haicheng area, Eastern Liaoning. *J. Changchun Univ. Earth Sci.* **1996**, *2*, 166–171.

60. Qu, H.; Zhang, Y.; Lei, G.; Wang, Z.; Cheng, P.; Zhang, F. On the Paleoproterozoic crust-mantle mixed complex in East Liaoning. *Land Resour.* **2000**, *3*, 199–205.

61. Chen, S.; Huan, Y.; Bing, Z. Characteristics of Paleo-Proterozoic intrusive rocks and continental dynamic evolution pattern of tectonomagma in east Liaoning. *Liaoning Geol.* **2001**, *1*, 43–51, (In Chinese with English Abstract).

62. Hao, D. *The Origin of Paleoproterozoic Granites in the Eastern Liaoning and Southern Jilin Provinces and the Crustal Evolution*; Ocean University of China: Qingdao, China, 2004.

63. Wu, F.Y.; Zhang, Y.B. Genesis of zircon and its constraints on interpretation of U-Pb age. *Chin. Sci. Bull.* **2004**, *49*, 1554–1569. [CrossRef]

64. Wan, Y.S.; Ma, M.Z.; Dong, C.Y.; Xie, H.Q.; Xie, S.W.; Ren, P.; Liu, D.Y. Widespread late Neoarchean reworking of Meso- to Paleoarchean continental crust in the Anshan-Benxi area, North China Craton, as documented by U–Pb–Nd–Hf–O isotopes. *Am. J. Sci.* **2015**, *315*, 620–670. [CrossRef]

65. Wan, Y.S.; Dong, C.Y.; Peng, R.; Bai, W.Q.; Xie, H.Q.; Liu, S.J. Spatial and temporal distribution, compositional characteristics and formation and evolution of Archean TTG rocks in the North China Craton: A synthesis. *Acta Petrol. Sin.* **2017**, *33*, 1405–1419.

66. Wan, Y.S.; Dong, C.Y.; Xie, H.Q.; Xie, S.W.; Liu, S.J.; Bai, W.Q.; Ma, M.Z.; Liu, D.Y. Formation Age of BIF-Bearing Anshan Group Supracrustal Rocks in Anshan-Benxi Area: New Evidence from SHRIMP U-Pb Zircon Dating. *Earth Sci.* **2017**, *43*, 57–81, (In Chinese with English Abstract).

67. Wen, F.; Tian, Z.H.; Liu, P.H.; Xu, W.; Liu, F.L.; Mitchell, N.R. Sedimentary evidence for the early Paleoproterozoic tectono-magmatic lull: Detrital zircon provenance of the 2.47-2.17 Ga Langzishan Formation, Liaohe Group, Eastern Block of the North China Craton. *J. Asian Earth Sci.* **2021**, *221*, 104939. [CrossRef]

68. Wu, D.; Liu, Y.J.; Li, W.M.; Chang, R.H. Ductile shear zone and uranium mineralization in the Lianshanguan area of eastern Liaoning uranium metallogenic belt. *Acta Petrol. Sin.* **2020**, *36*, 2571–2588, (In Chinese with English Abstract).

69. Wang, H.C.; Lu, S.N.; Chu, H.; Xiang, Z.Q.; Zhang, C.J.; Liu, H. Zircon U-Pb age and tectonic setting of meta-basalts of Liaohe Group in Helan area, Liaoyang, Liaoning Province. *J. Jilin Univ. (Earth Sci. Ed.)* **2010**, *41*, 1322–1334, (In Chinese with English Abstract).

70. Wu, D. *Early Precambrian Tectonic Evolution and Uranium Mineralizationin Lianshanguan Area, Eastern Liaoning Province*; Jilin University: Changchun, China, 2021; (In Chinese with English Abstract).

71. Yang, C.W.; Zhang, J.; Zhao, L.F.; Zhang, C.; Yang, H.X.; Lu, T.J. Detrital zircon constraints on tectonic evolution of the liaodong paleoproterozoic orogenic belt, north china craton. *Precambrian Res.* **2021**, *362*, 106152. [CrossRef]

72. Zhao, L.; Zou, H.; Qu, H.; Lv, X.; Hou, J. Preliminary study on the regional correlation of Liaohe group in eastern Liaoning. China. *Geol. Resour.* **2011**, *20*, 265–267, (In Chinese with English Abstract).

73. Zhai, F.; Liang, S.; Guo, H. The original formation-based discussion on Liaohe Group's environment. *Contrib. Geol. Miner. Resour. Res.* **2016**, *31*, 550–554, (In Chinese with English Abstract).

74. Andersen, T. Correction of common lead in u_pb analyses that do not report 204Pb. *Chem. Geol.* **2002**, *192*, 59–79. [CrossRef]

75. Blank, L.P.; Kamo, S.L.; Williams, I.S.; Mundil, R.; Davis, D.W.; Korsch, R.J.; Foudoulis, C. The application of SHEIMP to Phanerozoic geochronology: A critical appraisal of four zircon standards. *Chem. Geol.* **2003**, *200*, 171–188.

76. Ludwig, K.R. *ISOPLOT 3.00, A Geochronology Toolkit for Microsoft Excel*; Berkeley Geochronological Center Special Publication: Berkeley, CA, USA, 2003.

77. Diwu, C.R.; Sun, Y.; Guo, A.L.; Wang, H.L.; Liu, X.M. Crustal growth in the North China Craton at ~2.5 Ga: Evidence from in situ zircon U-Pb ages, Hf isotopes and whole-rock geochemistry of the Dengfeng Complex. *Gondwana Res.* **2011**, *20*, 149–170. [CrossRef]

78. Geng, Y.S.; Shen, Q.H.; Ren, L.D. Late Neoarchean to Early Paleoproterozoic magmatic events and tectono-thermal systems in the North China Craton. *Acta Petrol. Sin.* **2010**, *26*, 1945–1966.

79. Geng, Y.S.; Du, L.L.; Ren, L.D. Growth and reworking of the early Precambrian continental crust in the North China Craton: Constraints from zircon Hf isotopes. *Gondwana Res.* **2012**, *21*, 517–529. [CrossRef]

80. Guan, H.; Sun, M.; Wilde, S.A.; Zhou, X.H.; Zhai, M.G. SHRIMP U-Pb zircon geochronology of the Fuping complex: Implications for formation and assembly of the North China Craton. *Precambrian Res.* **2002**, *113*, 1–18. [CrossRef]

81. Liu, F.; Guo, J.H.; Lu, X.P.; Diwu, C.R. Crustal growth at ~2. 5Ga in the North China Craton: Evidence from whole-rock Nd and zircon Hf isotopes in the Huai'an gneiss terrane. *Chin. Sci. Bull.* **2009**, *24*, 4704–4713.

82. Liu, J.H.; Liu, F.L.; Ding, Z.J.; Liu, P.H.; Wang, F. The zircon Hf isotope characteristics of ~2.5Ga magmatic event, and implication for the crustal evolution in the Jiaobei terrane, China. *Acta Petrol. Sin.* **2012**, *28*, 2697–2704.

83. Luo, Y.; Sun, M.; Zhao, G.C.; Li, S.Z.; Xu, P.; Ye, K.; Xia, X.P. LA–ICP–MS U–Pb zircon ages of the Liaohe Group in the Eastern Block of the North China Craton: Constraints on the evolution of the Jiao-Liao-Ji Belt. *Precambrian Res.* **2004**, *134*, 349–371. [CrossRef]

84. Jahn, B.M.; Liu, D.Y.; Wan, Y.S.; Song, B.; Wu, J.S. Archean crustal evolution of the Jiaodong Peninsula, China, as revealed by zircon SHRIMP geochronology, elemental and Nd-isotope geochemistry. *Am. J. Sci.* **2008**, *308*, 232–269. [CrossRef]

85. Zhao, G.C.; Wilde, S.A.; Cawood, P.A.; Sun, M. SHRIMP U-Pb zircon ages of the Fuping complex: Implications for late archean to paleoproterozoic accretion and assembly of the North China Craton. *Am. J. Sci.* **2002**, *302*, 191–226. [CrossRef]

86. Li, S.Z.; Zhao, G.C.; Santosh, M.; Liu, X.; Dai, L.M. Paleoproterozoi Tectonothermal Evolution and Deep Crustal Processes in the JiaoLiao-Ji Belt, North China Craton: A Review. *Geol. J.* **2011**, *46*, 525–543. [CrossRef]

87. Li, S.Z.; Dai, L.; Zhang, Z.; Zhao, S.; Guo, L. Precambrian geodynamics (III): General features of Precambrian geology. *Earth Sci. Front.* **2015**, *22*, 27–45.

88. Wang, F.; Liu, F.L.; Liu, P.H.; Cai, J.; Schertl, H.P.; Ji, L.S.; Tian, Z.H. In situ zircon U-Pb dating and whole-rock geochemistry of metasedimentary rocks from South Liaohe Group, Jiao-Liao-Ji orogenic belt: Constraints on the depositional and metamorphic ages, and implications for tectonic setting. *Precambrian Res.* **2017**, *303*, 764–780. [CrossRef]

89. Wang, X.P.; Peng, P.; Wang, C.; Yang, S.Y. Petrogenesis of the 2115Ma Haicheng mafic sills from the eastern North China Craton: Implications for an intra-continental rifting. *Gondwana Res.* **2016**, *39*, 347–364. [CrossRef]

90. Liu, F.; Liu, C.; Itano, K.; Iizuka, T.; Cai, J.; Wang, F. Geochemistry, U-Pb dating, and Lu-Hf isotopes of zircon and monazite of porphyritic granites within the Jiao-Liao-Ji orogenic belt: Implications for petrogenesis and tectonic setting. *Precambrian Res.* **2017**, *300*, 78–106. [CrossRef]

91. Wan, Y.S.; Liu, D.Y.; Song, B.; Wu, J.S.; Yang, C.H.; Zhang, Z.Q.; Geng, Y.S. Geochemical and Nd isotopic compositions of 3.8 Ga meta-quartz dioritic and trondhjemitic rocks from the Anshan area and their geological significance. *J. Asian Earth Sci.* **2005**, *24*, 563–575. [CrossRef]

92. Wan, Y.S.; Liu, D.Y.; Nutman, A.; Zhou, H.Y.; Dong, C.Y.; Yin, X.Y.; Ma, M.Z. Multiple 3.8–3.1 Ga tectono-magmatic events in a newly discovered area of ancient rocks (the Shengousi Complex), Anshan, North China Craton. *J. Asian Earth Sci.* **2012**, *54–55*, 18–30. [CrossRef]

93. Wu, F.Y.; Zhang, Y.B.; Yang, J.H.; Xie, L.W.; Yang, Y.H. Zircon U–Pb and Hf isotopic onstraints on the Early Archean crustal evolution in Anshan of the North China Craton. *Precambrian Res.* **2008**, *167*, 339–362. [CrossRef]

94. Cui, P.L.; Sun, J.G.; Sha, D.M.; Wang, X.J.; Zhang, P.; Gu, A.L.; Wang, Z.Y. Oldest zircon xenocryst (4.17 Ga) from the North China Craton. *Int. Geol. Rev.* **2013**, *55*, 1902–1908. [CrossRef]

95. Liu, P.H.; Liu, F.L.; Wang, F. In-situ U-Pb dating of zircons from high-pressure granulites in Shandong Peninsula Eastern China and its geological significance. *Earth Sci. Front.* **2011**, *18*, 33–54.

96. Belousova, E.A.; Griffin, W.L.; O'Reilly, S.Y.; Fisher, N.I. Igneous zircon: Trace element composition as an indicator of source rock type. *Contrib. Mineral. Petrol.* **2002**, *143*, 602–622. [CrossRef]

97. EI-Bialy, M.Z.; Ali, K.A. Zircon trace element geochemical constraints on the evolution of the Ediacaran (600–614 Ma) post-collisional DokhanV olcanics and Y ounger Granites of SE Sinal, NE Arabian-Nubian Shield. *Chem. Geol.* **2013**, *360*, 54–73. [CrossRef]

98. Grimes, C.B.; John, B.E.; Kelemen, P.B.; Mazdab, F.K.; Wooden, J.L.; Cheadle, M.J.; HanghØj, K.; Schwartz, J.J. Trace element chemistry of zircons from oceanic crust: A method for distinguishing detrital zircon provenance. *Geology* **2007**, *35*, 643–646. [CrossRef]

99. Yang, J.H.; Gawood, P.A.; Du, Y.S.; Huang, H.; Huang, H.W.; Tao, P. Large Igneous Province and magmatic arc sourced Permian-Triassic volicanogenic sediments in China. *Sediment. Geol.* **2012**, *261*, 120–131. [CrossRef]

100. Chen, J.S.; Xing, D.H.; Liu, M.; Li, B.; Yang, H.; Tian, D.X.; Yang, F.; Wang, Y. Zircon U-Pb chronology and geological significance of felsic volcanic rocks in the Liaohe Group from the Liaoyang area, Liaoning Province. *Acta Petrol. Sin.* **2017**, *33*, 2792–2810.

101. Ludwig, K.R. User's Manual for Isoplot 3.70. In *A Geochronological Toolkit for Microsoft Excel*; Berkley Geochronology Centre Special Publication: Berkeley, CA, USA, 2008.

102. Tucker, R.T.; Roberts, E.M.; Hu, Y.; Kemp, A.I.S.; Salisbury, S.W. Detrital zircon age constraints for the Winton Formation, Queensland: Contextualizing Australia's Late Cretaceous dinosaur faunas. *Gondwana Res.* **2013**, *24*, 767–779. [CrossRef]

103. Song, Y.; Zhang, D.X.; Andrei, S.; Yuan, M.M.; Cong, X.R. Decomposition the detrital grain ages by Kernel Density Estimation and its applications: Determining the major tectonic events in the Songliao Basin, NE China. *Earth Sci. Front.* **2016**, *23*, 265–276, (In Chinese with English Abstract).

104. Wang, F.; Liu, F.; Schertl, H.P.; Xu, W.; Tian, Z. Detrital zircon U-Pb geochronology and hf isotopes of the Liaohe group, Jiao-Liao-ji belt: Implications for the Paleoproterozoic tectonic evolution. *Precambrian Res.* **2020**, *340*, 105633. [CrossRef]

105. Liu, J.; Zhang, J.; Liu, Z.; Yin, C.; Zhao, C.; Li, Z. Geochemical and geochronological study on the Paleoproterozoic rock assemblage of the Xiuyan region: New constraints on an integrated rift-and-collision tectonic process involving the evolution of the Jiao-Liao-Ji belt, North China Craton. *Precambrian Res.* **2018**, *310*, 179–197. [CrossRef]

106. Meng, E.; Liu, F.; Liu, P.; Liu, C.; Shi, J.; Kong, Q. Depositional ages and tectonic implications for South Liaohe group from Kuandian area in northeastern Liaodong Peninsula, northeast China. *Acta Petrol. Sin.* **2013**, *29*, 2465–2480.

107. Meng, E.; Wang, C.; Li, Y.; Li, Z.; Yang, H.; Cai, J.; Ji, L.; Jin, M. Zircon U–Pb–Hf 1278 isotopic and whole-rock geochemical studies of Paleoproterozoic metasedimentary rocks in 1279 the northern segment of the Jiao–Liao–Ji Belt, China: Implications for provenance and 1280 regional tectonic evolution. *Precambrian Res.* **2017**, *298*, 472–489. [CrossRef]

108. Qin, Y. *Geochronological Constraints on the Tectonic Evolution of the Liao-Ji 1297 Paleoproterozoic Rift Zone*; Jilin University: Changchun, China, 2013; (In Chinese with English Abstract).

109. Wang, F.; Liu, J.; Liu, C. Detrital zircon U-Pb geochronology of metasedimentary rocks from the Lieryu Formation of the South Liaohe Group in Sanjiazi area, the South Liaoning Province. *Acta Petrol. Sin.* **2017**, *33*, 2785–2791.

110. Wang, F.; Liu, F.L.; Liu, P.H.; Cai, J.; Ji, L.; Liu, L.S.; Tian, Z.H. Redefinition of the Gaixian formation of the South Liaohe group: Evidence from the detrital zircon U-Pb geochronology of metamorphosed sandstone in Huanghuadian-Suzigou area, the southern Liaoning Province. *Acta Petrol. Sin.* **2018**, *34*, 1219–1228.

111. Wang, H.C.; Ren, Y.W.; Lu, S.N. Stratigraphic Units and Tectonic Setting of the Paleoproterozoic Liao–Ji Orogen. *Acta Geosci. Sin.* **2015**, *36*, 583–598, (In Chinese with English Abstract).

112. Xu, W.; Liu, F.L.; Liu, P.H.; Tian, Z.H.; Cai, J.; Wang, W.; Ji, L. Paleoproterozoic transition in tectonic regime recorded by the Eastern Block of the North China Craton: Evidence from detrital zircons of the Langzishan Formation, Jiao-Liao-Ji Belt. *Int. Geol. Rev.* **2020**, *62*, 168–185. [CrossRef]

113. Xu, W.; Liu, F.L.; Wang, F.; Santosh, M.; Dong, Y.S.; Li, S. Palaeoproterozoic tectonic evolution of the Jiao-Liao-Ji Belt, North China Craton: Geochemical and isotopic evidence from ca. 2.17 Ga felsic tuff. *Geol. J.* **2020**, *55*, 409–424. [CrossRef]

114. Liu, P.H.; Liu, F.L.; Tian, Z.H.; Cai, J.; Ji, L.; Wang, F. Petrological and geochronological evidence for Paleoproterozoic granulite-facies metamorphism of the South Liaohe Group in the Jiao-Liao-Ji Belt, North China Craton. *Precambrian Res.* **2019**, *327*, 121–143. [CrossRef]

115. Wen, F.; Tian, Z.H. A metamorphic and deformational study of meta-pelites in the Liaohe Group located at Liaodong Peninsula: Significance to process of Paleoproterozoic orogenesis and exhumation. *Acta Petrol. Sinica* **2021**, *37*, 619–635.

116. Tian, Z.H.; Liu, F.L.; Windley, B.F.; Liu, P.H.; Wang, F.; Liu, C.H.; Wang, W.; Cai, J.; Xiao, W.J. Polyphase structural deformation of low-to medium-grade metamorphic rocks of the Liaohe Group in the Jiao-Liao-Ji Orogenic Belt, North China Craton: Correlations with tectonic evolution. *Precambrian Res.* **2017**, *303*, 641–659. [CrossRef]

117. Bi, J.H.; Ge, W.C.; Xing, D.H.; Yang, H.; Dong, Y.; Tian, D.X.; Chen, H.J. Paleoproterozoic meta-rhyolite and meta-dacite of the Liaohe Group, Jiao-Liao-Ji Belt, North China Craton: Petrogenesis and implications for tectonic setting. *Precambrian Res.* **2018**, *314*, 306–324. [CrossRef]

118. Lu, X.P.; Wu, F.Y.; Guo, J.H.; Yin, C.J. Late Paleoproterozoic granitic magmatism and crustal evolution in the Tonghua region, northeast China. *Acta Petrol. Sin.* **2005**, *21*, 721–736, (In Chinese with English Abstract).

119. Cawood, P.; Hawkesworth, C.J.; Dhuime, B.; Prave, T. Detrital zircon record and tectonic setting. *Geology* **2012**, *40*, 875–878. [CrossRef]

120. Li, Z.; Chen, B.; Wei, C. Hadean detrital zircon in the North China Craton. *J. Mineral. Petrol. Sci.* **2016**, *111*, 283–291. [CrossRef]

121. Zhu, K.; Liu, Z.; Xu, Z.; Wang, X.A.; Cui, W.; Hao, Y. Petrogenesis and tectonic implications of two types of Liaoji granitoid in the Jiao-Liao-Ji Belt, North China Craton. *Precambrian Res.* **2019**, *331*, 105369. [CrossRef]

122. Liu, P.H.; Cai, J.; Zou, L. Metamorphic P-T-t path and its geological implication of the Sanjiazi garnet amphibolites from the northern Liaodong Penisula, Jiao-Liao-Ji belt: Constraints on phase equilibria and Zircon U-Pb dating. *Acta Petrol. Sin.* **2017**, *33*, 2649–2674, (In Chinese with English Abstract).

123. Yang, C.; Liu, J.; Yang, H.; Zhang, C.; Feng, J.; Lu, T.; Sun, Y.; Zhang, J. Tectonics of the Paleoproterozoic Jiao-Liao-Ji orogenic belt in the Liaodong peninsula, North China Craton: A review. *J. Asian Earth Sci.* **2019**, *176*, 141–156. [CrossRef]

124. Xu, W.; Liu, F.; Tian, Z.; Liu, L.; Ji, L.; Dong, Y. Source and petrogenesis of Paleoproterozoic meta-mafic rocks intruding into the north Liaohe Group: Implications for back-arc extension prior to the formation of the Jiao-Liao-Ji belt, north china craton. *Precambrian Res.* **2018**, *307*, 66–81. [CrossRef]

125. Gerya, T. Future directions in subduction modeling. *J. Geodyn.* **2011**, *52*, 344–378. [CrossRef]

Article

Archean Crustal Evolution of the Alxa Block, Western North China Craton: Constraints from Zircon U-Pb Ages and the Hf Isotopic Composition

Pengfei Niu [1], Junfeng Qu [2,*], Jin Zhang [2], Beihang Zhang [2] and Heng Zhao [2]

[1] Chengdu Exploration and Development Research Institute of PetroChina Daqing Oilfield Co., Ltd., Chengdu 610000, China; niupf92@163.com

[2] Institute of Geology, Chinese Academy of Geological Sciences, Beijing 100044, China

* Correspondence: qujf@cags.ac.cn

Abstract: The Alxa Block is an important component of the North China Craton, but its metamorphic basement has been poorly studied, which hampers the understanding of the Alxa Block and the North China Craton. In this study, we conducted geochronological and geochemical studies on three TTG (tonalite–trondhjemite–granodiorite) gneisses and one granitic gneiss exposed in the Langshan area of the eastern Alxa Block to investigate their crustal evolution. The zircon U-Pb dating results revealed that the protoliths of the TTG and granitic gneisses were formed at 2836 ± 20 Ma, 2491 ± 18 Ma, 2540 ± 38 Ma, and 2763 ± 42 Ma, respectively, and were overprinted by middle–late Paleoproterozoic metamorphism (1962–1721 Ma). All gneiss samples had high Sr/Y ratios (41–274) and intermediate $Mg^{\#}$ values (44.97–55.78), with negative Nb, Ta, and Ti anomalies and moderately to strongly fractionated REE patterns $((La/Yb)_N = 10.6$–107.1), slight Sr enrichment, and positive Eu anomalies, displaying features of typical high-SiO_2 adakites and Archean TTGs. The magmatic zircons from the 2.84 Ga and 2.49 Ga TTG rocks had low $\varepsilon_{Hf}(t)$ values of -1.9–1.7, and -3.83–2.12 with two-stage model ages (T_{DMC}) of 3.24–3.11 Ga and 3.10–3.01 Ga, respectively, whereas those from the 2.54 Ga TTG rock exhibited $\varepsilon_{Hf}(t)$ values ranging from -1.1 to 3.46 and T_{DMC} from 3.0 Ga to 2.83 Ga, suggesting that the crustal materials of the basement rocks in the eastern Alxa Block were initially extracted from the depleted mantle during the late Paleoarchean to Mesoarchean era and were reworked in the late Mesoarchean and late Neoarchean era. By contrast, the Alxa Block probably had a relative younger crustal evolutionary history (<3.24 Ga) than the main North China (<3.88 Ga), Tarim (<3.9 Ga), and Yangtze (<3.8 Ga) Cratons and likely had a unique crustal evolutionary history before the early Paleoproterozoic era.

Keywords: Alxa Block; North China Craton; TTG rocks

Citation: Niu, P.; Qu, J.; Zhang, J.; Zhang, B.; Zhao, H. Archean Crustal Evolution of the Alxa Block, Western North China Craton: Constraints from Zircon U-Pb Ages and the Hf Isotopic Composition. *Minerals* **2023**, *13*, 685. https://doi.org/10.3390/min13050685

Academic Editors: Jin Liu, Jiahui Qian, Xiaoguang Liu and Alexandre V. Andronikov

Received: 24 March 2023
Revised: 4 May 2023
Accepted: 15 May 2023
Published: 17 May 2023

1. Introduction

Tonalite–trondhjemite–granodiorite (TTG) rocks constitute a major part of Archean continental crust and provide information about the composition, tectonic environment, and evolution of the early continental crust [1,2]. Studies have shown that the early Precambrian era was an important period of crustal growth, the continental crust formed between 3.0 Ga and 2.5 Ga accounted for 36% of the present continental crust, and the continental crust formed during 2.15–1.65 Ga accounted for 39% [3]. There are two views stating that Precambrian crustal growth was concentrated in three main stages: either 3.6 Ga, 2.7 Ga, and 1.8 Ga or 2.7 Ga, 1.9 Ga, and 1.2 Ga [3,4].

The China continent mainly consists of three early Precambrian nuclei, including the Yangtze Craton (YC), Tarim Craton (TC), and North China Craton (NCC) (Figure 1) [5–10]. In recent decades, major progress has been made in the reconstruction of the crustal growth history of the YC and TC [11–16]. Paleoarchean (3437–3262 Ma) TTG rocks have been found in the YC with two-stage Hf model ages from Hadean to Eoarchean [17–21].

Recently, Eoarchean (ca. 3.7 Ga) TTG rocks have been identified from the TC [22]. Additionally, the oldest detrital zircons from the basement rocks of the two cratons were dated 3.8–3.2 Ga with Eoarchean to Paleoarchean two-stage Hf model ages from the Eoarchean to Paleoarchean era [23–25], which suggest that the crustal evolution of the two cratons had already begun before the Paleoarchean era.

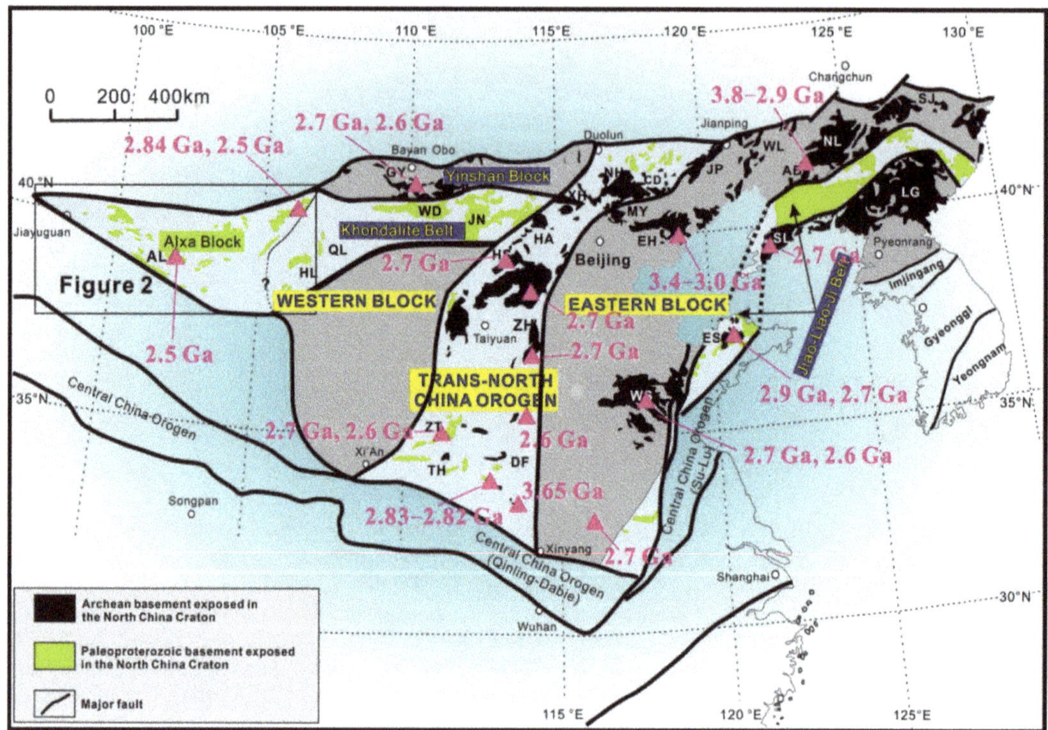

Figure 1. Distribution of Precambrian basement and subdivision of the North China Craton [26–28]. Additionally, shown are the locations of Archean TTGs with rock ages.

As one of the oldest cratons in China, the North China Craton (NCC) experienced a long and complicated geological history [29–32]. Most present models divide the NCC into the Eastern Block, Western Block, and Trans North China Orogen (TNCO) [27,33] (Figure 1). The early Precambrian tectonic pattern of the NCC remains controversial. Some scholars have believed that cratonization was completed ca. 2.5 Ga, marked by the "Wutai Movement" [34,35], followed by regional extension at the end of the Paleoproterozoic era that resulted in the destruction of the NCC (called activation) [36], while others have believed that the basement of the NCC had not been completely consolidated until ca. 1.9 Ga [37–40], and the "Lvliang Movement" ca. 1.8 Ga caused the Eastern Block and Western Block of the NCC to join together along the TNCO [7,41].

The Alxa Block is located in the westernmost part of the NCC. Compared with the main NCC and the YC and TC, the Precambrian basement of the Alxa Block is relatively less studied, which restricts a better understanding of the evolution of the NCC. In this paper, we report new geochronological and geochemical results for Meso-Neoarchean rocks from the Langshan area, which confirm the existence of the Archean basement in the Alxa Block and reveal evidence for the crustal evolution of the NCC.

2. Geological Background and Sample Descriptions

The Alxa Block, as the westernmost part of the NCC, is adjacent to the Central Asian Orogenic Belt in the north, the TC in the west, and the North Qilian Orogenic Belt in the south. The Precambrian basement of the Alxa Block is mainly exposed in the Longshoushan, Beidashan, Yabulaishan, Bayanwulashan, and Langshan areas (Figure 2), and the Bayanwula–Langshan Fault in the east is considered the eastern boundary of the Alxa Block [42,43]. The wide distribution of deserts and limited basement outcrops in the Alxa Block hamper the understanding of its tectonic pattern. The NE-oriented Langshan Mountains, located on the northeastern margin of the Alxa Block (Figure 2), are key areas in unravelling the early Precambrian geological evolution between the NCC and Alxa Block.

As the oldest basement in the Langshan area, the Diebusige Complex is mainly composed of banded biotite plagioclase gneiss, amphibolite gneiss, magnetite quartzite, marble, intrusive k-feldspar granite, and amphibolite. Some chronological studies have been conducted on the Diebusige Complex, but the formation age of the Diebusige Complex remains uncertain. Yang et al. (1988) [44] determined that the Rb-Sr isochron age of amphibolite was 3219 Ma, while Li et al. (2006) [45] obtained a Sm-Nd isochron age of 3081 ± 49 Ma, suggesting that the Diebusige Complex was formed in the Paleo-Mesoarchean era. The 2.75–3.5 Ga ages of detrital zircons and 2.5–2.69 Ga, 1.9–1.95 Ga, and 1.8–1.85 Ga ages of metamorphic zircons that Geng et al. (2006, 2007, 2010) [46–49] obtained from the Diebusige gneisses indicated that they were formed in the Neoarchean and underwent tectono-thermal events in the late Neoarchean and late Paleoproterozoic era. Dan et al. (2012) [50] suggested that the supracrustal rocks of the Diebusige Complex were deposited at 2.0–2.45 Ga and experienced metamorphism at 1.89 Ga and 1.79 Ga. Based on the chronology of the metamorphic basement, they believed that there was no Archean rock exposed in the eastern Alxa Block. However, Gong et al. (2012) [51] and Zhang et al. (2013) [52] found ca. 2.5 Ga TTG rock in the Beidashan area, providing evidence for the existence of exposed Archean rocks in the Alxa Block.

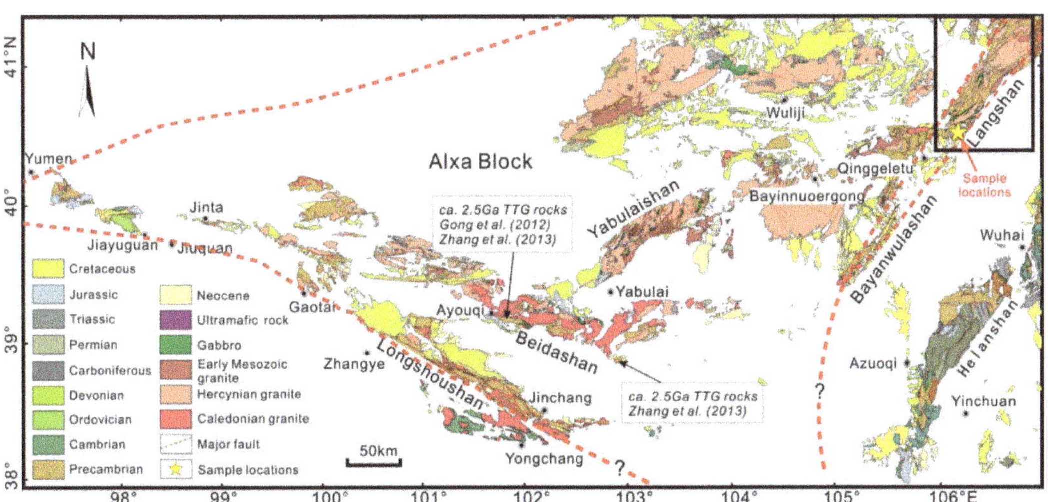

Figure 2. Geological map of the Alxa Block, westernmost North China Craton, and basement distribution of the Alxa Block. The ca. 2.5 Ga TTG rocks exposed in Beidashan area are from references [51,52]

Three TTG gneiss samples and one granitic gneiss sample were selected from the Diebusige Complex for this study; their detailed locations are shown in Table 1. All samples are gray–white, fine-to-medium grained, and generally show gneissic fabrics. Mineral grains are subhedral to anhedral, and some show serrated boundaries (Figure 3). The tonalitic gneiss (1810-1) mainly consists of plagioclase, quartz, hornblende, and minor biotite. Hornblende grains show recrystallization fronts, and quartz grains show irregular and crenulated margins, which suggests dynamic recrystallization. The trondhjemitic gneiss (1814-3) is mainly composed of plagioclase, quartz, hornblende, minor biotite, and K-feldspar. It was strongly affected by later deformation, and plagioclase and hornblende showed different degrees of fragmentation and alteration. Another tonalitic gneiss (1816-1) is characterized by a typical mineral assemblage of plagioclase, quartz, hornblende and biotite. Plagioclase grains show polysynthetic twinning, and quartz occurs as fine-grained and anhedral grains. Hornblende is the major dark mineral, and recrystallization is apparent in the surrounding area. The granitic gneiss (D798) is mainly composed of quartz, plagioclase, and biotite with minor accessory minerals of apatite, and mylonitization occurs under the superposition of later tectonic events.

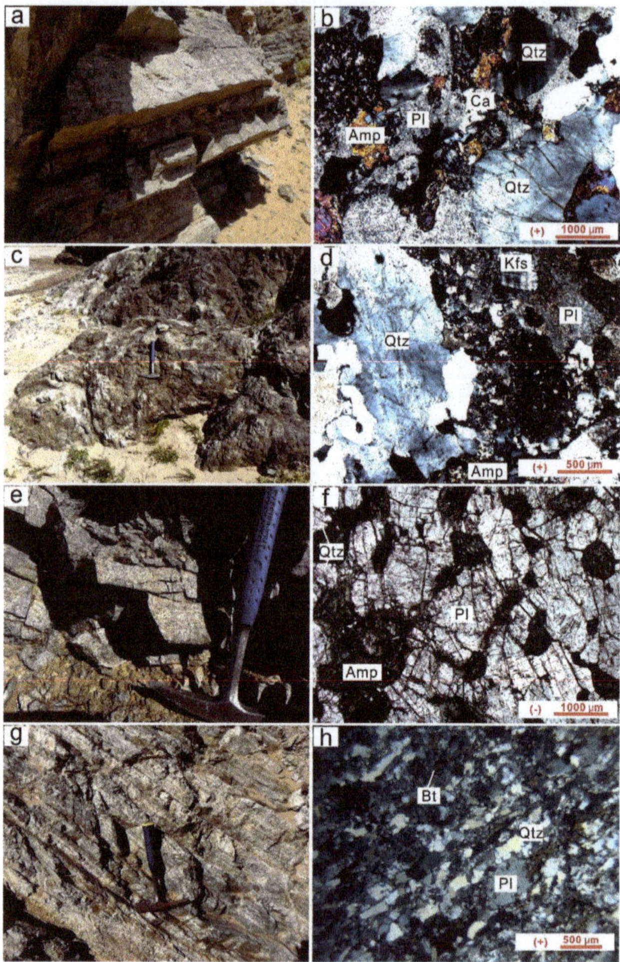

Figure 3. Field photos and representative photomicrographs of the TTG rocks and granitic samples from the Diebusige Complex, eastern Alxa Block. (**a**,**b**) Tonalitic gneiss sample 1810-1; (**c**,**d**) trondhjemitic gneiss sample 1814-3; (**e**,**f**) tonalitic gneiss sample 1816-1; and (**g**,**h**) granitic gneiss sample D798. Qtz: quartz; Pl: plagioclase; Bt: biotite; Kfs: K-feldspar; Amp: amphibole; and Ca: calcite.

Table 1. GPS locations and lithology of the representative samples.

Sample	GPS Location	Lithology	Mineral Assemblage
1810-1	40°35′55.15″ N 106°16′42.56″ E	Tonalitic gneiss	Qtz (25%) + Pl (65%) + Hb (5%) + Bt (5%)
1814-3	40°35′53.69″ N 106°16′53.29″ E	Trondhjemitic gneiss	Qtz (35%) + Pl (45%) + Hb (10%) + Bt (5%) + Kfs (5%)
1816-1	43°34′20.01″ N 106°12′29.44″ E	Tonalitic gneiss	Qtz (20%) + Pl (60%) + Hb (15%) + Bt (5%)
D798	40°35′19.14″ N 106°15′38.23″ E	Granitic gneiss	Qtz (30%) + Pl (55%) + Bt (15%)

3. Analytical Methods

3.1. Geochemistry

Whole-rock major and trace element analyses were completed at Wuhan Sample Solution Analytical Technology Co., Ltd. (Wuhan, China), and external standards and repeated samples were used to comprehensively control the analytical quality. Whole-rock major elements were measured using XRF, and five standards, BHVO-2, GSP-2, W-2A, GBW07103 and GBW07316, were determined in parallel. Trace elements were analyzed by inductively coupled plasma–mass spectrometry (ICP–MS) on an Agilent 7700e instrument with a shielded torch, and four standards, AGV-2, BHVO-2, BCR-2 and RGM-2, were used to monitor the analytical quality. The relative standard deviations for the whole-rock major and trace elements are within ±5%.

3.2. Zircon U-Pb Dating and Hf Isotopic Composition

SHRIMP zircon U-Pb dating was performed using a sensitive high-resolution ion microprobe (SHRIMP-II) at the Beijing SHRIMP Center, Institute of Geology, Chinese Academy of Geological Sciences, Beijing. The analytical procedure was the same as that of Williams. (1998) [53]. The primary flow intensity was 4.5 nA, and the spot size was 25–30 μm. Standard zircon TEM (417 Ma) was used for the age corrections [54]. Data processing was carried out using the ISOPLOT program [55], and uncertainties for individual analyses were quoted at 1σ, whereas those for weighted mean ages were quoted at a 2σ and 95% confidence level.

Zircon in situ Lu-Hf analyses were carried out using an NU plasma II MC-ICP–MS at the School of Earth and Space Sciences, Peking University. An ArF-excimer laser ablation system of Geolas HD (193 nm) was used with a 44 μm spot size. The analytical procedure was the same as that of Zhang et al. (2016) [56]. Data reduction was conducted using the IOLITE program [57]. Zircon 91,500 was used as an internal standard with a reference value of $^{176}Hf/^{177}Hf = 0.282307 \pm 31$ (2SD) [58], zircon Plešovice was used as the monitoring standard, and the value of $^{176}Hf/^{177}Hf = 0.282483$ (2SD) was obtained, which was consistent with the suggested value of 0.282482 ± 13 (2SD) [59].

4. Results

4.1. Whole-Rock Major and Trace Elements

Whole-rock major and trace element analyses were performed on four samples, including three TTG gneisses (1810-1, 1814-3, and 1816-1) and one granitic gneiss (D798). The analytical results are given in Table 2.

4.1.1. Major Element Geochemistry

The TTG gneisses from the Diebusige Complex in the Langshan area had SiO_2 contents of 60.55–77.80 wt.%, Al_2O_3 contents of 10.90–19.11 wt.%, and Na_2O/K_2O ratios of 0.93–6.38. In the normative An-Ab-Or diagram (Figure 4a), the samples 1810-1 (2.84 Ga) and 1816-1 (2.54 Ga) plotted in the tonalite field, except for sample 1814-3 (2.49 Ga), which plotted in the trondhjemite field. According to the A/NK vs. A/CNK classification (Figure 4b), the two tonalitic gneisses were weakly metaluminous, while the trondhjemitic gneiss was weakly peraluminous, which was in line with the TTG rocks with corresponding ages in the NCC. All TTG gneisses showed features of subalkaline series in the TAS diagram (Figure 4c). They mainly plotted in the medium-to-low K fields of the calc-alkaline and tholeiitic series in the K_2O vs. SiO_2 diagram (Figure 4d). In addition, they had $Mg^{\#}$ values ranging between 44.97 and 52.23, with an average of 49.96 (Table 2), slightly higher than those of Archean high-Al TTGs (42 on average).

Table 2. Analytical results of major (.wt%) and trace (ppm) elements for TTG and granitic gneisses in the eastern Alxa Block.

Sample	D798								1810-1	1814-3			1816-1		
Rock Type	Granitic Gneiss								Tonalitic Gneiss	Trondhjemitic Gneiss			Tonalitic Gneiss		
SiO_2	69.84	71.54	72.30	71.67	64.18	66.78	68.86	71.88	77.80	69.41	60.55	66.49	66.14	62.04	68.71
TiO_2	0.23	0.18	0.21	0.36	0.30	0.41	0.30	0.20	0.05	0.42	0.08	0.18	0.64	0.81	0.41
Al_2O_3	15.10	13.68	12.11	12.58	15.99	15.64	16.1	13.71	10.90	12.17	19.11	14.64	14.86	14.32	13.13
TFe_2O_3	1.71	1.48	1.60	2.61	2.90	3.28	2.06	1.72	1.39	4.66	4.06	4.80	5.26	7.73	4.32
MnO	0.03	0.03	0.03	0.04	0.05	0.05	0.03	0.02	0.02	0.07	0.06	0.07	0.09	0.11	0.07
MgO	1.09	0.71	0.76	1.26	1.55	1.64	0.88	0.88	0.76	2.54	2.24	2.56	2.17	3.21	2.37
CaO	1.30	1.87	2.33	2.13	2.61	1.96	2.99	1.12	3.73	1.79	1.50	1.73	4.19	5.00	4.94
Na_2O	3.89	3.05	2.85	4.14	3.68	4.13	5.35	3.28	2.81	4.03	6.77	4.46	3.66	3.44	3.61
K_2O	4.07	4.62	4.81	2.75	5.08	3.48	1.80	5.14	1.33	2.02	2.94	1.90	1.20	0.84	0.57
P_2O_5	0.02	0.01	0.07	0.08	0.28	0.10	0.13	0.01	0.06	0.07	0.13	0.08	0.14	0.11	0.14
LOI	2.56	2.66	2.74	2.22	3.03	2.32	1.17	1.77	1.47	2.60	2.30	2.90	1.46	2.19	1.61
total	99.84	99.83	99.80	99.84	99.64	99.77	99.74	99.72	100.32	99.77	99.73	99.81	99.81	99.80	99.87
Li	8.38	6.49	5.89	17.09	15.10	10.34	7.04	8.92	3.86	19.48	16.84	19.23	17.20	22.83	14.37
Be	0.84	0.72	0.50	1.28	0.93	1.08	1.18	0.75	3.66	19.48	16.84	19.23	1.06	1.07	14.37
V	20.76	21.38	22.70	48.02	33.44	40.58	30.73	19.20	14.24	40.84	25.75	36.85	79.20	147.61	66.26
Cr	10.99	15.67	9.71	25.59	26.64	18.78	11.87	9.86	8.93	34.20	55.55	29.44	111.93	201.04	42.63
Co	3.44	3.81	3.75	6.80	5.59	7.18	4.99	3.68	3.29	10.89	7.79	7.77	14.03	37.06	13.86
Ni	5.23	9.68	8.25	16.56	27.86	17.08	5.13	4.33	9.45	12.14	8.13	5.94	20.84	82.17	12.01
Cu	3.02	5.04	19.99	5.27	7.77	3.55	11.61	3.05	11.95	32.24	3.33	2.06	13.22	101.59	4.38
Zn	36.91	29.61	28.43	51.61	47.10	60.55	51.85	40.81	36.47	68.66	61.95	81.60	71.27	128.54	70.22
Ga	19.03	16.48	13.63	19.91	19.15	18.89	22.54	15.49	13.52	15.19	20.46	15.60	18.19	19.69	17.97
Rb	93.85	111.34	111.73	54.82	115.10	79.24	22.07	110.95	44.49	39.21	70.00	44.53	14.48	14.39	5.93
Sr	278.77	222.19	205.53	330.25	428.20	449.88	869.58	377.05	373.38	456.51	677.51	442.67	474.87	415.79	474.31
Zr	113.94	136.68	56.92	57.47	50.75	40.80	24.83	33.28	70.93	248.84	26.99	253.78	81.80	30.48	21.72
Nb	4.32	2.75	5.05	4.39	5.72	3.51	3.75	5.18	1.30	5.62	1.86	11.25	13.10	9.78	4.27
Sn	0.65	0.82	0.62	0.45	0.53	0.41	0.87	0.46	0.31	0.61	0.46	0.47	0.87	0.77	0.77
Cs	0.62	1.03	0.53	0.63	0.60	0.75	0.18	0.75	0.37	0.62	0.59	0.65	1.30	1.48	1.01
Ba	842.77	988.31	1338.22	856.66	2428.99	1319.21	1248.15	1951.79	320.81	1152.90	1509.41	768.83	772.23	579.27	371.54
La	18.41	14.32	19.41	14.81	34.00	19.66	25.73	11.03	14.10	19.33	26.71	16.84	27.88	18.44	16.61
Ce	27.52	21.03	26.58	21.87	57.76	27.93	45.31	14.38	23.64	28.00	42.47	27.73	47.66	30.38	27.16
Pr	2.61	1.88	2.64	2.21	6.03	2.65	4.90	1.10	2.03	2.64	4.04	2.70	5.19	3.57	3.35
Nd	8.45	5.89	8.95	7.79	21.12	8.71	17.59	3.01	5.50	8.58	13.57	9.39	18.77	13.46	12.59
Sm	1.23	0.71	1.40	1.24	3.19	1.26	2.45	0.31	0.59	1.02	1.75	1.29	2.88	2.00	2.12
Eu	0.75	0.84	0.89	0.92	1.79	1.20	0.98	1.00	0.22	1.15	1.62	1.19	1.18	1.36	1.04
Gd	0.89	0.54	1.19	1.04	2.41	1.02	1.56	0.23	0.41	0.82	1.29	1.04	2.55	1.85	1.90
Tb	0.12	0.07	0.17	0.15	0.33	0.14	0.18	0.04	0.07	0.11	0.17	0.14	0.38	0.27	0.28
Dy	0.48	0.43	0.75	0.76	1.48	0.70	0.60	0.16	0.32	0.49	0.72	0.68	2.03	1.51	1.50
Ho	0.10	0.11	0.14	0.16	0.27	0.14	0.09	0.03	0.06	0.11	0.14	0.15	0.42	0.33	0.31
Er	0.25	0.45	0.36	0.44	0.74	0.35	0.25	0.12	0.20	0.31	0.37	0.45	1.25	1.02	0.87
Tm	0.04	0.09	0.05	0.07	0.09	0.05	0.03	0.02	0.03	0.06	0.05	0.08	0.20	0.17	0.13
Yb	0.27	0.75	0.29	0.44	0.54	0.32	0.16	0.15	0.21	0.42	0.36	0.59	1.31	1.17	0.86
Lu	0.05	0.14	0.04	0.07	0.08	0.05	0.03	0.03	0.04	0.07	0.06	0.11	0.21	0.19	0.13
Y	2.95	3.46	4.35	4.68	7.84	3.84	3.18	1.75	2.09	3.23	3.54	4.05	11.57	8.70	8.33
Sc	2.47	3.73	1.63	6.32	3.36	5.26	2.69	2.68	2.23	4.77	4.41	4.95	11.37	17.85	9.83
Hf	3.99	4.34	1.70	1.67	1.68	1.24	0.74	1.13	1.89	7.30	0.84	6.55	2.34	1.62	0.67
Ta	0.45	0.38	0.44	0.60	0.49	0.29	0.31	0.35	0.04	0.75	0.79	2.94	2.74	0.80	0.68
Tl	0.53	0.58	0.55	0.36	0.55	0.39	0.14	0.51	0.29	0.26	0.34	0.25	0.13	0.11	0.07
Pb	23.65	21.86	20.55	17.63	22.81	18.43	19.37	20.22	15.96	7.48	15.20	5.78	9.76	10.42	7.81
Th	5.43	0.27	2.68	0.42	1.36	0.31	0.14	0.18	4.59	0.40	0.43	5.01	0.51	0.22	0.10

Minerals **2023**, 13, 685

Table 2. *Cont.*

Sample	D798								1810-1	1814-3			1816-1		
Rock Type	Granitic Gneiss								Tonalitic Gneiss	Trondhjemitic Gneiss			Tonalitic Gneiss		
U	0.53	0.37	0.19	0.17	0.50	0.14	0.07	0.09	0.65	0.31	0.09	0.51	0.26	0.26	0.06
$Mg^{\#}$	55.78	48.55	48.45	48.85	51.43	49.73	45.89	50.25	52.04	51.87	52.23	51.39	44.97	45.13	52.05
Eu_N/Eu_N^*	2.08	3.99	2.05	2.41	1.89	3.13	1.43	10.97	1.29	3.73	3.15	3.03	1.30	2.13	1.55
Sr/Y	94.66	64.29	47.27	70.51	54.65	117.03	273.54	215.45	178.30	141.29	191.28	109.33	41.03	47.80	56.91
La/Yb	68.20	19.02	66.24	33.57	62.62	61.23	158.81	76.08	67.36	46.47	75.03	28.49	21.23	15.70	19.32
Nb/Ta	9.55	7.25	11.56	7.33	11.58	12.05	12.17	14.96	30.37	7.52	2.36	3.83	4.78	12.20	6.25
Zr/Sm	92.33	192.51	40.72	46.46	15.89	32.33	10.13	108.04	119.91	243.96	15.44	196.73	28.40	15.21	10.24
Gd/Yb	3.31	0.71	4.06	2.35	4.44	3.18	9.62	1.61	1.98	1.96	3.62	1.76	1.94	1.58	2.21
Ce/Sr	0.10	0.09	0.13	0.07	0.13	0.06	0.05	0.04	0.06	0.06	0.06	0.06	0.10	0.07	0.06
$(La/Yb)_N$	45.98	12.82	44.66	22.63	42.22	41.28	107.07	51.29	45.41	31.33	50.58	19.21	14.31	10.59	13.02

$Mg^{\#} = 100 \times Mg/(Mg + Fe^{2+})$; $TFeO = TFe_2O_3 \times 0.8998$; $Eu_N/Eu_N^* = 2 \times Eu_N/(Sm_N + Gd_N)$; N: chondrite normalized; LOI: loss on ignition.

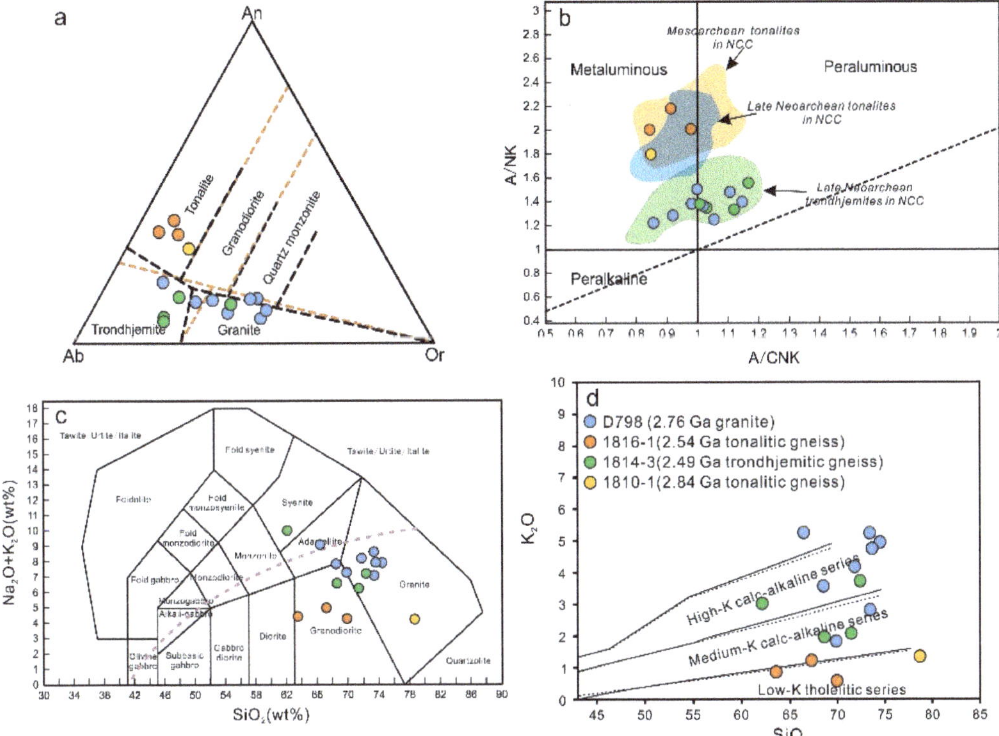

Figure 4. Geochemical discrimination diagrams for the TTG and granitic gneisses from the Diebusige Complex, eastern Alxa Block. (**a**) An-Ab-Or diagram of the gneiss samples after O'Conner (1965) and Barker (1979); (**b**) ANK vs. ACNK diagram; (**c**) SiO$_2$ vs. total alkali (Na$_2$O + K$_2$O) content diagram (Middlemost, 1994); and (**d**) SiO$_2$ vs. K$_2$O diagram.

The 2.76 Ga granitic gneiss (D798) had SiO$_2$ contents of 64.18–72.30 wt.% (69.63 wt.% on average), Al$_2$O$_3$ contents of 12.11–16.18 wt.% (14.37 wt.% on average), and Na$_2$O/K$_2$O ratios of 0.59–2.97 (1.15 on average). It displayed high-K calc-alkaline features in the K$_2$O vs. SiO$_2$ diagram (Figure 4d) and showed similar characteristics to the 2.5 Ga trondhjemitic gneiss (1814-3) in the TAS diagram and the A/NK vs. A/CNK diagram (Figure 4b,c). This rock had Mg$^{\#}$ values of 45.89–55.78 (49.87 on average), which are in accordance with the TTG gneisses (Table 2).

4.1.2. Trace Element Geochemistry

In the chondrite-normalized REE diagrams (Figure 5a,c), the TTG gneisses show similar characteristics to the TTG rocks in the NCC. Three TTG gneisses exhibit broadly similar REE distribution patterns and different LREE and HREE fractionation degrees. They all have positive Eu anomalies with Eu$_N$/Eu$_N$* > 1.29. The 2.84 Ga tonalitic gneiss (La$_N$/Yb$_N$ = 45.41) and the 2.54 Ga trondhjemitic gneiss (La$_N$/Yb$_N$ = 19.21–50.58, 32.06 on average) show high fractionation between LREEs and HREEs, while the 2.49 Ga tonalitic gneiss shows relatively lower fractionations (La$_N$/Yb$_N$ = 10.59–14.31, 12.64 on average).

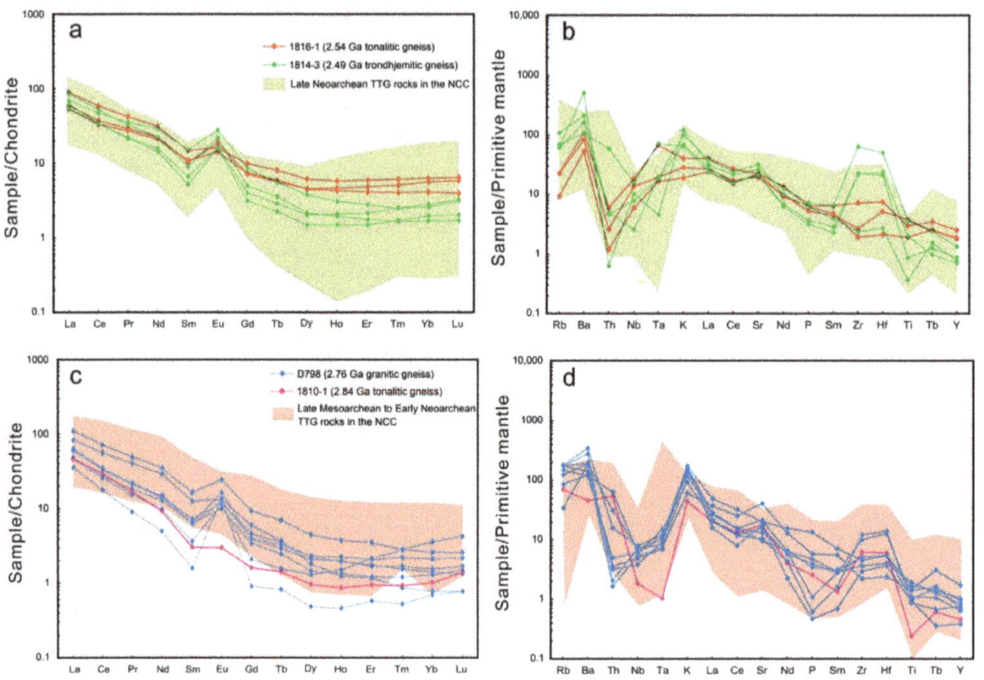

Figure 5. Chondrite-normalized REE patterns (**a**) and (**c**) and primitive mantle-normalized spider diagrams (**b**) and (**d**) for the TTG and granitic gneisses in the eastern Alxa Block.

In the primitive-normalized trace element diagrams (Figure 5b,d), three TTG gneisses show similar features in enrichment of LILEs (e.g., Rb, Ba, and Sr) and depletion of HFSEs (e.g., Nb, Ta, and Ti). They have variable contents of Cr and Ni. The 2.49 Ga tonalitic gneiss shows much higher Cr (118.5 ppm on average) and Ni (38.3 ppm on average) contents than the 2.54 Ga trondhjemitic gneiss (30.2 ppm and 7.8 ppm on average, respectively) and 2.84 Ga tonalitic gneiss (8.9 ppm and 9.4 ppm, respectively). The TTG gneisses and granitic gneisses are characterized by high Sr and low Y contents with high Sr/Y ratios (>41), analogous to average high-SiO2 adakites and Archean TTG [60].

The 2.76 Ga granitic gneiss exhibits characteristics similar to those of TTG gneisses in REE and trace element patterns (Figure 5c,d) and shows high REE fractionation (La$_N$/Yb$_N$ = 12.82–107.07, 45.99 on average) and distinctly positive Eu anomalies (Eu$_N$/Eu$_N$* = 1.43–10.97, 3.49 on average). It also shows concentrations of Rb, Ba, and Sr contents and depletions of Nb, Ta, and Ti contents and has low contents of Cr (16.1 ppm on average) and Ni (11.8 ppm on average) that are similar to 2.84 Ga tonalitic gneiss.

4.2. Zircon U-Pb Dating and Hf Isotopic Results

The zircon U-Pb dating results of the TTG and granitic gneiss samples (1810-1, 1814-3, 1816-1, and D798) are presented in Table 3, and representative zircon features are presented in Figure 6. All tested samples contain subhedral–euhedral zircon grains with near oval shapes and arc-shaped terminations. The diameter of zircons from samples 1810-1, 1814-3, 1816-1, and D798 are between 200 μm and 400 μm. Cathodoluminescence (CL) imaging of most zircons reveals core–mantle–rim textures of oscillatory zoned cores overprinted by broad (<80 μm) or thin mantle (<50 μm) and rim (<15 μm) domains (Figure 6). The oscillatory zoned zircon cores are characterized by lower CL brightness values than the rims. Overgrowth mantles and rims are commonly narrow in all samples, with rare grains that are bright gray, homogeneous, and internally structureless. We interpret the zircon cores to have a magmatic origin, with mantles and rims resulting from metamorphic recrystallization [61]. In situ zircon Hf isotope analyses were conducted on the representative zircons of the three TTG gneisses (Figure 6), and the results are listed in Table 3. For the 2.54 Ga tonalitic gneiss (1816-1), the Hf isotopic compositions of the six inherited zircon cores were calculated based on the weighted mean age of 2616 ± 11 Ma, while the other nineteen magmatic zircon cores were calculated based on the crystallization age of 2540 ± 38 Ma. Similarly, the Hf isotopic compositions of ten magmatic zircon cores from the 2.49 Ga trondhjemitic gneiss (1814-3) were calculated based on the crystallization age of 2491 ± 18 Ma.

4.2.1. Tonalitic Gneiss Sample 1810-1

Twenty-four analyses were obtained from the tonalitic gneiss sample (1810-1), and three analyses were discordant (spots 1.1, 7.1 and 11.1; Table 3). Two concordant analyses from zircon cores with (spots 6.1 and 15.3) well-preserved oscillatory zoning yield ^{207}Pb/^{206}Pb ages of 2826 ± 16 Ma and 2842 ± 13 Ma, respectively, with a weighted mean ^{207}Pb/^{206}Pb age of 2836 ± 20 Ma (MSWD = 0.63), which was proposed to be the crystallization age of the protolith (Figure 7a). In addition, there were two ^{207}Pb/^{206}Pb age groups from the inherited zircon cores that yielded mean ^{207}Pb/^{206}Pb ages of 2880 ± 17 Ma (MSWD = 0.04) and 2918 ± 8 Ma (MSWD = 0.80). The weighted mean ^{207}Pb/^{206}Pb ages of the two Paleoproterozoic age groups obtained from the unzoned rim domains were 1951 ± 12 Ma (MSWD = 0.95; 1962–1935 Ma) and 1867 ± 12 Ma (MSWD = 1.3; 1915–1843 Ma) (Figure 7a). We considered these two age groups, ca. 1.87 Ga and ca. 1.95 Ga, to represent the ages of metamorphic events [61].

Eight magmatic zircon cores from the 2.84 Ga tonalitic gneiss (1810-1) had ^{176}Hf/^{177}Hf ratios between 0.280948 and 0.281053 (Table 4), age-corrected $\varepsilon_{Hf}(t)$ values ranging from 1.89 to 1.71, with two-stage Hf model ages (T_{DMC}) of 3111–3242 Ma, respectively. Ten metamorphic zircon mantles or rims had relatively higher ^{176}Hf/^{177}Hf ratios between 0.281086 and 0.281181 and lower εHf(t) values from -18.90 to -13.49, with two-stage Hf model ages (T_{DMC}) of 3143–3336 Ma.

4.2.2. Trondhjemitic Gneiss Sample 1814-3

Of the twenty-seven analyses of zircons from the trondhjemitic gneiss sample (1814-3), fourteen from the oscillatory zoned cores yield ^{207}Pb/^{206}Pb ages ranging between 2191 ± 24 Ma and 2577 ± 21 Ma. Ten analyses (2443–2555 Ma) yield a weighted mean ^{207}Pb/^{206}Pb age of 2491 ± 18 Ma (MSWD = 0.99; Table 3; Figure 7b). Eleven analyses were obtained from the oscillatory unzoned rims and yield ^{207}Pb/^{206}Pb ages ranging between 1702 Ma and 1943 Ma, with a weighted mean ^{207}Pb/^{206}Pb age of 1834 ± 45 Ma (MSWD = 0.58; Figure 7b). Together with the intercept ages of 1819 ± 120 Ma and 2422 ± 59 Ma (MSWD = 0.66) (Figure 7b), we consider the mean ^{207}Pb/^{206}Pb age of 2491 ± 18 Ma obtained from the oscillatory zoned cores to represent the crystallization age of the trondhjemitic gneiss and the mean ^{207}Pb/^{206}Pb age of 1834 ± 45 Ma obtained from the unzoned rims to represent the age of metamorphism overprinted on the trondhjemitic gneiss [61].

Table 3. Zircon U-Pb isotopic data obtained by SHRIMP for TTG and granitic gneisses in the eastern Alxa Block.

Spot No	$^{206}Pb_c$ (%)	U ppm	Th ppm	Th/U	$^{206}Pb^*$ ppm	$^{207}Pb/^{206}Pb$	Error in %	$^{207}Pb/^{235}U$	Error in %	$^{206}Pb/^{238}Pb$	Error in %	Error corr	$^{207}Pb/^{206}Pb$ Age (Ma)	1σ	$^{206}Pb/^{238}U$ Age (Ma)	1σ	Discordant (%)
								D798: Granitic gneiss									
1.1	0.94	56	24	0.44	24.1	0.1831	2.7	12.56	3.2	0.4977	1.7	0.533	2681	45	2604	37	3
2.1	0.37	470	112	0.25	168	0.1600	0.51	9.14	1.2	0.4145	1.1	0.911	2456	8.6	2236	21	9
2.2	25.34	51	30	0.61	22.9	0.1610	14	7.70	15	0.3520	4.1	0.274	2446	240	1943	67	21
3.1	0.31	92	77	0.86	35.6	0.1609	1.3	9.98	1.9	0.4499	1.4	0.753	2465	21	2395	29	3
4.1	0.65	69	77	1.16	29.7	0.1848	1.7	12.69	2.3	0.4983	1.6	0.671	2696	28	2607	33	3
5.1	2.29	36	24	0.68	9.64	0.1087	6.5	4.52	6.9	0.3018	2.1	0.307	1777	120	1700	31	4
6.1	1.36	37	19	0.53	12.5	0.1381	5	7.41	6.4	0.3890	3.9	0.616	2202	87	2119	70	4
7.1	0.52	140	110	0.81	50	0.1599	1.2	9.10	1.8	0.4130	1.4	0.765	2454	20	2229	26	9
8.1	14.76	52	38	0.74	16.3	0.1010	17	4.19	17	0.3009	2.6	0.149	1643	320	1696	38	−3
9.1	0.69	74	31	0.43	27.4	0.1567	1.7	9.24	2.3	0.4278	1.6	0.695	2420	29	2296	31	5
10.1	3.97	31	15	0.51	12	0.1180	7.2	6.94	7.7	0.4270	2.4	0.317	1924	130	2293	46	−19
11.1	2.33	18	12	0.69	5.71	0.1160	5.9	5.68	6.5	0.3549	2.5	0.389	1895	110	1958	42	−3
11.2	10.14	53	14	0.27	10.9	0.1350	31	3.90	33	0.2108	4.5	0.139	2159	560	1233	44	43
12.1	1.16	55	45	0.84	15.3	0.1064	3	4.63	3.4	0.3159	1.7	0.497	1738	54	1769	26	−2
13.1	1.48	45	29	0.65	13.5	0.1133	3.5	5.34	4	0.3417	1.8	0.464	1852	64	1895	30	−2
14.1	18.37	31	17	0.56	8.15	0.0810	36	2.80	36	0.2480	3.6	0.098	1211	710	1428	44	−18
15.1	8.97	165	105	0.66	45.3	0.0990	13	3.91	13	0.2873	1.7	0.133	1597	230	1628	24	−2
16.1	2.34	46	20	0.45	14.7	0.1090	12	5.45	12	0.3616	2.5	0.201	1788	220	1990	40	−11
17.1	0.23	210	203	1.00	63.4	0.1234	1.2	5.97	1.9	0.3512	1.4	0.768	2005	21	1940	24	3
18.1	0.19	121	91	0.78	45.1	0.1730	0.81	10.31	1.6	0.4323	1.4	0.867	2587	14	2316	28	10
19.1	0.52	147	103	0.72	40.4	0.1073	1.4	4.70	1.9	0.3180	1.3	0.692	1753	25	1780	21	−2
20.1	0.11	177	146	0.85	81.2	0.1921	0.61	14.11	1.4	0.5326	1.3	0.900	2760	10	2752	28	0
21.1	0.16	224	222	1.03	90.9	0.1753	0.6	11.39	1.4	0.4712	1.2	0.900	2609	10	2489	25	5
22.1	0.26	170	1	0.00	81.4	0.1970	0.61	15.09	1.4	0.5557	1.3	0.902	2801	10	2849	29	−2
23.1	2.51	39	26	0.69	11.6	0.0956	4.7	4.41	5.1	0.3345	1.9	0.372	1539	89	1860	30	−21
24.1	0.67	140	102	0.75	39.8	0.1054	1.5	4.78	2	0.3286	1.3	0.669	1721	27	1832	21	−6
25.1	1.23	61	27	0.46	21.3	0.1531	2.8	8.40	3.3	0.3981	1.6	0.499	2380	49	2160	30	9
26.1	6.87	72	22	0.31	18	0.1220	11	4.47	12	0.2661	2.2	0.192	1985	200	1521	28	23
								1810–1: Tonalitic gneiss									
1.1	0.07	99	39	0.41	42.4	0.1849	0.78	12.68	1.8	0.4974	1.6	0.902	2697	13	2603	35	4
2.1	0.01	453	107	0.24	138	0.1186	0.85	5.80	1.4	0.3549	1.2	0.809	1935	15	1958	20	−1
3.1	0.01	47	102	2.23	13.5	0.1172	1.7	5.35	2.3	0.3310	1.6	0.702	1915	30	1843	26	4
3.2	0.03	307	176	0.59	88.8	0.1133	0.69	5.26	1.4	0.3368	1.2	0.872	1853	12	1871	20	−1
3.3	0.12	164	141	0.89	48	0.1158	0.94	5.43	1.6	0.3403	1.3	0.810	1892	17	1888	21	0
4.1	0.02	184	56	0.31	83.7	0.1935	0.57	14.09	1.4	0.5280	1.3	0.912	2772	9.3	2733	28	1
4.2	0.01	1102	3	0.00	337	0.1204	0.87	5.90	1.6	0.3557	1.4	0.846	1962	16	1962	23	0
5.1	0.10	52	49	0.98	14.8	0.1157	2	5.28	2.6	0.3312	1.7	0.640	1890	36	1844	27	2
6.1	0.01	338	168	0.51	160	0.2000	0.98	15.22	1.5	0.5521	1.2	0.773	2826	16	2834	27	0
7.1	0.01	771	548	0.74	370	0.2120	0.28	16.32	1.2	0.5584	1.1	0.970	2921	4.5	2860	26	2
7.2	0.09	76	28	0.38	39.7	0.2073	1.4	17.42	2.4	0.6090	1.9	0.803	2885	23	3068	47	−6
8.1	0.06	80	56	0.72	21.5	0.1156	1.3	4.95	1.9	0.3104	1.4	0.756	1890	23	1742	22	8
9.1	0.00	594	127	0.22	165	0.1141	0.56	5.08	1.3	0.3230	1.2	0.900	1866	10	1804	18	3
10.1	0.00	1373	743	0.56	531	0.1901	0.48	11.80	1.3	0.4503	1.2	0.925	2743	7.8	2397	23	13
11.1	0.07	91	36	0.41	40.5	0.2098	1.9	14.99	3.7	0.5180	3.2	0.854	2904	31	2691	70	7
12.1	0.04	123	48	0.40	62.2	0.2185	0.8	17.80	1.7	0.5908	1.5	0.875	2970	13	2993	35	−1
13.1	0.02	204	77	0.39	97.9	0.2101	0.69	16.17	1.6	0.5581	1.4	0.900	2906	11	2859	33	2
13.2	0.01	480	205	0.44	238	0.2067	0.56	16.45	1.6	0.5772	1.5	0.938	2880	9.1	2937	36	−2

Table 3. *Cont.*

Spot No	$^{206}Pb_c$ (%)	U ppm	Th ppm	Th/U	$^{206}Pb^*$ ppm	$^{207}Pb/^{206}Pb$	Error in %	$^{207}Pb/^{235}U$	Error in %	$^{206}Pb/^{238}Pb$	Error in %	Error corr	$^{207}Pb/^{206}Pb$ Age (Ma)	1σ	$^{206}Pb/^{238}U$ Age (Ma)	1σ	Discordant (%)
14.1	0.03	224	112	0.52	62.8	0.1127	0.95	5.07	1.7	0.3263	1.4	0.818	1843	17	1820	22	1
14.2	0.09	274	94	0.35	83.1	0.1187	0.86	5.78	1.6	0.3529	1.3	0.835	1937	15	1949	22	−1
15.1	0.09	94	36	0.40	26.8	0.1130	1.3	5.15	2	0.3304	1.5	0.760	1849	23	1840	24	0
15.2	0.02	775	11	0.01	239	0.1200	0.46	5.94	1.3	0.3588	1.2	0.931	1957	8.2	1976	20	−1
15.3	0.02	718	185	0.27	338	0.2020	0.78	15.25	2.5	0.5480	2.4	0.949	2842	13	2816	54	1
16.1	0.43	26	27	1.07	7.65	0.1148	2.5	5.42	3.2	0.3423	2	0.620	1876	46	1898	33	−1
							1814–3: Trondhjemitic gneiss										
1.1	0.19	31	31	1.05	12.8	0.1588	1.5	10.68	2.4	0.4879	1.8	0.773	2443	25	2561	38	−5
2.1	0.12	25	6	0.24	10.6	0.1661	1.7	11.43	2.6	0.4994	2	0.761	2518	28	2611	42	−4
3.1	0.02	84	86	1.06	34.7	0.1646	0.93	10.87	1.7	0.4788	1.4	0.837	2504	16	2522	30	−1
3.2	1.15	5	3	0.62	1.56	0.1172	6.3	5.29	7.4	0.3270	4	0.533	1914	110	1825	63	5
3.3	0.20	20	12	0.62	8.42	0.1698	2.8	11.44	3.6	0.4890	2.3	0.629	2555	47	2564	48	0
4.1	0.09	26	29	1.13	10.3	0.1557	1.7	9.75	2.6	0.4540	2	0.755	2410	29	2413	39	0
5.1	0.00	6	2	0.35	1.88	0.1101	4.5	5.39	6	0.3550	4	0.662	1802	82	1959	67	−9
6.1	0.17	32	18	0.58	13.3	0.1637	1.5	10.97	2.4	0.4861	1.8	0.767	2495	26	2554	39	−2
6.2	0.28	12	9	0.75	3.44	0.1171	3.5	5.22	4.4	0.3232	2.7	0.611	1913	63	1805	42	6
7.1	0.36	14	3	0.26	3.91	0.1103	3.3	5.05	4.2	0.3318	2.6	0.611	1805	60	1847	41	−2
8.1	0.36	10	4	0.49	2.77	0.1081	4.2	5.03	5.1	0.3374	2.9	0.574	1768	76	1874	47	−6
9.1	0.00	2	1	0.60	0.563	0.1191	7.8	5.93	10	0.3610	6.3	0.631	1943	140	1987	110	−2
9.2	0.33	15	13	0.89	6.52	0.1645	2.2	11.33	3.5	0.5000	2.7	0.769	2503	37	2612	58	−4
10.1	0.02	54	41	0.79	18.5	0.1371	1.4	7.52	2.1	0.3981	1.6	0.753	2191	24	2160	29	1
10.2	−0.23	16	4	0.25	4.66	0.1095	2.8	5.07	3.7	0.3360	2.4	0.657	1791	51	1867	39	−4
11.1	0.00	33	32	0.98	13.1	0.1593	3.7	10.07	4.2	0.4587	1.9	0.458	2448	63	2434	39	1
12.1	0.12	40	28	0.73	18.8	0.1811	2.5	15.71	3	0.5493	1.6	0.552	2663	41	2822	37	−6
13.1	0.09	37	26	0.72	16.1	0.1720	1.3	12.07	2.1	0.5089	1.7	0.803	2577	21	2652	37	−3
16.1	0.08	103	87	0.87	39	0.1517	0.93	9.2	1.8	0.4397	1.5	0.851	2365	16	2349	30	1
16.2	0.28	15	10	0.72	4.23	0.1136	3.6	5.22	4.5	0.3332	2.7	0.607	1858	64	1854	44	0
17.1	1.12	8	3	0.39	2.35	0.1043	7.9	4.8	9.5	0.3340	5.3	0.556	1702	150	1856	85	−9
18.1	3.93	7	3	0.40	1.85	0.1150	11	4.89	12	0.3070	5.2	0.423	1886	200	1727	79	8
19.1	0.47	39	20	0.53	15.7	0.1632	2.2	10.57	3.2	0.4700	2.3	0.722	2489	38	2483	48	0
20.1	0.15	31	22	0.73	14	0.1732	1.5	12.51	2.4	0.5240	1.9	0.795	2588	24	2716	42	−5
21.1	0.00	22	24	1.15	9.07	0.1601	1.8	10.61	3	0.4800	2.4	0.802	2457	30	2529	50	−3
22.1	0.00	15	3	0.21	4.15	0.1136	3.2	5.04	4.2	0.3217	2.7	0.646	1857	57	1798	42	3
23.1	0.27	24	18	0.78	10.2	0.1627	2.8	10.92	3.5	0.4870	2.1	0.613	2484	47	2557	45	−3
							1816–1: Tonalitic gneiss										
1.1	0.05	80	44	0.56	25.4	0.1365	2	6.96	2.5	0.3699	1.6	0.615	2029	27	2183	35	7
2.1	0.02	158	131	0.86	47.6	0.1259	1.8	6.11	2.8	0.3516	2.1	0.762	1942	36	2042	32	5
3.1	0.01	149	116	0.80	52	0.1481	2.2	8.3	2.6	0.4066	1.4	0.523	2200	25	2324	38	5
4.1	0.03	154	113	0.76	66.8	0.1754	1.1	12.19	1.8	0.5038	1.4	0.776	2630	30	2610	18	−1
4.2	0.13	140	71	0.53	59.9	0.1740	0.72	11.94	1.5	0.4975	1.3	0.879	2603	29	2597	12	0
5.1	0.09	85	56	0.69	32.7	0.1593	1	9.86	1.9	0.4488	1.6	0.839	2390	31	2448	17	2
6.1	0.04	108	54	0.52	44	0.1651	0.87	10.8	2.2	0.4746	2	0.918	2504	42	2508	15	0
7.1	0.01	355	342	1.00	138	0.1630	0.92	10.17	1.5	0.4523	1.2	0.791	2406	24	2487	16	3
8.1	0.00	153	88	0.59	67.1	0.1771	0.68	12.47	1.5	0.5105	1.3	0.889	2659	29	2626	11	−1
9.1	0.02	322	272	0.87	118	0.1524	0.88	8.93	1.8	0.4248	1.6	0.873	2282	30	2373	15	4

Table 3. *Cont.*

Spot No	^{206}Pb$_c$ (%)	U ppm	Th ppm	Th/U	^{206}Pb* ppm	^{207}Pb/^{206}Pb	Error in %	^{207}Pb/^{235}U	Error in %	^{206}Pb/^{238}Pb	Error in %	Error corr	^{207}Pb/^{206}Pb Age (Ma)	1σ	^{206}Pb/^{238}U Age (Ma)	1σ	Discordant (%)
10.1	0.12	110	71	0.67	30.3	0.1080	1.2	4.756	1.9	0.3194	1.4	0.758	1787	22	1766	22	−1
11.1	0.04	88	50	0.58	34	0.1598	1	9.88	1.8	0.4483	1.5	0.824	2388	29	2454	17	3
12.1	0.04	134	107	0.83	55.2	0.1762	0.79	11.67	1.6	0.4804	1.3	0.863	2529	28	2618	13	3
13.1	0.28	82	35	0.44	23.4	0.1141	1.5	5.18	2.1	0.3295	1.5	0.712	1836	24	1866	27	2
14.1	0.04	268	186	0.72	82.6	0.1264	0.88	6.254	1.6	0.3588	1.3	0.826	1977	22	2049	16	4
15.1	0.03	155	123	0.82	63.9	0.1634	1.4	10.8	1.9	0.4794	1.3	0.698	2525	28	2491	23	−1
16.1	0.00	44	27	0.64	12	0.1101	1.9	4.8	2.6	0.3163	1.8	0.691	1772	28	1801	34	2
17.1	0.04	79	50	0.66	35.7	0.1815	0.9	13.2	1.7	0.5275	1.5	0.853	2731	33	2667	15	−2
18.1	0.06	159	120	0.78	49.4	0.1365	1.1	6.8	1.7	0.3615	1.3	0.776	1989	22	2183	19	9
18.2	−0.03	110	61	0.57	46.7	0.1697	1	11.58	1.8	0.4949	1.5	0.822	2592	32	2555	17	−1
19.1	0.00	54	53	1.02	14.8	0.1091	1.7	4.81	2.4	0.3199	1.7	0.700	1789	26	1784	31	0
20.1	0.08	82	42	0.53	34	0.1731	2.2	11.54	2.7	0.4837	1.5	0.552	2543	31	2588	37	2
21.1	0.02	267	208	0.81	92.8	0.1420	2.3	7.92	2.7	0.4047	1.4	0.511	2191	26	2251	40	3
22.1	0.13	65	28	0.44	18	0.1096	1.5	4.85	2.2	0.3210	1.6	0.719	1794	25	1793	28	0
23.1	0.00	159	80	0.52	60.4	0.1578	0.74	9.64	1.6	0.4430	1.4	0.882	2364	27	2432	12	3
24.1	0.11	120	98	0.84	48.5	0.1630	0.85	10.51	1.6	0.4679	1.4	0.851	2474	28	2487	14	0
25.1	0.03	94	67	0.74	34.7	0.1579	1.8	9.33	2.3	0.4286	1.4	0.634	2299	28	2434	30	6
26.1	0.04	199	151	0.78	74.5	0.1590	1.9	9.54	2.5	0.4354	1.7	0.671	2330	33	2445	32	5
27.1	0.39	62	28	0.47	16.9	0.1069	2.1	4.68	2.7	0.3177	1.6	0.613	1778	25	1748	39	−2
28.1	0.04	131	75	0.59	56.5	0.1773	0.75	12.28	1.6	0.5024	1.4	0.888	2624	31	2628	12	0
29.1	0.35	67	33	0.50	18.2	0.1112	1.8	4.8	2.4	0.3132	1.6	0.672	1756	24	1820	32	3
30.1	0.66	265	187	0.73	86	0.1366	1.3	7.08	2.6	0.3757	2.3	0.875	2056	40	2185	22	6

Errors are 1-sigma; Pbc and Pb* indicate the common and radiogenic portions, respectively. Error in Standard calibration was 0.22% (not included in above errors but required when comparing data from different mounts). Common Pb corrected using measured 204Pb.

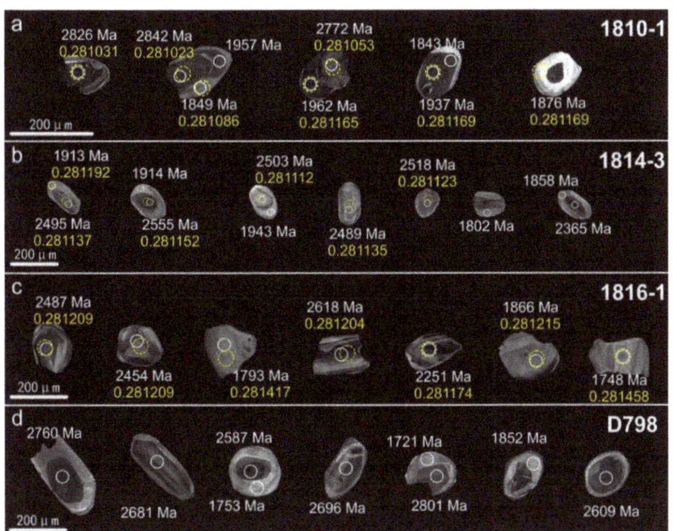

Figure 6. Representative cathodoluminescence images of dated zircons. The white solid-line circle and white number represent the analytical spot of U-Pb dating and dating result, respectively. The yellow dashed-line circle and yellow number represent the the analytical spot of Hf isotope and its corrected ^{176}Hf/^{177}Hf value, respectively.

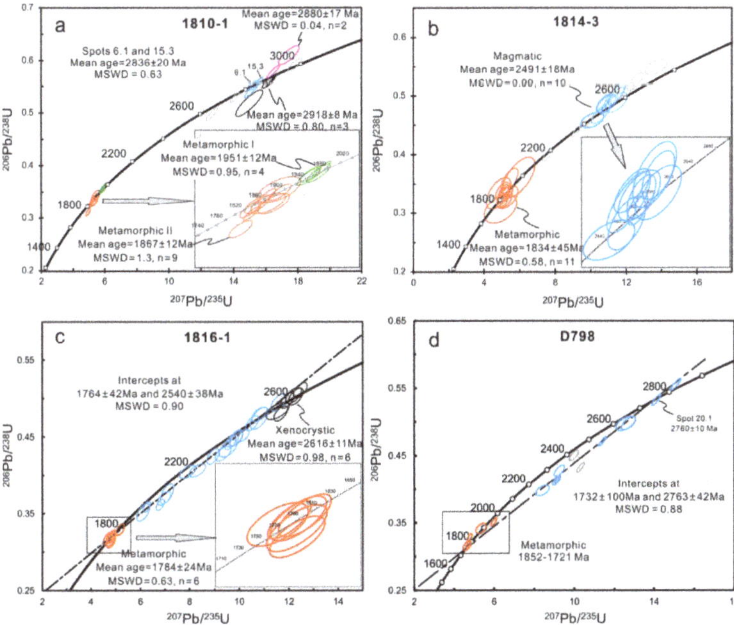

Figure 7. U–Pb concordia diagrams for zircons from the TTG rocks (**a–c**) and granitic gneiss (**d**) in the eastern Alxa Block. The purple and black ellipses represent analyses for inherited or xenocrystic zircons; The gray ellipses represent discordant analyses; The blue ellipses represent analyses for magmatic zircons, while the red and green ellipses represent analyses for metamorphic zircons.

Table 4. Lu-Hf isotopic data for zircons from TTG rocks in the eastern Alxa Block.

No.	Measured Age (Ma)	Used Age (Ma)	^{176}Yb/^{177}Hf	^{176}Lu/^{177}Hf	^{176}Hf/^{177}Hf (corr)	2σ	$\varepsilon_{Hf}(0)$	$\varepsilon_{Hf}(t)$	2σ	T_{DM}	T_{DMC}	$f_{Lu/Hf}$
					1810–1: Tonalitic gneiss							
2.1	1935	1935	0.008069	0.000277	0.281181	0.000013	−56.27	−13.49	0.46	2831	3134	−0.99
3.1	1915	1915	0.011251	0.000387	0.281154	0.000014	−57.22	−15.05	0.50	2875	3196	−0.99
3.3	1892	1892	0.009593	0.000331	0.281128	0.000012	−58.13	−16.41	0.44	2905	3246	−0.99
4.1	2772	2772	0.015876	0.000621	0.281053	0.000014	−60.78	0.42	0.48	3027	3111	−0.98
4.2	1962	1962	0.010297	0.000390	0.281165	0.000013	−56.84	−13.62	0.47	2861	3162	−0.99
5.1	1890	1890	0.005287	0.000177	0.281153	0.000013	−57.26	−15.37	0.44	2861	3192	−0.99
6.1	2826	2826	0.029679	0.001094	0.281031	0.000017	−61.55	−0.04	0.61	3094	3178	−0.97
8.1	1890	1890	0.007648	0.000272	0.281133	0.000013	−57.98	−16.23	0.45	2895	3235	−0.99
9.1	1866	1866	0.014119	0.000632	0.281168	0.000013	−56.73	−15.96	0.49	2874	3203	−0.98
10.1	2743	2743	0.032849	0.001155	0.281035	0.000015	−61.42	−1.89	0.54	3094	3203	−0.97
11.1	2904	2904	0.011499	0.000435	0.280979	0.000018	−63.42	1.15	0.64	3112	3182	−0.99
12.1	2970	2970	0.009732	0.000366	0.280948	0.000015	−64.51	1.71	0.53	3148	3208	−0.99
13.1	2906	2906	0.029166	0.001132	0.281010	0.000018	−62.32	0.92	0.63	3127	3195	−0.97
13.2	2880	2880	0.016740	0.000676	0.280963	0.000021	−63.98	−0.45	0.77	3153	3242	−0.98
14.2	1937	1937	0.008957	0.000322	0.281169	0.000014	−56.69	−13.93	0.49	2850	3158	−0.99
15.1	1849	1849	0.008050	0.000352	0.281086	0.000014	−59.61	−18.90	0.50	2962	3336	−0.99
15.3	2842	2842	0.041481	0.001464	0.281023	0.000014	−61.84	−0.68	0.49	3136	3223	−0.96
16.1	1876	1876	0.008185	0.000293	0.281169	0.000015	−56.69	−15.28	0.52	2848	3176	−0.99
					1814–3: Trondhjemitic gneiss							
1.1	2443	2491	0.023648	0.000801	0.281170	0.000014	−56.65	−2.12	0.49	2884	3011	−0.98
2.1	2518	2491	0.012920	0.000479	0.281123	0.000015	−58.32	−3.25	0.54	2923	3068	−0.99
3.1	2504	2491	0.010837	0.000401	0.281112	0.000016	−58.69	−3.49	0.55	2931	3079	−0.99
3.3	2555	2491	0.023738	0.000807	0.281152	0.000016	−57.29	−2.77	0.55	2909	3043	−0.98
6.1	2495	2491	0.016543	0.000587	0.281137	0.000016	−57.82	−2.93	0.58	2912	3051	−0.98
6.2	1913	1913	0.010601	0.000413	0.281192	0.000018	−55.89	−13.79	0.64	2826	3131	−0.99
8.1	1768	1768	0.008023	0.000293	0.281137	0.000015	−57.81	−18.83	0.54	2890	3269	−0.99
9.2	2503	2491	0.017259	0.000587	0.281112	0.000016	−58.71	−3.83	0.56	2946	3096	−0.98
11.1	2448	2491	0.020314	0.000703	0.281152	0.000015	−57.28	−2.59	0.53	2901	3034	−0.98
19.1	2489	2491	0.011975	0.000419	0.281135	0.000017	−57.90	−2.73	0.61	2903	3041	−0.99
21.1	2457	2491	0.021810	0.000711	0.281134	0.000017	−57.93	−3.26	0.62	2926	3068	−0.98
22.1	1857	1857	0.009303	0.000381	0.281235	0.000019	−54.35	−13.45	0.67	2766	3070	−0.99
23.1	2484	2491	0.017408	0.000619	0.281132	0.000017	−57.99	−3.15	0.62	2921	3062	−0.98
					1816–1: Tonalitic gneiss							
1.1	2183	2540	0.008120	0.000341	0.281196	0.000014	−55.73	0.71	0.51	2815	2909	−0.99
2.1	2042	2540	0.018306	0.000719	0.281205	0.000015	−55.40	0.39	0.52	2830	2925	−0.98
3.1	2324	2540	0.019608	0.000785	0.281240	0.000017	−54.16	1.52	0.61	2788	2868	−0.98
4.1	2610	2616	0.019010	0.000764	0.281190	0.000017	−55.94	1.48	0.59	2854	2931	−0.98
4.2	2597	2616	0.007257	0.000312	0.281223	0.000015	−54.77	3.46	0.54	2777	2832	−0.99
5.1	2448	2540	0.008014	0.000324	0.281208	0.000016	−55.30	1.17	0.58	2798	2885	−0.99
6.1	2508	2540	0.007759	0.000320	0.281183	0.000014	−56.18	0.29	0.51	2831	2929	−0.99
7.1	2487	2540	0.037669	0.001519	0.281252	0.000019	−53.76	0.65	0.67	2826	2911	−0.95
8.1	2626	2616	0.008833	0.000360	0.281186	0.000015	−56.09	2.05	0.52	2830	2903	−0.99
9.1	2373	2540	0.025353	0.001020	0.281249	0.000017	−53.86	1.41	0.62	2793	2873	−0.97
10.1	1766	1766	0.008814	0.000354	0.281229	0.000017	−54.58	−15.70	0.59	2773	3110	−0.99
11.1	2454	2540	0.010133	0.000398	0.281209	0.000017	−55.29	1.05	0.61	2803	2891	−0.99
12.1	2618	2616	0.024690	0.000973	0.281204	0.000018	−55.44	1.61	0.63	2851	2925	−0.97
13.1	1866	1866	0.007226	0.000292	0.281215	0.000017	−55.07	−13.87	0.60	2787	3098	−0.99
14.1	2049	2540	0.029831	0.001212	0.281251	0.000017	−53.79	1.15	0.61	2805	2886	−0.96
15.1	2491	2540	0.013793	0.000541	0.281188	0.000017	−56.02	0.07	0.60	2841	2941	−0.98
16.1	1801	1801	0.002951	0.000119	0.281388	0.000015	−48.93	−8.96	0.52	2544	2799	−1.00
17.1	2667	2616	0.010778	0.000429	0.281184	0.000018	−56.16	1.86	0.64	2838	2912	−0.99
18.1	2183	2540	0.017517	0.000689	0.281208	0.000018	−55.30	0.54	0.65	2824	2917	−0.98
18.2	2555	2540	0.011776	0.000483	0.281162	0.000015	−56.95	−0.76	0.55	2872	2982	−0.99
19.1	1784	1784	0.004382	0.000165	0.281405	0.000018	−48.35	−8.82	0.65	2525	2779	−1.00
20.1	2588	2540	0.014708	0.000577	0.281166	0.000015	−56.81	−0.78	0.54	2873	2983	−0.98
21.1	2251	2540	0.023336	0.000939	0.281174	0.000017	−56.49	−1.10	0.62	2888	2999	−0.97
22.1	1793	1793	0.006664	0.000281	0.281417	0.000018	−47.92	−8.32	0.65	2516	2761	−0.99
23.1	2432	2540	0.007641	0.000319	0.281159	0.000016	−57.04	−0.57	0.59	2863	2973	−0.99
24.1	2487	2540	0.017488	0.000671	0.281209	0.000019	−55.26	0.61	0.67	2821	2913	−0.98
25.1	2434	2540	0.011524	0.000450	0.281171	0.000017	−56.63	−0.39	0.60	2857	2964	−0.99
26.1	2445	2540	0.031465	0.001271	0.281214	0.000019	−55.09	−0.26	0.68	2859	2957	−0.96
27.1	1748	1748	0.004465	0.000183	0.281458	0.000022	−46.48	−7.78	0.77	2455	2697	−0.99
28.1	2628	2616	0.014402	0.000576	0.281162	0.000017	−56.95	0.80	0.62	2879	2965	−0.98
29.1	1820	1820	0.007359	0.000304	0.281517	0.000020	−44.39	−4.20	0.71	2383	2574	−0.99
30.1	2185	2540	0.012065	0.000470	0.281174	0.000017	−56.52	−0.31	0.60	2855	2960	−0.99

Note: measured age (Ma) represents the measured age of SHRIMP U-Pb dating for the analyses, and the used age (Ma) represents the ages that are used during the calculation of $\varepsilon_{Hf}(t)$ and model age. The crustal model ages (T_{DMC}) were calculated by assuming ^{176}Lu/^{177}Hf ratio of 0.010 for the upper crust.

Ten magmatic zircons from the 2.49 Ga trondhjemitic gneiss (1814-3) have ^{176}Hf/^{177}Hf ratios between 0.281112 and 0.281170 (Table 4), corresponding to age-corrected εHf(t) values between -3.38 and -2.12, slightly lower than those of magmatic zircons from sample 1810-1. The two-stage Hf model ages (T_{DMC}) range from 3011 Ma to 3096 Ma. Three analyses from the metamorphic zircon grains or rims present $^{176}Hf/^{177}Hf$ ratios varying from 0.281137 to 0.281235, and their age-corrected $\varepsilon Hf(t)$ values are between -18.83 and -13.45. The corresponding two-stage Hf model ages (T_{DMC}) range from 3070 to 3269 Ma.

4.2.3. Tonalitic Gneiss Sample 1816-1

Thirty-two analyses were obtained from the tonalitic gneiss sample (1816-1) (Table 3; Figure 7c). Six analyses from the inherited cores yield $^{207}Pb/^{206}Pb$ ages of 2588–2667 Ma, with a weighted mean age of 2616 ± 11 Ma (MSWD = 0.98; Figure 7c). Eighteen analyses from magmatic zircon cores show variable degrees of Pb loss with $^{207}Pb/^{206}Pb$ ages of 2042–2555 Ma, and seven analyses from metamorphic zircons or rims yield $^{207}Pb/^{206}Pb$ ages of 1748–1866 Ma. All analyses except for those from inherited zircons define a discordia line with an upper concordia intercept age of 2540 ± 38 Ma and a lower concordia intercept age of 1764 ± 42 Ma (MSWD = 0.90; Figure 7c). Six concordant analyses from unzoned rim domains (spots 10.1, 16.1, 19.1, 22.1, 27.1, and 29.1) that yield $^{207}Pb/^{206}Pb$ ages between 1820 ± 32 Ma and 1748 ± 39 Ma have a weighted mean age of 1784 ± 24 Ma (MSWD = 0.63), which is identical to the intercept ages within errors. Therefore, ages of 2540 ± 38 Ma and 1784 ± 24 Ma are proposed to reflect the crystallization and metamorphic ages of the tonalitic gneiss, respectively.

Nineteen magmatic zircon cores from the 2.54 Ga tonalitic gneiss have $^{176}Hf/^{177}Hf$ ratios ranging from 0.281159 to 0.281252 (Table 4), age-corrected $\varepsilon Hf(t)$ values from -1.10 to 1.52, with two-stage Hf model ages (T_{DMC}) ranging from 2868 Ma to 2999 Ma, respectively. Seven metamorphic zircon grains exhibit $^{176}Hf/^{177}Hf$ ratios between 0.281215 and 0.281517 and relatively lower $\varepsilon Hf(t)$ values between -15.70 and -4.20. The corresponding two-stage Hf model ages (T_{DMC}) are 2574–3110 Ma, which are mainly concentrated in the ranges of 2697–2799 Ma, respectively. In addition, six inherited zircon cores show a similar $^{176}Hf/^{177}Hf$ ratio range of 0.281162–0.281223, with $\varepsilon Hf(t)$ values of 0.80–3.46 and corresponding two-stage Hf model ages (T_{DMC}) of 2832–2965 Ma.

4.2.4. Granitic Gneiss Sample D798

Twenty-eight analyses were obtained from the granitic gneiss sample (D798) (Table 3; Figure 7d). Twelve analyses show a large error (≥ 87 Ma), which is useless for age determination, and two analyses are discordant (spots 3.1 and 18.1). Nine analyses show variable degrees of Pb loss with $^{207}Pb/^{206}Pb$ ages of 2380–2801 Ma and define a discordia line with an upper concordia intercept age of 2763 ± 42 Ma (MSWD = 0.88; Figure 7d). The only analysis (spot 20.1) close to the concordia line that yields a $^{207}Pb/^{206}Pb$ age of 2760 ± 10 Ma responds well to the upper intercept age, reflecting the protolith emplacement age of the granitic gneiss. Four analyses from unzoned rims yield $^{207}Pb/^{206}Pb$ ages of 1852–1721 Ma and are interpreted as the metamorphic ages of the granitic gneiss [61].

5. Discussion

5.1. Petrogenesis of the TTG Rocks and Granitic Gneiss

The TTG gneisses from the Diebusige Complex are characterized by high SiO_2 contents (>62 wt.%), Sr/Y ratios (41–191) and intermediate $Mg^{\#}$ values (44.97–52.23) (Table 2), with negative Nb, Ta, and Ti anomalies and enrichment in Sr (Figure 5b,d). These characteristics are similar to those of Archean TTGs and high-SiO_2 adakites, which are consistent with the results in the Sr vs. $(CaO + Na_2O)$ diagram (Figure 8a). The low $\varepsilon_{Hf}(t)$ values (-3.83 to 3.46) and the presence of old xenocrystic zircons (2.84 Ga and 2.54 Ga tonalitic gneiss samples) suggest a crustal origin for the protoliths. The moderately to strongly fractionated REE patterns $((La/Yb)_N - 10.59–50.58)$, low Sr contents (373.38–677.51 ppm) and positive Eu anomalies $(Eu_N/Eu_N^* = 1.29–3.73))$ suggest partial melting in the garnet stability field and the absence of plagioclase in the residue. Generally, rutile has a lower Nb/Ta ratio than chondrite, and its residue in the source or separation and differentiation during magmatic crystallization led to a higher Nb/Ta ratio of the melts [62]. Element Nb has a higher distribution coefficient than Ta in hornblende [63,64], and the presence of hornblende in the residue led to lower Nb/Ta and Dy/Yb ratios and higher Zr/Sm ratios for the corresponding melts [65]. The 2.49 Ga trondhjemitic gneiss and 2.54 Ga tonalitic gneiss in the Langshan area have Nb/Ta ratios lower than those of chondrite (17.6 [66]; 19.9 [67]), indicating that the negative Nb, Ta, and Ti anomalies were not caused by residual rutile in the source but were more likely related to the residues of hornblende in the source. The 2.84 Ga tonalitic gneiss has high Nb/Ta ratios; thus, its negative Nb, Ta, and Ti anomalies may have been controlled by rutile residues in the source. Together with the classification proposed by Moyen (2011) [68], the 2.49 Ga trondhjemitic gneiss and 2.54 Ga tonalitic gneiss were derived from partial melting of garnet-bearing amphibolite under high-to-medium pressure conditions, while the 2.84 Ga tonalitic gneiss was derived from partial melting of rutile-bearing eclogite under high-pressure conditions (Figure 8b,c).

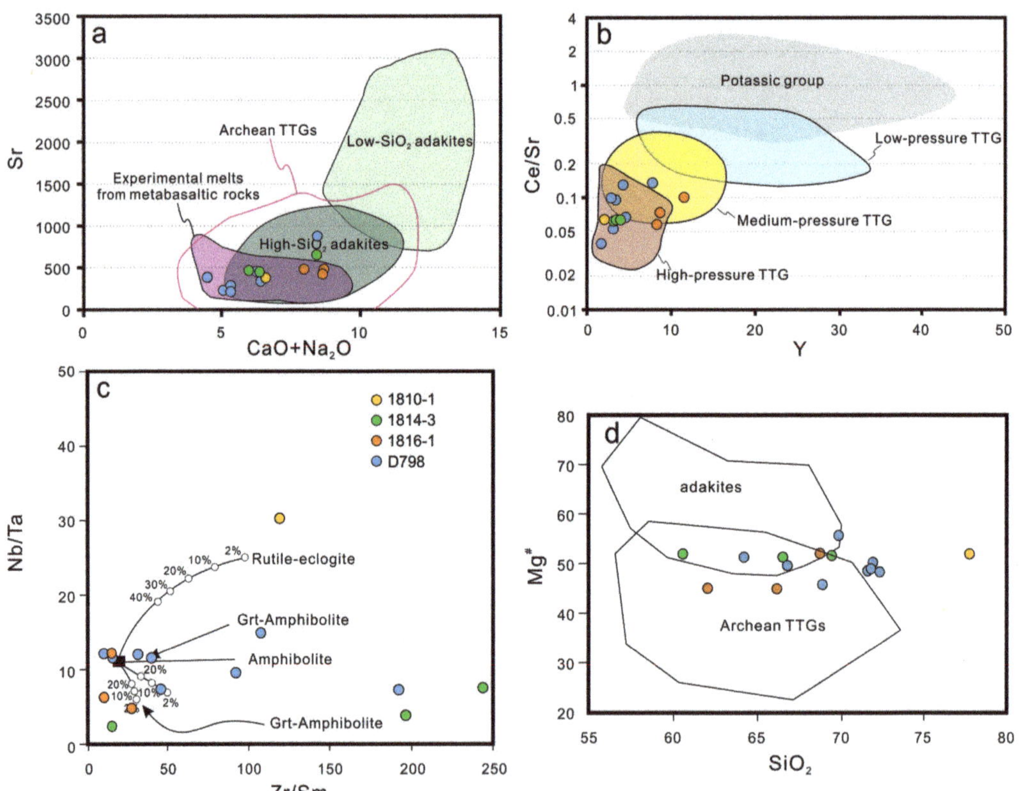

Figure 8. Geochemical modeling results for the TTG and granitic gneisses in the eastern Alxa Block. (**a**) Sr vs. (CaO + Na₂O) diagram after references [60,64,69]; (**b**) Ce/Sr vs. Y diagram after reference [68]; and (**c**) Nb/Ta vs. Zr/Sm diagram and melting curves after reference [70]; (**d**) Mg# vs. SiO₂ diagram after reference [71].

The 2.76 Ga granitic gneiss from the Diebusige Complex has SiO₂ contents (66.43–74.49 wt.%), Sr/Y ratios (47–274, 117 on average), and intermediate Mg# values (45.89–55.78, 49.87 on average) (Table 2 and Figure 8d), with significantly negative Nb and Ta anomalies and slight Ti anomalies (Figure 5d), which are similar to ca. 2.5 Ga TTG gneisses discussed above and show the features of high-SiO2 adakites (Figure 8a). High Sr/Y ratios, moderate to strong REE fractionations and positive Eu anomalies (Eu$_N$/Eu$_N$*

= 1.43–10.97) suggest that the granitic gneiss was probably derived from partial melting of a subducted basaltic slab with garnet in the residue. The granitic gneiss shows low Nb/Ta (7.25–14.96, 10.81 on average) and Zr/Sm ratios (10.13–192.51, 67.30 on average), indicating that it was derived from partial melting of garnet-bearing amphibolite (Figure 8c).

Previous studies have suggested that the potential source of Archean TTGs and modern adakites may have been the melting of subducting oceanic crust [2,60,72,73], thickened lower crust [74–76], or delaminated lower crust [74,77]. Generally, TTG or adakitic melts with low $Mg^\#$ values and Cr and Ni concentrations can be generated by the partial melting of mafic rocks underplating the lower crust [69,78], while those generated from the partial melting of a subducting slab and delaminated thickened lower crust would have higher $Mg^\#$ values and MgO, Cr, and Ni contents on account of the interaction with the overlying mantle wedge during ascent [60,74,79,80]. Additionally, TTG melts produced by the partial melting of the delaminated lower crust would have higher contents of MgO (>3 wt.%), TiO_2 (>0.9 wt.%) and compatible elements [81–83]. The low MgO (<3.21 wt.%) and TiO2 (<0.81 wt.%) contents of TTG gneisses from the Langshan area can rule out the origin of partial melting of the delaminated lower crust. The $Mg^\#$ value can be used as a marker to reflect whether the mafic rock was contaminated by the mantle during the melting process; generally, the $Mg^\#$ value of a typical mid-oceanic ridge basalt is <60 (51 on average), and the $Mg^\#$ value of the melt formed by its partial melting is <45 [69,76]. All TTG gneisses and the granitic gneiss in this study show similar $Mg^\#$ values (approximately 50) (Figure 8d), indicating that a certain degree of mantle contamination may have occurred during the ascent of the TTG melts. However, they have different compatible element compositions: the 2.84 Ga tonalitic gneiss and 2.76 Ga granitic gneiss have low Cr and Ni contents, and the 2.49 Ga trondhjemitic and 2.54 Ga tonalitic gneisses have relatively high contents of Cr and Ni. Therefore, the 2.84 Ga tonalitic gneiss and 2.76 Ga granitic gneiss might have formed by the partial melting of the thickened lower crust, whereas the 2.49 Ga trondhjemitic and the 2.54 Ga tonalitic gneisses are probably related to the partial melting of the subducted oceanic slab.

5.2. Archean to Late Paleoproterozoic Crustal Evolution in the Alxa Block

The Diebusige Complex is one of the oldest metamorphic series in the eastern Alxa Block. The results show that the 2.84 Ga and 2.54 Ga tonalitic gneisses, the 2.49 Ga trondhjemitic gneiss, and the 2.76 Ga granitic gneiss are components of the Diebusige Complex. The magmatic zircon age populations at ca. 2.8 Ga and ca. 2.5 Ga indicate that the eastern Alxa Block experienced at least two magmatic events in the late Mesoarchean to late Neoarchean era. Zircon Hf isotope analysis shows that all magmatic zircons from the TTG rocks have $\varepsilon_{Hf}(t)$ values ranging from −3.83 to 3.46, which suggest a crustal origin for the protoliths. Previous studies have suggested that the two-stage zircon Hf model age (T_{DMC}) can accurately reflect the extraction time of source materials from depleted mantle [84]. Magmatic zircons from the 2.84 Ga tonalitic gneiss have T_{DMC} values between 3.24 Ga and 3.11 Ga, while those from the 2.54 Ga tonalitic gneiss and 2.49 Ga trondhjemitic gneiss have T_{DMC} values of 3.0–2.83 Ga and 3.10–3.01 Ga, respectively, indicating that the Langshan TTG gneisses were derived from reworking of Paleo-Mesoarchean crust and mixed with mantle materials to different degrees during migration. The 2.84 Ga tonalitic gneiss is the oldest rock currently exposed in the Alxa Block. Recently, Gong et al. (2012) and Zhang et al. (2013b) [51,52] recognized 2.5 Ga TTG rocks from the Beidashan Complex in the western Alxa Block. Zircon Hf isotopic features suggested that the western Alxa Block experienced a mostly 2.8–2.7 Ga crustal growth and a ca. 2.5 Ga magmatic–metamorphic event. The T_{DMC} values of 3.59–3.02 Ga obtained from the ca. 2.8 Ga-inherited zircons also implied the existence of Paleo-Mesoarchean crustal materials in the western Alxa Block [52]. Combined datasets show that the eastern and western Alxa Block probably had the same Paleo-Mesoarchean crust, and the Alxa Block experienced Paleo-Mesoarchean crustal growth, a ca. 2.8 Ga magmatic event, and a ca. 2.5 Ga magmatic–metamorphic event.

The Langshan TTG gneisses and granitic gneiss recorded continuous metamorphic ages of 1962–1721 Ma with peaks at ca. 1.95 Ga and ca. 1.85 Ga. Paleoproterozoic metamorphic events were widely developed in every Precambrian basement in the Alxa Block, such as the Bayanwulashan Complex in the eastern Alxa Block and the Beidashan Complex and Longshoushan Complex in the western Alxa Block [50,52,85,86]. The remaining NCC also recorded these two metamorphic events, and previous studies have suggested that ca. 1.95 Ga and ca. 1.85 Ga corresponded to the formation ages of the Khondalite Belt and the TNCO [27,87–90], respectively. However, whether the formation of the TNCO could have affected the Alxa Block located in the westernmost part of the NCC is still uncertain. A few models suggested the ca. 1.95 Ga and ca. 1.85 Ga events were related to the assembly and breakup of the Paleoproterozoic Columbia supercontinent since they have been identified globally (e.g., Laurentia, Baltica, Amazonia, and India [91–96]).

5.3. Early Geological History of the Alxa Block in Comparison with the YC, TC, and Main NCC

The NCC, YC, and TC are three old cratons in China that constitute the main nucleus of the Chinese continent [5,6,27,97]. Despite considerable progress over recent decades in understanding the Precambrian evolution of these three cratons, limited work has been conducted on comparing their early geological histories [7,8,98–102].

As an important component of the Archean crust, TTG rocks play an important role in the study of Precambrian crustal evolution. In terms of the YC, previous studies show that Paleo-Neoarchean TTG rocks were well developed in the YC [11,13,103–107]. The oldest TTG rocks were formed during 3437–3262 Ma, and zircon Hf isotope studies have suggested that these rocks with $\varepsilon_{Hf}(t)$ values of −4.7–1.2 were sourced from the Hadean to Eoarchean crust [17–21,103,104] (Figure 9). Additionally, detrital zircons with ages of 3.8–3.2 Ga have been identified in the YC [24,25], indicating that the beginning of crustal evolution of the YC was as early as the Eoarchean to Paleoarchean era (Figures 9 and 10d). Mesoarchean TTG rocks from the YC have $\varepsilon_{Hf}(t)$ values of −10-3, which yield T_{DMC} values ranging between ca. 3.8 Ga and ca. 3.0 Ga [20,105] (Figure 10f). This suggests that Eoarchean to Paleoarchean crustal materials in the YC were reworked in the Mesoarchean. The Neoarchean rocks were also recognized from the YC [15], and their zircon Hf isotopic features suggest a derivation from Mesoarchean reworked components (Figure 10f).

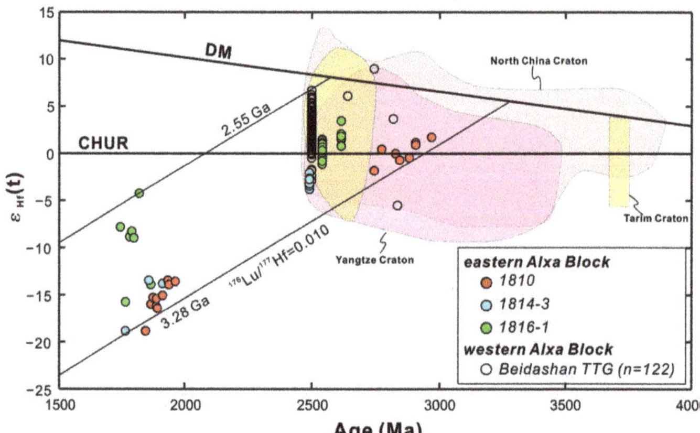

Figure 9. Diagram of $\varepsilon_{Hf}(t)$ values vs. $^{207}Pb/^{206}Pb$ ages for zircons from the basement rocks in the Alxa Block and the main North China, Tarim, and Yangtze cratons. Data for the Alxa Block are from references [51,52] and this study; (2) data for the main North China Craton are from references [30,65,106,108–115]; (3) data for the Tarim Craton are from references [22,23,98,99,116,117]; and (4) data for the Yangtze Craton are from references [11,14,17,19,21,24,102,103,107,118–120].

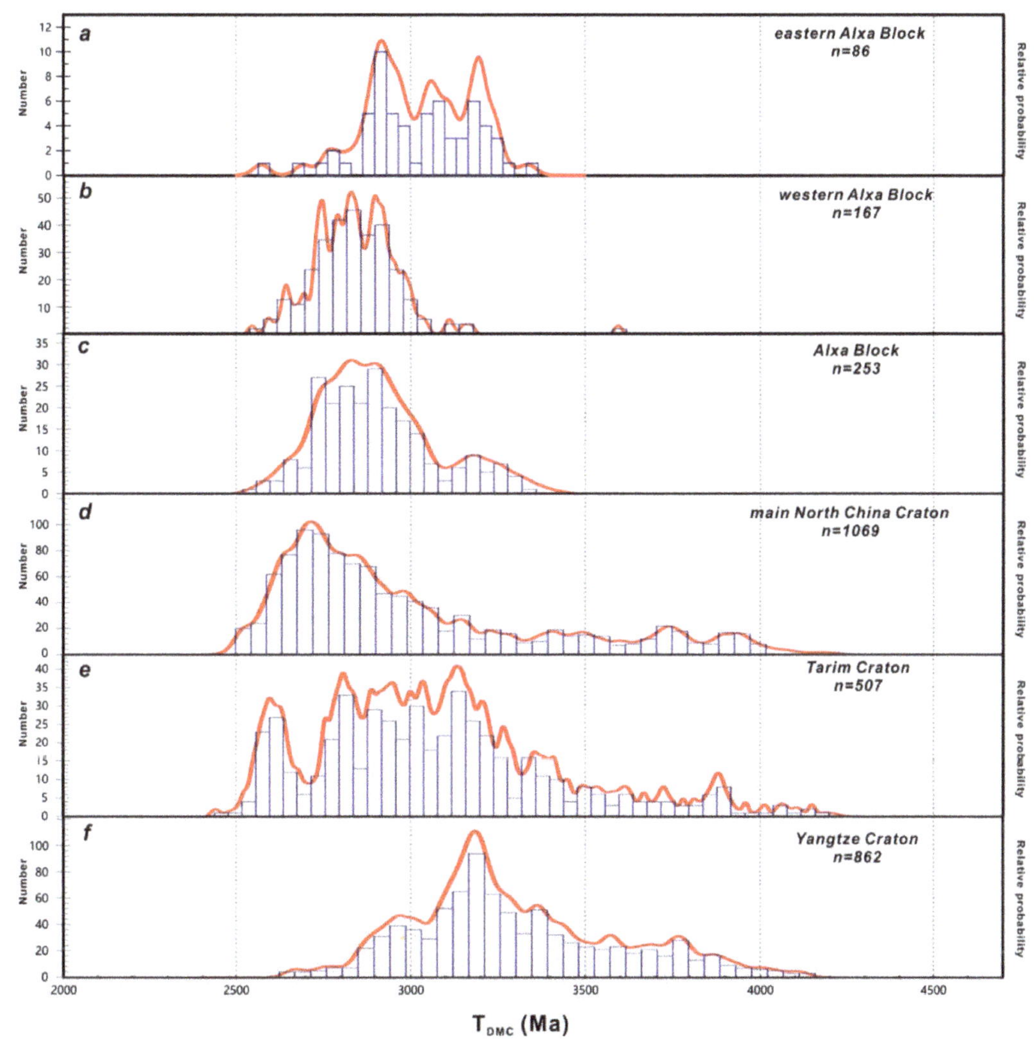

Figure 10. Zircon Hf isotope model age (T$_{DMC}$) histogram for basement rocks of the eastern Alxa Block (**a**), western Alxa Block (**b**), whole Alxa Block (**c**), and the main North China (**d**), Tarim (**e**), and Yangtze (**f**) cratons. Data sources are the same as those in Figure 9.

The discovery of ca. 3.7 Ga tonalitic gneisses with a mean $\varepsilon_{Hf}(t)$ value of -0.7 ± 2.6 suggests that the crustal evolution of the TC began before the Eoarchean era [22] (Figures 9 and 10c). Detrital zircons from metasedimentary rocks in the northern TC were dated ca. 2.5 Ga to ca. 3.5 Ga with T$_{DMC}$ values from ca. 3.9 Ga to ca. 3.7 Ga [23], similarly indicating that the crustal components in the TC may have been generated as early as ca. 3.9 Ga. Neoarchean orthogneisses and mafic–ultramafic rocks are widely exposed in the TC [12,16,95,98,99]. Previous studies have shown that continuous magmatic events occurred in the Neoarchean era, and zircon Hf isotopes yield a large range of $\varepsilon_{Hf}(t)$ values (ca. -8–10) with two-stage model ages from the Paleoarchean to Neoarchean era (ca. 3.4–2.8 Ga) [16,23,98,116,117] (Figures 9 and 10c). This suggests that basement rocks in the TC involved synchronous crustal growth and reworking during the Paleo-Neoarchean era.

Numerous detrital and inherited zircons from a variety of metasedimentary rocks in the eastern NCC have been dated 3.88–3.6 Ga, suggesting that the beginning of crustal evolution of the NCC was as early as before the Eoarchean era [30,121–124] (Figure 10d). Recently, ca. 3.8 Ga TTG rocks and granulite enclaves were also identified in the NCC [125,126], further confirming the existence of Eoarchean continental materials and the initial time of crustal evolution of the NCC. Paleoarchean zircons from the eastern NCC have $\varepsilon_{Hf}(t)$ values ranging from −5 to 3 and give Eo-Paleoarchean two-stage model ages varying from 3.9 Ga to 3.4 Ga (Figures 9 and 10b). Zircons from the Mesoarchean and Neoarchean TTG rocks have variable $\varepsilon_{Hf}(t)$ values of −5.5–10.2 (3.6 on average) and −7.8–12.6 (4.1 on average) with two-stage model ages of 4.2–2.8 Ga and 3.9–2.5 Ga, respectively (Figures 9 and 10b). These results indicate that basement rocks in the main NCC involved synchronous crustal growth and reworking, similar to the TC during the Paleo-Neoarchean era.

The Alxa Block is the westernmost component of the NCC. The existence of Archean rocks has long been controversial until ca. 2.5 Ga TTG rocks were identified from the western Alxa Block [51,52]. Mesoarchean to early Paleoproterozoic granitic gneisses (2.76 Ga) and TTG gneisses (2.84 Ga, 2.54 Ga, and 2.49 Ga) from the Langshan area in the eastern Alxa Block are reported in this study. As mentioned above, magmatic zircons from the Langshan TTG gneisses have $\varepsilon_{Hf}(t)$ values ranging from −3.83 to 3.46 and two-stage model ages ranging from 3.3 Ga to 2.9 Ga with peaks at ca. 3.2 Ga and ca. 3.0 Ga. Magmatic zircons from the ca. 2.5 Ga TTG rocks in the western Alxa Block have $\varepsilon_{Hf}(t)$ values varying from -5.54 to 8.98, and two-stage model ages mainly vary from 3.0 Ga to 2.6 Ga, with a peak at ca. 2.8 Ga [51,52]. By contrast (Figures 9 and 10), the eastern Alxa Block recorded a 3.3–2.9 Ga crustal growth and 2.8–2.7 Ga and ca. 2.5 Ga crustal reworking, which have been extensively recorded in most ancient cratons worldwide [31,75,126,127], whereas the western Alxa Block recorded a 2.8–2.7 Ga crustal growth and ca. 2.5 Ga crustal growth and reworking. This indicates that the eastern Alxa Block has older crustal materials than the western Alxa Block, and crustal growth and reworking simultaneously occurred 2.8–2.7 Ga in the Alxa Block (Figure 10a,b). The combination of available datasets suggests that the oldest basement exposed in the Alxa Block formed in the Paleo-Mesoarchean era and that crustal evolution began in the Paleoarchean era, which was younger than those of the main NCC, TC, and YC (Figures 9 and 10). Therefore, we suggest that the Alxa Block probably has its unique crustal evolutionary history before the early Paleoproterozoic.

6. Conclusions

Based on geological, geochronological, geochemical, and zircon Lu-Hf isotope data from the Langshan area in the eastern Alxa Block, we reach the following conclusions:

(1). The granitic gneiss and three TTG gneisses from the Langshan area were mainly emplaced 2.76 Ga, 2.84 Ga, 2.54 Ga, and 2.49 Ga, respectively, supporting the existence of Archean rocks in the eastern Alxa Block.

(2). Zircon Lu-Hf isotope data indicated that the Langshan TTG gneisses were derived from partial melting of crustal materials extracted from depleted mantle during the Paleoarchean to Mesoarchean era (3.24–2.83 Ga).

(3). The eastern Alxa Block experienced an important period of crustal growth during ca. 3.24–2.83 Ga, followed by crustal reworking of ca. 2.8 Ga and ca. 2.5 Ga, and the Alxa Block probably had its unique crustal evolution history from before the early Paleoproterozoic era, which was younger than that of the main NCC, TC, and YC.

Author Contributions: Conceptualization, J.Q. and J.Z.; methodology, B.Z.; software, H.Z.; investigation, P.N., J.Q. and J.Z.; resources, J.Q.; data curation, P.N., B.Z. and H.Z.; writing—original draft preparation, P.N.; writing—review and editing, J.Q. and J.Z.; visualization, P.N., B.Z. and H.Z.; supervision, J.Q.; funding acquisition, J.Q. and J.Z. All authors have read and agreed to the published version of the manuscript.

Funding: This research was funded by: Basic Scientific Research Fund of the Institute of Geology, Chinese Academy of Geological Sciences, grant number J2103; China Geological Survey, grant number DD20230217; National Natural Science Foundation of China, grant number 41972224.

Data Availability Statement: The data presented in this study are available on reasonable request from the corresponding authors.

Acknowledgments: We would like to thank the Key Laboratory of Mineral Resources Evaluation in Northeast Asia, Ministry of Land and Resources of China, the Wuhan Sample Solution Analytical Technology Co., Ltd., and the Beijing SHRIMP Center, Institute of Geology, Chinese Academy of Geological Sciences, Beijing, for their support and assistance on sample processing, zircon U–Pb dating, Hf isotope and major and trace element analyses. The authors are grateful for the critical comments from the anonymous reviewers, which profoundly enhanced the quality of this manuscript.

Conflicts of Interest: The authors declare no conflict of interest.

References

1. Jahn, B.; Glikson, A.Y.; Peucat, J.J.; Hickman, A.H. REE geochemistry and isotopic data of Archaean silicic volcanics and granitoids from the Pilbara block, western Australia: Implications for early crustal evolution. *Geochim. Cosmochim. Acta* **1981**, *45*, 1633–1652. [CrossRef]
2. Moyen, J.-F.; Martin, H. Forty years of TTG research. *Lithos* **2012**, *148*, 312–336. [CrossRef]
3. Condie, K.C. Episodic growth and supercontinents: A mantle avalanche connection? *Earth Planet. Sci. Lett.* **1998**, *163*, 91–108. [CrossRef]
4. Condie, K.C. Episodic continental growth models: After thoughts and extensions. *Tectonophysics* **2000**, *322*, 153–162. [CrossRef]
5. Lu, S.; Zhao, G.; Wang, H.; Hao, G. Precambrian metamorphic basement and sedimentary cover of the North China Craton: A review. *Precambrian Res.* **2008**, *160*, 77–93. [CrossRef]
6. Zhao, G.; Wilde, S.A.; Cawood, P.A.; Lu, L. Thermal Evolution of Archean Basement Rocks from the Eastern Part of the North China Craton and Its Bearing on Tectonic Setting. *Int. Geol. Rev.* **1998**, *40*, 706–721. [CrossRef]
7. Zhao, G.C. Paleoproterozoic assembly of the North China Craton. *Geol. Mag.* **2001**, *138*, 89–91. [CrossRef]
8. Zhao, G.; Wilde, S.A.; Cawood, P.A.; Sun, M. Archean blocks and their boundaries in the North China Craton: Lithological, geochemical, structural and P–T path constraints and tectonic evolution. *Precambrian Res.* **2001**, *107*, 45–73. [CrossRef]
9. Zhao, G.; Sun, M.; Wilde, S.A.; Li, S. Late Archean to Paleoproterozoic evolution of the North China Craton: Key issues revisited. *Precambrian Res.* **2005**, *136*, 177–202. [CrossRef]
10. Zhai, M.G.; Guo, J.H.; Liu, W.J. Neoarchaean to Paleoproterozoic continental evolution and tectonic history of the North China Craton: A review. *J. Asian Earth Sci.* **2005**, *24*, 547–561. [CrossRef]
11. Chen, Q.; Sun, M.; Zhao, G.; Zhao, J.; Zhu, W.; Long, X.; Wang, J. Episodic crustal growth and reworking of the Yudongzi terrane, South China: Constraints from the Archean TTGs and potassic granites and Paleoproterozoic amphibolites. *Lithos* **2019**, *326–327*, 1–18. [CrossRef]
12. Chen, S.; Guo, Z.; Huang, W. Traces of Archean Subduction in the Tarim Craton, Northwest China: Evidence from the Tuoge Complex and Implications for Basement Formation. *J. Geol.* **2019**, *127*, 323–341. [CrossRef]
13. Huang, M.; Cui, X.; Ren, G.; Guo, J.; Ren, F.; Pang, W.; Chen, F. The 2.73 Ga I-type granites in the Lengshui Complex and implications for the Neoarchean tectonic evolution of the Yangtze Craton. *Int. Geol. Rev.* **2019**, *62*, 649–664. [CrossRef]
14. Wu, Y.B.; Zhou, G.Y.; Gao, S.; Liu, X.C.; Zheng, Q.; Wang, H.; Yang, J.Z.; Yang, S.H. Petrogenesis of Neoarchean TTG rocks in the Yangtze Craton and its implication for the formation of Archean TTGs. *Precambrian Res.* **2014**, *254*, 73–86. [CrossRef]
15. Wang, K.; Li, Z.X.; Dong, S.W.; Cui, J.J.; Han, B.F.; Zheng, T.; Xu, Y.L. Early crustal evolution of the Yangtze Craton, South China: New constraints from zircon U-Pb-Hf isotopes and geochemistry of ca. 2.9–2.6 Ga granitic rocks in the Zhongxiang Complex. *Precambrian Res.* **2018**, *314*, 325–352. [CrossRef]
16. Cai, Z.; Jiao, C.; He, B.; Qi, L.; Ma, X.; Cao, Z.; Xu, Z.; Chen, X.; Liu, R. Archean–Paleoproterozoic tectonothermal events in the central Tarim Block: Constraints from granitic gneisses revealed by deep drilling wells. *Precambrian Res.* **2020**, *347*, 105776. [CrossRef]
17. Chen, K.; Gao, S.; Wu, Y.; Guo, J.; Hu, Z.; Liu, Y.; Zong, K.; Liang, Z.; Geng, X. 2.6–2.7 Ga crustal growth in Yangtze craton, South China. *Precambr. Res.* **2013**, *224*, 472–490. [CrossRef]
18. Chen, W.T.; Zhou, M.-F.; Zhao, X.-F. Late Paleoproterozoic sedimentary and mafic rocks in the Hekou area, SW China: Implication for the reconstruction of the Yangtze Block in Columbia. *Precambrian Res.* **2013**, *231*, 61–77. [CrossRef]
19. Jiao, W.; Wu, Y.; Yang, S.; Peng, M.; Wang, J. The oldest basement rock in the Yangtze Craton revealed by zircon U-Pb age and Hf isotope composition. *Sci. China Ser. D Earth Sci.* **2009**, *52*, 1393–1399. [CrossRef]
20. Li, Y.H.; Zheng, J.P.; Ping, X.Q.; Xiong, Q.; Xiang, L.; Zhang, H. Complex growth and reworking processes in the Yangtze cratonic nucleus. *Precambrian Res.* **2018**, *311*, 262–277. [CrossRef]
21. Zhang, S.B.; Zheng, Y.F.; Wu, Y.B.; Zhao, Z.F.; Gao, S.; Wu, F.Y. Zircon U–Pb age and Hf isotope evidence for 3.8 Ga crustal remnant and episodic reworking of Archean crust in South China. *Earth Planet. Sci. Lett.* **2006**, *252*, 56–71. [CrossRef]

22. Ge, R.; Zhu, W.; Wilde, S.A.; Wu, H. Remnants of Eoarchean continental crust derived from a subducted proto-arc. *Sci. Adv.* **2018**, *4*, eaao3159. [CrossRef] [PubMed]

23. Ge, R.; Zhu, W.; Wilde, S.A.; He, J. Zircon U–Pb–Lu–Hf–O isotopic evidence for ≥3.5Ga crustal growth, reworking and differentiation in the northern Tarim Craton. *Precambrian Res.* **2014**, *249*, 115–128. [CrossRef]

24. Zhang, S.B.; Zheng, Y.F.; Wu, Y.B.; Zhao, Z.F.; Gao, S.; Wu, F.Y. Zircon isotope evidence for ≥3.5Ga continental crust in the Yangtze craton of China. *Precambrian Res.* **2006**, *146*, 16–34. [CrossRef]

25. Zheng, J.; Griffin, W.; O'Reilly, S.Y.; Zhang, M.; Pearson, N.; Pan, Y. Widespread Archean basement beneath the Yangtze craton. *Geology* **2006**, *34*, 417–420. [CrossRef]

26. Wan, Y.S.; Liu, D.Y.; Dong, C.Y.; Xie, H.Q.; Kröner, A.; Ma, M.Z.; Liu, S.J.; Xie, S.W.; Ren, P. Formation and Evolution of Archean Continental Crust of the North China Craton. In *Precambrian Geology of China*; Zhai, M., Ed.; Springer: Berlin/Heidelberg, Germany, 2015; pp. 59–136. [CrossRef]

27. Shen, Q.; Geng, Y.; Song, B.; Wan, Y. New information from the surface outcrops and deep crust of Archean rocks of the North China and Yangtze blocks, and Qinling-Dabie orogenic belt. *Acta Geol. Sin.* **2005**, *79*, 616–627.

28. Niu, P.F.; Qu, J.F.; Zhang, J. Precambrian tectonic affinity of the southern Langshan area, northeastern margin of the Alxa Block: Evidence from zircon U-Pb dating and Lu-Hf Isotopes. *Acta Geol. Sin.* **2022**, *96*, 1516–1533. [CrossRef]

29. Jahn, B.; Auvray, B.; Shen, Q.; Liu, D.; Zhang, Z.; Dong, Y.; Ye, X.; Zhang, Q.; Cornichet, J.; Mace, J. Archean crustal evolution in China: The Taishan complex, and evidence for juvenile crustal addition from long-term depleted mantle. *Precambrian Res.* **1988**, *38*, 381–403. [CrossRef]

30. Wu, F.Y.; Zhang, Y.B.; Yang, J.H.; Xie, L.W.; Yang, Y.H. Zircon U–Pb and Hf isotopic constraints on the Early Archean crustal evolution in Anshan of the North China Craton. *Precambrian Res.* **2008**, *167*, 339–362. [CrossRef]

31. Zhai, M.G.; Santosh, M. The early Precambrian odyssey of the North China Craton: A synoptic overview. *Gondwana Res.* **2011**, *20*, 6–25. [CrossRef]

32. Zhao, G.; Sun, M.; Wilde, S.A.; Li, S. A Paleo-Mesoproterozoic supercontinent: Assembly, growth and breakup. *Earth-Sci. Rev.* **2004**, *67*, 91–123. [CrossRef]

33. Zhao, G.; Cawood, P.A.; Li, S.; Wilde, S.A.; Sun, M.; Zhang, J.; He, Y.; Yin, C. Amalgamation of the North China Craton: Key issues and discussion. *Precambrian Res.* **2012**, *222–223*, 55–76. [CrossRef]

34. Li, J.H.; Qian, X.L.; Huang, C.N.; Liu, S.W. Tectonic framework of North China Block and its cratonization in the Early Precambrian. *Acta Petrol. Sin.* **2000**, *16*, 1–10. (In Chinese)

35. Lu, L.Z.; Xu, X.C.; Liu, F.L. *Early Precambrian Khondalite Series in North China*; Changchun Publishing House: Changchun, China, 1996. (In Chinese)

36. Zhai, M.G. Precambrian tectonic evolution of the North China Craton. *Geol. Soc. Lond. Spec. Publ.* **2004**, *226*, 57–72. [CrossRef]

37. Bai, J.; Huang, X.G.; Dai, F.Y.; Wu, C.H. *The Precambrian Evolution of China*, 2nd ed.; Geological Publishing House: Beijing, China, 1996; pp. 36–38. (In Chinese)

38. Ma, X.H. *Precambrian Tectonic Framework and Research Methods in China*; Geological Publishing House: Beijing, China, 1987. (In Chinese)

39. Sun, D.Z.; Li, H.M.; Lin, Y.X.; Zhou, H.F.; Zhao, F.Q.; Tang, M. Precambrian geochronology, chronological framework and model of chronocrustal structures of Zhongtiao Mountains. *Acta Geol. Sin.* **1991**, *3*, 20–35. (In Chinese)

40. Wu, C.H. Metasedimentary rocks and tectonic division of North China Craton. *Geol. J. China Univ.* **2007**, *13*, 442–457. (In Chinese)

41. Zhao, G.; Wilde, S.; Cawood, P.A.; Sun, M. SHRIMP U-Pb zircon ages of the Fuping Complex: Implications for Late Archean to Paleoproterozoic accretion and assembly of the North China Craton. *Am. J. Sci.* **2002**, *302*, 191–226. [CrossRef]

42. Wang, Z.-Z.; Han, B.-F.; Feng, L.-X.; Liu, B.; Zheng, B.; Kong, L.-J. Tectonic attribution of the Langshan area in western Inner Mongolia and implications for the Neoarchean–Paleoproterozoic evolution of the Western North China Craton: Evidence from LA-ICP-MS zircon U–Pb dating of the Langshan basement. *Lithos* **2016**, *261*, 278–295. [CrossRef]

43. Zhang, J.; Li, J.; Xiao, W.; Wang, Y.; Qi, W. Kinematics and geochronology of multistage ductile deformation along the eastern Alxa block, NW China: New constraints on the relationship between the North China Plate and the Alxa block. *J. Struct. Geol.* **2013**, *57*, 38–57. [CrossRef]

44. Yang, Z.D.; Pan, X.S.; Yang, Y.F. *Geological Structure Characteristics and Deposits of Alxa Blocks and Adjacent Region*; Science Press: Beijing, China, 1988; pp. 1–254. (In Chinese)

45. Li, J.J. Regional Metallogenic System of Alashan Block in Inner Monolia Autonomous Region. Ph.D. Thesis, China University of Geoscience, Beijing, China, 2006. (In Chinese).

46. Geng, Y.S.; Wang, X.S.; Shen, Q.H.; Wu, C.M. Redefinition of the Alxa Group complex (Precambrian metamorphic basement) in the Alxa area, Inner Mongolia. *Geol. China* **2006**, *33*, 138–145. (In Chinese)

47. Geng, Y.S.; Wang, X.S.; Shen, Q.H.; Wu, C.M. Chronology of the Precambrian metamorphic series in the Alxa area, Inner Mongolia. *Geol. China* **2007**, *34*, 251–261. (In Chinese)

48. Geng, Y.S.; Zhou, X.W. Early Neoproterozoic granite events in Alxa area of Inner Mongolia and their geological significance: Evidence from geochronology. *Acta Petrol. Mineral.* **2010**, *29*, 779–795. (In Chinese)

49. Geng, Y.S.; Shen, Q.H.; Ren, L.D. Late Neoarchean to Early Paleoproterozoic magmatic events and tectonothermal systems in the North China Craton. *Acta Petrol. Sin.* **2010**, *26*, 1945–1966. (In Chinese)

50. Dan, W.; Li, X.H.; Guo, J.H.; Liu, Y.; Wang, X.C. Paleoproterozoic evolution of the eastern Alxa Block, westernmost North China: Evidence from in situ zircon U-Pb dating and Hf-O isotopes. *Gondwana Res.* **2012**, *21*, 838–864. [CrossRef]
51. Gong, J.; Zhang, J.; Yu, S.; Li, H.; Hou, K. Ca. 2.5 Ga TTG rocks in the western Alxa Block and their implications. *Chin. Sci. Bull.* **2012**, *57*, 4064–4076. [CrossRef]
52. Zhang, J.X.; Gong, J.H.; Yu, S.Y.; Li, H.K.; Hou, K.J. Neoarchean-Paleoproterozoic multiple tectonothermal events in the western Alxa block, North China Craton and their geological implication: Evidence from zircon U-Pb ages and Hf isotopic composition. *Precambrian Res.* **2013**, *235*, 36–57. [CrossRef]
53. Williams, I.S. U-Th-Pb geochronology by ion microprobe. In *Applications of Microanalytical Techniques to Understanding Mineralizing Processes*; McKibben, M.A., Shanks, W.C., Ridley, W.I., Eds.; Economic Geology Publishing Co.: Littleton, CO, USA, 1998; Volume 7, pp. 1–35.
54. Black, L.P.; Kamo, S.L.; Allen, C.M.; Aleinikoff, J.N.; Davis, D.W.; Korsch, R.J.; Foudoulis, C. TEMORA 1: A new zircon standard for Phanerozoic U–Pb geochronology. *Chem. Geol.* **2003**, *200*, 155–170. [CrossRef]
55. Ludwig, K.R. Manual for Isoplot 3.0: A Geochronological, Toolkit for Microsoft Excel. *Berkeley Geochronol. Cent. Spec. Publ.* **2003**, *4*, 1–71.
56. Zhang, S.-H.; Zhao, Y.; Ye, H.; Hu, G.-H. Early Neoproterozoic emplacement of the diabase sill swarms in the Liaodong Peninsula and pre-magmatic uplift of the southeastern North China Craton. *Precambrian Res.* **2016**, *272*, 203–225. [CrossRef]
57. Paton, C.; Hellstrom, J.; Paul, B.; Woodhead, J.; Hergt, J. Iolite: Freeware for the visualisation and processing of mass spectrometric data. *J. Anal. At. Spectrom.* **2011**, *26*, 2508–2518. [CrossRef]
58. Wu, F.-Y.; Yang, Y.-H.; Xie, L.-W.; Yang, J.-H.; Xu, P. Hf isotopic compositions of the standard zircons and baddeleyites used in U–Pb geochronology. *Chem. Geol.* **2006**, *234*, 105–126. [CrossRef]
59. Sláma, J.; Košler, J.; Condon, D.J.; Crowley, J.L.; Gerdes, A.; Hanchar, J.M.; Horstwood, M.S.A.; Morris, G.A.; Nasdala, L.; Norberg, N.; et al. Plešovice zircon–A new natural reference material for U-Pb and Hf isotopic microanalyses. *Chem. Geol.* **2008**, *249*, 1–35. [CrossRef]
60. Martin, H.; Smithies, R.; Rapp, R.; Moyen, J.-F.; Champion, D. An overview of adakite, tonalite–trondhjemite–granodiorite (TTG), and sanukitoid: Relationships and some implications for crustal evolution. *Lithos* **2005**, *79*, 1–24. [CrossRef]
61. Grant, M.L.; Wilde, S.A.; Wu, F.Y.; Yang, J.H. The application of zircon cathodoluminescence imaging, Th-U-Pb chemistry and U-Pb ages in interpreting discrete magmatic and high-grade metamorphic events in the North China Craton at the Archean/Proterozoic boundary. *Chem. Geol.* **2009**, *261*, 155–171. [CrossRef]
62. Xiong, X.; Adam, J.; Green, T. Rutile stability and rutile/melt HFSE partitioning during partial melting of hydrous basalt: Implications for TTG genesis. *Chem. Geol.* **2005**, *218*, 339–359. [CrossRef]
63. Green, T.H. Significance of Nb/Ta as an indicator of geochemical processes in the crust-mantle system. *Chem. Geol.* **1995**, *120*, 347–359. [CrossRef]
64. Rapp, R.P.; Shimizu, N.; Norman, M.D. Growth of early continental crust by partial melting of eclogite. *Nature* **2003**, *425*, 605–609. [CrossRef]
65. Huang, X.L.; Niu, Y.; Xu, Y.G.; Yang, Q.J.; Zhong, J.W. Geochemistry of TTG and TTG-like gneisses from Lushan-Taihua complex in the southern North China Craton: Implications for late Archean crustal accretion. *Precambrian Res.* **2010**, *182*, 43–56. [CrossRef]
66. Jochum, K.P.; Stolz, A.J.; McOrist, G. Niobium and tantalum in carbonaceous chondrites: Constraints on the solar system and primitive mantle niobium/tantalum, zirconium/niobium, and niobium/uranium ratio. *Meteorit. Planet. Sci.* **2000**, *35*, 229–235. [CrossRef]
67. Münker, C.; Pfänder, J.A.; Weyer, S.; Büchl, A.; Kleine, T.; Mezger, K. Evolution of Planetary Cores and the Earth-Moon System from Nb/Ta Systematics. *Science* **2003**, *301*, 84–87. [CrossRef]
68. Moyen, J.F. The composite Archean grey gneisses: Petrological significance, and evidence for a non-unique tectonic setting for Archean crustal growth. *Lithos* **2011**, *123*, 21–36. [CrossRef]
69. Rapp, R.; Shimizu, N.; Norman, M.; Applegate, G. Reaction between slab-derived melts and peridotite in the mantle wedge: Experimental constraints at 3.8 GPa. *Chem. Geol.* **1999**, *160*, 335–356. [CrossRef]
70. Shan, H.; Zhai, M.; Dey, S. Petrogenesis of Two Types of Archean TTGs in the North China Craton: A Case Study of Intercalated TTGs in Lushan and Non-intercalated TTGs in Hengshan. *Acta Geol. Sin.—Engl. Ed.* **2016**, *90*, 2049–2065. [CrossRef]
71. Condie, K.C. TTGs and adakites: Are they both slab melts? *Lithos* **2005**, *80*, 33–44. [CrossRef]
72. Drummond, M.S.; Defant, M.J.; Kepezhinskas, P.K.; Brown, M.; Candela, P.; Peck, D.; Stephens, W.; Walker, R.; Zen, E.-A. Petrogenesis of Slab-Derived Trondhjemite–Tonalite-Dacite/Adakite Magmas. *Trans. R. Soc. Edinb.—Earth Sci.* **1996**, *87*, 205–215. [CrossRef]
73. Martin, H. Adakitic magmas: Modern analogues of Archaean granitoids. *Lithos* **1999**, *46*, 411–429. [CrossRef]
74. Bédard, J.H. A catalytic delamination-driven model for coupled genesis of Archaean crust and sub-continental lithospheric mantle. *Geochim. Cosmochim. Acta* **2006**, *70*, 1188–1214. [CrossRef]
75. Condie, K.C.; Beyer, E.; Belousova, E.; Griffin, W.; O'reilly, S.Y. U–Pb isotopic ages and Hf isotopic composition of single zircons: The search for juvenile Precambrian continental crust. *Precambrian Res.* **2005**, *139*, 42–100. [CrossRef]
76. Smithies, R. The Archaean tonalite–trondhjemite–granodiorite (TTG) series is not an analogue of Cenozoic adakite. *Earth Planet. Sci. Lett.* **2000**, *182*, 115–125. [CrossRef]

77. Zegers, T.E.; van Keken, P.E. Middle Archean continent formation by crustal de-lamination. *Geology* **2001**, *29*, 1083–1086. [CrossRef]

78. Rapp, R.P.; Watson, E.B. Dehydratation melting of metabasalt at 8–32 kbar: Implications for continental growth and crust-mantle recycling. *J. Petrol.* **1995**, *36*, 891–931. [CrossRef]

79. Smithies, R.H.; Champion, D.C. The Archaean high-Mg diorite suite: Links to tonalite–trondhjemite–granodiorite magmatism and implications for early Archaean crustal growth. *J. Petrol.* **2000**, *41*, 1653–1671. [CrossRef]

80. Moyen, J.F. High Sr/Y and La/Yb ratios: The meaning of the "adakitic signature". *Lithos* **2009**, *112*, 556–574. [CrossRef]

81. Castillo, P.R. An overview of adakite petrogenesis. *Chin. Sci. Bull.* **2006**, *51*, 257–268. [CrossRef]

82. Wang, Q.; Wyman, D.A.; Xu, J.; Zhao, Z.; Jian, P.; Zi, F. Partial Melting of Thickened or Delaminated Lower Crust in the Middle of Eastern China: Implications for Cu-Au Mineralization. *J. Geol.* **2007**, *115*, 149–161. [CrossRef]

83. Xu, J.F.; Shinjo, R.; Defant, M.J.; Wang, Q.; Rapp, R.P. Origin of Mesozoic adakitic intrusive rocks in the Ningzhen area of east China: Partial melting of delaminated lower continental crust? *Geology* **2002**, *30*, 1111–1114. [CrossRef]

84. Wu, F.Y.; Li, X.H.; Zheng, Y.F.; Gao, S. Lu–Hf isotopic systematics and their applications in petrology. *Acta Petrol. Sin.* **2007**, *23*, 185–220. (In Chinese)

85. Dong, C.Y.; Liu, D.Y.; Li, J.J.; Wan, Y.S.; Zhou, H.Y.; Li, C.D.; Yang, Y.H.; Xie, L.W. New evidences from the Bayanwulashan-Helanshan region for the formation age of Khondalite Belt: SHRIMP zircon dating and Hf isotopic composition. *Chin. Sci. Bull.* **2007**, *52*, 1913–1922. (In Chinese) [CrossRef]

86. Dong, C.; Wan, Y.; Wilde, S.A.; Xu, Z.; Ma, M.; Xie, H.; Liu, D. Earliest Paleoproterozoic supracrustal rocks in the North China Craton recognized from the Daqingshan area of the Khondalite Belt: Constraints on craton evolution. *Gondwana Res.* **2014**, *25*, 1535–1553. [CrossRef]

87. Krŏner, A.; Wilde, S.A.; Li, J.H.; Wang, K.Y. Age and evolution of a late Archean to early Palaeozoic upper to lower crustal section in the Wutaishan/Hengshan/Fuping terrain of northern China. *J. Asian Earth Sci.* **2005**, *24*, 577–595. [CrossRef]

88. Santosh, M.; Tsunogae, T.; Ohyama, H.; Sato, K.; Li, J.; Liu, S. Carbonic metamorphism at ultrahigh-temperatures: Evidence from North China Craton. *Earth Planet. Sci. Lett.* **2008**, *266*, 149–165. [CrossRef]

89. Wilde, S.A.; Zhao, G.C.; Sun, M. Development of the North China Craton during late Archaean and its final amalgamation at 1.8 Ga: Some speculations on its position within a global palaeoproterozoic supercontinent. *Gondwana Res.* **2002**, *5*, 85–94. [CrossRef]

90. Zhao, G.C.; Wilde, S.A.; Sun, M.; Li, S.Z.; Li, X.P.; Zhang, J. SHRIMP U-Pb zircon ages of granitoid rocks in the Lvliang Complex: Implications for the accretion and evolution of the Trans-North China Orogen. *Precambrian Res.* **2008**, *160*, 213–226. [CrossRef]

91. Condie, K.C.; Aster, R.C. Episodic zircon age spectra of orogenic granitoids: The supercontinent connection and continental growth. *Precambrian Res.* **2010**, *180*, 227–236. [CrossRef]

92. Rogers, J.J.; Santosh, M. Configuration of Columbia, a Mesoproterozoic Supercontinent. *Gondwana Res.* **2002**, *5*, 5–22. [CrossRef]

93. Rogers, J.J.; Santosh, M. Tectonics and surface effects of the supercontinent Columbia. *Gondwana Res.* **2009**, *15*, 373–380. [CrossRef]

94. Santosh, M. Assembling North China Craton within the Columbia supercontinent: The role of double-sided subduction. *Precambrian Res.* **2010**, *178*, 149–167. [CrossRef]

95. Zhang, C.L.; Li, H.K.; Santosh, M.; Li, Z.X.; Zou, H.B.; Wang, H.; Ye, H. Precambrian evolution and cratonization of the Tarim Block, NW China: Petrology, geochemistry, Nd isotopes and U-Pb zircon geochronology from Archaean gabbro-TTG-potassic granite suite and Paleoproterozoic metamorphic belt. *J. Asian Earth Sci.* **2012**, *47*, 5–20. [CrossRef]

96. Zhao, G.C.; Cawood, P.A. Precambrian geology of China. *Precambrian Res.* **2012**, *222–223*, 13–54. [CrossRef]

97. Qiu, Y.M.; Gao, S.; McNaughton, N.J.; Groves, D.I.; Ling, W.L. First evidence of >3.2 Ga continental crust in the Yangtze craton of south China and its implications for Archean crustal evolution and Phanerozoic tectonics. *Geology* **2000**, *28*, 11–14. [CrossRef]

98. Ge, R.F.; Zhu, W.B.; Wilde, S.A.; Wu, H.L.; He, J.W.; Zheng, B.H. Archean magmatism and crustal evolution in the northern Tarim Craton: Insights from zircon U-Pb-Hf-O isotopes and geochemistry of ~2.7 Ga orthogneiss and amphibolite in the Korla Complex. *Precambrian Res.* **2014**, *252*, 145–165. [CrossRef]

99. Long, X.P.; Chao, Y.; Sun, M.; Zhao, G.C.; Xiao, W.J.; Wan Yu, J.; Yang, Y.H.; Hu, A.Q. Archean crustal evolution of the northern Tarim craton, NW China: Zircon U-Pb and Hf isotopic constraints. *Precambrian Res.* **2010**, *180*, 272–284. [CrossRef]

100. Wan, Y.S.; Zhang, Z.Q.; Wu, J.S.; Song, B.; Liu, D.Y. Geochemical and Nd isotopic characteristic of some rocks from the Paleoarchean Chentaigou supracrustal, Anshan area, NE China. *Cont. Dyn.* **1997**, *2*, 39–46.

101. Wan, Y.S.; Dong, C.Y.; Ren, P.; Bai, W.Q.; Xie, H.Q.; Liu, S.J.; Xie, S.W.; Liu, D.Y. Spatial and temporal distribution, compositional characteristics and formation and evolution of Archean TTG rocks in the North China Craton: A synthesis. *Acta Petrol. Sin.* **2017**, *33*, 1405–1419. (In Chinese)

102. Zhang, S.-B.; Zheng, Y.-F.; Wu, Y.-B.; Zhao, Z.-F.; Gao, S.; Wu, F.-Y. Zircon U-Pb age and Hf-O isotope evidence for Paleoproterozoic metamorphic event in South China. *Precambrian Res.* **2006**, *151*, 265–288. [CrossRef]

103. Guo, J.-L.; Gao, S.; Wu, Y.-B.; Li, M.; Chen, K.; Hu, Z.-C.; Liang, Z.-W.; Liu, Y.-S.; Zhou, L.; Zong, K.-Q.; et al. 3.45 Ga granitic gneisses from the Yangtze Craton, South China: Implications for Early Archean crustal growth. *Precambrian Res.* **2014**, *242*, 82–95. [CrossRef]

104. Guo, J.-L.; Wu, Y.-B.; Gao, S.; Jin, Z.-M.; Zong, K.-Q.; Hu, Z.-C.; Chen, K.; Chen, H.-H.; Liu, Y.-S. Episodic Paleoarchean-Paleoproterozoic (3.3–2.0 Ga) granitoid magmatism in Yangtze Craton, South China: Implications for late Archean tectonics. *Precambrian Res.* **2015**, *270*, 246–266. [CrossRef]

105. Qiu, X.-F.; Ling, W.-L.; Liu, X.-M.; Lu, S.-S.; Jiang, T.; Wei, Y.-X.; Peng, L.-H.; Tan, J.-J. Evolution of the Archean continental crust in the nucleus of the Yangtze block: Evidence from geochemistry of 3.0 Ga TTG gneisses in the Kongling high-grade metamorphic terrane, South China. *J. Asian Earth Sci.* **2018**, *154*, 149–161. [CrossRef]
106. Wu, M.; Zhao, G.; Sun, M.; Li, S.; Bao, Z.; Tam, P.Y.; Eizenhöefer, P.R.; He, Y. Zircon U–Pb geochronology and Hf isotopes of major lithologies from the Jiaodong Terrane: Implications for the crustal evolution of the Eastern Block of the North China Craton. *Lithos* **2014**, *190–191*, 71–84. [CrossRef]
107. Zhou, G.; Wu, Y.; Li, L.; Zhang, W.; Zheng, J.; Wang, H.; Yang, S. Identification of ca. 2.65 Ga TTGs in the Yudongzi complex and its implications for the early evolution of the Yangtze Block. *Precambrian Res.* **2018**, *314*, 240–263. [CrossRef]
108. Diwu, C.; Sun, Y.; Guo, A.; Wang, H.; Liu, X. Crustal growth in the North China Craton at ~2.5Ga: Evidence from in situ zircon U–Pb ages, Hf isotopes and whole-rock geochemistry of the Dengfeng complex. *Gondwana Res.* **2011**, *20*, 149–170. [CrossRef]
109. Geng, Y.S.; Yang, C.H.; Du, L.L.; Ren, L.D.; Song, H.X. Late Neoarchean magmatism and crustal growth in eastern Hebei: Constraint from geochemistry, zircon U-Pb ages and Hf isotope. *Acta Petrol. Sin.* **2018**, *34*, 1058–1082. (In Chinese)
110. Kang, J.L.; Wang, H.C.; Xiao, Z.B.; Sun, Y.W.; Zeng, L.; Pang, Z.B.; Li, J.R. Neoarchean crustal accretion of the North China Craton: Evidence from the TTG gneisses and monzogranitic gneisses in Yunzhong Mountain area, Shanxi. *Acta Petrol. Sin.* **2017**, *33*, 2881–2898.
111. Liu, J.H.; Liu, F.L.; Ding, Z.J.; Liu, C.H.; Yang, H.; Liu, P.H.; Wang, F.; Meng, N. The growth, reworking and metamorphism of early Precambrian crust in the Jiaobei terrane, the North China Craton: Constraints from U-Th-Pb and Lu-Hf isotopic systematics, and REE concentrations of zircon from Archean granitoid gneisses. *Precambrian Res.* **2013**, *224*, 287–303. [CrossRef]
112. Luo, Z.X.; Huang, X.L.; Wang, Y.; Yang, F.; Han, L. Geochronology and Geochemistry of the TTG Gneisses from the Taihua Group in the Xiaoshan Area, North China Craton: Constraints on Petrogenesis. *Geotecton. Metallog.* **2018**, *42*, 332–347. (In Chinese)
113. Song, H.X.; Yang, C.H.; Du, L.L.; Ren, L.D.; Geng, Y.S. Delineation of the ~2.7Ga TTG gneisses in Zanhuang Complex, Hebei Province, and its geological significance. *Acta Petrol. Sin.* **2018**, *34*, 1599–1611. (In Chinese)
114. Wang, W.; Zhai, M.G.; Li, T.S.; Santosh, M.; Zhao, L.; Wang, H.Z. Archean-Paleoproterozoic crustal evolution in the eastern North China Craton: Zircon U-Th-Pb and Lu- Hf evidence from the Jiaobei terrane. *Precambrian Res.* **2014**, *241*, 146–160. [CrossRef]
115. Zhang, S.-B.; Tang, J.; Zheng, Y.-F. Contrasting Lu–Hf isotopes in zircon from Precambrian metamorphic rocks in the Jiaodong Peninsula: Constraints on the tectonic suture between North China and South China. *Precambrian Res.* **2014**, *245*, 29–50. [CrossRef]
116. Zhang, J.X.; Yu, S.Y.; Gong, J.H.; Li, H.K.; Hou, K.J. The latest Neoarchean-Paleoproterozoic evolution of the Dunhuang block, eastern Tarim craton, northwestern China: Evidence from zircon U–Pb dating and Hf isotopic analyses. *Precambrian Res.* **2013**, *226*, 21–42. [CrossRef]
117. Zong, K.; Liu, Y.; Zhang, Z.; He, Z.; Hu, Z.; Guo, J.; Chen, K. The generation and evolution of Archean continental crust in the Dunhuang block, northeastern Tarim craton, northwestern China. *Precambrian Res.* **2013**, *235*, 251–263. [CrossRef]
118. Zhou, G.; Wu, Y.; Gao, S.; Yang, J.; Zheng, J.; Qin, Z.; Wang, H.; Yang, S. The 2.65 Ga A-type granite in the northeastern Yangtze craton: Petrogenesis and geological implications. *Precambrian Res.* **2015**, *258*, 247–259. [CrossRef]
119. Hui, B.; Dong, Y.; Cheng, C.; Long, X.; Liu, X.; Yang, Z.; Sun, S.; Zhang, F.; Varga, J. Zircon U–Pb chronology, Hf isotope analysis and whole-rock geochemistry for the Neoarchean-Paleoproterozoic Yudongzi complex, northwestern margin of the Yangtze craton, China. *Precambrian Res.* **2017**, *301*, 65–85. [CrossRef]
120. Wang, Z.; Deng, Q.; Duan, T.; Yang, F.; Du, Q.; Xiong, X.; Liu, H.; Cao, B. 2.85 Ga and 2.73 Ga A-type granites and 2.75 Ga trondhjemite from the Zhongxiang Terrain: Implications for early crustal evolution of the Yangtze Craton, South China. *Gondwana Res.* **2018**, *61*, 1–19. [CrossRef]
121. Liu, D.Y.; Nutman, A.P.; Compston, W.; Wu, J.S.; Shen, Q.H. Remnants of ≥3800 Ma crust in the Chinese part of the Sino-Korean craton. *Geology* **1992**, *20*, 339–342. [CrossRef]
122. Liu, D.Y.; Wan, Y.S.; Wu, J.S.; Wilde, S.A.; Dong, C.Y.; Zhou, H.Y.; Yin, X.Y. Archean Crustal evolution and the oldest rocks in the North China Craton. *Geol. Bull. China* **2007**, *26*, 1131–1138. (In Chinese)
123. Liu, D.; Wilde, S.; Wan, Y.; Wu, J.; Zhou, H.; Dong, C.; Yin, X. New U-Pb and Hf isotopic data confirm Anshan as the oldest preserved segment of the North China Craton. *Am. J. Sci.* **2008**, *308*, 200–231. [CrossRef]
124. Song, B.; Nutman, A.P.; Liu, D.Y.; Wu, J.S. 3800 to 2500 Ma crustal evolution in the Anshan area of Liaoning Province, northern China. *Precambrian Res.* **1996**, *78*, 79–94. [CrossRef]
125. Wan, Y.S.; Xie, H.Q.; Wang, H.C.; Li, P.C.; Chu, H.; Xiao, Z.B.; Dong, C.Y.; Liu, S.J.; Li, Y.; Hao, G.M.; et al. Discovery of ~3.8 Ga TTG rocks in eastern Hebei, North China Croton. *Acta Geol. Sin.* **2021**, *95*, 1321–1333. (In Chinese)
126. Condie, K.C.; Belousova, E.; Griffin, W.; Sircombe, K.N. Granitoid events in space and time: Constraints from igneous and detrital zircon age spectra. *Gondwana Res.* **2009**, *15*, 228–242. [CrossRef]
127. Jiang, N.; Guo, J.H.; Zhai, M.G.; Zhang, S.Q. 2.7 Ga crust growth in the North China craton. *Precambrian Res.* **2010**, *179*, 37–49. [CrossRef]

Article

Metamorphic Ages and *PT* Conditions of Amphibolites in the Diebusige and Bayanwulashan Complexes of the Alxa Block, North China Craton

Feng Zhou [1], Longlong Gou [1,*], Xiaofei Xu [1,2] and Zhibo Tian [1]

1   State Key Laboratory of Continental Dynamics, Department of Geology, Northwest University, Xi'an 710069, China; xuxiaofei@kust.edu.cn (X.X.)
2   Faculty of Land Resources Engineering, Kunming University of Science and Technology, Kunming 650093, China
*   Correspondence: LLgou@nwu.edu.cn

Abstract: The metamorphism and geological significance of amphibolites in the Diebusige and Bayanwulashan Complexes of the eastern Alxa Block, North China Craton, were poorly understood until now. This study presents the results of petrology, laser ablation inductively coupled plasma mass spectrometry (LA-ICP-MS) zircon U–Pb analysis, phase equilibrium modeling and geothermobarometry for these rocks. The peak mineral assemblage of clinopyroxene + hornblende + plagioclase + K-feldspar + ilmenite + quartz + melt is inferred for amphibolite sample ALS2164 in the Diebusige Complex. Correspondingly, the peak mineral assemblage of clinopyroxene + hornblende + plagioclase ± K-feldspar + ilmenite + quartz + melt is identified for amphibolite sample ALS2191 in the Bayanwulshan Complex. Phase equilibrium modelling constrained the peak metamorphic condition of amphibolite sample ALS2164 in the Diebusige Complex to be 825–910 °C/7.2–10.8 kbar, which is similar to that (800–870 °C/7.0–10.7 kbar) of amphibolite sample ALS2191 in the Bayanwulashan Complex. Hbl–pl–qz thermobarometry yielded the metamorphic *PT* conditions of 732–810 °C/3.0–6.7 kbar for these amphibolites, which are consistent with the average temperatures of 763 °C, 768 °C and 780 °C calculated by Ti-zircon thermometry. As a result, phase equilibrium modelling yielded wide *PT* condition ranges of 800–910 °C/7.0–10.8 kbar, the lower limit of which is consistent with the upper limit of estimates by the hbl–pl–qz thermobarometer. In addition, LA-ICP-MS U–Pb analysis on metamorphic zircons yielded weighted mean $^{207}Pb/^{206}Pb$ ages of 1901 ± 22–1817 ± 21 Ma, which represent the timing of amphibolite-facies metamorphism. As a whole, the *PT* estimates display a high geothermal gradient, which is consistent with coeval ultrahigh-temperature metamorphism and associated mantle-derived mafic-ultramafic rocks in the Diebusige Complex. Combing this information with the previously published data from the Diebusige Complex, an extensional setting after continental collision is inferred for the eastern Alxa Block during the late Paleoproterozoic. The HREE enrichment patterns of metamorphic zircons from the amphibolites in this study are in agreement with that these amphibolites formed at relatively shallower crust than the garnet-bearing mafic granulites in the Diebusige Complex.

Keywords: amphibolites; the Alxa Block; the Khondalite Belt; zircon U-Pb age; extensional setting

Citation: Zhou, F.; Gou, L.; Xu, X.; Tian, Z. Metamorphic Ages and *PT* Conditions of Amphibolites in the Diebusige and Bayanwulashan Complexes of the Alxa Block, North China Craton. *Minerals* **2023**, *13*, 1426. https://doi.org/10.3390/min13111426

Academic Editors: Jin Liu, Jiahui Qian, Xiaoguang Liu and Silvana Martin

Received: 29 August 2023
Revised: 31 October 2023
Accepted: 3 November 2023
Published: 9 November 2023

1. Introduction

Amphibolites, one of the common metabasic rocks in metamorphic terranes from Precambrian to Phanerozoic, can provide valuable information for the study of geological processes [1,2]. Previous experimental studies suggest that amphibolites can be stable in a relatively wide *PT* range of 0–15 kbar/500–950 °C [3–7]. Afterwards, some researchers proposed that the upper temperature limit of amphibolite-facies metamorphism is 700–850 °C [8–10], which is lower than the results determined by experimental studies. In terms of origin, amphibolite can be formed by metamorphism of mafic igneous rock (ortho-amphibolite) or sedimentary tuff (para-amphibolite) (two different origins of amphibolites in the Xiangshan area of Jiangxi province [11]). Amphibolite may be prograde products of Barrow-type metamorphism [4,12,13] with a clockwise *P–T* path, reflecting the collision thickening environment, such as the amphibolite in the Wutai–Hengshan area [14]. It can also be retrograde products of overprinted high-pressure/ultrahigh-pressure (HP/UHP) eclogites, where the *P–T* path is dominated by decompression [7,15,16], such as the amphibolite in the Qingyouhe area of Qinling Complex [17]. The formation of amphibolite may be related to the underplating of mantle-derived magmas, which is characterized by an anticlockwise *P–T* path (amphibolite in the northern Liaoning Complex [18]). In addition, amphibolite with a decompressional heating *P–T* segment can reflect an extensional setting that developed on a previous orogen (amphibolite in the Xilingol Complex of Inner Mongolia [19]).

The Alxa Block is located in the westernmost part of the North China Craton (NCC), adjacent to the Khondalite Belt (KB) (Figure 1a). In recent decades, some researchers considered that the Alxa Block has an affinity to the Yangtze Craton or the Tarim Craton due to the identification of the Neoproterozoic magmatism [20,21], whereas others thought that the Alxa Block was the western extension of the Yinshan Block, NCC [22,23]. Additionally, some scholars argued that the Alxa Block is an independent block or a microcontinental block in a Paleozoic orogenic belt [24–26]. Recently, more and more evidence has supported the Alxa Block as the westward extension of the KB due to the discovery of ca. 1.96–1.83 Ga metamorphism, ca. 2.3–2.0 Ga magmatism and small-scale Neoarchean magmatism in this block [27–35]. Therefore, the relationship between the Alxa Block and the NCC has become the focus of controversy, and research on metamorphism can provide important insights for solving this dispute. Furthermore, the Diebusige and Bayanwulashan Complexes distributed discontinuously along the eastern Alxa Block (Figure 1b) are composed of felsic orthogneiss, amphibolite lenses or interlayers, and a few metasedimentary rocks [36,37]. Previous work revealed that these amphibolite lenses or interlayers had metamorphic ages of ca. 1.93–1.81 Ga [36,37], which are comparable to published metamorphic ages of ca. 1.96–1.83 Ga for high- to ultrahigh-temperature (HT–HUT) metamorphic rocks in the Diebusige Complex [33–35]. Due to the lack of detailed metamorphic *PT* condition calculations for these amphibolites, their relationship with HT–UHT metamorphic rocks, and the implications for tectonic evolution of the Alxa Block are unclear until now.

In this study, we describe amphibolites in the Diebusige and Bayanwulashan Complexes of the eastern Alxa Block, NCC. Detailed studies of petrology, U–Pb ages and trace elements of metamorphic zircons, and metamorphic *PT* condition calculations using phase equilibrium modelling and conventional thermobarometries are used to reveal the metamorphic ages and characteristics of these amphibolites. Metamorphic ages and *PT* conditions constrained for amphibolites in this study, coupled with previously published data in the region, provide us new insight to elucidate their tectonic significance of the Alxa Block.

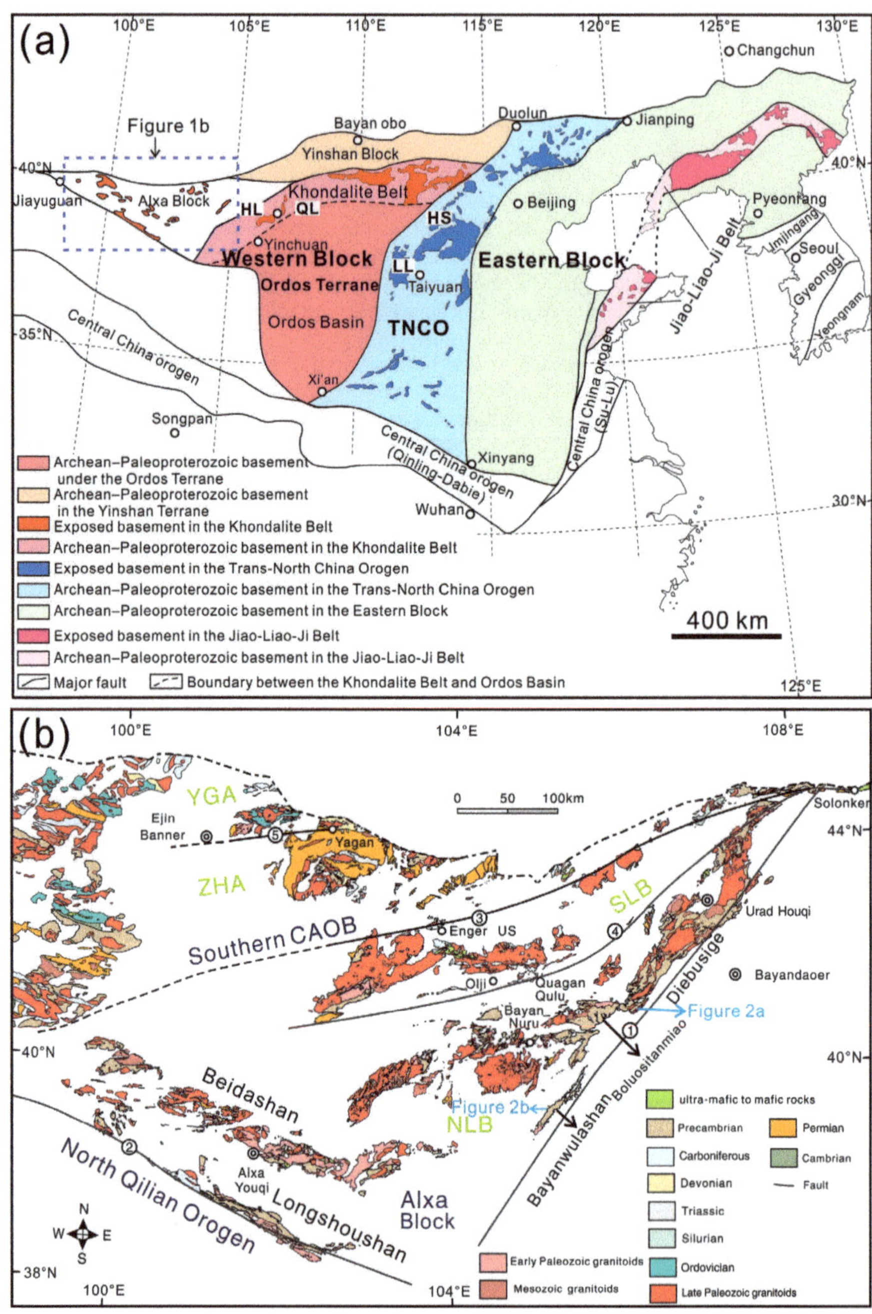

Figure 1. (**a**) Simplified tectonic sketch map of the North China Craton (NCC) showing the location of the Alxa Block and (**b**) geological map of the Alxa Block (revised from BGMRIM (1991) [38]). CAOB, Central Asian Orogenic Belt; YGA, the Yagan arc; ZHA, the Zhusileng–Hangwula arc; SLB, the Shalazhashan belt; NLB, the Nuru–Langshan belt. ① The Langshan fault, ② the Longshoushan fault, ③ the Enger US fault, ④ the Badain Jaran fault, ⑤ the Yagan fault.

2. Geological Setting

The westernmost part of the NCC, termed as the Alxa block, is bounded by the Central Asian Orogenic Belt and the Qilian Orogen to the south, respectively (Figure 1a,b). The Alxa Block is mainly covered by Cenozoic sediments, and the outcrops of Precambrian metamorphic basement rocks are sporadically exposed in the western and eastern parts of the block (Figure 1b). The Precambrian basement of the western Alxa Block is mainly composed of the Longshoushan and Beidashan Complexes [27,30,39,40] (Figure 1b), and the Precambrian basement of the eastern Alxa Block consists of the Diebusige, Bayanwulashan and Boluositanmiao Complexes [29,36,37,41,42] (Figure 1b).

2.1. The Western Alxa Block

The Longshoushan Complex is composed of felsic gneisses, amphibolites and metasedimentary rocks [30,31,43–46]. The crystallization ages for the protolith of felsic gneisses were constrained to be ca. 2.35–2.01 Ga, and their metamorphic ages vary from ca. 1.97 to 1.84 Ga [30,31,43–45,47]. The metamorphic ages of amphibolites have been determined at ca. 1.85 Ga [31]. Metasedimentary rocks have detrital zircon ages ranging from ca. 3.0 to 1.98 Ga, and their metamorphic ages are ca. 1.96–1.92 Ga [31,45,46].

The Beidashan Complex is predominantly composed of granodioritic–trondhjemitic gneisses, whose protolith crystallization ages are ca. 2.84–2.50 Ga, and their metamorphic ages are constrained at ca. 2.53–2.47 Ga and ca. 1.87–1.83 Ga [27,39]. The sequence of metamorphic events is similar to those of the NCC, thereby the Alxa Block was considered as the western extension of the Khondalite Belt or an integrated component of the NCC [27,45].

2.2. The Eastern Alxa Block

The Diebusige Complex mainly consists of felsic gneisses, amphibolite, and minor metasedimentary rocks (e.g., banded iron formations and pelitic gneisses) [29,32–36,42]. SHRIMP and SIMS zircon U–Pb dating yielded protolith crystallization ages of ca. 1.98–1.97 Ga and the metamorphic ages of ca. 1.93 and 1.80 Ga for felsic gneisses [29,36,42]. The metamorphic ages of 1.84 and 1.85 Ga were obtained for magnetite quartzites, and peak metamorphic temperature was estimated at ~860–800 °C by the two-pyroxene geothermometer [33]. The metamorphic ages of corundum-bearing garnet–sillimanite gneisses are constrained to be ca. 1.95–1.83 Ga, which recorded a clockwise *P–T* path with peak metamorphic conditions of ~890–940 °C at ~7.5–9.8 kbar [34]. Wang et al. (2023) [35] obtained a clockwise *P–T* path with the peak metamorphic condition of 948–1048 °C/11–14 kbar, and metamorphic ages of ca. 1.96–1.86 Ga for mafic granulites.

The Boluositanmiao Complex is mainly composed of felsic gneisses, and the protolith crystallization ages of these felsic gneisses are determined at ca. 1.84–1.82 Ga [30,48,49].

In the Bayanwulashan Complex, there are felsic gneisses, amphibolites and metasedimentary rocks [28,36,37,42,48,50,51]. Zircon U–Pb dating indicates that the crystallization age for the protolith of felsic gneisses is ca. 2.40 to 2.23 Ga [48,50], whereas the metamorphic ages of these rocks have been constrained to be ca. 1.96–1.77 Ga [28,36,37]. The crystallization ages for the protolith of amphibolites were dated at ca. 2.30–2.34 Ga, which have metamorphic ages of ca. 1.94–1.80 Ga [36,51]. The detrital zircons from metasedimentary rocks (e.g., felsic paragneiss) range from ca. 2.50 to 2.27 Ga, which underwent ca. 1.90–1.80 Ga metamorphism [36,42]. These late Paleoproterozoic metamorphic ages in the Diebusige and Bayanwulashan Complexes indicate that the Alxa Block may be the westward extension of the KB [27,30,33–35].

3. Analytical Methods

The mineral compositions (Tables S1–S3) were analyzed using the JEOL JXA-8230 electron microprobe (EMP) at the State Key Laboratory of Continental Dynamics (SKLCD), Northwest University, Xi'an. The operating conditions are a 15 kV accelerating voltage, a beam current of 10 nA, and a beam diameter of 2 μm.

The whole-rock compositions were determined by X-ray fluorescence (XRF) analyses (Table 1) at the SKLCD, Northwest University, Xi'an. The FeO compositions were measured by a titration method at the Regional Geology and Mineral Resources Institute, Langfang, Hebei province.

Separated zircon grains were mounted in epoxy resin and polished to expose grain centers. Cathodoluminescence (CL) imaging was carried out at Tuoyan Analytical Technology Co., Ltd., Guangzhou, which was used to identify internal structures of zircons. Zircon U–Pb dating and rare earth element (REE) analysis were simultaneously carried out by using laser ablation inductively coupled plasma mass spectrometry (LA-ICP-MS) at the Wuhan Sample Solution Analytical Technology Co., Ltd., Wuhan, China. The laser spot was 24 μm in size, and a laser frequency of 5 Hz was used. Harvard zircon 91500 and glass NIST610 were selected as external standards. Detailed operating conditions, analytical procedures, and data processing methods can be found in Liu et al. (2010) [52].

4. Field Occurrence and Petrology

4.1. Field Occurrence

The Diebusige and Bayanwulashan Complexes mainly comprise felsic gneiss with amphibolite lenses or interlayers (Figure 2a,b). Amphibolites display a homogeneous distribution of leucosomes (Figure 3a,e,f), suggesting they have experienced partial melting (Figure 3a–f). Amphibolite samples ALS2164 and ALS2166 were collected from the Diebusige Complex, whereas amphibolite samples ALS2191 and ALS2156 are from the Bayanwulashan Complex. Detailed petrography and mineral chemistry were carried out for three amphibolite samples (i.e., ALS2164, ALS2166 and ALS2191).

4.2. Petrology and Mineral Chemistry

Petrographic features are showed in Figures 4 and 5, whereas classifications of mineral composition are presented in Figure 6. The mineral abbreviations in this study follow Whitney and Evans (2010) [53]. Hornblende compositions are plotted on the Leake classification diagram [54].

4.2.1. Amphibolite Sample ALS2164

Amphibolite sample ALS2164 displays a porphyroblastic texture and is composed of clinopyroxene (~15%), plagioclase (~45%), hornblende (~30%), K-feldspar (~8%), minor quartz (~2%), with accessory zircon, apatite, ilmenite and hematite (Figure 4a–f). Plagioclase occurs as coarse-grained crystals and is in textural equilibrium with clinopyroxene, hornblende and K-feldspar (Figure 4a–f). Minor quartz appears as interstitial crystals between hornblende and clinopyroxene (Figure 4e,f). Ilmenite and hematite are present as interstitial grains (Figure 4d,e). Clinopyroxenes are diopside in composition (Table S1; Figure 6a). Plagioclases have X_{An} of 0.32–0.34 and are andesine in composition (Table S1; Figure 6b). Hornblendes are pargasite and magnesio-hastingsite in composition, with Ti = 0.22–0.28 cpfu. and X_{Mg} = 0.60–0.62 (Table S1; Figure 6d).

4.2.2. Amphibolite Sample ALS2166

Amphibolite sample ALS2166 is composed of clinopyroxene (~15%), plagioclase (~55%), hornblende (~30%) with accessory ilmenite, apatite and zircon (Figure 4g,h). Ilmenite is present as interstitial grains (Figure 4g,h). Clinopyroxenes in this sample are diopside in composition (Table S2; Figure 6a), which are similar to those of clinopyroxene from sample ALS2164. Plagioclases display andesine composition with homogeneous X_{An} value (0.34–0.37) (Table S2; Figure 6b). Coarse-grained hornblendes are ferropargasite or pargasite, with X_{Mg} values of 0.45–0.49 and Ti contents of 0.20–0.29 cpfu. (Table S2; Figure 6d).

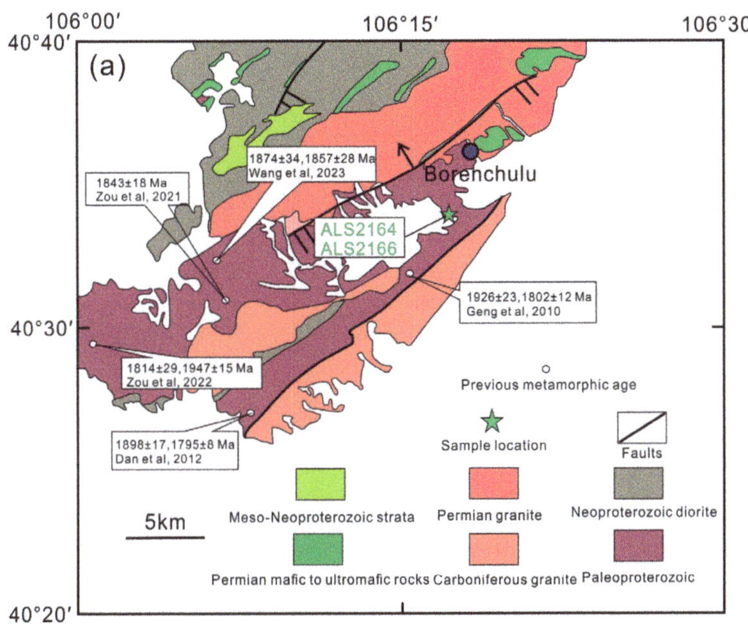

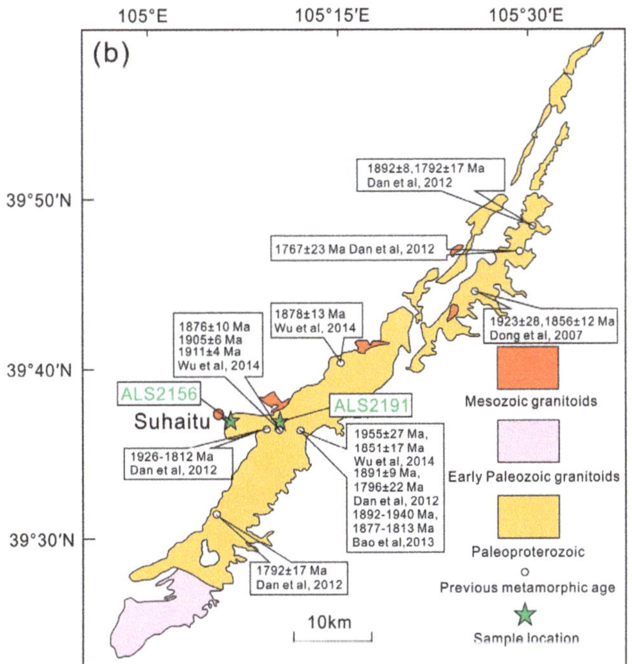

Figure 2. Geological maps of the Diebusige Complex [29,33–36] (**a**) and the Bayanwulashan Complex [28,36,37,51] (**b**) in the Alxa Block (revised from BGMRIM (1991) [38]).

Figure 3. Field photographs of the amphibolites: (**a–c**) amphibolite interlayers and felsic gneiss in the Diebusige Complex; (**d–f**) amphibolite interlayers and felsic gneiss in the Bayanwulashan Complex.

4.2.3. Amphibolite Sample ALS2191

Amphibolite sample ALS2191 consists of clinopyroxene (~10%), plagioclase (~47%), hornblende (~40%), K-feldspar (~2%) and minor quartz (~3%), with accessory ilmenite, sphene and zircon (Figure 5a–d). Clinopyroxene, hornblende, plagioclase and quartz are in contact with each other. Clinopyroxene grains have been transformed to the composite of quartz and actinolite, which was termed as pseudomorph of Cpx (Figure 5a–d). This phenomenon has been observed in clinopyroxene from HP–HT metamorphic rocks, which was considered to be formed by retrogression with water ingress [55–61]. K-feldspar was observed as veins in the fractures (Figure 5b), which may be crystallized from K-bearing fluid during retrogression. Ilmenite is present as interstitial grains (Figure 5a).

Sphene occurs as corona around ilmenite (Figure 5f). Plagioclases have slight compositional variation with X_{An} = 0.36–0.39 and belong to andesine (Figure 6b; Table S3). Hornblendes belong to pargasite and magnesio-hastingsite, with X_{Mg} values of 0.50–0.52 and Ti contents of 0.18–0.31 (Figure 6d; Table S3), whereas actinolites have higher X_{Mg} values of 0.62–0.72 but lower Ti contents of 0.00–0.03 (Figure 6c; Table S3).

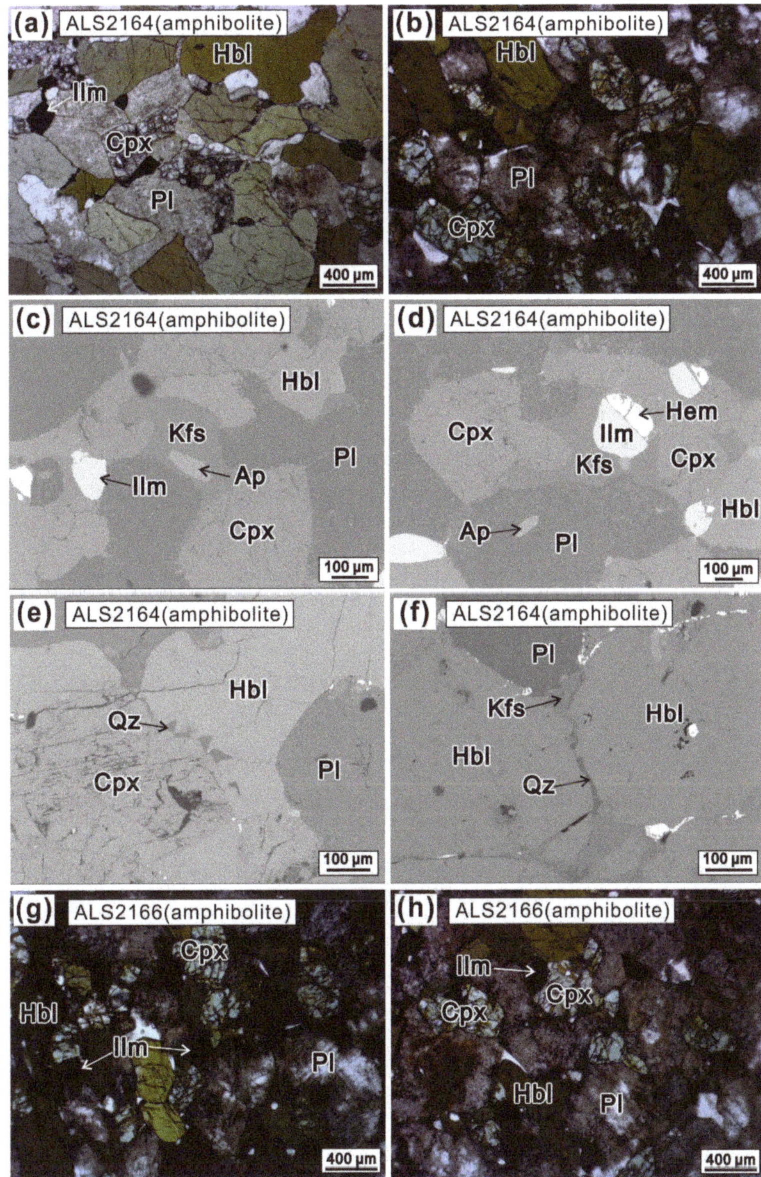

Figure 4. Photomicrographs for the amphibolite samples ALS2164 and ALS2166 in the Diebusige Complex: (**a**,**b**) clinopyroxene, hornblende, plagioclase and ilmenite (plane polarized light); (**c**–**f**) back-scattered electron (BSE) image shows K-feldspar, quartz, apatite, ilmenite and hematite; (**g**,**h**) clinopyroxene, hornblende and plagioclase (plane polarized light).

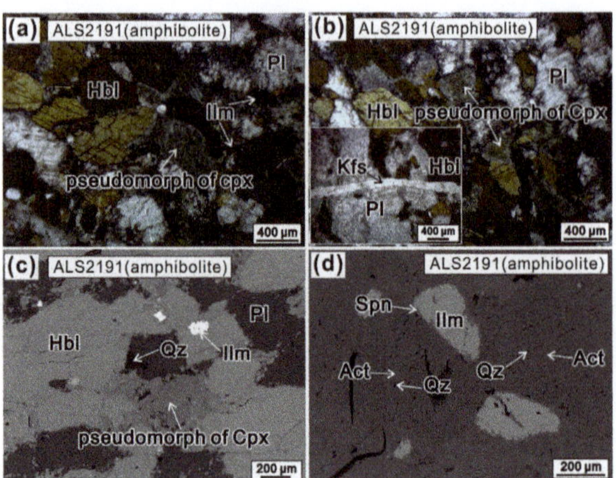

Figure 5. Photomicrographs for the amphibolite sample ALS2191 in the Bayanwulashan Complex: (**a**,**b**) hornblende, plagioclase, K-feldspar and pseudomorph of clinopyroxene (plane polarized light); (**c**,**d**) BSE image shows quartz, sphene, and actinolite and quartz in the pseudomorph of clinopyroxene.

4.3. Mineral Assemblage Evolution

Based on observations of field outcrops and petrography described above, the peak mineral assemblage of cpx + hbl + pl + kfs + qz + ilm + melt is identified for sample ALS2164, and the peak mineral assemblage of cpx + hbl + pl ± kfs + ilm + qz + melt is inferred for sample ALS2191. Actinolites in the pseudomorph of Cpx (Figure 5d; sample ALS2191) are a typical greenschist facies mineral, indicating later hydrothermal alteration. Sphenes only occur around ilmenite in the pseudomorph of Cpx and have not been found in the other domain (i.e., Figure 5a,c). Likely, sphene coronas around ilmenite in the pseudomorph of Cpx (Figure 5d; sample ALS2191) are also the product of hydrothermal alteration.

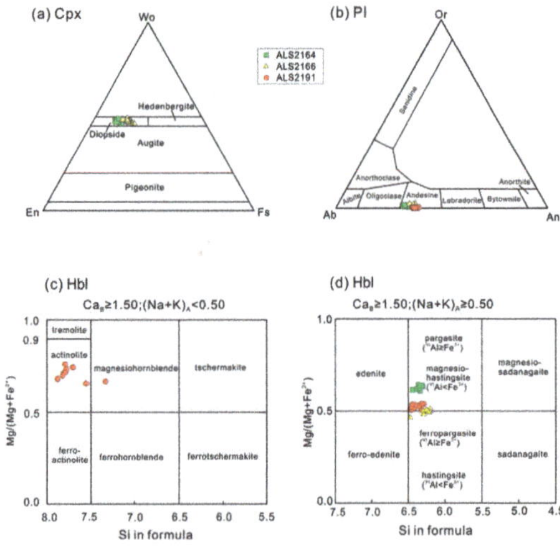

Figure 6. Compositions of clinopyroxene, plagioclase and hornblende: (**a**) the Wo–En–Fs classification diagram for the clinopyroxene; (**b**) the X_{Or}–X_{Ab}–X_{An} diagram for the plagioclase; (**c**,**d**) the classification diagram for the hornblende.

5. Calculation of Metamorphic Conditions

5.1. Phase Equilibrium Modelling

Phase equilibrium modelling was carried out for amphibolite samples ALS2191 and ALS2164 in the system NCKFMASHTO (Na$_2$O–CaO–K$_2$O–FeO–MgO–Al$_2$O$_3$–SiO$_2$–H$_2$O–TiO$_2$–O$_2$) using the Theriak-Domino software (ver.11.03.2020; [62]), with the dataset file ds62_mafic (adopted from the Thermocalc database ds62 by Holland and Powell, 1998 [63], 2011 [64]). The adopted activity–composition (a–x) models are listed in the following: liquid (melt), clinopyroxene and hornblende [65]; garnet and orthopyroxene [66]; olivine [64]; plagioclase [67]; magnetite–spinel [68]; and ilmenite–hematite [69]. Pure phases include quartz and rutile. The amount of FeO was analyzed by titration and then Fe$_2$O$_3$ was calculated by the difference. The T–X_{H2O} pseudosection was used to evaluate the appropriate H$_2$O content for the final P–T pseudosection modelling. The bulk-rock compositions for modelling are listed in Table 1.

5.1.1. Amphibolite Sample ALS2164

A T–X_{H2O} pseudosection constructed at 8 kbar (Figure 7a) using the oxygen content obtained by the Fe^{2+} titration was used to select a suitable H$_2$O content, so that the final retrograde mineral assemblage is stable just above the post-melt-loss solidus [70]. The adopted pressure of 8 kbar is consistent with the pressure range for the peak mineral assemblage of cpx–hbl–pl–kfs–qz–ilm–liq in the P–T pseudosection (Figure 7b). The variations in X_{H2O} correspond to a range of H$_2$O contents from 0 to 5 mol.% (Table 1; Figure 7b,c). A X_{H2O} value of 0.65 (equivalent to 3.25 mol.% H$_2$O) was chosen, which crosses the liquid-out line of the peak mineral assemblage (Table 1; Figure 7b).

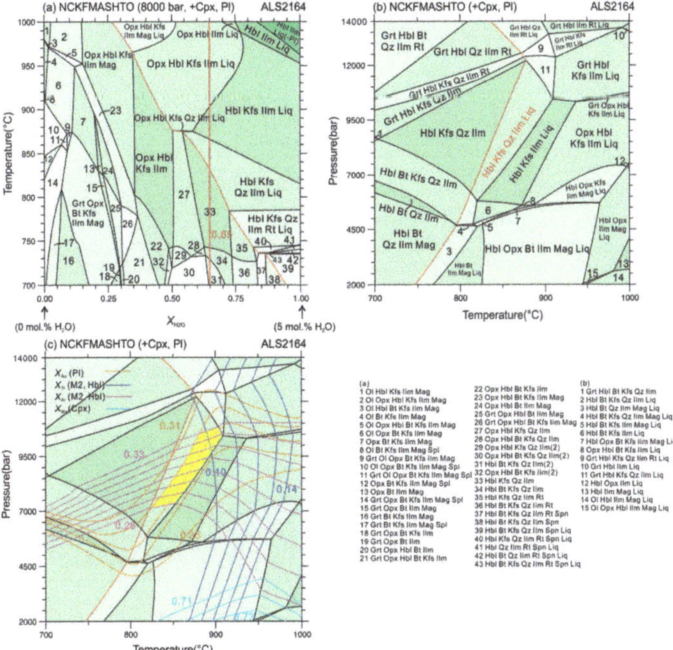

Figure 7. T–X_{H2O} and P–T pseudosections for the amphibolite sample ALS2164 in the Diebusige Complex: (**a**) T–X_{H2O} pseudosection at 8 kbar; (**b**) P–T pseudosection; (**c**) P–T pseudosection with isopleths of X_{Ti}(M2) and X_{Al}(M2) in hornblende, X_{Mg} in clinopyroxene and X_{An} in plagioclase. The field of peak mineral assemblage is marked by cpx + hbl + pl + kfs + ilm + liq in red type. Red bars in (**a**) denote H$_2$O content used for subsequent modelling.

The *P–T* pseudosection was calculated over the *P–T* window of 2–14 kbar/700–1000 °C (Figure 7b). The *P–T* pseudosection is presented with isopleths of $X_{Ti}(M2)$ and $X_{Al}(M2)$ in hornblende, X_{Mg} in clinopyroxene and X_{An} in plagioclase (Figure 7c). The inferred peak mineral assemblage of cpx + hbl + pl + kfs + qz + ilm + liq was stable over a *P–T* range of 810–910 °C/5.8–12.2 kbar (Figure 7b). The peak metamorphic condition is furtherly constrained to be 825–910 °C/7.2–10.8 kbar by the measured X_{An} = 0.32–0.34 of plagioclase and $X_{Al}(M2)$ = 0.26–0.33 of hornblende (Figure 7c). However, the measured $X_{Ti}(M2)$ of 0.11–0.14 in hornblende and X_{Mg} of 0.74–0.80 in clinopyroxene fall outside of the range of the peak phase field (Table S1; Figure 7c).

5.1.2. Amphibolite Sample ALS2191

The H_2O content was determined using the $T–X_{H2O}$ pseudosection (Figure 8a) to ensure that the final mineral assemblage is stable just above the solidus [70]. The $T–X_{H2O}$ pseudosection was constructed at 9 kbar, which is within pressure range for the peak assemblage of cpx–hbl–pl–kfs–qz–ilm–liq in the *P–T* pseudosection (Figure 8b). The variations in X_{H2O} correspond to a range of H_2O contents from 0 to 5.0 mol.% (Table 1; Figure 8a). The peak mineral assemblage of cpx–hbl–pl–kfs–qz–ilm–liq occurs at >0.55 X_{H2O} and 772–868 °C in the $T–X_{H2O}$ pseudosection (Figure 8a). A X_{H2O} value of 0.64 (equivalent to 3.20 mol.% H_2O) was selected for the subsequent *P–T* pseudosection calculation, which crosses the liquid-out line for the peak mineral assemblage (Table 1; Figure 8a).

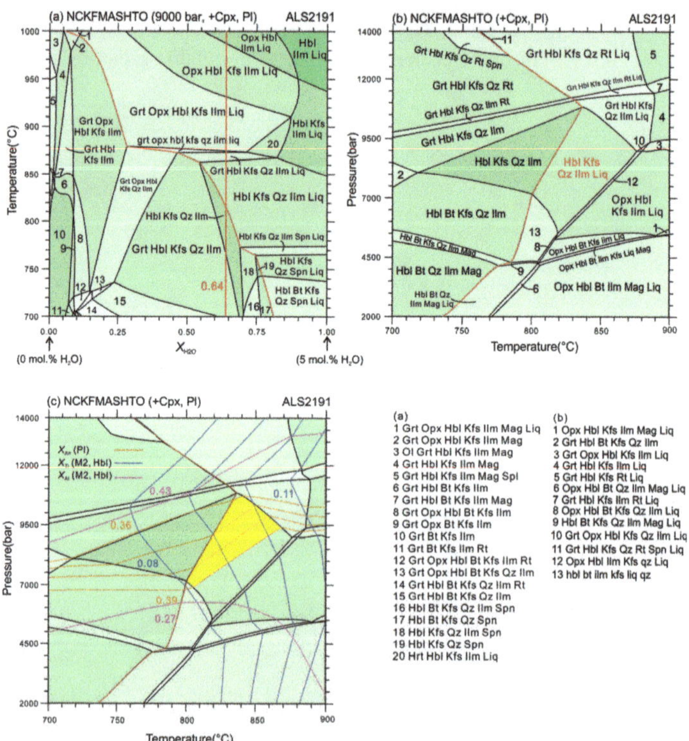

Figure 8. $T–X_{H2O}$ and *P–T* pseudosections for the amphibolite sample ALS2191 in the Bayanwulashan Complex: (**a**) $T–X_{H2O}$ pseudosection at 9 kbar; (**b**) *P–T* pseudosection; (**c**) *P–T* pseudosection with isopleths of $X_{Ti}(M2)$ and $X_{Al}(M2)$ in hornblende, and X_{An} in plagioclase. The field of peak mineral assemblage is marked by cpx + hbl + pl + kfs + qz + ilm + liq in red type. Red bars in (**a**) denote H_2O content used for subsequent modelling.

The P–T pseudosection was calculated over the P–T window of 700–900 °C and 2–14 kbar (Figure 8c). The solidus was modelled at temperature between 737 and 837 °C. Isopleths of X_{Ti}(M2) and X_{Al}(M2) in hornblende, and X_{An} in plagioclase, were contoured on the fields of relevant assemblages in the P–T pseudosection (Figure 8c). The inferred peak mineral assemblage of cpx + pl $\pm$ kfs + hbl + ilm + qz + liq occurs in the range of 5.5–10.8 kbar/800–877 °C (Table S1; Figure 8b). The measured X_{An} = 0.36–0.39 of plagioclase furtherly constrained the peak metamorphic condition to be 800–870 °C/7.0–10.7 kbar. As the X_{Al}(M2) in hornblende ranges from 0.27 to 0.43 (Figure 8c; Table S3), it cannot furtherly constrain the P–T condition of the peak metamorphism. Hornblendes have a high X_{Ti}(M2) of 0.09–0.16, of which the Ti-rich composition (X_{Ti}(M2) = 0.11–0.16) of hornblende (Table S3) and the corresponding X_{Ti}(M2) isopleths fall outside of the field of the peak mineral assemblage.

Table 1. Bulk compositions used for phase equilibrium modelling.

Whole Rock Compositions (wt.%)													
Sample	SiO_2	TiO_2	Al_2O_3	Fe_2O_3	FeO	MnO	MgO	CaO	Na_2O	K_2O	P_2O_5	LOI	Total
ALS2164	45.23	2.35	11.85	6.36	7.21	0.19	8.81	10.38	2.53	1.76	0.65	1.49	99.61
ALS2191	48.65	1.27	14.76	3.32	7.75	0.17	6.38	9.49	2.78	2.10	0.44	1.66	99.63

Normalized Molar Proportion Used for Phase Equilibria Modelling	Figures		H_2O	SiO_2	Al_2O_3	CaO	MgO	FeO	K_2O	Na_2O	TiO_2	O
ALS2164	Figure 7a	$x = 0$	0.00	48.07	7.42	10.84	13.96	11.49	1.19	2.61	1.88	2.54
		$x = 1$	5.00	45.66	7.05	10.30	13.26	10.92	1.13	2.48	1.78	2.41
	Figure 7b,c		3.25	46.51	7.18	10.49	13.50	11.12	1.15	2.52	1.82	2.46
ALS2191	Figure 8a	$x = 0$	0.00	53.10	9.49	10.42	10.38	9.80	1.46	2.94	1.04	1.36
		$x = 1$	5.00	50.45	9.02	9.90	9.86	9.31	1.39	2.79	0.99	1.29
	Figure 8b,c		3.20	51.40	9.19	10.09	10.05	9.48	1.42	2.85	1.01	1.32

Abbreviation: LOI, loss on ignition.

5.2. Conventional Thermobarometry

Hbl–pl–qz thermobarometry [71,72] yielded metamorphic PT conditions of 777 °C/3.0 kbar–785 °C/4.0 kbar, 732 °C/6.6 kbar–810 °C/6.7 kbar and 738 °C/6.1 kbar–759 °C/3.0 kbar for samples ALS2164, ALS2166 and ALS2191, respectively. Ti-in-amphibole thermometry [73] yielded temperatures of 825–881 °C (an average of 861 °C), 808–917 °C (an average of 866 °C) and 783–913 °C (an average of 851 °C) for samples ALS2164, ALS2166 and ALS2191, respectively (Tables S1–S3; Figure 9a). The revised Ti-in-zircon thermometry was also used to estimate the metamorphic temperatures [74]. The coexisting Ti-phase in our samples is ilmenite, which means the TiO_2 activity was 0.6 [75]. The SiO_2 activity was set as 1.0 due to the presence of quartz. After excluding abnormal spot >20 ppm that may be caused by Ti-rich mineral inclusions, the metamorphic temperatures recorded by zircons were estimated to be 679–850 °C with a mean value of 763 °C (2.80–16.0 ppm), 728–857 °C with a mean value of 780 °C (4.96–17.40 ppm), and 689–853 °C with a mean value of 768 °C (3.15–16.80 ppm), for samples ALS2164, ALS2166 and ALS2191, respectively (Table S4; Figure 9b).

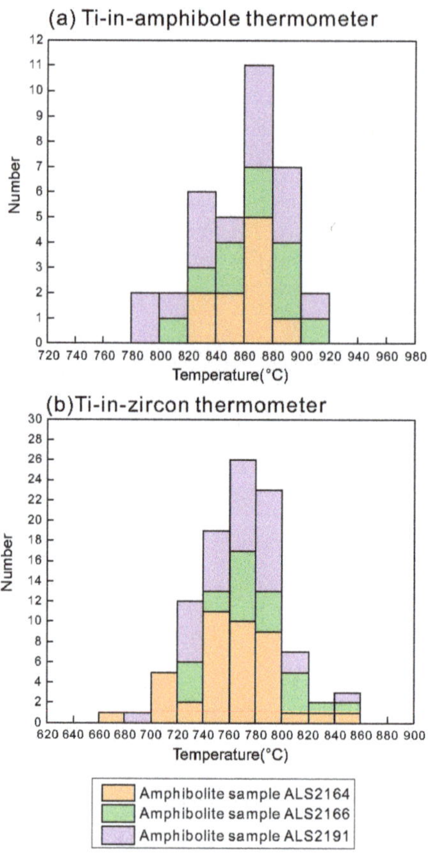

Figure 9. Metamorphic temperatures calculated for amphibolite samples ALS2164, ALS2166 and ALS2191 using Ti-in-amphibole (**a**) and Ti-in-zircon (**b**) thermometers.

6. Zircon U–Pb Dating

6.1. Amphibolite Sample ALS2164

Zircon grains from this sample are rounded or stubby in shape, with a size of 100–200 μm, and exhibit planar or no zoning, indicating metamorphic origin (Figure 10). Fifty-three spot analyses were conducted on metamorphic zircons, of which fifty-two spot analyses were plotted on or close to the concordia curve (Figure 11a). After excluding twelve spots with old apparent data (may be carried out on inherited igneous zircons), the remaining concordant data yielded a weighted mean ^{207}Pb/^{206}Pb age of 1847 ± 25 Ma (n = 40, MSWD = 1.4) (Table S4; Figure 11a). On the chondrite-normalized REE patterns (data of chondrite are cited from Sun and McDonough (1989) [76]), all zircons show variable Eu anomalies (Eu/Eu* = 0.25–6.02) and an enriched HREE pattern ((Gd/Lu)$_N$ = 0.02–0.08) (Table S5; Figure 11b). Th vs U diagram (Figure 11b) displays relatively low Th (27.91–221.54 ppm) and high U (31.48–268.31 ppm) contents with Th/U ratios of 0.45–2.41 (Figure 11b; Table S4).

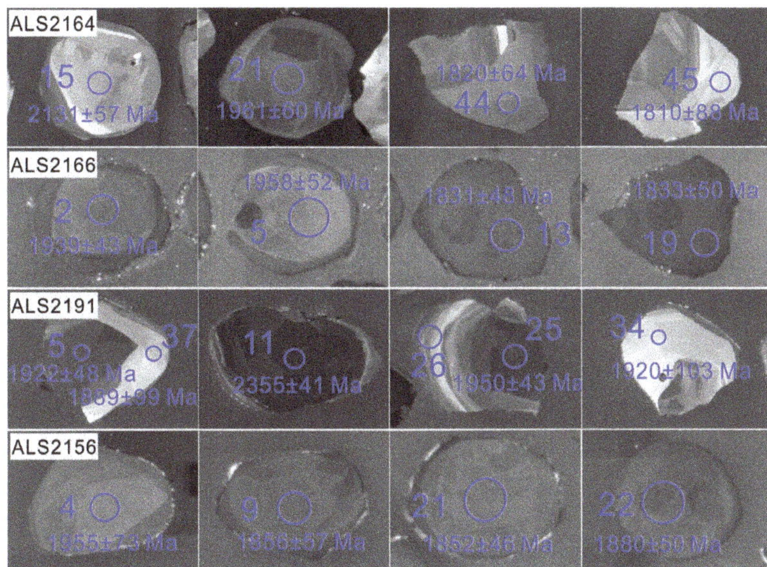

Figure 10. CL images of representative zircons within amphibolite samples ALS2164, ALS2166, ALS2191 and ALS2156. The blue circles represent analytical spot with a size of ~24 μm.

6.2. Amphibolite Sample ALS2166

Zircon grains from this sample are generally stubby to rounded in shape, with a diameter of 70–150 μm. Features under CL images suggest they are metamorphic rather than igneous in origin (Figure 10). Twenty-three spot analyses were performed on metamorphic zircons, and all spot analyses yielded concordant data (Table S4). These concordant data yielded a weighted mean ^{207}Pb/^{206}Pb age of 1817 ± 21 Ma Ma (n = 23, MSWD = 2.0) (Figure 11c). Obviously, all zircons exhibit variable Eu anomalies (Eu/Eu* = 0.26–1.19) and an enriched HREE pattern ((Gd/Lu)$_N$ = 0.01–0.03) (Figure 11d; Table S5). The Th vs U diagram displays that all zircons have relatively low Th (68.58–244.17 ppm) and high U (268.19–811.76 ppm) contents with low Th/U ratios (0.24–0.44) (Figure 11d).

6.3. Amphibolite Sample ALS2191

Zircon grains from this sample are rounded or stubby in shape, with a diameter of 100–200 μm, and display sector and patchy zoning, reflecting metamorphic origin (Figure 10). Forty spot analyses were performed on metamorphic zircons, of which thirty-eight spot analyses yielded concordant data (Table S4). After excluding three old apparent age of 2355 ± 41, 2300 ± 39 and 2080 ± 68 Ma (i.e., spots 3, 4 and 11), the remaining concordant data yielded a weighted mean ^{207}Pb/^{206}Pb age of 1901 ± 22 Ma (n = 35, MSWD = 0.4) (Table S4; Figure 11e). Notably, all metamorphic zircons display an enriched HREE pattern ((Gd/Lu)$_N$ = 0.02–0.08) with negative Eu anomalies (Eu/Eu* = 0.13–0.60)(Table S5; Figure 11f)). The U vs Th diagram shows that metamorphic zircons have relatively low Th (66.83–714.27 ppm) and high U (25.03–791.68 ppm) contents, with variable Th/U ratios (0.38–4.82) (Figure 11f; Table S4).

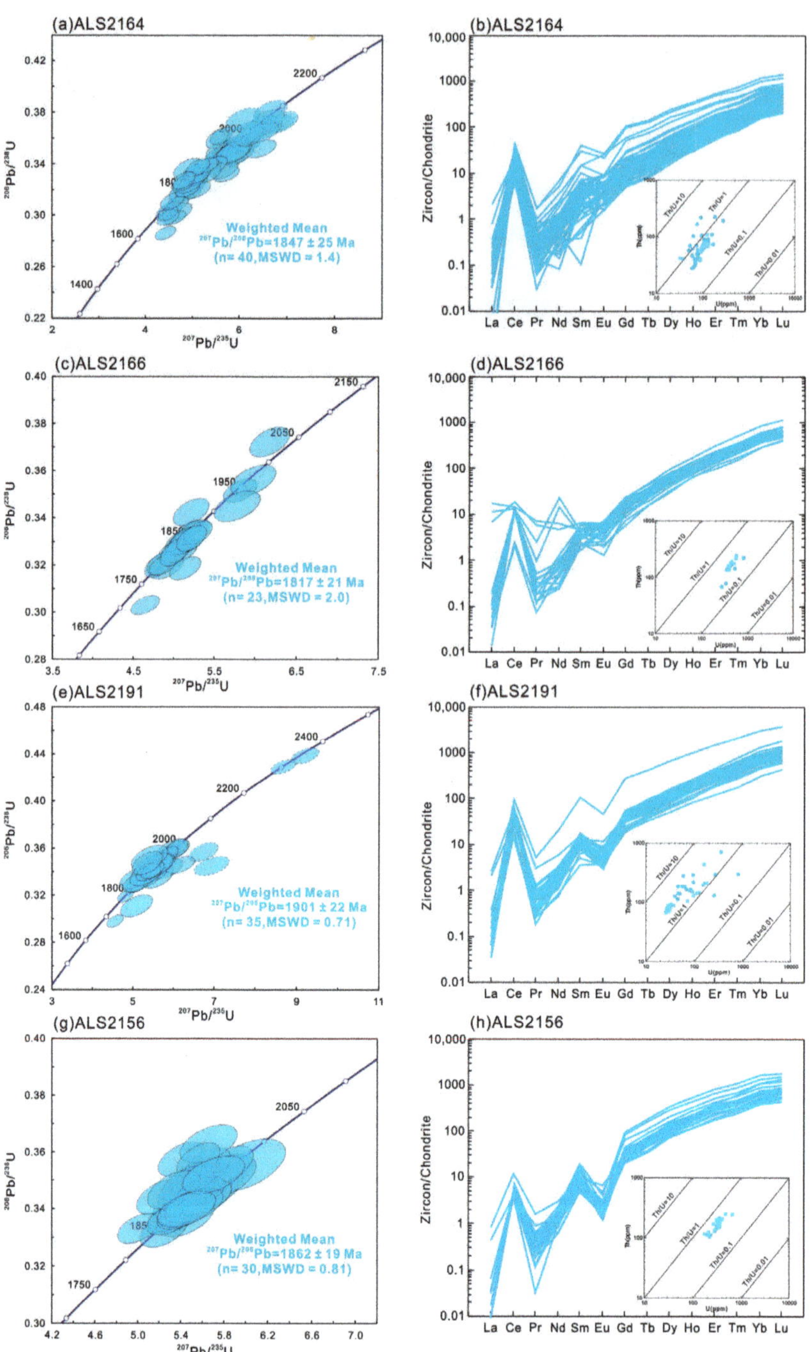

Figure 11. Concordia diagrams, chondrite-normalized rare earth element patterns and Th vs. U diagram with Th/U isopleths for the metamorphic zircons of amphibolite samples ALS2164 (**a,b**), ALS2166 (**c,d**), ALS2191 (**e,f**) and ALS2156 (**g,h**).

6.4. Amphibolite Sample ALS2156

Zircon grains from this sample are stubby to rounded in shape, with a size of 60–120 μm, and display fir-tree or patchy zoning, suggesting metamorphic origin (Figure 10). All thirty spot analyses on metamorphic zircons yielded concordant data (Table S4). These concordant data yielded a weighted mean ^{207}Pb/^{206}Pb age of 1862 ± 19 Ma (n = 30, MSWD = 0.81) (Table S4; Figure 11g). On the chondrite-normalized REE patterns (Figure 11h), all zircons display a enriched heavy REE pattern ((Gd/Lu)$_N$ = 0.02–0.06) with negative Eu anomalies (Eu/Eu* = 0.06–0.38) (Table S5; Figure 11h). The U versus (vs.) Th diagram shows that all zircons have relatively low Th (104.69–251.08 ppm) and high U (172.68–645.94 ppm) contents, with low Th/U ratios (0.38–0.68) (Table S4; Figure 11h).

7. Discussion

7.1. Age Interpretation of Amphibolites

In the eastern Alxa Block, a large number of metamorphic ages have been reported in the Bayanwulashan and Diebusige Complexes [19,20,24,25,30,31,34,41,45]. For example, SHRIMP and LA-ICP-MS zircon U–Pb dating yielded two metamorphic age groups of 1.95–1.90 Ga and 1.85–1.80 Ga for granitic gneiss and amphibolites from the Bayanwulashan Complex [19,45]. Wu et al. (2014) [31] reported two groups of metamorphic ages of 1.95–1.90 Ga and 1.85–1.80 Ga by LA-ICP-MS zircon U–Pb dating for several orthogneiss samples in the Bayanwulashan Complex. Additionally, SIMS zircon U–Pb dating determined that orthogneisses and paragneisses from both the Diebusige and Bayanwulashan Complexes recorded two groups of metamorphic ages of ca. 1.90 Ga and ca. 1.80 Ga [30]. In the Diebusige Complex, Geng et al. (2007, 2010) [34,35] constrained the timing of metamorphism of felsic gneiss to be ca. 2.00–1.90 Ga and ca. 1.93 Ga, respectively. Recently, Zou et al. (2021) [24] demonstrated that the granulite-facies metamorphism of magnetite quartzite from the Diebusige Complex occurred at ca. 1.85 Ga by LA-ICP-MS zircon U–Pb dating. Moreover, a corundum-bearing garnet–sillimanite gneiss sample from the Diebusige Complex was determined to firstly undergo granulite-facies metamorphism at ca. 1.95 Ga, then achieved UHT metamorphic condition, and finally cooled to the solidus at ca. 1.83 Ga [25]. Wang et al. (2023) [35] reported metamorphic ages of ca. 1.96–1.86 Ga for UHT mafic granulites in the Diebusige Complex. Therefore, both the Bayanwulashan and Diebusige Complexes underwent metamorphism at ca. 1.95–1.90 Ga and ca. 1.85–1.80 Ga.

In this study, LA-ICP-MS zircon U–Pb dating yielded four weighted mean ^{207}Pb/^{206}Pb ages of 1847 ± 25, 1817 ± 21, 1901 ± 22 and 1862 ± 19 Ma for metamorphic zircons of amphibolite samples ALS2164, ALS2166, ALS2191 and ALS2156, respectively (Figure 11). These metamorphic chronological data (ca. 1901–1817 Ma) are consistent with reported zircon U–Pb ages in the Alxa Block (Table S6). The calculated average values of Ti-in-zircon temperatures of 763, 768 and 780 °C suggest that the amphibolite-facies metamorphism occurred at ca. 1901–1817 Ma.

7.2. Metamorphic Style and Tectonic Implications

Metamorphic conditions calculated by *P–T* pseudosection modelling and hbl–pl–qz thermobarometry in this study, and previous results in the Alxa Block, are summarized in Figure 12, in which the boundaries of metamorphic facies are from Brown (2014) [9]. As described in Section 5.1, the metamorphic *P–T* conditions of 825–910 °C/7.2–10.8 kbar were obtained for amphibolite sample ALS2164, whereas that of 800–870 °C/7.0–10.7 kbar were constrained for amphibolite sample ALS2191. The dated zircons display the relatively low crystallization temperatures with averages of 763 °C, 768 °C and 780 °C, which are consistent with the metamorphic temperatures constrained by the hbl–pl–qz thermobarometry in Figure 12. In addition, the Ti-in-amphibole thermometer yielded high metamorphic temperatures with averages of 861 °C, 866 °C and 851 °C, which agree with the peak metamorphic temperatures constrained by *P–T* pseudosection modelling (Figure 12). Our samples preserved typical amphibolite-facies mineral assemblage (Opx-free mineral assemblage), which are consistent with temperature and pressure estimates from hbl–pl–qz thermobarometry and the Ti-in-zircon thermometer. Such a large difference in metamorphic conditions calculated by different methods may be a result of the following reasons: (1) The absolute *PT* conditions of the *P–T* pseudosection modelling may have a large error (~0.1 GPa and ~50 °C; [77,78]). Thus, considering this uncertainty range, the results obtained from conventional thermobarometry can closely align with those derived from *P–T* pseudosection modelling. (2) These amphibolites may have experienced granulite-facies metamorphism although they are characterized by opx-free mineral assemblage. This can be attributed to the effect of bulk-rock composition, as interpreted by opx-bearing phase fields occur at high temperatures in these samples (Figures 7 and 8). Nevertheless, the lower limit of metamorphic conditions constrained by *P–T* pseudosection modelling are consistent with the upper limit of metamorphic conditions yielded by the hbl–pl–qz thermobarometer (732–810 °C/3.0–6.7 kbar) in uncertainty (Figure 12).

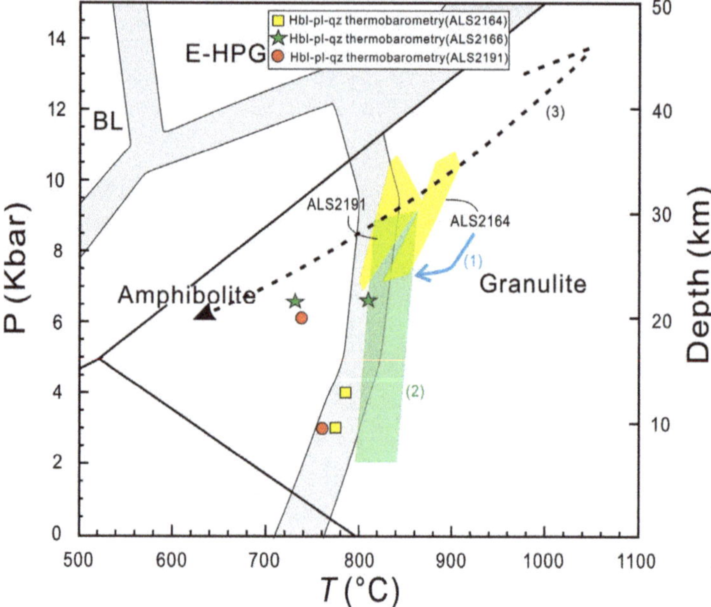

Figure 12. Summary of previously published metamorphic *P–T* paths and conditions (1–3), and metamorphic *PT* condition for amphibolites in the Diebusige (sample ALS2164) and Bayanwulashan (sample ALS2191) Complexes of the Alxa Block, NCC in this study: (1) corundum-bearing garnet–sillimanite gneiss in the Diebusige Complex [25]; (2) garnet-bearing magnetite quartzite in the Diebusige Complex [24]; (3) garnet-bearing mafic granulites in the Diebusige Complex [35]. Bl, blueschist facies; E–HPG, eclogite–high-pressure granulite-facies.

The NCC display two characteristics, including ca. 2.70–2.50 Ga magmatic-metamorphic events and ca. 1.95–1.85 Ga metamorphic events [22,79]. Although the Neoarchean TTG gneiss in the Alxa Block [27,30,31] is similar to those of the Yinshan Block, Dan et al. (2012) [36] considered that the Alxa Block was not the western extension of the Yinshan Block due to the absence of the Neoarchean greenstone belt and supracrustal rocks in the Alxa Block. Therefore, the extant data does not support the Alxa Block as the western extension of the Yinshan Block. The KB, a Paleoproterozoic orogenic belt in the Western Block of the NCC, mainly consists of the khondalite series (a set of upper amphibolite- to granulite-facies metasedimentary rocks, mafic granulite, charnockite, S-type granite, and a small amount of TTG gneiss) [22,79]. By comparing metamorphic age characteristics of the Alxa Block and the Western Block (Yinshan Block, KB, and Ordos Block) of the NCC (Figure 1a), Zou et al. (2021) [33] pointed out the metamorphic age characteristics of the Alxa Block is consistent with those in the KB, indicating that they may have a similar metamorphic evolutionary history, at least in the Neoarchean–late Paleoproterozoic. Recently, several studies have reported HT–UHT metamorphic rocks with metamorphic ages of ca. 1.96–1.83 Ga (garnet-bearing magnetite quartzite, garnet-sillimanite-biotite-plagioclase gneiss, corundum-bearing gneiss and garnet-bearing mafic granulite) in the Diebusige Complex of the eastern Alxa Block [33–35], which is similar to the khondalite series of the KB.

The tectonic evolution of the KB includes the following processes: (1) the crustal thickening led to the burial of the protolith of khondalite series to the depth of the lower crust during syn-collision stage (ca. 1.96–1.95 Ga), which is characterized by formation of HP pelitic granulites [79,80]; (2) the underplating or emplacement of mantle-derived magmas resulted in the formation of UHT metamorphic rocks during the post-collisional extension stage (ca. 1.93–1.85 Ga) [81–83]. In this study, amphibolites have metamorphic ages of ca. 1.90–1.82 Ga, which are similar to previously reported metamorphic ages of ca. 1.96–1.83 Ga for HT–UHT rocks in the Diebusige Complex [32–34]. This indicates that the eastern Alxa Block likely experienced a shared HT–UHT metamorphism during the late Paleoproterozoic. Amphibolites in the Diebusige Complex display higher peak temperature but similar peak pressure, compared with those of amphibolites in the Bayanwulashan Complex, which indicates there is an extra heat source in the Diebusige Complex. However, both amphibolites from the two complexes recorded a higher geothermal gradient (~30 °C/km) than Phanerozoic continental crust (~20 °C/km) [84,85]. This high geothermal gradient is also supported by reported UHT metamorphic rocks and associated mantle-derived mafic-ultramafic rocks in the Diebusige Complex [33,34]. Additionally, the numerical modeling of Sizava et al. (2014) [85] showed that these high apparent thermal gradients can be achieved during the Paleoproterozoic hot orogensis. The HREE enrichment patterns of metamorphic zircons from the amphibolites in this study are in agreement with that these amphibolites formed under *PT* conditions where garnet is not stable (Figure 11) and at relatively shallower crust than those of the garnet-bearing mafic granulites in the Diebusige Complex [35]. Combining clockwise *P–T* paths of UHT metamorphic rocks in the Diebusige Complex (Figure 12) and metamorphic ages, an extensional setting after continental collision is inferred for these amphibolites in the eastern Alxa Block [84–86].

As discussion above, the Alxa Block is similar to the KB in that both have similar metamorphic rock types and experienced the late Paleoproterozoic tectonic thermal events. As several researchers have proposed, the late Paleoproterozoic metamorphic data from the Diebusige Complex [33–35] and this study suggest that the Alxa Block may be the westward extension of the KB [27,30,32–35]. Based on current speculation, there are still some issues which require clarification in future work, such as the occurrence of multiple magmatic-metamorphic events since the Neoproterozoic in the Alxa Block, has been served as an indicator that distinguishes the NCC [20,21,24–26].

8. Conclusions

(1) LA-ICP-MS zircon U–Pb dating yielded the metamorphic ages of ca. 1901–1817 Ma for amphibolites in the Diebusige and Bayanwulashan Complexes, representing the timing of amphibolite-facies metamorphism.

(2) Phase equilibrium modelling and conventional thermobarometries yielded a relatively large *PT* range for amphibolites in the Diebusige and Bayanwulashan Complexes, with a high geothermal gradient, which may correspond to an extensional setting following continental collision.

(3) The HREE enrichment patterns of metamorphic zircons are consistent with that these amphibolites formed under *PT* conditions where garnet is not stable and at relatively shallower crust than the garnet-bearing mafic granulites in the Diebusige Complex.

Supplementary Materials: The following supporting information can be downloaded at: https://www.mdpi.com/article/10.3390/min13111426/s1, Table S1: Major chemical compositions of clinopyroxene, hornblende, plagioclase and K-feldspar in the amphibolite sample ALS2164. Table S2: Major chemical compositions of clinopyroxene, hornblende, plagioclase and K-feldspar in the amphibolite sample ALS2166. Table S3: Major chemical compositions of hornblende, plagioclase and K-feldspar in the amphibolite sample ALS2191. Table S4: LA-ICP-MS zircon U–Pb isotope data of amphibolite samples ALS2164, ALS2166 and ALS2191. Table S5: LA-ICP-MS zircon trace element data of amphibolite samples ALS2164, ALS2166 and ALS2191. Table S6: Summary of Paleoproterozoic magmatic and metamorphic ages in the Alxa Block. References [27–30,33,35–37,39,42–51,53] are cited in the supplementary materials.

Author Contributions: Conceptualization, L.G.; Investigation, X.X. and Z.T.; Writing—original draft, F.Z. All authors have read and agreed to the published version of the manuscript.

Funding: This work was jointly supported by the National Natural Science Foundation of China (Grants No. 41772051 and 41890831) and a research grant from the State Key Laboratory of Continental Dynamics (SKLCD-04).

Data Availability Statement: Data is contained within the article or Supplementary Material. The data presented in this study are available in [Tables 1 and S1–S6].

Acknowledgments: The authors thank the editor and managing editor very much for their editorial work, and three anonymous reviewers for their constructive and insightful comments and suggestions.

Conflicts of Interest: The authors declare that the research was conducted in the absence of any commercial or financial relationships that could be construed as a potential conflict of interest.

References

1. Barker, A.J. *Introduction to Metamorphic Textures and Microstructures*; Blackie: Glasgow, UK, 1990.
2. Stowell, H.H.; Stein, E. The significance of plagioclase-dominant coronas on garnet, Wenatchee Block, northern Cascades, Washington, U. *S.A. Can. Mineral.* **2005**, *43*, 367–385. [CrossRef]
3. Kohn, M.J.; Spear, F. Two new barometers for garnet amphibolites with applications to eastern Vermont. *Am. Mineral.* **1990**, *75*, 89–96.
4. Surour, A.A. Medium- to high-pressure garnet-amphibolites from Gebel Zabara and Wadi Sikait, South Eastern Desert, Egypt. *J. Afr. Earth Sci.* **1995**, *21*, 443–457. [CrossRef]
5. Liu, J.; Bohlen, S.R.; Ernst, W.G. Stability of hydrous phases in subducting oceanic crust. *Earth Planet. Sci. Lett.* **1996**, *143*, 161–171. [CrossRef]
6. Dale, J.; Holland, T.; Powell, R. Hornblende–garnet–plagioclase thermobarometry: A natural assemblage calibration of the thermodynamics of hornblende. *Contrib. Mineral. Petrol.* **2000**, *140*, 353–362. [CrossRef]
7. Lopez, S.V.; Gomez, P.M.Y.; Azor, A.; Fernandez, S.J.M. Phase diagram sections applied to amphibolites: A case study from the Ossa_morena/Central Iberian Variscan suture (Southwestern Iberian Massif). *Lithos* **2003**, *68*, 1–21.
8. Winter, J.D. *An Introduction to Igneous and Metamorphic Petrology*, 2nd ed.; Prentice Hall: Upper Saddle River, NJ, USA, 2010; pp. 1–702.
9. Brown, M. The contribution of metamorphic petrology to understanding lithosphere evolution and geodynamics. *Geosci. Front.* **2014**, *5*, 553–569. [CrossRef]
10. Wei, C.J.; Guan, X.; Dong, J. HT-UHT metamorphism of metabasites and the petrogenesis of TTGs. *Acta Petrol. Sin.* **2017**, *33*, 1381–1404.
11. Zhang, B.T.; Ling, H.; Chen, P.; Hu, G.; Jiang, Y.; Yu, J. New recognition criteria for ortho and paraamphibolites: The Comparative study on mineral petrochemical characteristics of the precambrian orthoparaamphibolites from Xiangshan, Central Jiangxi Province. *Contrib. Geol. Miner. Resour. Res.* **2005**, *20*, 223–232.
12. Zhao, G.C.; Cawood, P.A.; Lu, L.Z. Petrology and P-T history of the Wutai amphibolites: Implications for tectonic evolution of the Wutai complex, China. *Precambrian Res.* **1999**, *93*, 181–199. [CrossRef]
13. Miyashiro, A. *Metamorphic Petrology*; UCL Press Limited: London, UK, 1994.
14. Qian, J.H.; Wei, C.J. P–t–t evolution of garnet amphibolites in the wutai–hengshan area, north china craton: Insights from phase equilibria and geochronology. *J. Metamorph. Geol.* **2016**, *34*, 423–446. [CrossRef]
15. Bader, T.; Zhang, L.F.; Li, X.W. Is the Songshugou Complex, Qinling Belt, China, an Eclogite Facies Neoproterozoic Ophiolite? *J. Earth Sci.* **2019**, *30*, 460–475. [CrossRef]
16. Lou, Y.; Wei, C.; Liu, X.; Zhang, C.; Tian, Z.; Wang, W. Metamorphic evolution of garnet amphibolite in the western Dabieshan eclogite belt, Central China: Evidence from petrography and phase equilibria modeling. *J. Asian Earth Sci.* **2013**, *63*, 130–138. [CrossRef]
17. Wang, Y.W.; Liu, L.; Liao, X.Y.; Gai, Y.S.; Yang, W.Q.; Kang, L. Multi-metamorphism of amphibolite in the Qinling complex,Qingyouhe area: Revelation from trace elements and mineral inclusions in zircons. *Acta Petrol. Sin.* **2016**, *32*, 1467–1492.

18. Wu, K.K.; Zhao, G.; Sun, M.; Yin, C.; He, Y.; Tam, P.Y. Metamorphism of the northern Liaoning Complex: Implications for the tectonic evolution of Neoarchean basement of the Eastern Block, North China Craton. *Geosci. Front.* **2013**, *4*, 305–320. [CrossRef]
19. Zhang, J.; Wei, C.J.; Chu, H. High-T and low-P metamorphism in the xilingol complex of central inner mongolia, China: An indicator of extension in a previous orogeny. *J. Metamorph. Geol.* **2018**, *36*, 393–417. [CrossRef]
20. Ge, X.H.; Liu, J.L. Broken "Western China Craton". *Acta Petrol. Sin.* **2000**, *16*, 59–66. (In Chinese with English Abstract)
21. Li, X.H.; Su, L.; Song, B.; Liu, D.Y. SHRIMP U-Pb zircon age of the Jinchuan ultramafic intrusion and its geological significance. *Chin. Sci. Bull.* **2004**, *49*, 420–422. [CrossRef]
22. Zhao, G.C.; Sun, M.; Wilde, S.A.; Li, S.Z. Late Archean to Paleoproterozoic evolution of the North China Craton: Key issues revisited. *Precambrian Res.* **2005**, *136*, 177–202. [CrossRef]
23. Zhao, G.C. Metamorphic evolution of major tectonic units in the basement of the north China craton: Key issues and discussion. *Acta Petrol. Sin.* **2009**, *25*, 1772–1792.
24. Dan, W.; Li, X.H.; Wang, Q.; Wang, X.C.; Wyman, D.A.; Liu, Y. Phanerozoic amalgamation of the Alxa block and North China Craton: Evidence from Paleozoic granitoids, U-Pb geochronology and Sr-Nd-Pb-Hf-O isotope geochemistry. *Gondwana Res.* **2016**, *32*, 105–121. [CrossRef]
25. Yuan, W.; Yang, Z.Y. The Alashan terrane was not part of north China by the late Devonian: Evidence from detrital zircon U-Pb geochronology and Hf isotopes. *Gondwana Res.* **2015**, *27*, 1270–1282. [CrossRef]
26. Liu, S.J.; Tsunogae, T.; Li, W.; Shimizu, H.; Santosh, M.; Wan, Y.S.; Li, J.H. Paleoproterozoic granulites from Heling'er: Implications for regional ultrahigh temperature metamorphism in the north China craton. *Lithos* **2012**, *148*, 54–70. [CrossRef]
27. Zhang, J.X.; Gong, J.H.; Yu, S.; Li, H.; Hou, K. Neoarchean–Paleoproterozoic multiple tectonothermal events in the western Alxa block, North China Craton and their geological implication: Evidence from zircon U–Pb ages and Hf isotopic composition. *Precambrian Res.* **2013**, *235*, 36–57. [CrossRef]
28. Dong, C.; Liu, D.Y.; Li, J.J.; Wang, Y.S.; Zhou, H.Y.; Li, C.D.; Yang, Y.H.; Xie, L.W. Palaeoproterozoic Khondalite Belt in the western North China Craton: New evidence from SHRIMP dating and Hf isotope composition of zircons from metamorphic rocks in the Bayan Ul-Helan Mountains area. *Chin. Sci. Bull.* **2007**, *52*, 2984–2994. (In Chinese with English Abstract) [CrossRef]
29. Geng, Y.S.; Wang, X.S.; Wu, C.M.; Zhou, X.W. Late-Paleoproterozoic tectonothermal events of the metamorphic basement in Alxa area: Evidence from geochronology. *Acta Petrol. Sin.* **2010**, *26*, 1159–1170. (In Chinese with English Abstract)
30. Zhang, J.X.; Gong, J.H. Revisiting the nature and affinity of the Alxa Block. *Acta Petrol. Sin.* **2018**, *34*, 940–962. (In Chinese with English Abstract)
31. Gong, J.H.; Zhang, J.X.; Yu, S.Y. The origin of Longshoushan Group and associated rocks in the southern part of the Alxa block: Constraint from LA-ICP-MS U-Pb zircon dating. *Acta Petrol. Mineral.* **2011**, *30*, 795–818. (In Chinese with English Abstract)
32. Zou, L.; Liu, P.H.; Liu, L.S.; Wang, W.; Tian, Z.H. Diagenetic and Metamorphic Timing of the Diebusige Complex, the Eastern Alxa Block: New Evidence from Zircon LA-ICP-MS U-Pb Dating of Biotite-Plagioclase Gneiss. *Earth Sci.* **2020**, *45*, 3313–3329. (In Chinese with English Abstract)
33. Zou, L.; Guo, J.H.; Liu, L.S.; Ji, L.; Liu, P.H. Palaeoproterozoic granulite-facies metamorphism in the eastern Alxa block: New petrological and geochronological evidence from the diebusige complex. *Precambrian Res.* **2021**, *354*, 106051. [CrossRef]
34. Zou, L.; Guo, J.H.; Jiao, S.J.; Huang, G.Y.; Tian, Z.H.; Liu, P.H. Paleoproterozoic ultrahigh-temperature metamorphism in the Alxa Block, the Khondalite Belt, North China Craton: Petrology and phase equilibria of quartz-absent corundum-bearing pelitic granulites. *J. Metamorph. Geol.* **2022**, *40*, 1159–1187. [CrossRef]
35. Wang, Y.; Ge, R.; Si, Y. Late Paleoproterozoic ultrahigh-temperature mafic granulite from the eastern Alxa Block, North China Craton. *Precambrian Res.* **2023**, *397*, 107172. [CrossRef]
36. Dan, W.; Li, X.H.; Guo, J.; Liu, Y.; Wang, X.C. Paleoproterozoic evolution of the eastern Alxa Block, westernmost North China: Evidence from in situ zircon U–Pb dating and Hf–O isotopes. *Gondwana Res.* **2012**, *21*, 838–864. [CrossRef]
37. Wu, S.; Hu, J.; Ren, M.; Gong, W.; Liu, Y.; Yan, J. Petrography and zircon U–Pb isotopic study of the Bayanwulashan Complex: Constrains on the Paleoproterozoic evolution of the Alxa Block, westernmost North China Craton. *J. Asian Earth Sci.* **2014**, *94*, 226–239. [CrossRef]
38. BGMRIM (Bureau of Geology and Mineral Resources of Inner Mongolia Autonomous Region). *Regional Geology of Inner Mongol Autonomous Region*; Geological Publishing House: Beijing, China, 1991. (In Chinese)
39. Gong, J.H.; Zhang, J.X.; Yu, S.Y.; Li, H.K.; Hou, K.J. Ca. 2.5 Ga TTG rocks in the western Alxa Block and their implications. *Chin. Sci. Bull.* **2012**, *57*, 4064–4076. (In Chinese with English Abstract) [CrossRef]
40. Gong, J.H.; Zhang, J.X.; Yu, S.Y. Redefinition of the" Longshoushan Group" outcropped in the eastern segment of Longshoushan on the southern margin of Alxa block: Evidence from detrital zircon U-Pb dating results. *Acta Petrol. Mineral.* **2013**, *32*, 1–22. (In Chinese with English Abstract)
41. Geng, Y.S.; Wang, X.S.; Shen, Q.H.; Wu, C.M. Redefinition of the Alxa Group-complex (Precambrian metamorphic basement) in the Alxa area, Inner Mongolia. *Geol. China* **2006**, *33*, 138–145.
42. Geng, Y.S.; Wan, X.S.; Shen, Q.H.; Wu, C.M. Chronology of the Precambrian metamorphic series in the Alxa area, Inner Mongolia. *Geol. China* **2007**, *34*, 251–261. (In Chinese with English Abstract)
43. Xiu, Q.; Yu, H.; Li, Q.; Zuo, G.; Li, J.; Cao, C. Discussion on the petrogenic time of Longshoushan Group, Gansu Province. *Acta Geol. Sin.* **2004**, *78*, 366–373. (In Chinese with English Abstract)

44. Liu, Y. Characteristics and Their Geological Significance of Paleoproterozoic Granite in Jinchuan, Gansu Province. Master's Thesis, University of Geosciences, Beijing, China, 2008. (In Chinese with English Summary)

45. Gong, J.; Zhang, J.; Wang, Z.; Yu, S.; Li, H.; Li, Y. Origin of the Alxa Block, western China: New evidence from zircon U–Pb geochronology and Hf isotopes of the Longshoushan Complex. *Gondwana Res.* **2016**, *36*, 359–375. [CrossRef]

46. Liu, J.; Yin, C.; Zhang, J.; Qian, J.; Li, S.; Xu, K.; Wu, S.; Xia, Y. Tectonic evolution of the Alxa Block and its affinity: Evidence from the U-Pb geochronology and Lu-Hf isotopes of detrital zircons from the Longshoushan Belt. *Precambrian Res.* **2020**, *344*, 105733. [CrossRef]

47. Zeng, R.; Lai, J.; Mao, X.; Li, B.; Zhang, J.; Bayless, R.C.; Yang, L. Paleoproterozoic multiple Tectonothermal events in the Longshoushan Area, Western North China Craton and their geological implication: Evidence from geochemistry, Zircon U–Pb geochronology and Hf Isotopes. *Minerals* **2018**, *8*, 361. [CrossRef]

48. Shen, Q.H.; Geng, Y.S.; Wang, X.S.; Wu, C.M. Petrology, geochemistry, formation environment and ages of Precambrian amphibolites in Alxa region. *Acta Petrol. Mineral.* **2005**, *24*, 21–31. (In Chinese with English Abstract)

49. Zhou, H.Y.; Mo, X.X.; Li, J.J.; Li, H.M. The U-Pb isotopic dating age of single zircon from biotite plagioclase gneiss in the Qinggele area, Alashan, western Inner Mongolia. *Bull. Mineral. Petrol. Geochem.* **2007**, *26*, 221–223. (In Chinese with English Abstract)

50. Li, J.J.; Shen, B.F.; Li, H.M.; Zhou, H.Y.; Guo, L.J.; Li, C.Y. Single-zircon U–Pb age of granodioritic gneiss in the Bayan Ul area, western Inner Mongolia. *Geol. Bull. China* **2004**, *23*, 1243–1245. (In Chinese with English Abstract)

51. Bao, C.; Chen, Y.L.; Li, D.P. LA-MC-ICP-MS zircons U-Pb dating and Hf isotopic compositions of the Paleoproterozoic amphibolite in Bayan Ul area, Inner Mongolia. *Geol. Bull. China* **2013**, *32*, 1513–1524.

52. Liu, Y.S.; Gao, S.; Hu, Z.C.; Gao, C.G.; Zong, K.Q.; Wang, D.B. Continental and oceanic crust recycling-induced melt-peridotite interactions in the Trans-North China Orogen: U-Pb dating, Hf isotopes and trace elements in zircons of mantle xenoliths. *J. Petrol.* **2010**, *51*, 537–571. [CrossRef]

53. Whitney, D.L.; Evans, B.W. Abbreviations for names of rock-forming minerals. *Am. Mineral.* **2010**, *95*, 185–187. [CrossRef]

54. Leake, B.E.; Woolley, A.R.; Arps CE, S.; Birch, W.D.; Gilbert, M.C.; Grice, J.D.; Guo, Y.Z. Nomenclature of amphiboles: Report of the subcommittee on amphiboles of the International Mineralogical Association, Commission on New Minerals and Mineral Names. *Can. Mineral.* **1997**, *35*, 219–246.

55. Anderson, E.D.; Moecher, D.P. Omphacite breakdown reactions and relation to eclogite exhumation rates. *Contrib. Mineral. Petrol.* **2007**, *154*, 253–277. [CrossRef]

56. Gayk, T.; Kleinschrodt, R.; Langosch, A.; Seidel, E. Quartz exsolution in clinopyroxene of high-pressure granulite from the Munchberg Massif. *Eur. J. Mineral.* **1995**, *7*, 1217–1220. [CrossRef]

57. Nakano, N.; Osanai, Y.; Owada, M. Multiple breakdown and chemical equilibrium of silicic clinopyroxene under extreme metamorphic conditions in the Kontum Massif, central Vietnam. *Am. Mineral.* **2007**, *92*, 1844–1855. [CrossRef]

58. Page, F.Z.; Essene, E.J.; Mukasa, S.B. Prograde and retrograde history of eclogites from the Eastern Blue Ridge, North Carolina, USA. *J. Metamorph. Geol.* **2003**, *21*, 685–698. [CrossRef]

59. Page, F.Z.; Essene, E.J.; Mukasa, S.B. Quartz exsolution in clinopyroxene is not proof of ultrahigh pressures: Evidence from eclogites from the Eastern Blue Ridge, Southern Appalachians, USA. *Am. Mineral.* **2005**, *90*, 1092–1099. [CrossRef]

60. Proyer, A.; Krenn, K.; Hoinkes, G. Orientated precipitates of quartz and amphibole in clinopyroxene of metabasites from the Greek Rhodope: A product of open system precipitation during eclogite-granulite-amphibolite transition. *J. Metamorph. Geol.* **2009**, *27*, 639–654. [CrossRef]

61. Faryad, S.W.; Fisera, M. Olivine-bearing symplectites in fractured garnet from eclogite Moldanubian zone (Bohemian massif)—A short-lived, granulite facies event. *J. Metamorph. Geol.* **2015**, *33*, 597–612. [CrossRef]

62. De Capitani, C.; Petrakakis, K. The computation of equilibrium assemblage diagrams with Theriak/Domino software. *Am. Mineral.* **2010**, *95*, 1006–1016. [CrossRef]

63. Holland, T.J.B.; Powell, R. An internally consistent thermodynamic data set for phases of petrological interest. *J. Metamorph. Geol.* **1998**, *16*, 309–343. [CrossRef]

64. Holland, T.J.B.; Powell, R. An improved and extended internally consistent thermodynamic dataset for phases of petrological interest, involving a new equation of state for solids. *J. Metamorph. Geol.* **2011**, *29*, 333–383. [CrossRef]

65. Green, E.C.R.; White, R.W.; Diener, J.F.A.; Powell, R.; Holland, T.J.B.; Palin, R.M. Activity–composition relations for the calculation of partial melting equilibria in metabasic rocks. *J. Metamorph. Geol.* **2016**, *34*, 845–869. [CrossRef]

66. White, R.W.; Powell, R.; Holland, T.J.B.; Johnson, T.E.; Green, E.C.R. New mineral activity–composition relations for thermodynamic calculations in metapelitic systems. *J. Metamorph. Geol.* **2014**, *32*, 261–286. [CrossRef]

67. Holland, T.J.B.; Powell, R. Activity–composition relations for phases in petrological calculations: An asymmetric multicomponent formulation. *Contrib. Mineral. Petrol.* **2003**, *145*, 492–501. [CrossRef]

68. White, R.W.; Powell, R.; Clarke, G.L. The interpretation of reaction textures in Fe-rich metapelitic granulites of the Musgrave Block, central Australia: Constraints from mineral equilibria calculations in the system $K_2O–FeO–MgO–Al_2O_3–SiO_2–H_2O–TiO_2–Fe_2O_3$. *J. Metamorph. Geol.* **2002**, *20*, 41–55. [CrossRef]

69. White, R.W.; Powell, R.; Holland TJ, B.; Worley, B.A. The effect of TiO_2 and Fe_2O_3 on metapelitic assemblages at greenschist and amphibolite facies conditions: Mineral equilibria calculations in the system $K_2O–FeO–MgO–Al_2O_3–SiO_2–H_2O–TiO_2–Fe_2O_3$. *J. Metamorph. Geol.* **2000**, *18*, 497–511. [CrossRef]

70. Korhonen, F.J.; Brown, M.; Clark, C.; Bhattacharya, S. Osumilite–melt interactions in ultrahigh temperature granulites: Phase equilibria modelling and implications for the P–T–t evolution of the Eastern Ghats Province, India. *J. Metamorph. Geol.* **2013**, *31*, 881–907. [CrossRef]

71. Holland, T.J.B.; Blundy, J.D. Non-ideal interactions in calcic amphiboles and their bearing on amphibole-plagioclase thermometry. *Contrib. Mineral. Petrol.* **1994**, *116*, 433–447.

72. Bhadra, S.; Bhattacharya, A. The barometer tremolite + tschermakite + 2 albite = 2 pargasite + 8 quartz: Constraints from experimental data at unit silica activity, with application to garnet-free natural assemblages. *Am. Mineral.* **2007**, *92*, 491–502. [CrossRef]

73. Liao, Y.; Wei, C.J.; Rehman, H.U. Titanium in calcium amphibole: Behavior and thermometry. *Am. Mineral.* **2021**, *106*, 180–191.

74. Ferry, J.M.; Watson, E.B. New thermodynamic models and revised calibrations for the Ti-in-zircon and Zr-in-rutile thermometers. *Contrib. Mineral. Petrol.* **2007**, *154*, 429–443. [CrossRef]

75. Gao, X.Y.; Zheng, Y.F. On the Zr-in-rutile and Ti-in-zircon geothermometers. *Acta Petrol. Sin.* **2011**, *27*, 417–432. (In Chinese with English Abstract)

76. Sun, S.S.; McDonough, W.F. Chemical and isotopic systematics of oceanic basalts: Implications for mantle composition and processes. *Geol. Soc. Lond. Spec. Publ.* **1989**, *42*, 313–345. [CrossRef]

77. Guevara, V.E.; Caddick, M.J. Shooting at a moving target: Phase equilibria modelling of high-temperature metamorphism. *J. Metamorph. Geol.* **2016**, *34*, 209–235.

78. Palin, R.M.; Weller, O.M.; Waters, D.J.; Dyck, B. Quantifying geological uncertainty in metamorphic phase equilibria modelling: A Monte Carlo assessment and implications for tectonic interpretations. *Geosci. Front.* **2016**, *7*, 591–607.

79. Zhao, G.C.; Cawood, P.A.; Li, S.Z.; Wilde, S.A.; Sun, M.; Zhang, J.; He, Y.H.; Yin, C.Y. Amalgamation of the North China Craton: Key issues and discussion. *Precambrian Res.* **2012**, *222–223*, 56–76.

80. Yin, C.Q.; Zhao, G.C.; Wei, C.J.; Sun, M.; Guo, J.H.; Zhou, X.W. Metamorphism and partial melting of high-pressure pelitic granulites from the Qianlishan Complex: Constraints on the tectonic evolution of the Khondalite Belt in the North China Craton. *Precambrian Res.* **2014**, *242*, 172–186.

81. Santosh, M.; Wilde, S.; Li, J. Timing of Paleoproterozoic ultrahigh-temperature metamorphism in the North China Craton: Evidence from SHRIMP U-Pb zircon geochronology. *Precambr. Res.* **2007**, *159*, 178–196. [CrossRef]

82. Gou, L.L.; Zi, J.W.; Dong, Y.P.; Liu, X.M.; Li, Z.H.; Xu, X.F.; Zhang, C.L.; Liu, L.; Long, X.P.; Zhao, Y.H. Timing of two separate granulite-facies metamorphic events in the Helanshan complex, North China Craton: Constraints from monazite and zircon U-Pb dating of pelitic granulites. *Lithos* **2019**, *350–351*, 105216.

83. Jiao, S.J.; Fitzsimons, I.C.W.; Guo, J.H. Paleoproterozoic UHT metamorphism in the Daqingshan Terrane, North China Craton: New constraints from phase equilibria modeling and SIMS U-Pb zircon dating. *Precambrian Res.* **2017**, *303*, 208–227.

84. Zheng, Y.F.; Zhao, G.C. Two styles of plate tectonics in Earth's history. *Sci. Bull.* **2020**, *65*, 329–334. [CrossRef]

85. Sizova, E.; Gerya, T.; Brown, M. Contrasting styles of Phanerozoic and Precambrian continental collision. *Gondwana Res.* **2014**, *25*, 522–545.

86. Thompson, A.B.; England, P.C. Pressure–temperature–time paths of regional metamorphism II. *Their inference and interpretation using mineral assemblages in metamorphic rocks. J. Petrol.* **1984**, *25*, 929–955.

 *minerals*

Article

Tracing the Early Crustal Evolution of the North China Craton: New Constraints from the Geochronology and Hf Isotopes of Fuchsite Quartzite in the Lulong Area, Eastern Hebei Province

Chen Zhao [1,2,3,*], Jian Zhang [3,4], Xiao Wang [3,4], Chao Zhang [1,2], Guokai Chen [4], Shuhui Zhang [4] and Minjie Guo [4]

[1] Shenyang Center of Geological Survey, China Geological Survey, Shenyang 110034, China; congray@163.com
[2] Northeast Geological S&T Innovation Center of China Geological Survey, Shenyang 110034, China
[3] Department of Earth Sciences, The University of Hong Kong, Pokfulam Road, Hong Kong, China; jian@hku.hk (J.Z.); wx20160117@163.com (X.W.)
[4] School of Earth Sciences and Engineering, Sun Yat-sen University, Zhuhai 519082, China; chengk8@mail2.sysu.edu.cn (G.C.); zhangshh35@mail2.sysu.edu.cn (S.Z.); guomj5@mail2.sysu.edu.cn (M.G.)
* Correspondence: aaron198807@163.com

Abstract: Understanding the composition, formation and evolution of the oldest continental crust is crucial for comprehending the mechanism and timing of crustal growth and differentiation on early Earth. However, the preservation of the ancient continental crust is limited due to extensive reworking by later tectonothermal events. In the Lulong area of eastern Hebei, abundant ca. 3.8–3.4 Ga detrital zircons of the fuchsite quartzite have been previously identified. Nonetheless, the provenance and Hf isotopic compositions of the fuchsite quartzite remain unclear. In this study, we present new detrital zircon ages and Hf isotopic for the fuchsite quartzite in the Lulong area to establish the timing of deposition, the provenance and the regional stratigraphic relationship. Zircon U-Pb dating indicates that the fuchsite quartzite was deposited between 3.3–3.1 Ga and most grains were sourced from the 3.8 Ga TTG gneisses and Paleoarchean magmas. Field investigations and regional correlations reveal that the fuchsite quartzite from the Lulong area is equivalent to that of the Caozhuang area. Zircon Hf isotopic data from eastern Hebei Province (Lulong and Caozhuang areas) and Anshan and Xinyang areas indicate that the oldest crustal growth event of North China Craton occurred in the Hadean.

Keywords: early earth; eastern Hebei Province; North China Craton; detrital zircon; crustal growth

Citation: Zhao, C.; Zhang, J.; Wang, X.; Zhang, C.; Chen, G.; Zhang, S.; Guo, M. Tracing the Early Crustal Evolution of the North China Craton: New Constraints from the Geochronology and Hf Isotopes of Fuchsite Quartzite in the Lulong Area, Eastern Hebei Province. *Minerals* **2023**, *13*, 1174. https://doi.org/10.3390/min13091174

Academic Editor: Nigel J. Cook

Received: 25 July 2023
Revised: 30 August 2023
Accepted: 4 September 2023
Published: 7 September 2023

1. Introduction

The composition, formation and evolution of the early continental crust during Hadean and Archean is a fundamental issue of Precambrian geology. It encompasses a series of essential topics, such as determining the age and tectonic framework of the oldest continental crust and identifying the crustmantle interactions that existed during that time [1,2]. The most direct method for investigating these topics is the study of the crustal materials formed during the early stages of the Earth. However, ancient crustal rocks or materials are only found in limited amounts in certain old cratons, including the ~4.02 Ga Acasta gneiss in northern Labrador, Quebec [3]; The Itsaq Gneiss Complex, Greenland [4]; Eastern Antarctica [5]; The Pilbara and Yilgarn Cratons in Western Australia [6]; And the Anshan and Lulong areas in the North China Craton [7–10]. However, the preservation of the ancient crustal rocks is limited due to the extensive reworking by later tectonothermal events. The occurrence of the ancient zircons in younger rocks is more abundant than ancient rocks, making the study of the ancient zircon grains a key component in understanding the early evolution of the Earth. Currently, the oldest zircons on Earth (4.4 Ga) have been identified in the Jack Hills, Western Australia [11]. Meanwhile, ancient zircon grains of Hadean age have also been reported in ancient craton of China.

The North China Craton (NCC) is the largest and the oldest craton in China. It is subdivided into the Eastern and Western Blocks, which are separated by a Paleoproterozoic collisional belt, named the Trans-North China Orogen (Figure 1) [12]. The Western Block was formed by amalgamation of the Ordos and Yinshan Blocks along the EW-trending Khondalite Belt at 1.95–1.92 Ga (Figure 1) [13–15], whereas the Eastern Block underwent a Paleoproterozoic rifting event along its eastern margin at 2.2–1.9 Ga and formed the Jiao-Liao-Ji mobile belt at ca. 1.9 Ga (Figure 1) [16–20]. The final amalgamation between the Eastern and Western Blocks occurred in 1.95–1.85 Ga (Figure 1) [13,21–26], which formed the uniform basement of the NCC. Subsequently, the NCC was covered by abundant Proterozoic to Paleozoic sediments [27,28].

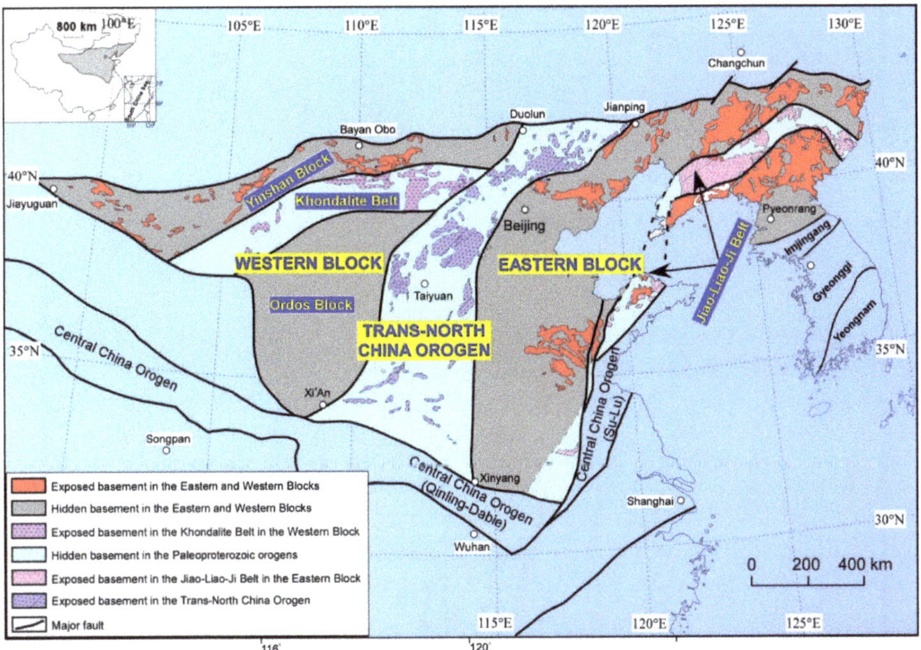

Figure 1. Tectonic subdivision of the North China Craton proposed by [12,13,29], showing the spatial distribution of basement rocks in the Neoarchean Eastern and Western Blocks, as well as the intervened Paleoproterozoic Trans-North China Orogen.

The discoveries of ancient rocks and zircons can provide evidence for understanding the early evolution of NCC. Early Precambrian rocks are widely distributed in the Eastern Block, wherea ca. 3.8 Ga quartzofeldspathic gneiss complex has been reported in Anshan, Liaoning Province and Lulong, eastern Hebei Province (Figure 2) [7–10]. The original oldest rock identified was a 3.8 Ga mylonitized trondhjemitic gneiss in the Anshan area [7]. Recently, Wan et al. [10] reported 3.8 Ga granodioritic rocks in the Labashan, Lulong area. For Xinyang, the southern part of the NCC, Ma et al. [30] reported that 3.8 Ga granulite xenoliths were hosted in the Mesozoic volcanic rocks. A large number of 3.8–3.4 Ga detrital zircons have been found in fuchsite quartzite and paragneiss from the Caozhuang area of eastern Hebei Province, indicating that the provenance of these meta-supracrustal rocks mainly consists of Eo- to Paleoarchean igneous rocks [2,31–33]. Likewise, abundant ca. 3.8–3.4 Ga detrital zircons have been identified in the fuchsite quartzite in the Lulong area, eastern Hebei Province [34]. However, their provenance and Hf isotopic compositions are poorly constrained. In this paper, we present new geochronological and Hf isotopic data of the ancient detrital zircons from the fuchsite quartzites in Labashan, in the Lulong area, to determine the timing of deposition, the provenance and the regional stratigraphic relationship and further discuss the early crustal growth of the NCC.

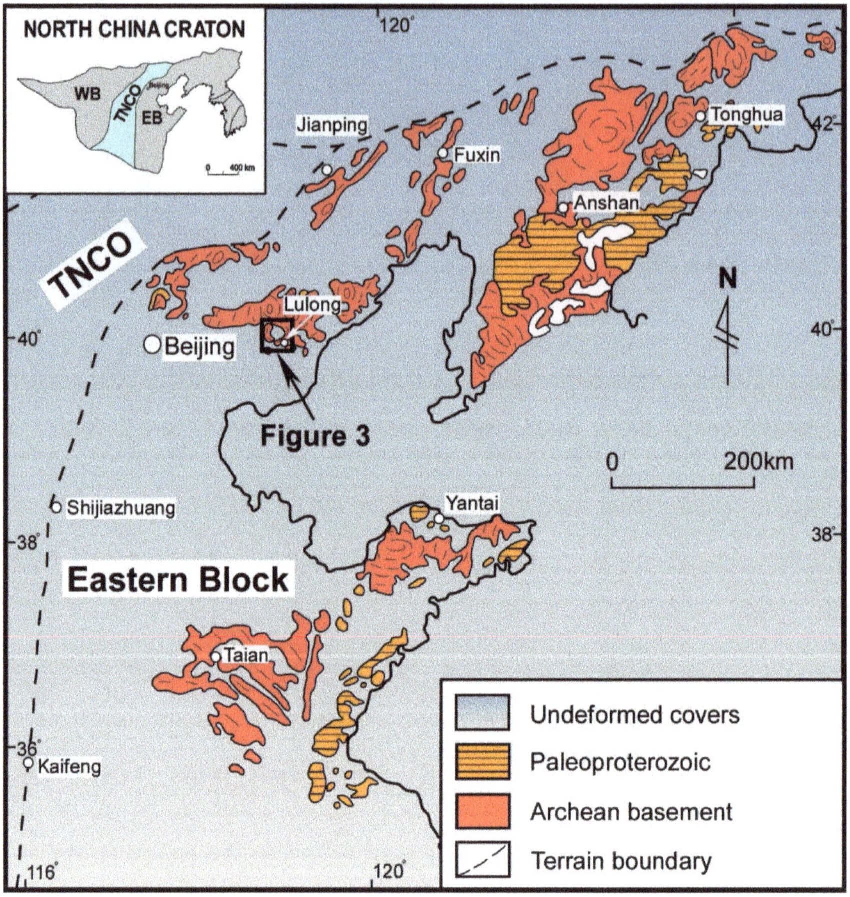

Figure 2. Geological map of the Eastern Block, showing the distribution of the Archean basements (revised in accordance with [13]).

2. Geological Setting and Sample Collection

The eastern Hebei Province is located in the core of the Eastern Block (Figure 2). The Precambrian terrain in eastern Hebei Province is mainly occupied by Neoarchean rock assemblages which comprise up to 85% of the exposed basement [35,36]. Due to the extensive reworking by later tectonothermal events, the areal extent and the tectonic history of the early Archean basement has been obliterated [37]. The Eoarchean to Mesoarchean rocks only crop out in Caozhuang, Qian' an and Labashan, Lulong areas (Figure 3) [2,7,31,35]. The Caozhuang supracrustal sequences, which are located in the southwest of the Qian' an TTG granitic dome, are mainly composed of quartzite, amphibolite, garnet-biotite gneiss, calc–silicate rock, marble, fine-grained biotite gneiss and banded iron formation (Figure 3) [38]. Some scholars proposed that the amphibolites from the Caozhuang area have an age of ca. 3.5 Ga, as determined using the whole rock Sm–Nd isochron method, which may represent the time of the protolithic basalt eruption [39–41], although this age may be an inconclusive result [35]. Detrital zircons of 3.8–3.4 Ga have been reported from fuchsite quartzite within the Caozhuang area [2,7,31]. Similar to Caozhuang in the

Qian' an area, Eo- to Paleoarchean materials have also been reported in Labashan in the Lulong area, such as the 3.8 Ga TTG gneisses and 4.0–3.2 Ga detrital zircons in paragneisses (Figure 3) [10,33]. Fuchsite quartzites also developed in Labashan in theLulong area, which are temporally and spatially associated with TTG gneiss, paragneiss, potassium-rich granite and meta-gabbro (Figure 4a). The lithologic features of fuchsite quartzite are similar to those of the Caozhuang area, with a thickness of 5–8 m and an extensional length of more than 100 m (Figure 4b). The fuchsite quartzite, which is composed of quartz (80 vol%) and muscovite (20 vol%) with minor plagioclase and ilmenite (Figure 4c), is intruded by the meta-gabbro (Figure 4d) and is in tectonic contact with the gneisses (Figure 4e).

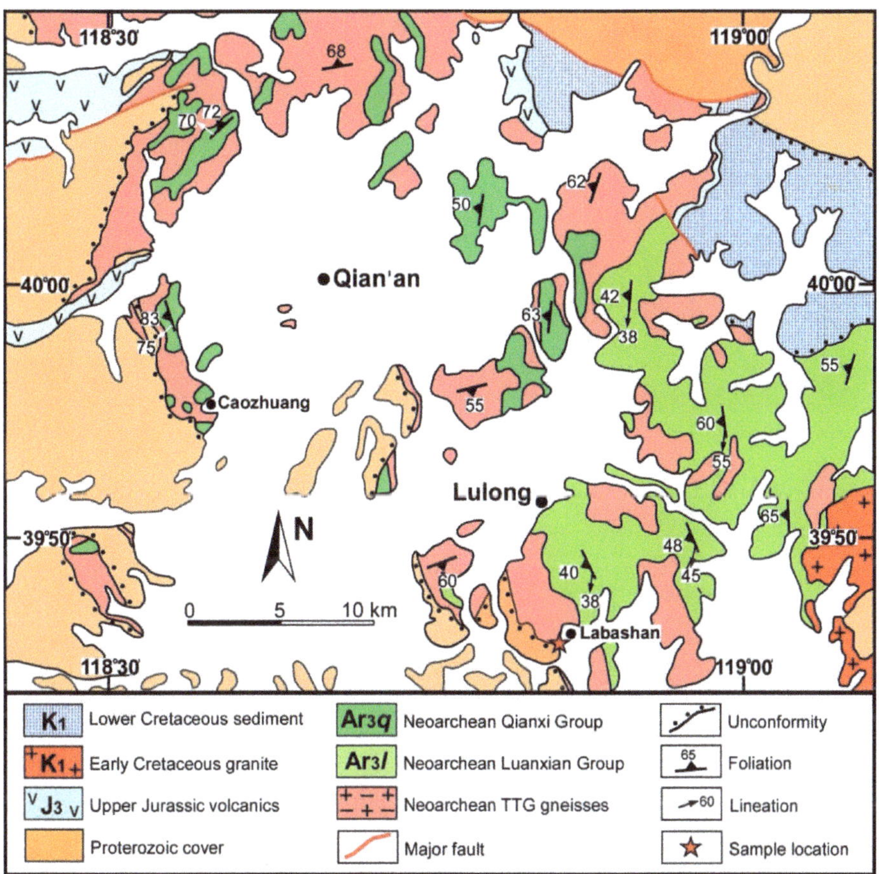

Figure 3. Detailed geological map of the Qian'an and Lulong areas, showing the spatial relationshios, lithology and structural features of the Archean supracrustal rock assemblage and granitic pluton.

In this study, two representative samples were collected from Labashan, Lulong to constrain the formation time of the fuchsite quartzite (Figure 4a,b,d). The youngest zircon of Sample 20JD03-1 (fuchsite quartzite) may represent the maximum sedimentation age. In the field, the meta-gabbro dyke clearly intrudes into the fuchsite quartzite (Figure 4a,d); Thus, it can be used to constrain the minimum depositional timing of fuchsite quartzite.

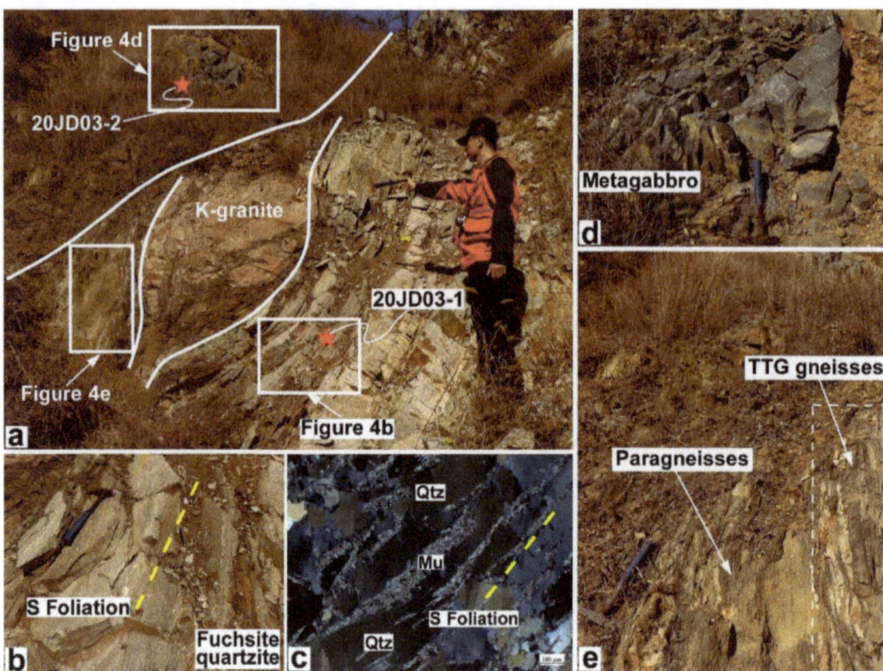

Figure 4. Representative outcrop and microscopic images of the fuchsite quartzite in the Labashan, Lulong, eastern Hebei Province. (**a**) The fuchsite quartzite was intruded by the meta-gabbro. (**b**) Field geological characteristics of the fuchsite quartzite (Sample 20JD03-1). (**c**) Microscopic images showed mineral composition of the fuchsite quartzite. Abbreviations for different minerals: Qtz—quartz; Mu—muscovite. (**d**) Field geological characteristics of the meta-gabbro (Sample 20JD03-2). (**e**) Field geological characteristics of the TTG gneisses and paragneisses.

3. Analytical Methods

Zircons were extracted from the samples and separated using conventional heavy liquid and magnetic techniques at the Tuoyan Analytical Technology Co., Ltd., testing center (Guangzhou, China). Representative zircons were hand-picked and mounted onto the epoxy resin surface and then polished to approximately half-thickness. Cathodoluminescence (CL) images of the zircons were acquired using CL spectroscopy on a Quanta 200F SEM (scanning electron microscope) at Nanjing Hongchuang GeoAnalysis, Nanjing, China. The U-Pb analyses of zircons were conducted by laser ablation inductively coupled plasmamass spectrometry (LA-ICP-MS) at the Chengpu Geological Testing Co., Ltd., Langfang, China. The analyses were performed using a Plasma Quant MS elite ICP-MS instrument designed by the Analytik Jena AG (Jena, Germany) in combination with an excimer 193 nm laser ablation system (New Wave, NWR193). Each analysis consisted of a background acquisition of approximately 30s followed by 30s of sample data acquisition. The spot size and frequency of the laser were set to 20 μm and 8 Hz, respectively, in this study. Helium was used as the carrier gas. NIST SRM 610 was applied to yield the highest sensitivity, lowest oxidation, and a stable signal to achieve the optimal condition [42]. A 15 ms dwell time was set for ^{204}Pb, ^{206}Pb, ^{207}Pb, ^{208}Pb, ^{232}Th and ^{238}U. Zircon Plešovice (~337 Ma) [43] and Qinghu (~159 Ma) [44] were used as external standards for U-Pb dating and as the blind sample for monitoring the instrument's condition, respectively, and were measured twice every 5 zircon analyses. The analytical data were reduced using GLITTER 4.0 software (Macquarie University, Sydney, Australia). The calculation of weighted average ages and plotting of concordia diagrams were completed with Isoplot 4.15 [45].

Zircon Lu–Hf isotopic ratios were determined following U-Pb analyses viaLA-ICP-MS at Chengpu Geological Testing Co., Ltd., Langfang, China. A Neptune Plus (MC-ICP-MS) instrument equipped with a New Wave UP21 LA system at a spot diameter of 44 μm was used with a laser repetition rate of 8 Hz at 6 J/cm^2 for 27 s. The detailed analytical procedures and interference corrections used were similar to those described by Wu et al. [46]. The $\varepsilon_{Hf}(t)$ values were calculated based on the chondrite values of ^{176}Hf/^{177}Hf (0.282772) and ^{176}Lu/^{177}Hf (0.0332) ratios [47].

4. Analytical Results

Cathodoluminescence (CL) images of representative zircons from each sample are shown in Figure 5, and zircon U-Pb dating results are shown in Figure 6 and listed in Supplementary Table S1.

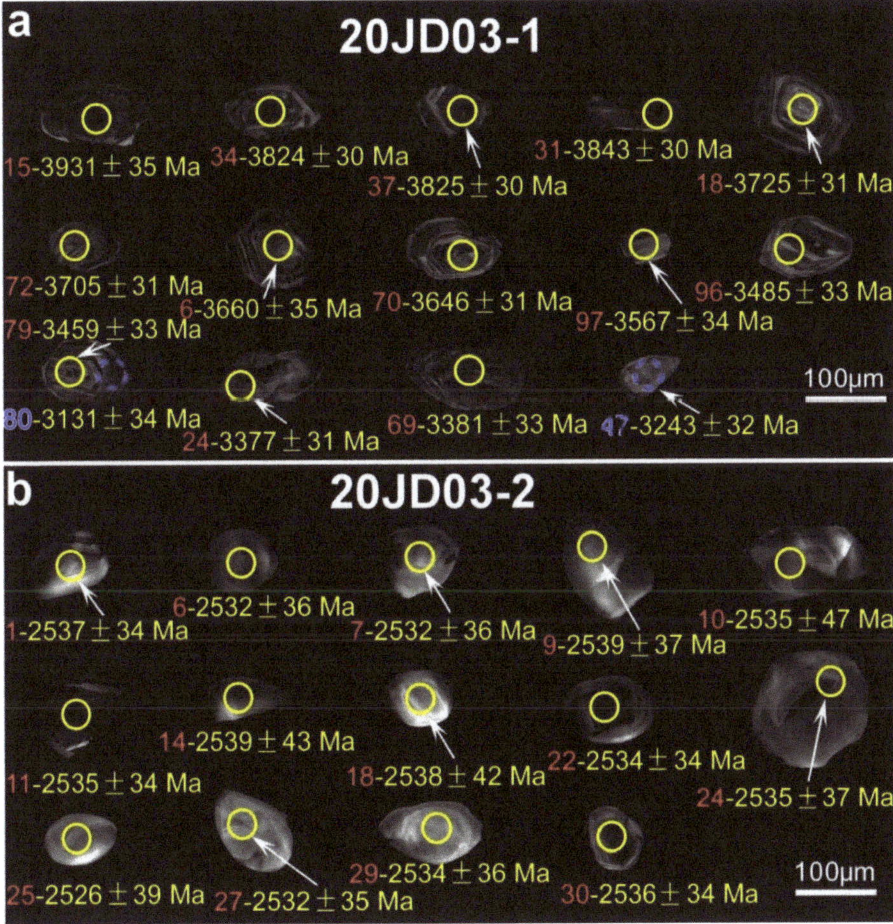

Figure 5. Representative selection of Cathodoluminescence (CL) images for the selected samples. (**a**) Sample 20JD03-1: fuchsite quartzite. (**b**) Sample 20JD03-2: meta-gabbro.

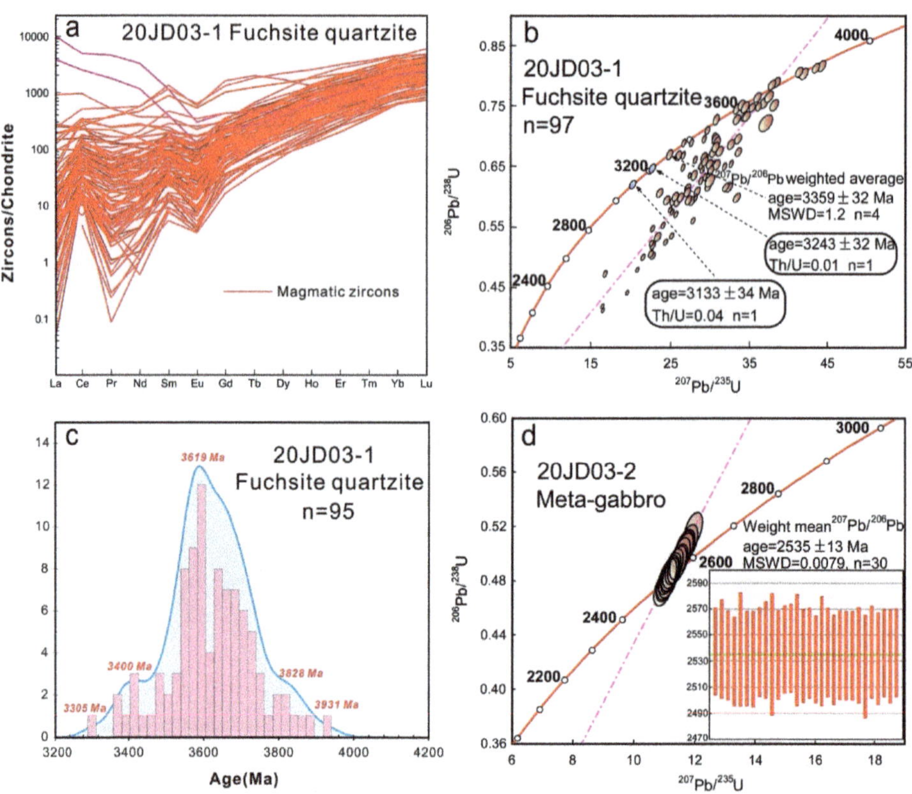

Figure 6. REE pattern, concordia diagrams and probability distribution of dating zircons. (**a**) Chondrite-normalized REE pattern for magmatic zircons from the fuchsite quartzite (20JD03-1); (**b**) U–Pb concordia diagram for zircons from the fuchsite quartzite; (**c**) Frequency distribution of the zircon age clusters of the detrital zircons from the fuchsite quartzite (20JD03-1); (**d**) U-Pb concordia diagram for zircons from the meta-gabbro (20JD03-2).

4.1. Zircon LA-ICP-MS U-Pb Dating

4.1.1. Fuchsite Quartzite (20JD03-1)

Sample 20JD03-1 was collected from a fuchsite quartzite in the Labashan, Lulong (Figures 3 and 4a). The fuchsite quartzite exhibits penetrative foliation S at the field and microstructure scales (Figure 4b,c). Zircon grains extracted from the fuchsite quartzite are mostly euhedral–subhedral crystal morphologies and 100–150 μm in grain size (Figure 5a). Most of the zircons exhibit clear oscillatory zoning structure (Figure 5a), indicative of a magmatic origin. Meanwhile, they exhibit enriched LREEs ($(Gd/Yb)_N$ = 0.01–1.11), mostly negative Eu anomalies (Eu/Eu^* = 0.02–0.34) and moderately to highly positive Ce anomalies (Ce/Ce^* = 0.37–216.78), which further confirm a magmatic origin for these zircons (Supplementary Table S2 and Figure 6a). Only one zircon shows thecore–rim structure, where the core (Spot #79) display concentric oscillatory zoning in the CL image (Figure 5a), indicative of a magmatic origin. In contrast, zircon rims (Spot #80 and Spot #47) are relatively luminescent and lack interior structure (Figure 5a), indicative of a metamorphic origin. This is consistent with the relatively low Th/U ratios of 0.01–0.04 (Supplementary Table S1). Ninety-seven zircon grains were analyzed and plotted on the U-Pb concordia diagram (Figure 6b). Of these, two young zircons were assigned the ages of 3133 ± 34 Ma and 3243 ± 32 Ma, respectively, which represented their metamorphic ages (Figure 6b). Most of the zircons fall into the range of 3305 Ma to 3931 Ma and form five distinct age peaks (Figure 6c). The youngest single magmatic zircon age of this sample is ca. 3305 Ma (Figure 6c). The age of the Paleoarchean zircons ranges from 3371 ± 35 Ma to 3796 ± 33 Ma with the most prominent age peak of ca. 3691 Ma and the subordinate age peak of ca. 3400 Ma (Figure 6c). The Eoarchean zircon cluster ranges from 3802 ± 35 Ma to 3882 ± 33 Ma forming a peak at ca. 3828 Ma (Figure 6c). The oldest zircon age in this sample was 3931 ± 35 Ma, with only one single grain (Figure 6c).

4.1.2. Meta-Gabbro (20JD03-2)

Sample 20JD03-2 (meta-gabbro) was collected from the same outcrop as Sample 20JD03-1 (Figure 4a,d). Regionally, the fuchsite quartzite is intruded by the meta-gabbro dyke (Figure 4a). Thus, this gabbro dyke can be used to determine the minimum depositional timing of the fuchsite quartzite. Zircon grains extracted from this sample are transparent and spherical-oval in shape, with a grain size ranging from 100 to 150 μm. CL images reveal that most zircons exhibit blurred oscillatory zoning (Figure 5b), indicating that these zircons could have been influenced by the subsequent anatexis. Th/U ratios for most analyzed zircons are 0.26–4.01 (Supplementary Table S1), which is indicative of a magmatic origin. Thirty zircon grains were analyzed and plotted on the U-Pb concordia diagram (Figure 6d). All the data are concordant and yield a ^{207}Pb/^{206}Pb weighted mean age of 2535 ± 13 Ma (MSWD = 0.0079) (Figure 6d), indicating the crystallization age of the meta-gabbro.

4.2. Zircon Hf Dating

Fifty zircon grains with concordant U–Pb ages from different age populations were analyzed for their Lu–Hf isotopic compositions. The zircon grain with an age of 3305 Ma has an $\varepsilon_{Hf}(t)$ value of −4.2 and a T_{DM2} model age of 3871 Ma (Supplementary Table S3). The zircon grains with ages in the 3785–3370 Ma range yielded similar compositions (^{176}Hf/^{177}Hf = 0.280253–0.280854, ^{176}Lu/^{177}Hf = 0.000394–0.000872, and ^{176}Yb/^{177}Hf = 0.010547–0.059906) and have $\varepsilon_{Hf}(t)$ values of −9.2 to +8.9, and T_{DM2} model ages of 3316–4346 Ma (Supplementary Table S3). The zircons of 3862–3802 Ma have similar compositions (^{176}Hf/^{177}Hf = 0.280311–0.280484, ^{176}Lu/^{177}Hf = 0.000425–0.001972, and ^{176}Yb/^{177}Hf = 0.011637–0.059104), and their $\varepsilon_{Hf}(t)$ values and T_{DM2} model ages are −1.5 to +1.2 and 4034 to 3903 Ma, respectively (Supplementary Table S3). The oldest zircon (3931 Ma) displayed asimilar composition (^{176}Hf/^{177}Hf = 0.280442, ^{176}Lu/^{177}Hf = 0.000889, and ^{176}Yb/^{177}Hf = 0.023647), and it has an $\varepsilon_{Hf}(t)$ value of +4.7, and a T_{DM2} model age of 3828 Ma (Supplementary Table S3).

5. Discussion

5.1. The Depositional Age of the Fuchsite Quartzite

Detrital zircon geochronology is a widely used method for constraining the depositional age of specific stratigraphic successions [48,49]. Although some geologists proposed that the fuchsite quartzite in the Lulong area was probably deposited after the Mesoarchean (ca. 3.2 Ga) [34], there is currently no confirmed geochronologicalevidence to support this hypothesis.

Given that a sedimentary unit can not be older than its youngest detrital zircon grain [50], our youngest concordant ^{207}Pb/^{206}Pb age group with a weight mean age of 3359 ± 32 Ma, provides a maximum age for the deposition of the fuchsite quartzite in this study (Figure 6b). The meta-gabbro that clearly intruded the fuchsite quartzite yielded an age of 2535 ± 13 Ma, indicating that the fuchsite quartzite must have been deposited sometime earlier than ca. 2535 Ma (Figure 4a,d and Figure 6d). Meanwhile, the metamorphic zircons from the fuchsite quartziteyielded the ages of 3243 ± 32 Ma and 3133 ± 34 Ma, which constrain the metamorphic event (Figure 6b). On the one hand, the ages of the zircon cores are obviously older than 3.1 Ga, indicating that the U–Pb isotopic system of these metamorphic zircons has reached a new equilibrium. On the other hand, the metamorphic zircons have prismatic shape, which can exclude the possibility of metamorphism in source areas [33]. Furthermore, 3.2–3.1 Ga probably represents the record of a tectono-thermal event that occurred after the fuchsite quartzite was deposited. For the same outcrop, Wan et al. [33] also reported the evidence of the Mesoarchean (ca. 3.1 Ga) metamorphic event in the paragneisses (Figure 4a,e). Combining this with the age of the youngest detrital zircon at 3.3 Ga, we can constrain the depositional age of the fuchsite quartzite to be 3.3–3.1 Ga.

Some scholars have proposed that the supracrustal rock assemblage in Labashan, Lulong, is similar to the Caozhuang supracrustal sequence (excluding biotite–granulite and BIF) in Huangbaiyu, Qian'an, with abundant Eoarchean–Paleoarchean zircons (specifically, 3.8 Ga to 3.3 Ga) [33,34]. Additionally, regional stratigraphic correlation indicates that much of the Caozhuang supracrustal sequence is equivalent to the supracrustal rock assemblage in Labashan, Lulong, and also has a depositional age of 3.3–3.1 Ga [33]. To sum up, we defined the whole area as a 3.3–3.1 Ga Caozhuang–Labashan sequence, which is mainly composed of fuchsite quartzite, feldspathic quartzite, biotite plagioclase gneiss, para–amphibolite, ortho–amphibolite and meta–ultramafic rock.

5.2. Determination of Provenance

Determining the source of sedimentary rocks can be achieved through dating of detrital zircon grains [51,52]. In this study, detrital zircons obtained from the fuchsite quartzite sample (20JD03-1) yielded the ages ranging between 3931 and 3305 Ma, which were grouped into five populations (Supplementary Table S1 and Figure 6c). The weighted mean ages of these populations are 3931 Ma, 3828 Ma, 3619 Ma, 3400 Ma and 3305 Ma, respectively (Figure 6c). Most of these ages are consistent with the known tectonothermal events, indicating the potential source information for the fuchsite quartzite.

It is notable that the zircon grains from thefuchsite quartzite exhibit clear oscillatory zoning texture andeuhedral crystal shape (Figure 5a), suggesting no or short distance of sedimentary transportation. Sixzircon grains from the fuchsite quartzite were found to be older than 3800 Ma, with the oldest age of 3931 ± 35 Ma (Supplementary Table S1 and Figure 6c). This is nearly the oldest zircon recorded in the Lulong area. Given that the 3.8 Ga TTG gneiss was reported in the same outcrop of the Labashan area (Figure 4a,e) [10], these zircons were likely derived from this Eoarchean gneiss. The ca. 3.6 Ga, ca. 3.4 Ga and ca. 3.3 Ga detrital zircon grains from the fuchsite quartzite were most likely derived from the Paleoarchean magmatic events in the Caozhuang and Huangbaiyu areas [33,34]. These zircon geochronological data provided valuable information for identifying the Eo- to Paleoarchean rocks in the eastern Hebei Province [33].

5.3. Early Crustal Evolution of the NCC

The high closure temperature of Hf in zircon reduces the influence of later metamorphism or magmatic processes, except for the formation of overgrowths [53]. Therefore, zircon Lu–Hf isotopic data can be used to determine magmatic sources and petrogenetic processes, and to constrain the history of crustal growth [54].

The U-Pb age results of this study provide evidence of the existence of 3.9–3.3 Ga old crustal fragments in the Lulong area, eastern Hebei Province, NCC. Most of the detrital zircon grains of this age display negative $\varepsilon_{Hf}(t)$ values as low as -9.2 (Supplementary Table S3 and Figure 7), while a significant number of zircon grains have positive $\varepsilon_{Hf}(t)$ values as high as $+8.9$ (Supplementary Table S3 and Figure 7). As shown in Figure 7, the 3305–3524 Ma zircons from the fuchsite quartzite show negative $\varepsilon_{Hf}(t)$ values that are distinct from the positive $\varepsilon_{Hf}(t)$ value of the 3931 Ma. The Hf model ages also suggest that these zircon grains were mainly derived from old recycled basement rocks rather thanjuvenile materials (Supplementary Table S3). Zircon grains with ages between 3540 and 3862 Ma have both positive and negative $\varepsilon_{Hf}(t)$ values (Supplementary Table S3 and Figure 7), indicating that these zircon grains were derived from both recycled old basements and juvenile materials. T_{DM2} model ages for the 3.9–3.3 Ga zircon grains range from 4.2 to 3.4 Gawith a peak at ca. 3.8 Ga (Supplementary Table S3 and Figure 8), which are significantly older than their U–Pb ages, suggesting that these grains were derived from the reworking of Hadean to Paleoarchean crust. It is worth noting that zircon grains in fuchsite quartzite from the Caozhuang, Qian' an area show a broadly similar distribution of U–Pb ages spectrum and

$\varepsilon_{Hf}(t)$ values (Figure 7). The $\varepsilon_{Hf}(t)$ values of zircons from both fuchsite quartzites decrease regularly with the decreasing ages (Figure 7). Petrologiccharacteristics indicate that the fuchsite quartzite from the Caozhuang area is equivalent to the fuchsite quartzite from the Lulong area.

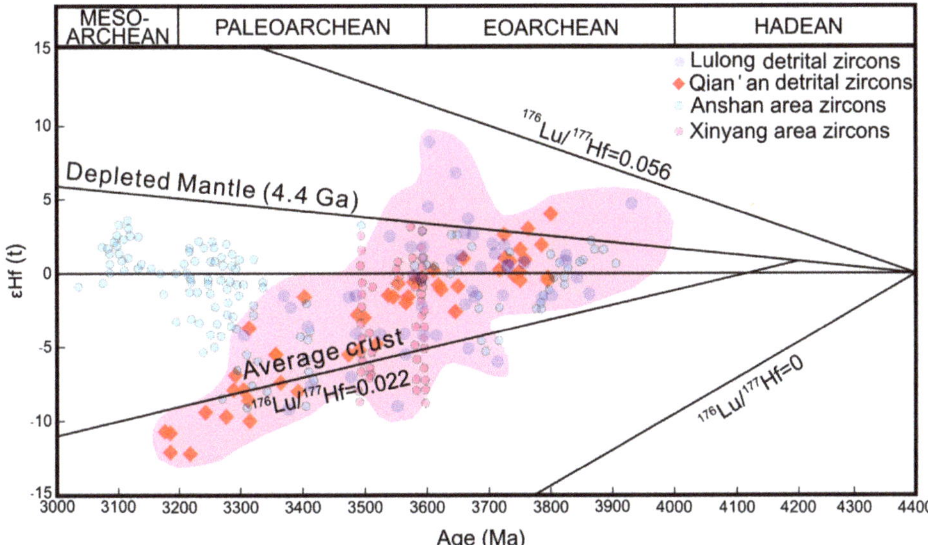

Figure 7. $\varepsilon_{Hf}(t)$ vs. U–Pb age diagram. The published data of the Qian'an, Anshan and Xinyang are taken from [2,55,56], respectively.

To address the issues regarding the early crustal evolution of NCC, we compared 3.8–3.0 Ga zircon grains from the Anshan and Xinyang areas (Figure 7). The Anshan area contains the oldest (ca. 3.8 Ga) igneous rocks in the northern NCC [7], as well as abundant ca. 3.45–2.90 Ga granitoids and minor Paleo- to Mesoarchean supracrustal rocks [57,58]. Zircon Hf isotopes indicate that both ancient and juvenile crustal materials contributed to the generation of the Eo- to Mesoarchean zircon grains in the Anshan area (Figure 7) [55]. Their zircon $T_{DM2(t)}$ ages range from 4.0 to 3.2 Ga with a peak at ca. 3.6 Ga (Figure 8b), suggesting significant crustal growth during the Hadean to Paleoarchean [55]. In the southern NCC, the oldest rocks and minerals are the ca. 3.6 Ga felsic granulite xenoliths [59,60] and ca. 3.6 Ga detrital zircons from the Songshan Group [61], respectively. The ca. 3.6 Ga zircons from the Xinyang xenoliths show variable $\varepsilon_{Hf}(t)$ (Figure 7). Zircons with negative $\varepsilon_{Hf}(t)$ values (-2.2 to -8.2) have $T_{DM2(t)}$ model ages of 3.97–4.31 Ga, suggesting ancient crustal components of the Hadean age (Figures 7 and 8c) [56]. Similarly, ca 3.5 Ga zircons in the Xinyang area also show a wide range of $\varepsilon_{Hf}(t)$ (Figure 7). Some of them display negative $\varepsilon_{Hf}(t)$ values of -1.2 to -7.8 and yield $T_{DM2(t)}$ model ages of 3.95–4.12 Ga (Figures 7 and 8c) [56]. This implies that the Hadean crustal materials were involved in the generation of the Paleoarchean felsic magmas. In summary, old (≥ 4.0 Ga) crustal components have been found in the Eastern Hebei Province (Caozhuang and Lulong areas), Anshan and Xinyang areas. This evidence indicates that the oldest crust in the NCC was formed in the Hadean.

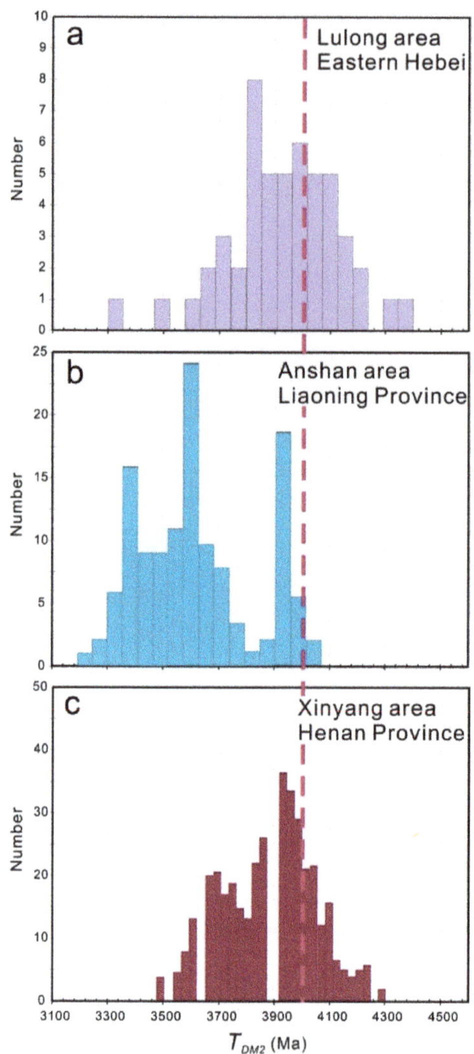

Figure 8. The comparison of the Hf two-stage model age of the zircons from (**a**) sample 20JD03-1 (fuchsite quartzite, Lulong area, eastern Hebei Province); (**b**) Anshan area; and (**c**) Xinyang area. The Hf isotopic data of the Anshan area and Xinyang area are taken from [55] and [56], respectively.

6. Conclusions

(1) Regional correlation reveals that the fuchsite quartzite from the Lulong area is equivalent to that of the Caozhuang area, with a depositional age of 3.3–3.1 Ga constrained by the detrital zircon ages and the metamorphic zircon ages.

(2) The primary sources of sediment in the fuchsite quartzite from the Lulong area were the 3.8 Ga TTG gneisses and the Paleoarchean magmas, specifically those from the ca. 3.6 Ga, 3.4 Ga, and 3.3 Ga.

(3) Zircon Hf model ages from eastern Hebei Province, specifically the Anshan and Xinyang areas, reinforce the theory of the earliest crustal growth event in NCC occurring in Hadean.

Supplementary Materials: The following supporting information can be downloaded at: https://www. mdpi.com/article/10.3390/min13091174/s1, Supplementary Table S1: LA-ICP-MS U–Pb data of zircons for the samples from the Lulong area; Supplementary Table S2: REE concentrations of zircons from the fuchsite quartzite in Lulong area; Supplementary Table S3: Zircon Hf isotopic data of the fuchsite quartzite in the Lulong area.

Author Contributions: Conceptualization, C.Z. (Chen Zhao) and J.Z.; methodology, C.Z. (Chen Zhao) and J.Z.; software, C.Z. (Chen Zhao); validation, J.Z. and C.Z. (Chen Zhao); formal analysis, J.Z.; investigation, C.Z. (Chen Zhao), J.Z., X.W., C.Z. (Chao Zhang), G.C., S.Z. and M.G.; data curation, C.Z. (Chen Zhao) and X.W.; writing—original draft preparation, C.Z. (Chen Zhao); writing—review and editing, J.Z.; project administration, J.Z. and C.Z. (Chao Zhang); funding acquisition, J.Z. and C.Z. (Chao Zhang). All authors have read and agreed to the published version of the manuscript.

Funding: This study was financially supported by the China Geological Survey (DD20230216) and the National Natural Science Foundation of China (42025204 and 42102271).

Institutional Review Board Statement: Not applicable.

Informed Consent Statement: Not applicable.

Data Availability Statement: Not applicable.

Acknowledgments: We sincerely thank three anonymous reviewers for their constructive and helpful comments. We also thank Yazhou Miao and Xiaoman Wang for their assistance in the field.

Conflicts of Interest: The authors declare no conflict of interest.

References

1. Smithies, R.H.; Champion, D.C.; Cassidy, K.F. Formation of Earth's early Archaean continental crust. *Precambrian Res.* **2003**, *127*, 89–101. [CrossRef]

2. Wu, F.Y.; Yang, J.H.; Liu, X.M.; Li, T.S.; Xie, L.W.; Yang, Y.H. Hf isotopes of the 3.8 Ga zircons in eastern Hebei Province, China: Implications for early crustal evolution of the North China Craton. *Chin. Sci. Bulletin.* **2005**, *50*, 2473–2480. [CrossRef]

3. Bowring, S.A.; Williams, I.S. Priscoan (4.00–4.03 Ga) orthogneisses from northwestern Canada. *Contrib. Mineral. Petrol.* **1999**, *134*, 3–16. [CrossRef]

4. Nutman, A.P.; McGregor, V.R.; Friend, C.R.L.; Bennett, V.C.; Kinny, P.D. The Itsaq Gneiss Complex of southern Western Greenland: The world's most extensive record of early crustal evolution (3900–3600 Ma). *Precambrian Res.* **1996**, *78*, 1–39. [CrossRef]

5. Black, L.P.; William, I.S.; Compston, W. Four zircon ages from one rock: The history of a 3930 Ma-old granulite from Mount Sones, Enderby Land, Antarctica. *Contrib. Mineral. Petrol.* **1986**, *94*, 427–437. [CrossRef]

6. Nutman, A.P.; Kinny, P.D.; Compston, W.; Williams, I.S. SHRIMP U–Pb zircon geochronology of the Narryer Gneiss Complex, Western Australia. *Precambrian Res.* **1991**, *52*, 275–300. [CrossRef]

7. Liu, D.Y.; Nutman, A.P.; Compston, W.; Wu, J.S.; Shen, Q.H. Remnants of ≥3800 Ma crust in the Chinese part of the Sino-Korean Craton. *Geology* **1992**, *20*, 339–342. [CrossRef]

8. Song, B.; Nutman, A.P.; Liu, D.Y.; Wu, J.S. 3800 to 2500 Ma crust in the Anshan area of Liaoning Province, northeastern China. *Precambrian Res.* **1996**, *78*, 79–94. [CrossRef]

9. Wan, Y.S.; Liu, D.Y.; Song, B.; Wu, J.S.; Yang, C.H.; Zhang, Z.Q.; Geng, Y.S. Geological and Nd isotopic compositions of 3.8 Ga meta-quartz dioritic and trondhjemitic rocks from the Anshan area and their geological significance. *J. Asian Earth Sci.* **2005**, *24*, 563–575. [CrossRef]

10. Wan, Y.S.; Xie, H.Q.; Wang, H.C.; Li, P.C.; Chu, H.; Xiao, Z.B.; Dong, C.Y.; Liu, S.J.; Li, Y.; Hao, G.M.; et al. Discovery of ~3.8 Ga TTG rocks in eastern Hebei, North China Craton. *Acta Geol. Sin.* **2021**, *95*, 1321–1333. (In Chinese with English Abstract)

11. Wilde, S.A.; Valley, J.W.; Peck, W.H.; Graham, C.M. Evidence from detrital zircons for the existence of continental crust and oceans on the Earth 4.4 Gyr ago. *Nature* **2001**, *409*, 175. [CrossRef] [PubMed]

12. Zhao, G.C.; Wilde, S.A.; Cawood, P.A.; Sun, M. Archean blocks and their boundaries in the North China Craton: Lithological, geochemical, structural and P-T path constraints and tectonic evolution. *Precambrian Res.* **2001**, *107*, 45–73. [CrossRef]

13. Zhao, G.C.; Sun, M.; Wilde, S.A.; Li, S.Z. Late Archean to Paleoproterozoic evolution of the North China Craton: Key issues revisited. *Precambrian Res.* **2005**, *136*, 177–202. [CrossRef]

14. Yin, C.Q.; Zhao, G.C.; Sun, M.; Xia, X.P.; Wei, C.J.; Zhou, X.W. LA-ICP-MS U-Pb zircon ages of the Qianlishan Complex: Constrains on the evolution of the Khondalite Belt in the Western Block of the North China Craton. *Precambrian Res.* **2009**, *174*, 78–94. [CrossRef]

15. Yin, C.Q.; Zhao, G.C.; Guo, J.H.; Sun, M.; Xia, X.P.; Zhou, X.W. U-Pb and Hf isotopic study of zircons of the Helanshan Complex: Constrains on the evolution of the Khondalite Belt in the Western Block of the North China Craton. *Lithos* **2011**, *122*, 25–38. [CrossRef]

16. Tam, P.Y.; Zhao, G.C.; Zhou, X.W.; Sun, M.; Guo, J.H.; Li, S.Z.; Yin, C.Q.; Wu, M.L.; He, Y.H. Metamorphic P-T path and imphcations of high-pressure politic granulites from the Jiaobei massif in the Jiao-Liao-Ji Belt, North China Craton. *Gondwana Res.* **2011**, *22*, 104–117. [CrossRef]
17. Tam, P.Y.; Zhao, G.C.; Sun, M.; Li, S.Z.; Wu, M.L.; Yin, C.Q. Petrology and metamorphic P-T path of high-pressure mafic granulites from the Jiaobei massif in the Jiao-Liao-Ji Belt, North China Craton. *Lithos* **2012**, *155*, 94–109. [CrossRef]
18. Liu, J.; Zhang, J.; Liu, Z.H.; Yin, C.Q.; Zhao, C.; Li, Z.; Yang, Z.J.; Dou, S.Y. Geochemical and geochronological study on the Paleoproterozoic rock assemblage of the Xiuyan region: New constraints on an integrated rift-and-collision tectonic process involving the evolution of the Jiao-Liao-Ji Belt, North China Craton. *Precambrian Res.* **2018**, *310*, 179–197. [CrossRef]
19. Liu, J.; Zhang, J.; Yin, C.Q.; Cheng, C.Q.; Liu, X.G.; Zhao, C.; Chen, Y.; Wang, X. Synchronous A-type and adakitic granitic magmatism at ca. 2.2 Ga in the Jiao-Liao-Ji belt, North China Craton: Implications for rifting triggered by lithospheric delamination. *Precambrian Res.* **2020**, *342*, 105629. [CrossRef]
20. Liu, J.; Zhang, J.; Liu, Z.H.; Yin, C.Q.; Xu, Z.Y.; Cheng, C.Q.; Zhao, C.; Wang, X. Late Paleoproterozoic crustal thickening of the Jiao-Liao-Ji belt, North China Craton: Insights from ca. 1.95-1.88 Ga syn-collisional adakitic granites. *Precambrian Res.* **2021**, *355*, 106120. [CrossRef]
21. Zhang, J.; Zhao, G.C.; Li, S.Z.; Sun, M.; Liu, S.W.; Wilde, S.A.; Kröner, A.; Yin, C.Q. Deformation history of the Hengshan complex: Implications for the tectonic evolution of the Trans-North China Orogen. *J. Struct. Geol.* **2007**, *29*, 933–949. [CrossRef]
22. Zhang, J.; Zhao, G.C.; Li, S.Z.; Sun, M.; Wilde, S.A.; Liu, S.W.; Yin, C.Q. Polyphase deformation of the Fuping complex, Trans-North china Orogen: Structures, SHRIMP U-Pb zircon ages and tectonic implications. *J. Struct. Geol.* **2009**, *31*, 177–193. [CrossRef]
23. Zhang, J.; Zhao, G.C.; Li, S.Z.; Sun, M.; Chan, L.S.; Shen, W.L. Structural pattern of the Wutai complex and its constraints on the tectonic framework of the Trans-North China Orogen. *Precambrian Res.* **2012**, *222*, 212–229. [CrossRef]
24. Qian, J.H.; Wei, C.J.; Zhou, X.W.; Zhang, Y.H. Metamorphic P-T paths and New Zircon U-Pb age data for garnet-mica schist from the Wutai Group, North China Craton. *Precambrian Res.* **2013**, *233*, 282–296. [CrossRef]
25. Qian, J.H.; Yin, C.Q.; Zhang, J.; Ma, L.; Wang, L.J. High-pressure granulites in the Fuping Complex of the central North China Craton: Metamorphic P-T-t evolution and tectonic implications. *J. Asian Earth Sci.* **2018**, *154*, 255–270. [CrossRef]
26. Qian, J.H.; Yin, C.Q.; Wei, C.J.; Zhang, J. Two phases of Paleoproterozoic metamorphism in the Zhujiafang ductile shear zone of the Hengshan Complex: Insights into the tectonic evolution of the North China Craton. *Lithos* **2019**, *330*, 35–54. [CrossRef]
27. Zhao, C.; Zhang, J.; Liu, J.; Yin, C.Q.; Yang, C.H.; Cui, Y.S.; Peng, Y.B.; Jiang, C.Y. Detrital zircon U–Pb and Hf isotopic study of the Yushulazi Formation in the Gaizhou-Zhuanghe area of the eastern Liaoning: Constraints on the crustal evolution of the North China Craton. *Acta Petrol. Sin.* **2019**, *35*, 2407–2432. (In Chinese with English Abstract)
28. Liu, X.G.; Zhang, J.; Yin, C.Q.; Li, S.Z.; Liu, J.; Qian, J.H.; Zhao, C. A synthetic geochemical and geochronological dataset of the Mesoproterozoic sediments along the southern margin of North China Craton: Unraveling a prolonged peripheral subduction involved in breakup of Supercontinent Columbia. *Precambrian Res.* **2021**, *357*, 106154. [CrossRef]
29. Zhao, G.C.; Wilde, S.A.; Cawood, P.A. Thermal evolution of Archean basement rocks from the eastern part of the North China craton and its bearing on tectonic setting. *Int. Geol. Rev.* **1998**, *40*, 706–721. [CrossRef]
30. Ma, Q.; Xu, Y.G.; Huang, X.L.; Zheng, J.P.; Ping, X.Q.; Xia, X.P. Eoarchean to Paleoproterozoic crustal evolution in the North China Craton: Evidence from U-Pb and Hf-O isotopes of zircons from deep-crustal xenoliths. *Geochim. Cosmochim. Acta* **2020**, *278*, 94–109. [CrossRef]
31. Wilde, S.A.; Valley, J.W.; Kita, N.T.; Cavosie, A.J.; Liu, D.Y. SHRIMP U-Pb and CAMECA 1280 Oxygen isotope results from ancient detrital zircons in the Caozhuang quartzite, eastern Hebei, North China Craton: Evidence for crustal reworking 3.8 Ga ago. *Am. J. Sci.* **2008**, *308*, 185–199. [CrossRef]
32. Wan, Y.S.; Xie, H.Q.; Dong, C.Y.; Kröner, A.; Wilde, S.A.; Bai, W.Q.; Liu, S.J.; Xie, S.W.; Ma, M.Z.; Li, Y. Hadean to Paleoarchean rocks and zircons in China. In *Earth's Oldest Rocks*, 2nd ed.; Van Kranendonk, M.J., Smithies, R.H., Bennett, V., Eds.; Elsevier: Amsterdam, The Netherlands, 2019; pp. 294–327.
33. Wan, Y.S.; Xie, H.Q.; Wang, H.C.; Liu, S.J.; Chu, H.; Xiao, Z.B.; Li, Y.; Hao, G.M.; Li, P.C.; Dong, C.Y. Discovery of early Eoarchean-Hadean zircons in eastern Hebei, North China Craton. *Acta Geol. Sin.* **2021**, *95*, 277–291. (In Chinese with English Abstract)
34. Chu, H.; Wang, H.C.; Rong, G.L.; Chang, Q.S.; Kang, J.L.; Jin, S.; Xiao, Z.B. The geological significance of the rediscovered fuchsite quartzite with abundant Eoarchean detrital zircons in eastern Hebei Province. *Chin. Sci. Bull.* **2016**, *61*, 2299–2308.
35. Nutman, A.P.; Wan, Y.S.; Du, L.L.; Friend, C.R.L.; Dong, C.Y.; Xie, H.Q.; Wang, W.; Sun, H.Y.; Liu, D.Y. Multistage late Neoarchean crustal evolution of the North China Craton, eastern Hebei. *Precambrian Res.* **2011**, *189*, 43–65. [CrossRef]
36. Zhao, G.C.; Zhai, M.G. Lithotectonic elements of Precambrian basement in the North China Craton: Review and tectonic implications. *Gondwana Res.* **2013**, *23*, 1207–1240. [CrossRef]
37. Liu, S.J.; Wan, Y.S.; Sun, H.Y.; Nutman, A.P.; Xie, H.Q.; Dong, C.Y.; Ma, M.Z.; Liu, D.Y.; Jahn, B. Paleo-to Eoarchean crustal evolution in eastern Hebei, North China Craton: New evidence from SHRIMP U-Pb dating and in-situ Hf isotopic study of detrital zircons from paragneisses. *J. Asian Earth Sci.* **2013**, *78*, 4–17. [CrossRef]
38. Zhao, C.; Zhang, J.; Zhao, G.C.; Yin, C.Q.; Chen, G.K.; Liu, J.; Liu, X.G.; Chen, W.L. Kinematics and structural evolution of the Anziling dome-and-keel architecture in east China: Evidence of Neoarchean vertical tectonism in the North China Craton. *Geol. Soc. Am. Bull.* **2022**, *134*, 2115–2129. [CrossRef]
39. Huang, X.; Zi, W.B.; DePaolo, D.J. Sm-Nd isotope study of early Archean rocks, Qianan, Hebei Province, China. *Geochim. Cosmochim. Acta* **1986**, *50*, 625–631. [CrossRef]

40. Jahn, B.M.; Auvray, B.; Cornichet, J.; Bai, Y.L.; Shen, Q.H.; Liu, D.Y. 3.5 Ga old amphibolites from eastern Hebei Province, China: Field occurrence, petrography, Sm-Nd isochron age and REE geochemistry. *Precambrian Res.* **1987**, *34*, 311–346. [CrossRef]
41. Cui, X.H.; Zhai, M.G.; Guo, J.H.; Zhao, L.; Zhu, X.Y.; Wang, H.Z.; Huang, G.Y.; Ge, S.S. Field occurrences and Nd isotopic characteristics of the meta-mafic ultramafic rocks from the Caozhuang Complex, eastern Hebei: Implications for early Archean crustal evolution of the North China Craton. *Precambrian Res.* **2018**, *310*, 425–442. [CrossRef]
42. Pearce, N.J.G.; Perkins, W.T.; Westgate, J.A.; Gorton, M.P.; Jackson, S.E.; Neal, C.R.; Chenery, S.P. A compilation of new and published major and trace element data for NIST SRM 610 and NIST SRM 612 Glass reference materials. *Geostand. Geoanal. Res.* **1997**, *21*, 115–144. [CrossRef]
43. Sláma, J.; Košler, J.; Condon, D.J.; Crowley, J.L.; Gerdes, A.; Hanchar, J.M.; Horstwood, M.S.A.; Morris, G.A.; Nasdala, L.; Norberg, N.; et al. Plesovice Zircon-Anew natural reference material for U-Pb and Hf isotopic microanalysis. *Chem. Geol.* **2008**, *249*, 1–35. [CrossRef]
44. Li, X.H.; Tang, G.Q.; Gong, B.; Yang, Y.H.; Hou, K.J.; Hu, Z.C.; Li, Q.L.; Liu, Y.; Li, W.X. Qinghu Zircon: A working reference for microbeam analysis of U-Pb age and Hf and O isotopes. *Chin. Sci. Bull.* **2013**, *58*, 4647–4654. (In Chinese with English Abstract) [CrossRef]
45. Ludwig, K.R. Isoplot/Ex Version 4.15: A Geochronological Toolkit for Microsoft Excel. *Berkeley Geochronol. Cent. Spec. Publ.* **2012**, *5*, 75.
46. Wu, F.Y.; Yang, Y.H.; Xie, L.W.; Yang, J.H.; Xu, P. Hf isotopic compositions of the standard zircons and baddeleyites used in U-Pb geochronology. *Chem. Geol.* **2006**, *234*, 105–126. [CrossRef]
47. Blichert-Toft, J.; Albarede, F. The Lu-Hf isotope geochemistry of chondrites and the evolution of the mantle-crust system. *Earth Planet. Sci. Lett.* **1997**, *148*, 243–258. [CrossRef]
48. Rainbird, R.H.; Stern, R.A.; Khudoley, A.K.; Kropachev, A.P.; Heaman, L.M.; Sukhorukov, V.I. U-Pb geochronology of Riphean sandstone and gabbro from southeast Siberia and its bearing on the Laurentia-Siberia connection. *Earth Planet. Sci. Lett.* **1998**, *164*, 409–420. [CrossRef]
49. Fedo, C.M.; Sircombe, K.N.; Rainbird, R.H. Detrital zircon analysis of the sedimentary record. *Rev. Mineral. Geochem.* **2003**, *53*, 277–303. [CrossRef]
50. Gehrels, G. Detrital Zircon U-Pb geochronology applied to tectonics. *Annu. Rev. Earth Planet. Sci.* **2014**, *42*, 127–149. [CrossRef]
51. Geslin, J.K.; Link, P.K.; Fanning, C.M. Highprecision provenance determination using detrital-zircon ages and petrography of Quaternary sands on the eastern Snake River Plain, Idaho. *Geology* **1999**, *27*, 295–298. [CrossRef]
52. Cawood, P.A.; Nemchin, A.A. Provenance record of a rift basin: U/Pb ages of detrital zircons from the Perth Basin, Western Australia. *Sediment. Geol.* **2000**, *134*, 209–234. [CrossRef]
53. Kinny, P.D.; Compston, W.; Williams, I.S. A reconnaissance ion–probe study of hafnium isotopes in zircons. *Geochim. Cosmochim. Acta* **1991**, *55*, 849–859. [CrossRef]
54. Wu, F.Y.; Li, X.H.; Zheng, Y.F.; Gao, S. Lu-Hf isotopic systematics and their applications in petrology. *Acta Petrol. Sin.* **2007**, *23*, 185–220, (In Chinese with English Abstract)
55. Wu, F.Y.; Zhang, Y.B.; Yang, J.H.; Xie, L.W.; Yang, Y.H. Zircon U–Pb and Hf isotopic constraints on the Early Archean crustal evolution in Anshan of the North China Craton. *Precambrian Res.* **2008**, *167*, 339–362. [CrossRef]
56. Ping, X.Q.; Zheng, J.P.; Tang, H.Y.; Ma, Q.; Griffin, W.L.; Xiong, Q.; Su, Y.P. Hadean continental crust in the southern North China Craton: Evidence from the Xinyang felsic granulite xenoliths. *Precambrian Res.* **2018**, *307*, 155–174. [CrossRef]
57. Wan, Y.S.; Wu, J.S.; Liu, D.Y.; Zhang, Z.Q.; Song, B. Geochemistry and Nd, Pb isotopic characteristics of 3.3 Ga Chentaigou granite in Anshan area. *Acta Geol. Sin.* **1997**, *18*, 382–388. (In Chinese with English Abstract)
58. Dong, C.Y.; Wan, Y.S.; Xie, H.Q.; Nutman, A.; Xie, S.W.; Liu, S.J.; Ma, M.Z.; Liu, D.Y. The Mesoarchean Tiejiashan-Gongchangling potassic granite in the Anshan Benxi area, North China Craton: Origin by recycling of Paleo- to Eoarchean crust from U-Pb-Nd-Hf-O isotopic studies. *Lithos* **2017**, *290*, 116–135. [CrossRef]
59. Zheng, J.P.; Griffin, W.L.; O'Reilly, S.Y.; Lu, F.X. 3.6 Ga lower crust in central China: New evidence on the assembly of the North China Craton. *Geology* **2004**, *32*, 229–232. [CrossRef]
60. Ping, X.Q.; Zheng, J.P.; Tang, H.Y.; Xiong, Q.; Su, Y.P. Paleoproterozoic multistage evolution of the lower crust beneath the southern North China Craton. *Precambrian Res.* **2015**, *269*, 162–182. [CrossRef]
61. Zhang, H.F.; Wang, J.L.; Zhou, D.W.; Yang, Y.H.; Zhang, G.W.; Santosh, M.; Yu, H.; Zhang, J. Hadean to Neoarchean episodic crustal growth: Detrital zircon records in Paleoproterozoic quartzites from the southern North China Craton. *Precambrian Res.* **2014**, *254*, 245–257. [CrossRef]

 *minerals*

Article

The Earliest Clastic Sediments of the Xiong'er Group: Implications for the Early Mesoproterozoic Sediment Source System of the Southern North China Craton

Yuan Zhang *, Guocheng Zhang and Fengyu Sun

School of Resources and Environment, Henan Polytechnic University, Jiaozuo 454003, China
* Correspondence: zy3519718@163.com

Abstract: The volcanic activity of the Xiong'er Group and its concomitant sedimentation are related to the stretching–breakup of the Columbia supercontinent. The Dagushi Formation overlies the Paleoproterozoic Shuangfang Formation with an angular unconformity. The Dagushi Formation, as the earliest clastic strata of the Xiong'er Group and the first stable sedimentary cover overlying the Archean crystalline basement in the southern margin of the North China Craton, provides tectonic evolution information that predates Xiong'er volcanic activity. By distinguishing lithologic characteristics and sedimentary structures, we identified that the sedimentary facies of the Dagushi Formation were braided river delta lake facies from bottom to top. The U–Pb ages of the detrital zircons of the Dagushi Formation can be divided into four groups: ~1905–1925, ~2154–2295, ~2529–2536, and ~2713–2720 Ma, indicating the provenance from the North China Craton basement. Based on the geochemical characteristics of the Dagushi Formation, we suggest that the sediments accumulated rapidly near the source, which were principally felsic in nature, and were supplemented by recycled materials. The provenance area pointed to the underlying metamorphic crystalline basement of the North China Craton as the main source area with an active tectonic background. The Chemical Index of Alteration (CIA) values of the Dagushi Formation sandstone samples ranged from 60.8 to 76.7, indicating that the source rocks suffered from slight to moderate chemic chemical weathering. The Index of Composition Variability (ICV) values ranged from 0.8 to 1.3, which indicates the first cyclic sediments. The vertical facies and provenance changes of the Dagushi Formation reflect a continuous crust fracturing process that occurred in the North China Craton.

Keywords: North China Craton; Dagushi Formation; sedimentary facies; detrital zircon; geochemistry

Citation: Zhang, Y.; Zhang, G.; Sun, F. The Earliest Clastic Sediments of the Xiong'er Group: Implications for the Early Mesoproterozoic Sediment Source System of the Southern North China Craton. *Minerals* **2023**, *13*, 971. https://doi.org/10.3390/min13070971

Academic Editors: Georgia Pe-Piper, Jin Liu, Jiahui Qian and Xiaoguang Liu

Received: 8 June 2023
Revised: 15 July 2023
Accepted: 17 July 2023
Published: 22 July 2023

1. Introduction

The North China Craton (NCC) is an ancient landmass with a long history covering 3.8 billion years, with evidence concerning many supercontinent events in geological history [1–6], including assembly and breakup records of the supercontinent Columbia [7–10]. Scholars think that the subduction between the eastern and western blocks occurred at ca. 1850 Ma, thus forming a unified NCC [11–14]. After that, a volcanic sedimentary succession, which is called the Xiong'er Group, was widely developed in the southern margin of the NCC. The Xiong'er Group can be divided from bottom to top into the Dagushi Formation, Xushan Formation, Jidanping Formation, and Majiahe Formation, which are a set of clastic rock and volcanic strata with low deformation and metamorphism. In recent years, scholars have conducted extensive research on the Xiong'er Group in the southern margin of the NCC using petrology, geochemistry, and chronology [6,9,12,15–35] (Figure 1a). The chronological data of zircon uranium-lead (U–Pb) isotopes show that most of the volcanic rocks of the Xiong'er Group were formed between 1800 and 1750 Ma [24,30]. However, the formation mechanism remains controversial [9,15–20]. To date, the main viewpoints include an Andean-type continental margin [9,22], a passive continental margin and rift [26–28,36,37], and an active continental margin and rift [22].

However, research on the Xiong'er Group has focused on its volcanic lavas, while the sedimentary rocks have not been a major subject of systematic research. Only a few scholars have investigated the geochemical characteristics of the clastic rocks from the Dagushi Formation [38]. Ref. [39] posited that the existence of primary glauconite in the sandstone of the Majiahe Formation proved that the Xiong'er Group was in a marine environment during its later stage. There is also controversy concerning the sedimentary source, which limits the correlation between the Mesoproterozoic strata in the southern margin of the NCC and other regions. Therefore, studies on the sedimentary environment, provenance characteristics, and tectonic setting of early Mesoproterozoic strata in the southern margin of the NCC are critical to reveal the evolution of paleogeography during the Proterozoic eon (Figure 1b,c). schematic diagram of the geographical location of the Dagushi Formation in the study area (modified from [40,41]).

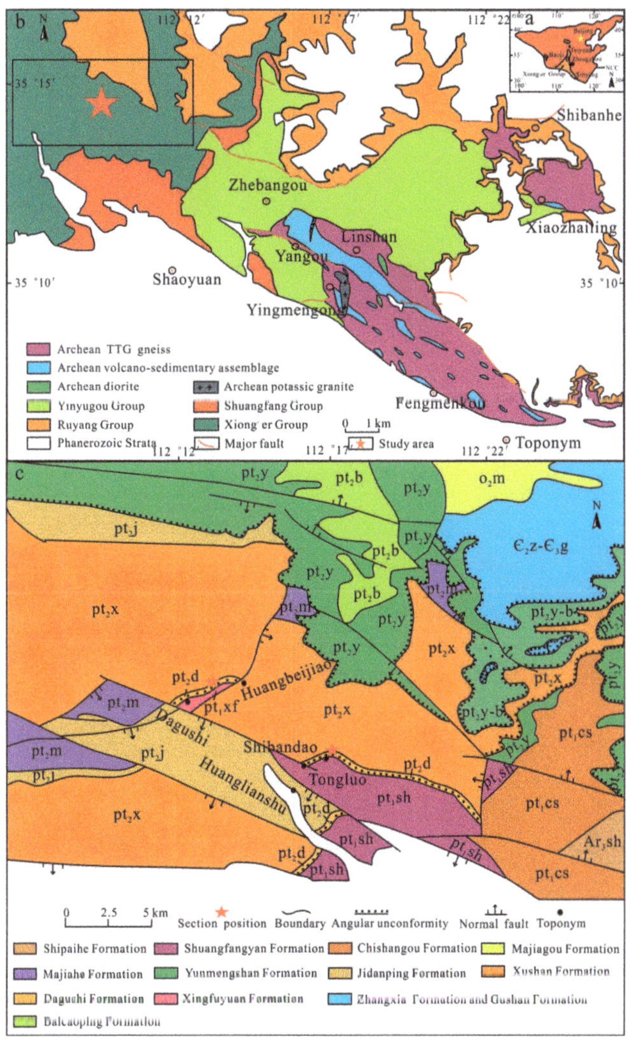

Figure 1. (**a**) Geological distribution of the Xiong'er Group in the NCC during the Precambrian era (modified from [19]); (**b**) simplified geological map of the study area (modified from [40]); (**c**).

2. Geological Background

The total area of the NCC is approximately 1.6 million km^2. Important components include northern China, Bohai Bay, and Inner Mongolia [42,43]. The NCC has a long history of geological evolution and has experienced several multi-stage tectonic evolution events [5,26,44–49]. Furthermore, the formation of the NCC has experienced three important geological events: the main continental crust growth at ~2.7 Ga, the cratonization event at ~2.5 Ga, and the final formation of the NCC at ~2.0–1.8 Ga [42,50,51].

The Xiong'er Group is primarily exposed in the Zhongtiao, Xiao, Xiong'er, and Waifang mountains [25,52]. The Xiong'er Group, as the cover of an ancient continental crust basement, overlies the Archean crystalline basement or early Proterozoic strata with an unconformity [53]. They represent the most extensive magmatic activity that occurred after the formation of the crystalline basement in the NCC [38,52]. The volcanic rocks are primarily andesite and basaltic andesite, with a small amount of dacite, rhyolite, and minor interlayered sedimentary rocks. Sedimentary rocks, sedimentary interbeds, and pyroclastic rocks are limited, covering only 4.3% of the total thickness of the layer, and mainly distributed in the Dagushi Formation at the bottom and the Majiahe Formation at the top, while only a small number of local intervals are exposed in the Xushan and Jidanping Formations [24,38,54].

The outcrop range of sedimentary rocks in the Dagushi Formation is small, and the thickness varies greatly. The representative section is located in Huangbeijiao and Xiaogoubei in Shaoyuan Town, Jiyuan City, Henan Province. The Dagushi Formation in this area overlays the biotite quartz schist of the late Neoarchean Shuangfang Formation with an angular unconformity, and its distribution is continuous, and its largest thickness is 189.5 m. The Dagushi Formation is also distributed in Moshigou, northern Luanchuan County, with a maximum thickness of 92 m. The main lithology is as follows: conglomerate, arkose sandstone, feldspathic quartz sandstone, and purple mudstone. Cross-bedding is developed in sandstone strata. In addition, the Dagushi Formation in eastern Yuanqu County of Shanxi Province and Luoning County of Henan Province outcrops sporadically. In this study, the Dagushi Formation, which is completely exposed in Huangbeijiao and Xiaogoubei north of Shaoyuan Town, Jiyuan City, was selected as the research object (Figure 1c).

3. Analytical Methods

Two coarse- to medium-grained sandstone samples (approximately 4 kg) were collected from the lowest and uppermost parts of the Dagushi Formation within the Huangbeijiao section in Jiyuan City, southern North China, for zircon separation (Figures 1 and 2). Zircon was separated using a heavy fluid and magnetic separator at the Hebei Institute of Regional Geology and Mineral Resources, Langfang, China. Approximately 400 zircons were hand-selected from each sample using a binocular microscope, and 177 zircons were selected to analyze. Zircon particles were attached to adhesive tape with M257 standard [55], sealed with epoxy resin, and polished to half of their thickness. The images of these zircons were taken using an optical microscope with transmitted and reflected light. High-resolution cathodoluminescence imaging was performed by a scanning electron microscope using the Gatan monoCL 3 + cathodoluminescence system from Wuhan SampleSolution Analytical Technology Co., Ltd. (Wuhan, China). Both imaging methods were used to identify internal structures and select targets for further U–Pb analysis.

U–Pb dating and zircon trace element analysis were conducted simultaneously by laser ablation inductively coupled plasma mass spectrometry (ICP-MS). Detailed operating parameters and procedures for laser ablation systems, ICP-MS instruments (Wuhan SampleSolution Analytical Technology Co., Ltd.), and data simplification can be found in [56–58]. Laser sampling was performed using a GeoLas 2005 instrument, and an Agilent 7900 ICP-MS system (Agilent Technologies, Palo Alto, CA, USA) was used to obtain the ion signal strength. A laser beam repetition rate of 5 Hz and an analysis point size of 24 μm diameter were used for each analysis. Helium was used as a carrier gas, and argon was used as a supplementary gas. Before entering the ICP, the sample was mixed with the carrier gas through a T-joint. The laser is equipped with an ablation system to smooth the signal, which can produce a smooth signal even at very low temperatures and frequencies [59] (i.e., down to 1 Hz). Each analysis consisted of approximately 20 s of background acquisition and 30 s of gas blank, followed by 50 s of sample data acquisition. The Excel-based software, ICPMSDataCal (Ver. 10.0), performed the offline selection, unified background and signal analysis, time trend correction of the records, quantitative element correction analysis, and U–Pb dating [56,57,60].

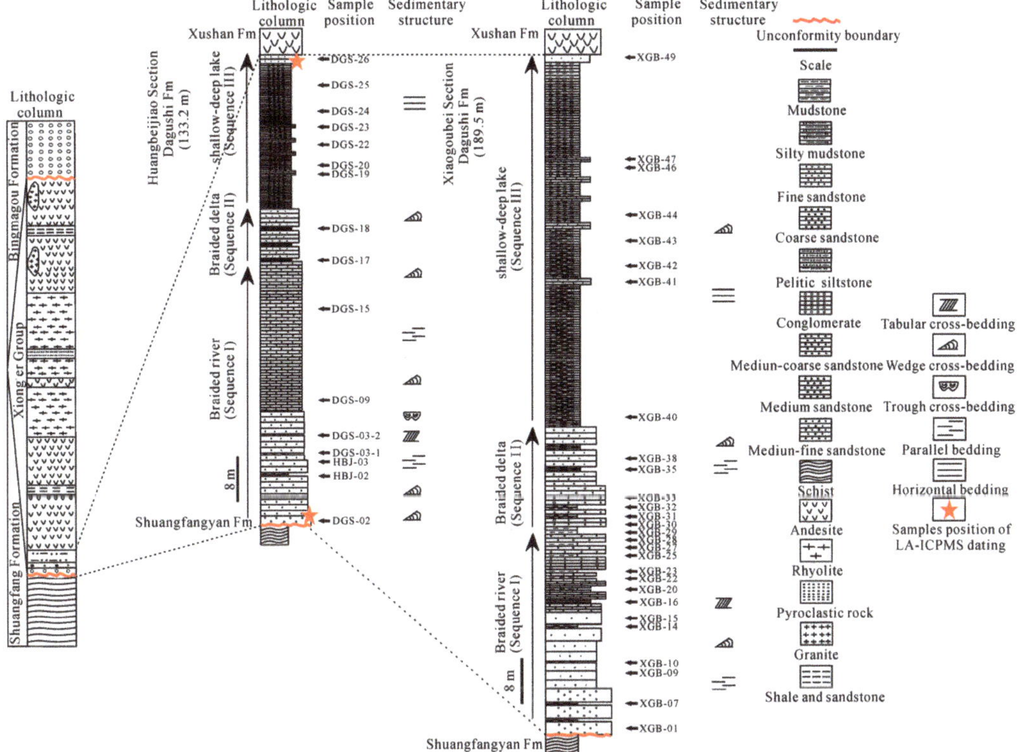

Figure 2. Generalized column of the Mesoproterozoic Dagushi Formation in Jiyuan (this study).

The standard zircon 91,500 was used for U–Pb dating, and two determinations were analyzed for every five measurements. The variation of the U–Th–Pb isotope ratio with time was corrected for every five analyses by linear interpolation [57,58]. The standard preferred U–Th–Pb isotope ratio for 91,500 was taken from [61]. The $^{207}Pb/^{206}Pb$ age was used for zircons greater than 1000 Ma [62]. For zircons less than 1000 Ma, the discordance was defined as $100\% \times abs\,[1 - (^{206}Pb/^{238}U)/(^{207}Pb/^{235}U)]$, and for zircons greater than 1000 Ma, the discordance was defined as $100\% \times abs\,[1 - (^{206}Pb/^{238}Pb)/(^{207}Pb/^{206}Pb)]$. Harmonic graphs were then generated, and the weighted average was determined using Isoplot/Ex_ver3 software [63].

Samples were collected from sandstone and argillaceous rocks from the Dagushi Formation in the Jiyuan area for geochemical analysis, and major elements in the sandstone and trace elements in the argillaceous rocks were determined. After the sample was naturally dried, it was crushed into a 200-mesh powder using a mortar. An Axios^{max} X-ray fluorescence (XRF) spectrometer (PANalytical BV, Almelo, the Netherlands) was used in a laboratory at ALS Minerals-ALS Chemex (Guangzhou, China) Co., Ltd. in Guangdong

Province. The whole rock major element described the detailed analysis procedure with errors between ±1% and ±2% [64].

Trace elements, including rare earth elements (REE), were detected by an Agilent 7900 ICP-MS Inductively Coupled Plasma Mass Spectrometer in ALS Minerals-ALS Chemex (Guangzhou, China) Co., Ltd. The analytical accuracy of trace elements exceeded 95%, and the detection range was less than or equal to 2 ppm in most cases. The detection range of Ba, Cr, Rb, Sr, and V was 5 ppm.

4. Results

4.1. Sedimentary Facies

We researched two sections of strata from the Mesoproterozoic Dagushi Formation in the Jiyuan area, divided their sedimentary facies, and further analyzed their sedimentary environment. Our measurements showed that the thickness of the Huangbeijiao Section from the Dagushi Formation in the Jiyuan area was approximately 133.2 m, while the thickness of the Xiaogoubei Section was 189.5 m (Figure 2). Our field observations showed that the Dagushi Formation rested upon the Paleoproterozoic crystalline basement unconformably and was the only sedimentary rock stratum in the Xiong'er Group. The color, composition, and structural characteristics of sediments in this stratum had distinct variations. In general, this stratum was a retrograding sequence which consisted of three different sedimentary sequences from bottom to top (Figure 2). Each sequence is summarized and described below.

Huangbeijiao Section (133.2 m).

Sequence I (71.4 m): the lithology of Sequence I in the Huangbeijiao Section was gray and purple pebbly sandstone and medium- to coarse-grained sandstone. From bottom to top, the grain size of sediments changed from coarse to fine with a positive rhythm, and the roundness changed from subangular to subrounded. Sequence I had parallel bedding formed by water ripple and large cross-bedding (Figure 3a–d) which occurred repeatedly from bottom to top. Sandstone layers with different grain sizes constituted multiple sedimentary cycles, and lenticular sand bodies were observed (Figure 3e). Based on the lithologic characteristics and recognizable sedimentary structures, Sequence I exhibited characteristics of a braided river deposit.

Sequence II (17.5 m): the lithology of Sequence II in the Huangbeijiao Section was primarily interbedding formed by purple medium- to fine-grained sandstone, siltstone, and mudstone (Figure 3f). The grain size was significantly smaller than that of Sequence I, and the main bedding was small wedge-shaped cross-bedding (Figure 3g). Irregular horizontal bedding was occasionally seen in the mudstone layers, and mud cracks were developed in the bedding of the local mudstone layers (Figure 3h). Mud cracks were most common in dry areas, and they were formed after silty or argillaceous sediments exposed their water surface, lost water, became dry, and shrank, which indicated exposure to a dry environment. Sequence II was thought to be a braided delta deposit.

Sequence III (44.3 m): the lithology of Sequence III in the Huangbeijiao Section was purple argillaceous siltstone and mudstone, which were typical products of overbank deposition. A significant amount of the rocks was severely weathered (Figure 3i), and there was no sedimentary structure of fluvial origin. The lithology in this sequence began to significantly change, and a great number of argillaceous sediments occurred. Compared to Sequence II, the grain size continued to decrease. The roundness of the sediment grains was mainly subrounded and occasionally subangular. Sequence III was thought to be a shore shallow lake deposit with weak hydrokinetics and without stagnancy.

Xiaogoubei Section (189.5 m).

Sequence I (71.6 m): the lithology of Sequence I in the Xiaogoubei Section was an interbedding of light purple medium- to thick-bedded, medium- to fine-grained sandstone and brown thin-bedded argillaceous rock. The basal sandstone included gravel with a diameter range from millimeters to centimeters which reflected a process in which sediments accumulated rapidly after short-distance transportation. From bottom to top, this sequence demonstrated multiple normal cycle rhythmites that were thick at the bottom and thin at the top. At the bottom of each cycle, there were pronounced scouring surfaces and coarse-grained imbricated pebbly sandstone layers (Figure 4a). The sandstone layer in the lower part of the cycle was characterized by large wedge- and trough-shaped cross-bedding, graded bedding, and parallel bedding, and lenticular sand bodies were commonly seen, which were flat at the top and convex at the bottom (Figure 4b–d). The siltstone or argillaceous siltstone in the upper part of the cycle was primarily small cross-bedding and wavy bedding, suggesting that it belonged to a braided river deposit with a strong hydrodynamic force that was controlled by directional flow in shallow water.

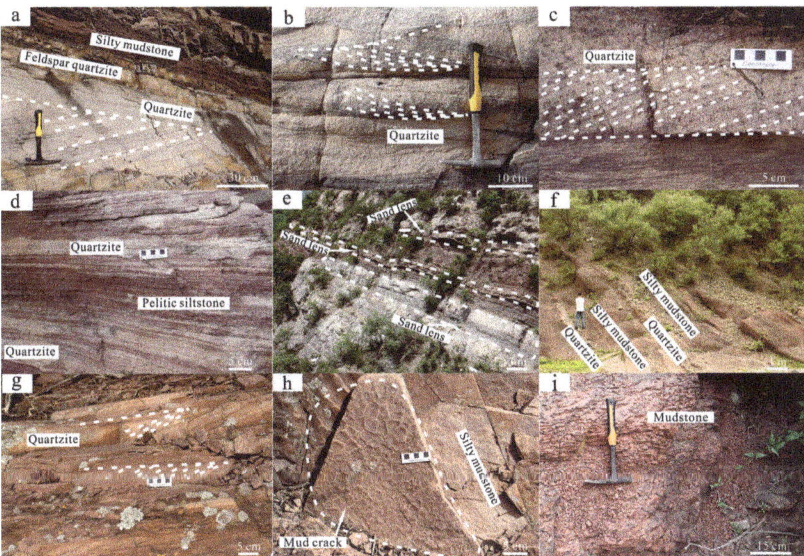

Figure 3. Lithologic characteristics and sedimentary structure of the Dagushi Formation in the Huangbeijiao section of Jiyuan. (**a**) Wedge-shaped cross-bedding; (**b**) trough cross-bedding; (**c**) tabular cross-bedding; (**d**) parallel bedding; (**e**) sand lens; (**f**) lithology characteristics of section II; (**g**) small wedge cross-bedding; (**h**) mud crack structure; (**i**) lithology characteristics of section III.

Sequence II (22 m): the lithology of Sequence II in the Xiaogoubei Section was a shallow conglomerate with medium thickness and pebbly coarse sandstone at the bottom, an interbedding of purple medium-grained sandstone and purple sandy mudstone in the middle, and purple and gray–green silty mudstone at the top. The basal scouring surface had a gentle slope, and the conglomerate, pebbled coarse sandstone, and sandstone all exhibited low compositional and textural maturity with mixed sizing (Figure 4e). The grain size of the sediments was significantly attenuated from bottom to top. Medium and small cross-bedding, parallel bedding, and lateral accretion cross-bedding were developed (Figure 4f,g). The sizes of various beds were smaller than those of Sequence I, suggesting that the hydrodynamics shifted from strong to weak and the water turbulence increased. Sequence II was determined to be a braided delta deposit.

Sequence III (95.9 m): the lithology included amaranth medium- to thick-bedded argillaceous siltstone at the bottom, the interbedding of thick-bedded purple argillaceous siltstone and mudstone in the middle, and gray–green coarse sandstone at the top. The compositional maturity, sizing, and roundness of the clastic material were better than those of Sequence II, and there was no sedimentary structure of fluvial origin; however, there was irregular horizontal bedding in local areas (Figure 4h), which indicated weak hydrodynamics. In addition to outcropped coarse sandstone at the top, the sediments of

Sequence III were mostly fine-grained silty and argillaceous sediments, indicating that the water gradually deepened. Thus, Sequence III was thought to be a shore shallow lake deposit.

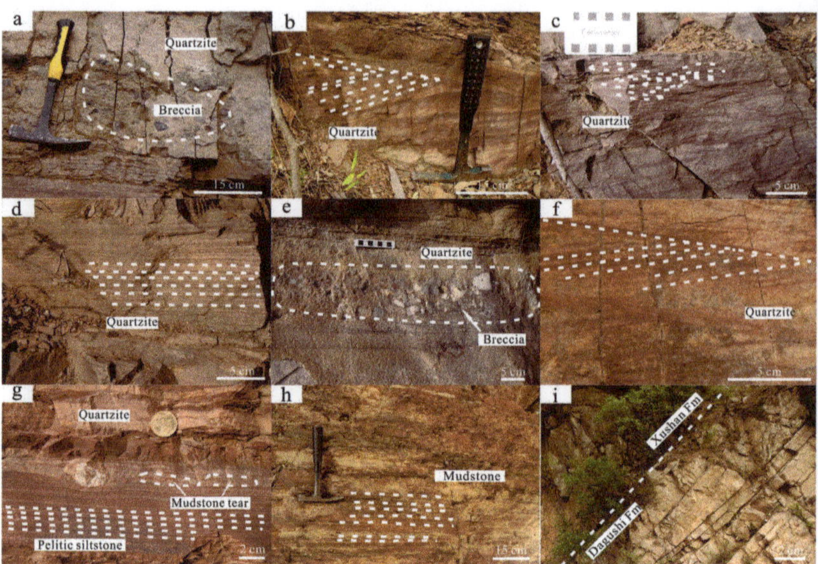

Figure 4. Lithologic characteristics and sedimentary structure of the Dagushi Formation in the Xiaogoubei section of Jiyuan. (**a**) Normal-grading bedding; (**b**) wedge-shaped cross-bedding; (**c**) trough cross-bedding; (**d**) parallel bedding; (**e**) conglomerate and trough cross-bedding; (**f**) wedge-shaped cross-bedding; (**g**) parallel bedding; (**h**) irregular horizontal bedding; (**i**) the boundary between the Dagushi and Xushan Formations.

4.2. Zircon U–Pb Geochronology

The CL images of representative zircons and their U–Pb ages are presented in Figures 5 and 6a,c. Most zircons are polyhedral in shape. Ref. [65] introduced that the calculation process of the age data. The analysis data are shown in Table S1, and the results show that the confidence intervals of zircon ranged from 90% to 100%. The ICP-MS U–Pb chronological test was performed on two sandstone samples collected from the top and bottom of the Dagushi Formation in the Huangbeijiao section of the Jiyuan area. The U–Pb age Concordia plot, age distribution histogram, and Th/U plot have the following characteristics (Figure 6):

Sample DGS-02 was collected from the bottom of the Dagushi Formation in Jiyuan (Figure 2). The age ranges of the 87 zircon grains were between 1784 and 2721 Ma. There were two main age peaks (1905 and 2154 Ma) and three secondary peaks (2295, 2536, and 2720 Ma) (Figure 6a,b). The youngest zircon U–Pb age measured from this sample was 1784 ± 43 Ma (concordant 99%). The Th/U value was between 0.31 and 1.40. Sample DGS-26 was collected from the top of the Dagushi Formation in Jiyuan (Figure 2). The age ranges of the 90 zircon grains were between 1832 and 2850 Ma, with three main peaks (2162, 2529, and 2713 Ma) and a secondary peak (1925 Ma) (Figure 6c,d). The youngest zircon U–Pb age measured from this sample was 1832 ± 34 Ma (concordant 99%). The Th/U value was between 0.07 and 1.22.

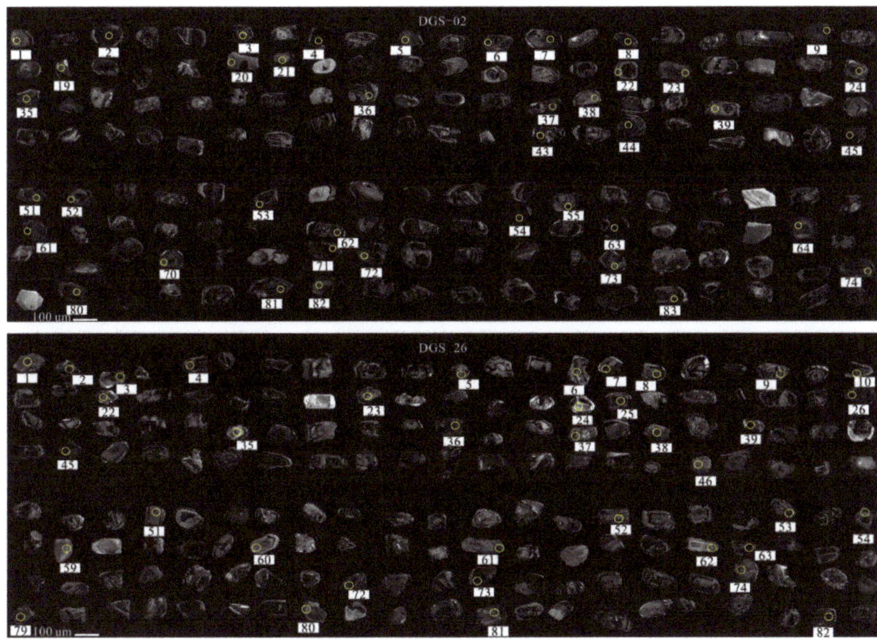

Figure 5. Cathodoluminescence (CL) images of representative zircons from the Dagushi Formation in Jiyuan (samples DGS-02 and DGS-26).

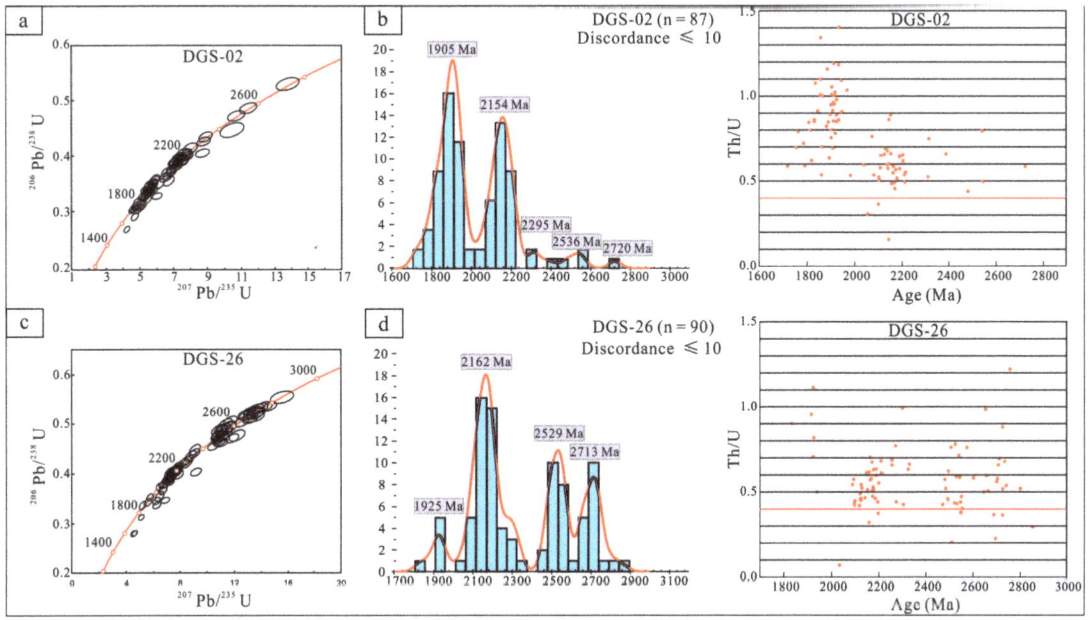

Figure 6. U–Pb age Concordia diagram, age distribution histogram, and Th/U age diagram of detrital zircons from the Dagushi Formation in Jiyuan; (**a,b**) DGS-02; (**c,d**) DGS-26.

4.3. Whole Rock Geochemistry

The major element compositions of the sandstones and trace element compositions of the argillaceous rocks are listed in Table S2. The sandstone samples from the Dagushi Formation were characterized by low SiO_2 contents (45.5%–77.9%), high $Fe_2O_3^T$ + MgO contents (2.7%–12.4%), and low TiO_2 contents (0.2%–0.8%). The Al_2O_3 and K_2O contents of the samples were 10.5%–20.4% and 1.7%–5.1%, respectively, while the MgO and Na_2O contents were 0.4%–4.1% and 0.2%–3.4%, respectively. We calculated the Chemical Index of Alteration (CIA) = 100 $[Al_2O_3/(Al_2O_3 + CaO^* + Na_2O + K_2O)]$ (Table S2) [66], which yielded values for the sandstone samples from the Dagushi Formation that varied between 60.8 and 76.7 (average value: 66.4) and are indicative of a slightly to moderately weathered source [66–68] (Figure 7). We also calculated the Index of Compositional Variability (ICV) = $(Fe_2O_3 + K_2O + Na_2O + CaO + MgO + TiO_2)/Al_2O_3$ (Table S2) in order to determine the proportion of primary source material relative to the weathered minerals that occurred in the sedimentary rocks [69,70]. Calculated ICV values for the Dagushi Formation sandstones generally varied between 0.8 and 1.3 (average value: 1). The ICV values of these samples were greater than 1 or close to 1, which indicates the first cyclic sediments [70].

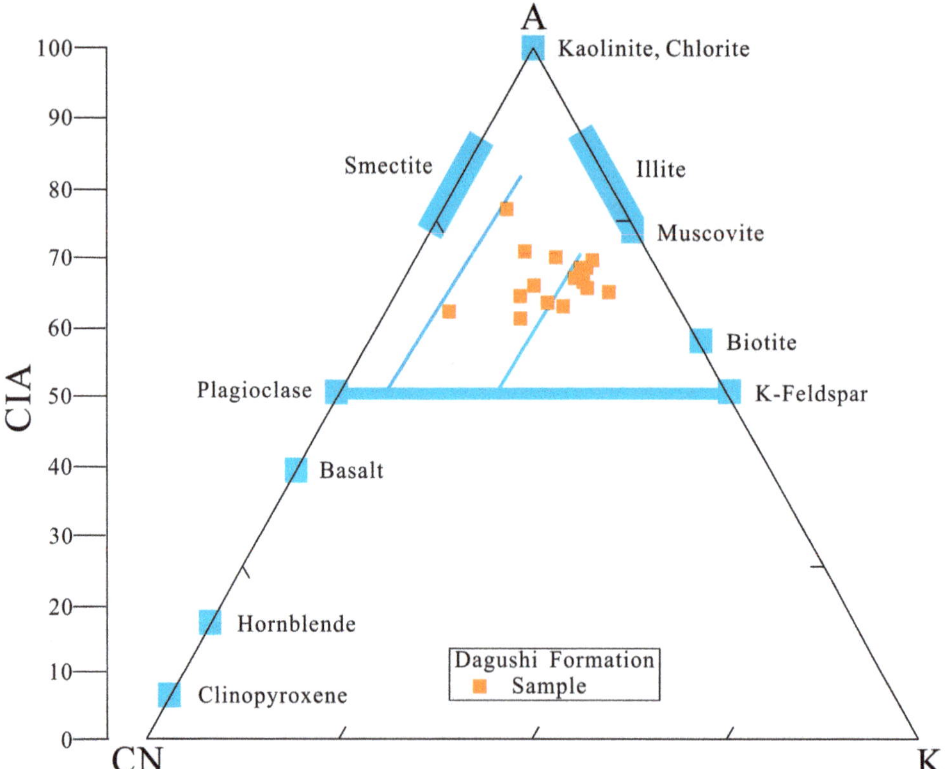

Figure 7. A–CN–K diagram of sandstones from the Dagushi Formation in Jiyuan (modified based on [66–68]). A = Al_2O_3; CN = $CaO^* + Na_2O$; K = K_2O; CaO^* refers only to CaO in the silicate minerals, which is the molar coefficient of oxide.

The REE test results ($\times 10^{-6}$) and partial characteristic indices (Table S2) of the mudstones sample from the Dagushi Formation indicated that the total REE contents of the mudstones ranged between 185 and 368 µg/g (average value: 258), which is greater than the averages for the North America shale (173 µg/g). Similarly, the ΣLREE/ΣHREE values varied from 9 to 13 (average value: 11), which is indicative of LREE enrichment. According to the chondrite-normalized REE patterns diagram [71] (Figure 8a), the La_N/Yb_N ratios ranged from 9.32 to 15.44, and the Gd_N/Yb_N ratios ranged from 1.52 to 1.89, indicating obvious fractionation of light and heavy REEs. δ Eu ranged from 0.56 to 0.74 (average value: 0.64), showing an obvious negative anomaly. δ Ce ranged from 0.99 to 1.12 (average value: 1.03), showing no obvious abnormality. Figure 8b shows that the La_A/Yb_A ratios ranged from 1.34 to 2.22 (average value: 1.84). LREEs were slightly enriched, and the overall content of REEs displayed a roughly synchronous change [72]. As shown in Figure 8c, the contents of the large ion lithophile elements Ba, Nb, and Sr were depleted, and Rb, Th, La, Ce, and Nd were enriched [71]. The contents of the transition elements, such as Ba, Rb, Y, Sc, V, Cr, Co, and Ni, were higher than the average values of the trace elements in the UCC, and the contents of terrigenous elements, such as Th, Zr, and Hf, were higher than their average values in the UCC [73] (Figure 8d).

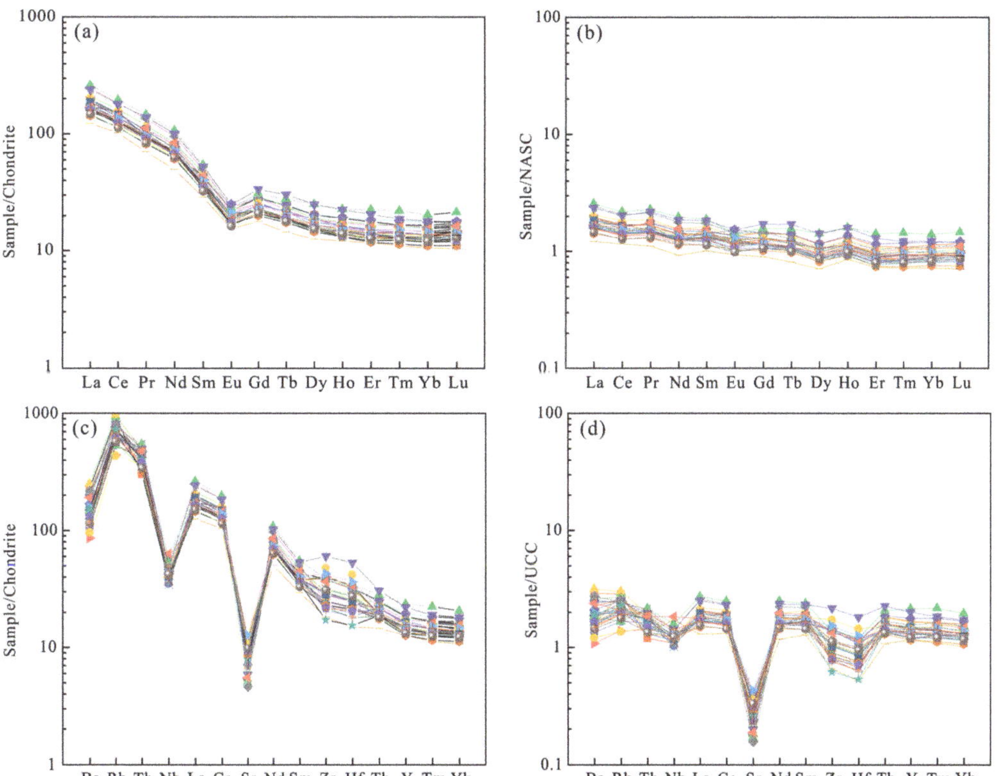

Figure 8. Distribution patterns of normalized rare earth elements (REEs) and normalized trace elements of argillaceous rock from the Mesoproterozoic Dagushi Formation in Jiyuan; (**a,c**) chondrite data from [71]; (**b**) North American shale (NASC) data from [72]; (**d**) UCC data from [73]).

5. Discussion

5.1. Sedimentary Provenance of the Dagushi Formation

The results of the detrital zircon from the sandstone samples (DGS-02 and DGS-26) at the bottom of the Dagushi Formation in the Jiyuan area indicated that there were two primary age peaks, 1908 and 2147 Ma, and three secondary peaks, 2291, 2517, and 2713 Ma (Figure 9). This implied that the provenance was dominated by Paleoproterozoic geological bodies, with a small number of Neoarchean ones. The age peak at 2713 Ma corresponded to the growth period of the Neoarchean crust [5,74–77]. In the Lushan area near Jiyuan, Trondhjemite, Tonalite, and Granodiorite (TTG) gneiss, plagioclase amphibolite, garnet two-pyroxene granulite, aluminum-rich and carbon-rich gneiss, marble, quartzite, etc. of the Taihua Group (2800–2700 Ma) were widely distributed [46,49,78] (Figure 9). The age peak at 2517 Ma corresponded to the tectonic–magmatic events of the late Neoarchean era, which was an important stage of continental crust accretion and cratonization in the NCC [79–83]. The detrital zircons between 2650 and 2500 Ma were likely derived from the TTG gneiss and supracrustal rock of the Dengfeng complex [5,42,84–89] (Figure 9). The age peaks of 2147 and 2291 Ma corresponded to multiple Paleoproterozoic active tectonic zones which developed during 2350–1950 Ma in the NCC; for example, the Shanxi-Henan active zone and the supracrustal rocks of the Songshan Group were formed between 2350–1960 Ma [28,51,90] (Figure 9). The age peak at 1908 Ma corresponded to tectonic-thermal events of collision and suturing between the eastern and western blocks of North China [2,4,91–95] (Figure 9). The ages of detrital zircons indicated that the provenance of the Dagushi Formation was primarily the NCC basement.

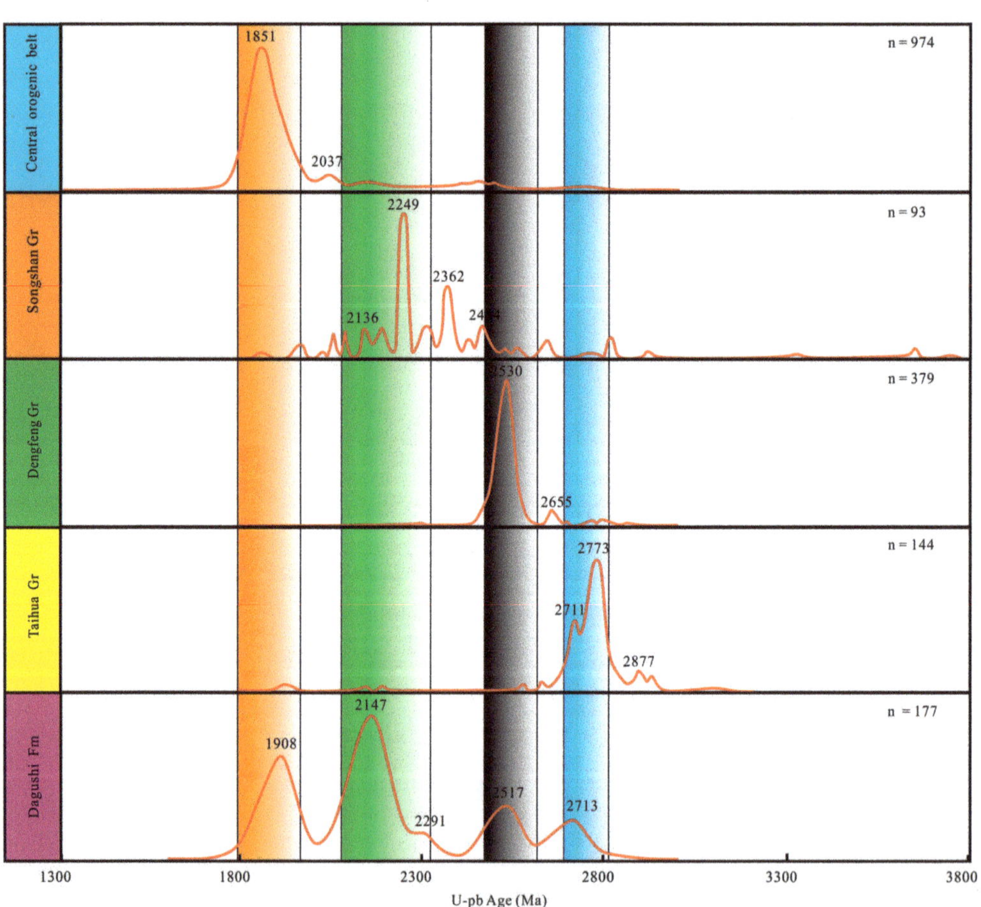

Figure 9. Histograms of concordant detrital zircon ^{207}Pb/^{206}Pb ages for samples from the Dagushi Formation, the Taihua, Dengfeng, and Songshan Groups, and the Central Orogenic Belt. Data sources: Dagushi Formation (this study); Taihua Group [46,49,78]; Dengfeng Group [86,88,89]; Songshan Group [90]; Central Orogenic Belt [2,4,91–95].

Based on the sandstone type discrimination diagram of log(SiO$_2$/Al$_2$O$_3$)-log(Fe$_2$O$_3$/K$_2$O) [96], the sandstones of the Dagushi Formation were found to be low-maturity wacke, litharenite sublitharenite, and arkose (Figure 10), which indicated that these sandstones in the Jiyuan area had undergone rapid proximal accumulation.

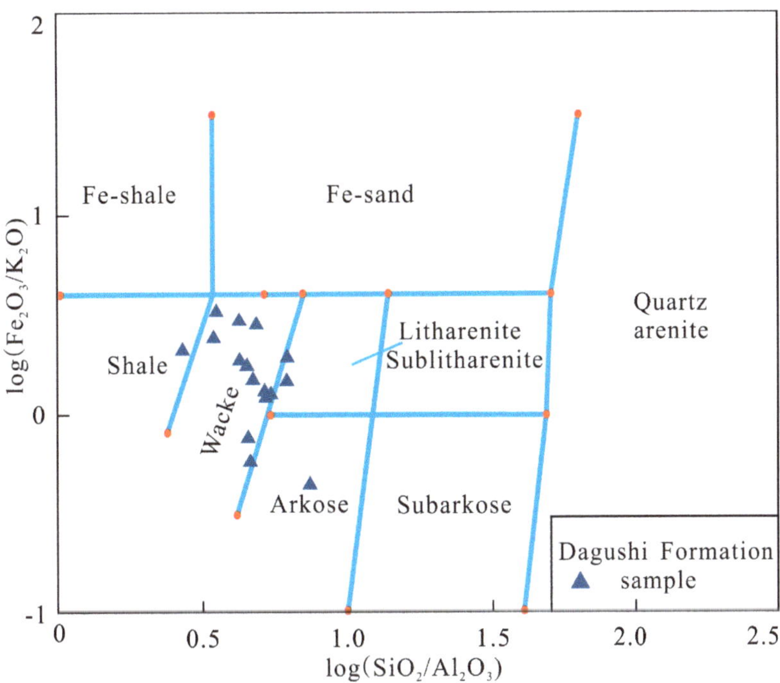

Figure 10. Discrimination plot of sandstone type for the Mesoproterozoic Dagushi Formation in Jiyuan (modified after [96]).

Several trace elements and REEs of clastic sedimentary rocks were inherited from their parent rocks and have been used to show their properties [97–100]. The discrimination diagram of the argillaceous rock of La/Yb–Ce [101] (Figure 11a) shows that the distribution of argillaceous rocks from the Dagushi Formation form a cluster located in the field for intermediate and silicate rocks. The diagram of La/Sc–Co/Th [101] (Figure 11b) shows that the argillaceous rock samples from the Dagushi Formation plot near felsic volcanic rocks. The La/Yb–$\sum$REE discrimination diagram of the source rock [102] (Figure 11c) reveals that the scatter plot distribution of argillaceous rocks in the Dagushi Formation was concentrated in the granite field. The discrimination diagram of La/Th–Hf [103] (Figure 11d) shows that the argillaceous rock samples of the Dagushi Formation were located near the felsic provenance area, with only a few in the passive continental provenance and mixed provenance areas of felsic and mafic rocks. The discrimination diagram of Zr/Sc–Th/Sc [104] (Figure 11e) shows that the projective points of argillaceous rock samples from the Dagushi Formation fell near the upper crust (felsic volcanic rocks). Based on regional geological data, the felsic metamorphic crystalline basement and granitic rocks in the late Archean–Paleoproterozoic era were widely distributed in the Wangwu Mountain area where Jiyuan is located [24,25,105,106]. For this reason, the sources of argillaceous rocks from the Dagushi Formation were primarily felsic in provenance. The provenance area pointed to is the underlying metamorphic crystalline basement of the NCC.

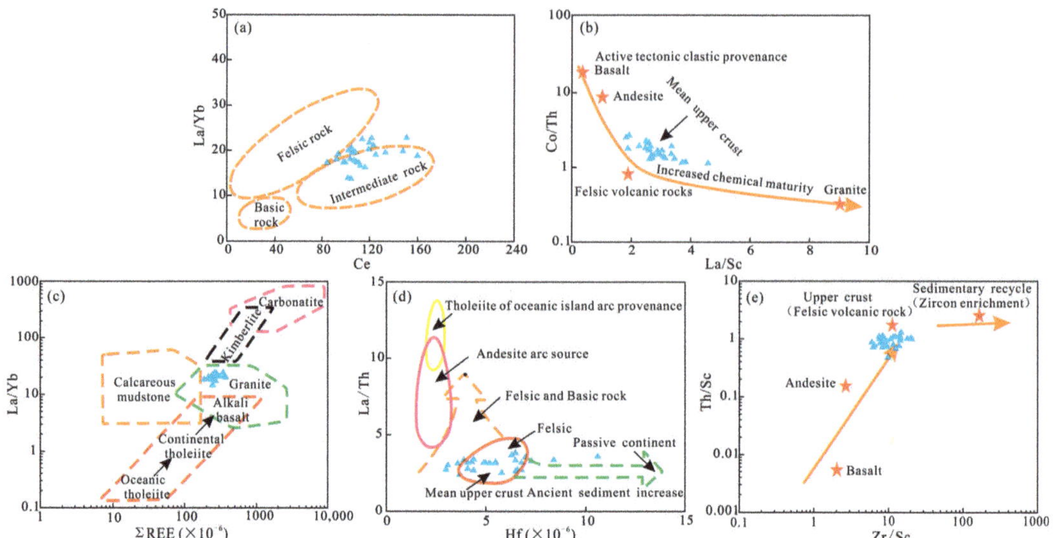

Figure 11. Source discrimination diagrams of argillaceous rock from the Mesoproterozoic Dagushi Formation in Jiyuan; (**a,b**) modified after [101]; (**c**) modified after [102]; (**d**) modified after [103]; (**e**) modified after [104].

The diagrams of trace elements Th–Sc–Zr/10 and Th–Co–Zr/10 (Figure 12a,b) and major elements wt(Fe$_2$O$_3$^{T} + MgO)%-wt(TiO$_2$)% and wt(Fe$_2$O^T + MgO)%-wt(Al$_2$O$_3$)% / wt(SiO$_2$)% (Figure 12c,d) have been used to distinguish tectonic environments [98,107–112]. Based on the discrimination diagrams of Th–Sc–Zr/10 and Th–Co–Zr/10 (Figure 12a,b), the argillaceous rock sample from the Dagushi Formation in the Jiyuan area primarily fell into the continent island arc and active continental margin, with a few samples in the nearby area. Based on discrimination diagrams of wt(Fe$_2$O$_3$^{T} + MgO)%-wt(TiO$_2$)% (Figure 12c), the sandstone samples from the Dagushi Formation in the Jiyuan area also primarily fell into the continent island arc and active continental margin fields, with a few near the ocean island arc field. The discrimination diagrams of wt(Fe$_2$O$_3$^{T} + MgO)%-wt(Al$_2$O$_3$)%/wt(SiO$_2$)% (Figure 12d) show that most samples fell into the active continental margin, ocean island arc field, and nearby areas, with a few in the ocean island arc field. Thus, the above findings indicated that the provenance area showed mixed provenance characteristics dominated by the active tectonic setting, supplemented by the settings of an active continental margin and island arc. The tectonic settings identified above only reflected that of the provenance area and did not represent the tectonic settings for the formation of the Dagushi Formation. The determination of the tectonic background during the sedimentary period of the Dagushi Formation in the Xionger Group should refer to evidence of coeval magmatic rocks in the Xionger Group. However, judgment of the tectonic background of the Dagushi Formation and even the Xionger Group is still controversial. To date, the main viewpoints include an Andean-type continental margin [9,22], passive continental margin and rift [26–28,36,37], and active continental margin and rift [22]. Thus far, the above viewpoints have not been unified to reach consensus.

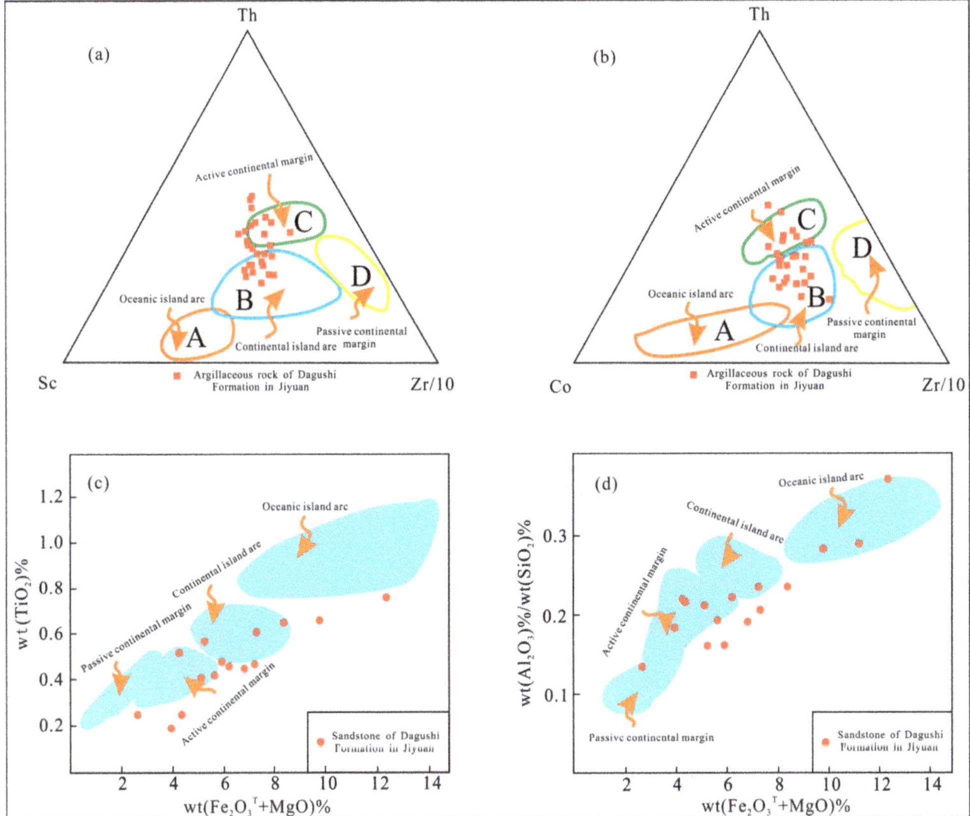

Figure 12. Discrimination plot of tectonic background for sandstone and argillaceous source rocks from the Mesoproterozoic Dagushi Formation in Jiyuan; (**a,b**) modified after [107]; (**c,d**) modified after [108]. (**a**) Oceanic island arc; (**b**) Continental island arc; (**c**) Active continental margin; (**d**) Passive continental margin.

5.2. Tectonic and Sedimentary Implications of Dagushi Formation for the Southern NCC during the Mesoproterozoic

The North China Craton underwent multiple periods of continuous rift development from ~18 Ga to the Neoproterozoic. Among them, the Xiong'er Rift, located at the southern margin of the North China Craton, represents a significant continental stretching and fracturing event that developed in response to the rifting process of the Mesoproterozoic Columbia supercontinent within the context of extensional tectonics. During this rifting process, subsidence initially occurred in the regions of Yuanqu County—Jiyuan City, Luoning City—Luanchuan County, and Ruzhou City, located at the southern margin of the North China Craton (Figure 13a). Because of the difference in terrain elevation, the felsic crystalline basement of the relatively uplifted NCC was denuded quickly, and the product of basement denudation carried by rivers was removed from the basin margin and deposited in the adjacent depression. The coarse clastic sediments at the bottom were gradually superimposed and extended to the rift center, forming a set of clastic sedimentary strata consisting of coarse clastic sandstone and argillaceous clastic rock, i.e., the Dagushi Formation of the Xiong'er Group (Figure 13a). This was followed by large-scale and continuous volcanism throughout the entire region, resulting in the formation of the giant thick volcanic rocks of the Xiong'er Group. The clastic rock filling of the Dagushi Formation in the Xiong'er Rift Basin and the subsequent eruption of the Xiong'er Group volcanic rocks originated from the tectonic setting of a continental margin rift [25,38,105,106] and represented the beginning of multi-stage fracturing events in the NCC from the end of the Paleoproterozoic era to the beginning of the Mesoproterozoic era, which was likely related to the transition of the Columbia supercontinent from collage and aggregation to stretching and fracturing.

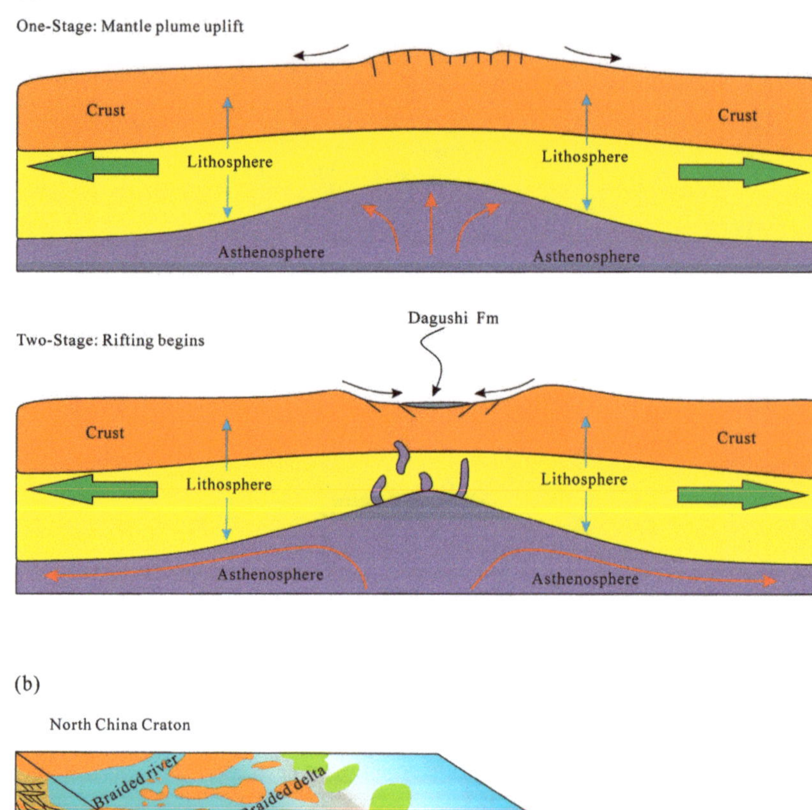

Figure 13. (**a**) Schematic section showing the location of our study area and the Dagushi Formation relative to tectonic units that might be provenance sources. (**b**) Proposed depositional model for the Dagushi Formation.

During the rapid sedimentary process of the Dagushi Formation in Jiyuan City, the initial stage was characterized by braided river facies, which resulted in the formation of sedimentary sandstones. As hydrodynamic conditions gradually weakened, the water body became deeper, leading to the development of delta facies with sand–mud interaction layers. This subsequently transformed upward into mudstone-based lake facies, and this vertical change marked the beginning of crustal fracturing and continuous sinking (Figure 13b). The Dagushi Formation, located at the base of the Xiong'er Group, is overlain by the unconformable Gaoshanhe Formation and Xiaogoubei Formation that consist of conglomerate and mixed sandstone. These sediments indicate several regional tectonic movements that occurred on the southern edge of North China before and after the eruption of the Xiong'er Volcano. Meanwhile, the Dagushi Formation represents the first regional subsidence since the Paleozoic era and distinguishes between the metamorphic basement of the Lower Paleoproterozoic Shuangfang Formation complex and the upper non-metamorphic Mesoproterozoic Xiong'er Group. It serves as a marker to delineate the earliest, non-metamorphic stable sedimentary cover on the crystalline basement of the North China Craton, which actually indicates the Paleoproterozoic and Mesoproterozoic boundary. The age of this study boundary is ~17.84 Ga, which is ~1.84 Ga earlier than the internationally recognized Paleoproterozoic and Mesoproterozoic boundary age.

6. Conclusions

By identifying the lithologic characteristics and sedimentary tectonics of the Dagushi Formation in the field, we determined that from bottom to top, the sedimentary facies of the Dagushi Formation are braided river, braid delta, and lake.

The CIA indicates a slightly to moderately weathered source, and ICV indicates the first cyclic sediments. The U–Pb ages of the detrital zircons of the Dagushi Formation indicate that the sediments were likely supplied by the Taihua, Dengfeng, and Songshan Groups and the Central Orogenic Belt.

The vertical facies and provenance changes of the Dagushi Formation reflect a continuous crust fracturing process that occurred in the North China Craton.

Supplementary Materials: The following supporting information can be downloaded at: https://www.mdpi.com/article/10.3390/min13070971/s1, Table S1: The ICP-MS U-Pb test data was performed on two sandstone samples collected from the DGS-02 and DGS-026 of the Dagushi Formation in the Jiyuan area; Table S2: For the major element assemblage of sandstone and trace element assemblage of argillaceous rock of Dagushi Formation in North China Craton.

Author Contributions: Conceptualization, Y.Z. and G.Z.; methodology, Y.Z.; software, Y.Z.; validation, Y.Z., G.Z. and F.S.; formal analysis, Y.Z.; investigation, Y.Z. and G.Z.; resources, Y.Z.; data curation, Y.Z.; writing—original draft preparation, Y.Z.; writing—review and editing, Y.Z.; visualization, Y.Z.; supervision, G.Z.; project administration, Y.Z. and F.S.; funding acquisition, Y.Z. and F.S. All authors have read and agreed to the published version of the manuscript.

Funding: This research was funded by the National Natural Science Foundation (grant no. 41872238). The APC was funded by Yuan Zhang.

Data Availability Statement: Not applicable.

Acknowledgments: Discussions with Deshun Zheng, Fengbo Sun, and Sicong Liu greatly improved the manuscript.

Conflicts of Interest: The authors declare no potential conflict of interest.

References

1. Chen, Y.J.; Zhai, M.G.; Jiang, S.Y. Significant achievements and open issues in study of orogenesis and metallogenesis surrounding the North China continent. *Acta Petrol. Sin.* **2009**, *25*, 2695–2726, (In Chinese wuth English abstract).
2. Zhao, G.C.; Wilde, S.A.; Guo, J.H.; Cawood, P.A.; Sun, M.; Li, X.P. Single zircon grains record two Paleoproterozoic collisional events in the North China Craton. *Precambrian Res.* **2010**, *177*, 266–276. [CrossRef]
3. Zhao, G.C.; Cawood, P.A. Precambrian geology of China. *Precambrian Res.* **2012**, *222*, 13–54. [CrossRef]

4.   Zhao, G.C.; Zhai, M.G. Lithotectonic elements of Precambrian basement in the North China Craton: Review and tectonic implications. *Gondwana Res.* **2013**, *23*, 1207–1240. [CrossRef]

5.   Zhai, M.G.; Santosh, M. The early Precambrian odyssey of the North China Craton: A synoptic overview. *Gondwana Res.* **2011**, *20*, 6–25. [CrossRef]

6.   Zhai, M.G.; Zhao, Y.; Zhao, T.P. *Main Tectonic Events and Metallogeny of the North China Craton*; Springer: Singapore, 2016.

7.   Kusky, T.M.; Li, J.H. Paleoproterozoic tectonic evolution of the North ChinaCraton. *J. Asian Earth Sci.* **2003**, *22*, 383–397. [CrossRef]

8.   Santosh, M.; Maruyama, S.; Yamamoto, S. The making and breaking of super- continents: Some speculations based on super-plumes, super downwelling and the role of tectosphere. *Gondwana Res.* **2009**, *15*, 324–341. [CrossRef]

9.   Zhao, G.C.; He, Y.H.; Sun, M. The Xiong'er volcanic belt at the southern margin of the North China Craton: Petrographic and geochemical evidence for its outboard position in the Paleo-Mesoproterozoic Columbia Supercontinent. *Gondwana Res.* **2009**, *16*, 170–181. [CrossRef]

10.   Yang, Q.Y.; Santosh, M.; Rajesh, H.M.; Tsunogae, T. Late paleoproterozoic chamockite suite within post-collisional setting from the North China Craton: Petrology, geochemistry, zircon U-Pb geochronology and Lu-Hf isotopes. *Lithos* **2014**, *208*, 34–52. [CrossRef]

11.   Zhao, G.C.; Wilde, S.A.; Cawood, P.A.; Sun, M. Thermal evolution of the Archaean basement rocks from the eastern part of the North China Craton and its bearing ontectonic setting. *Int. Geol. Rev.* **1998**, *40*, 706–721. [CrossRef]

12.   Zhao, G.C.; Wilde, S.A.; Cawood, P.A.; Sun, M. Archaean blocks and their boundaries in the North China Craton: Lithological, geochemical, structural and P-T path constraints and tectonic evolution. *Precambrian Res.* **2001**, *107*, 45–73. [CrossRef]

13.   Zhao, G.C.; Sun, M.; Wilde, S.A.; Li, S.Z. Assembly, accretion and break up of the Paleo-Mesoproterozoic Columbia supercontinent: Records in the North China Craton. *Gondwana Res.* **2003**, *6*, 417–434. [CrossRef]

14.   Zhao, G.C.; Kroner, A.; Wilde, S.A.; Sun, M.; Li, S.Z.; Li, X.P.; Zhang, J.; Xia, X.P.; He, Y.H. Lithotectonic elements and geological events in the Heng shan-Wutai-Fuping belt: A synthesis and implications for the evolution of the Trans-North China Orogen. *Geol. Mag.* **2007**, *144*, 753–775. [CrossRef]

15.   Rogers, J.J.W.; Santosh, M. Configuration of Columbia, a Mesoproterozoic supercont- inent. *Gondwana Res.* **2002**, *5*, 5–22. [CrossRef]

16.   Zhao, G.C.; Cawood, P.A.; Wilde, S.A.; Sun, M. Review of global 2.1–1.8 Gaorogens: Implications for a pre-Rodinia supercontinent. *Earth-Sci. Rev.* **2002**, *59*, 125–162. [CrossRef]

17.   Zhao, G.C.; Sun, M.; Wilde, S.A.; Li, S.Z. A Paleo-Mesoproterozoic sup- ercontinent: Assembly, growth and breakup. *Earth-Sci. Rev.* **2004**, *67*, 91–123. [CrossRef]

18.   He, Y.H.; Zhao, G.Z.; Sun, M.; Wilde, S.A. Geochemistry, isotope systema tics and petrogenesis of the volcanic rocks in the Zhongtiao Mountain: An alterna- tive inter-pretation for the evolution of the southern margin of the North China Craton. *Lithos* **2008**, *102*, 157–178. [CrossRef]

19.   Peng, P.; Zhai, M.G.; Ernst, R.E.; Guo, J.H.; Liu, F.; Hu, B. A 1.78 Ga large igneous province in the North China craton: The Xiong'er volcanic province and the North China dyke Swarm. *Lithos* **2008**, *101*, 260–280. [CrossRef]

20.   Geng, Y.S.; Kuang, H.W.; Du, L.L.; Liu, Y.Q.; Zhao, T.P. On the Paleo-Mesoproterozoic boundary from the breakup event of the Columbia supercontinent. *Acta Petrol. Sin.* **2019**, *35*, 2299–2324.

21.   Sun, S.; Zhang, G.W.; Chen, Z.M. *Precambrian Geological Evolution in the South of North China Fault Block*; Metallurgical Industry Press: Beijing, China, 1985. (In Chinese)

22.   Chen, Y.J.; Fu, S.G.; Qiang, L.Z. The Tectonic environment for the formation of the Xiong'er Group and the Xiyanghe Group. *Geol. Rev.* **1992**, *38*, 325–333, (In Chinese wuth English abstract).

23.   Zhao, T.P.; Zhou, M.F.; Jin, C.W.; Guan, H.; Li, H.M. Discussion on age of the Xiong'er Group in southern margin of North China Craton. *Sci. Geolo.-Gica. Sin.* **2001**, *36*, 326–334, (In Chinese with English abstract).

24.   Zhao, T.P.; Zhai, M.G.; Xia, B.; Li, H.M.; Zhang, Y.X.; Wan, Y.S. Zircon U-Pb SHRIMP dating for the volcanic rocks of the Xiong'er Group: Constrains on the initial formation age of the cover of the North China Craton. *Chinese Sci. Bull.* **2004**, *49*, 2495–2502. [CrossRef]

25.   Zhao, T.P.; Xu, Y.H.; Zhai, M.G. Petrogenesis and Tectonic Setting of the Paleoproterozoic Xiong'er Group in the Southern Part of the North China Craton: A Review. *Geol. J. China Univ.* **2007**, *13*, 191–206, (In Chinese with English abstract).

26.   Zhai, M.G. 2.1~1.7 Ga geological event group and its geotectonic sig- nificance. *Acta Petrol. Sin.* **2004**, *20*, 1343–1354, (In Chinese with English abstract).

27.   Zhai, M.G. Multi-stage crustal growth and cratonization of the North China Craton. *Geosci. Front.* **2014**, *5*, 457–469. [CrossRef]

28.   Zhai, M.G.; Peng, P. Paleoproterozoic events in North China Craton. *Acta Petrol. Sin.* **2007**, *23*, 2665–2682, (In Chinese with English abstract).

29.   Kusky, T.; Li, J.G.; Santosh, M. The Paleoproterozoic North Hebei orogen: North China craton's collisional suture with the Columbia supercontinent. *Gond-Wana Res. Int. Geosci. J.* **2007**, *12*, 4–28. [CrossRef]

30.   He, Y.H.; Zhao, G.C.; Sun, M.; Xia, X.P. SHRIMP and LA-ICP-MS zircon geochro-nology of the Xiong'er volcanic rocks: Implications for the Paleo-Meso- proterozoic evolution of the southern margin of the North China Craton. *Precam.-Brian Res.* **2009**, *168*, 213–222. [CrossRef]

31.   He, Y.H.; Zhao, G.C.; Sun, M.; Han, Y.G. Petrogenesis and tectonic setting of volcanic rocks in the Xiaoshan and Waifangshan areas along the southern mar- gin of the North China Craton: Constraints from bulk–rock geochemistry and Sr- Nd isotopic composition. *Lithos* **2010**, *114*, 186–199. [CrossRef]

32. Hu, G.H.; Zhao, T.P.; Zhou, Y.Y.; Yang, Y. Depositional age and provena- nce of the Wufoshan Group in the southern margin of the North China Craton: Evidence from detrital zircon U-Pb ages and Hf isotopic compositions. *Geochimica* **2012**, *41*, 326–342, (in Chinese with English abstract).

33. Hu, G.H.; Zhou, Y.Y.; Zhao, T.P. Geochemistry of Proterozoic Wufoshan Group sedimentary rocks in the Songshan area, Henan Province: Implications for prove-nance and tectonic setting. *Acta Petrol. Sin.* **2012**, *28*, 3692–3704, (In Chinese with English abstract).

34. Hu, G.H.; Zhao, T.P.; Zhou, Y.Y.; Wang, S.Y. Meso-Neoproterozoic sedim- entary formation in the southern margin of the North China Craton and its geolo- gical implications. *Acta Petrol. Sin.* **2013**, *29*, 2491–2507, (In Chinese with English abstract).

35. Hu, G.H.; Zhao, T.P.; Zhou, Y.Y. Depositional age, provenance and tect- onic setting of the Proterozoic Ruyang Group, southern margin of the North Ch- ina Craton. *Precambrian Res.* **2014**, *246*, 296–318. [CrossRef]

36. Zhai, M.G. Where is the north China-South China block boundary in eastern China? Comment. *Geology* **2002**, *30*, 667, (In Chinese with English abstract).

37. Zhai, M.G.; Hu, B.; Zhao, T.P.; Peng, P.; Meng, Q.R. Late Paleoproterozo ic-Neoproterozoic multi-rifting events in the North China Craton and their geolo- gical significance: A study advance and review. *Tectonophysics* **2015**, *662*, 153–166. [CrossRef]

38. Xu, Y.H.; Zhao, T.P.; Zhang, Y.X.; Chen, W. Geochemical Characteristics and Geological Significances of the Dagushi Formation Siliciclastic Rocks, the Paleoproterozoic Xiong'er Group from the Southern North China Craton. *Geol. Rev.* **2008**, *54*, 316–326, (in Chinese with English abstract).

39. Xu, Y.H.; Zhao, T.P.; Chen, W. The Discovery and Geological Significance of Glauconites from the Palaeoproterozoic Xiong'er Group in the Southern Part of the North China Craton. *Acta Sedimentol. Sin.* **2010**, *28*, 671–675, (In Chinese with English abstract).

40. Deng, H.; Kusky, T.M.; Polat, A.L.; Fu, H.Q.; Wang, L.; Wang, J.P.; Wang, S.J.; Zhai, W.J. A Neoarchean arc-backarc pair in the Linshan Massif, southern North China Craton. *Precambrian Res.* **2020**, *341*, 105649. [CrossRef]

41. Henan Bureau of Geology and Mineral Resources. *Regional Geology of Henan Province*; Geological Publishing House: Beijing, China, 1989. (In Chinese)

42. Zhai, M.G.; Guo, J.H.; Liu, W.J. Neoarchean to Paleoproterozoic conti- nental evolution and tectonic history of the North China Craton. *J. Asian Earth Sci.* **2005**, *24*, 547–561. [CrossRef]

43. Zhao, G.C.; Sun, M.; Wilde, S.A.; Li, S.Z. Late Archean to Paleoproter- ozoic evo-lution of the North China Craton: Key issues revisited. *Precambrian Res.* **2005**, *136*, 177–202. [CrossRef]

44. Jahn, B.M. Early crustal evolution as viewed from archean basic rocks of China. *Chem. Geol.* **1988**, *70*, 141. [CrossRef]

45. O'Neill, C.; Lenardic, A.; Moresi, L.; Torsvik, T.H.; Lee, C.T.A. Episodic Precam- brian subduction. *Earth Planet. Sci. Lett.* **2007**, *262*, 552–562. [CrossRef]

46. Huang, X.L.; Niu, Y.L.; Xu, Y.G.; Yang, Q.J.; Zhong, J.W. Geochemistry of TTG and TTG-like gneisses from Lushan-Taihua complex in the southern North China Craton: Implications for late Archean crustal accretion. *Precambrian Res.* **2010**, *182*, 43–56. [CrossRef]

47. Wan, Y.S.; Dong, C.Y.; Wang, W.; Xie, H.Q.; Liu, D.Y. Archean Basement and a Paleoproterozoic Collision Orogen in the Huoqiu Area at the Southeastern Margin of North China Craton: Evidence from Sensitive High Resolution Ion Micro-Probe U-Pb Zircon Geochronology. *Acta Geol. Sin.-Engl. Ed.* **2010**, *84*, 91–104. [CrossRef]

48. Wan, Y.S.; Liu, D.Y.; Wang, S.J.; Yang, E.X.; Wang, W.; Dong, C.Y.; Zhou, H.Y.; Du, L.L.; Yang, Y.H.; Diwu, C.R. ~2.7 Ga juvenile crust formation in the North China Craton (Taishan-Xintai area, western Shandong Province): Further evidence of an understated event from U-Pb dating and Hf isotopic composition of zircon. *Precambrian Res.* **2011**, *186*, 169–180. [CrossRef]

49. Diwu, C.R.; Sun, Y.; Lin, C.L.; Wang, H.L. LA-(MC)-ICPMS U-Pb zircon geochronology and Lu-Hf isotope compositions of the Taihua Complex on the southern margin of the North China Craton. *Chin. Sci. Bull.* **2010**, *55*, 2112–2123, (In Chinese with English abstract). [CrossRef]

50. Zhai, M.G.; Bian, A.G.; Zhao, T.P. The amalgamation of the supercontinent of North China Craton at the end of Neo-Archaean and its breakup during late Palaeoproterozoic and Meso-Proterozoic. *Sci. China (Seies D)* **2000**, *43*, 219–232. [CrossRef]

51. Zhai, M.G.; Liu, W. Palaeoproterozoic tectonic history of the North China Craton: A review. *Precambrian Res.* **2003**, *122*, 183–199. [CrossRef]

52. Zhao, T.P.; Zhou, M.F.; Zhai, M.G.; Xia, B. Palaeoproterozoic rift-related volcanism of the Xiong'er Group in the North China Craton:implications for the break- up of Columbia. *Int. Geol. Rev.* **2002**, *44*, 336–351. [CrossRef]

53. Guan, B.D. *The Precambrian-Lower Cambrian Geology and Metallogenesis in the South Border of the North China Platform in Henan Province*; Press of China University of Geosciences: Wuhan, China, 1996; pp. 1–328. (In Chinese)

54. Zhao, T.P.; Yuan, Z.L.; Guan, B.D. The characteristics and sedimentary environment of sedimentary interbeds of Xiong'er Group distributed in the juncture of Henan-Shanxi-Shaanxi Provinces. *Henan Geol.* **1998**, *16*, 261–272, (In Chinese with English abstract).

55. Nasdala, L.; Hofmeister, W.; Norberg, N.; Martinson, J.M.; Corfu, F.; Dörr, W.; Kamo, S.L.; Kennedy, A.K.; Kronz, A.; Reiners, P.W.; et al. Zircon M257: A homo-geneous natural re-ference material for the ion microprobe U-Pb analysis of zir-con. *Geostand. Geoanal. Res.* **2008**, *32*, 247–265. [CrossRef]

56. Liu, Y.S.; Hu, Z.C.; Gao, S.; Gunther, D.; Xu, J.; Gao, C.G.; Chen, H.H. In situ analysis of major and trace elements of anhydrous minerals by LA-ICP- MS without applying an internal standard. *Chem. Geol.* **2008**, *257*, 34–43. [CrossRef]

57. Liu, Y.S.; Gao, S.; Hu, Z.C.; Gao, C.G.; Zong, K.Q.; Wang, D.B. Continental and oceanic crust recycling-induced melt-peridotite interactions in the Trans-North China Orogen: U-Pb dating, Hf isotopes and trace elements in zircons of mantle xenoliths. *J. Petrol.* **2010**, *51*, 537–571. [CrossRef]

58. Liu, Y.S.; Hu, Z.C.; Zong, K.Q.; Gao, C.G.; Gao, S.; Xu, J.; Chen, H.H. Reappraisement and refinement of zircon U-Pb isotope and trace element analyses by LA-ICP-MS. *Chin. Sci. Bull.* **2010**, *55*, 1535–1546. [CrossRef]

59. Hu, Z.C.; Liu, Y.S.; Gao, S.; Xiao, S.Q.; Zhao, L.S.; Günther, D.; Li, M.; Zhang, W.; Zong, K.Q. A "wire" signal smoothing device for laser ablation inductively coupled plasma mass spectrometry analysis. *Spectrochim. Acta* **2012**, *78*, 50–57. [CrossRef]

60. Liu, Y.S.; Zong, K.Q.; Kelemen, P.B.; Gao, S. Geochemistry and magmatic history of eclogites and ultramafic rocks from the Chinese continental scientific drill hole: Subduction and ultrahigh-pressure metamorphism of lower crustal cumulates. *Chem. Geol.* **2008**, *247*, 133–153. [CrossRef]

61. Wiedenbeck, M.; Alle, P.; Corfu, F.; Griffin, W.L.; Meier, M.; Oberli, F.; Quadt, A.V.; Roddick, J.C.; Spiegel, W. Three natural zircon standards for U-Th-Pb, Lu-Hf, trace element and REE analyses. *Geostand. Geoanal. Res.* **1995**, *19*, 1–23. [CrossRef]

62. Compston, W.; Williams, I.S.; Kirschvink, J.L.; Zhang, Z.; Ma, G. Zircon U-Pb ages for the early cambrian time-scale. *J. Geol. Soc.* **1992**, *149*, 171–184. [CrossRef]

63. Ludwig, K.R. *User's Manual for Isoplot 3.00: A Geochronological Toolkit for Microsoft Excel*; Berkeley Geochronology Center: Alameda, Berkeley, CA, USA, 2003; Volume 39.

64. Li, X.H.; Li, Z.X.; Wingate, M.T.D.; Chung, S.L.; Liu, T.; Lin, G.C.; Li, W.X. Geochemistry of the 755 Ma Mundine Well dyke swarm, northwestern Australia: Part of a Neoproterozoic mantle superplume beneath Rodinia? *Precambrian Res.* **2006**, *146*, 1–15 [CrossRef]

65. Sircombe, K.N. Tracing provenance through the isotope ages of littoral and sedi-mentary detrital zircon, eastern Australia. *Sediment. Geol.* **1999**, *124*, 47–67. [CrossRef]

66. Nesbitt, H.W.; Young, G.M. Early Proterozoic climates and plate motions inferred from major element chemistry of lutites. *Nature* **1982**, *299*, 715–717. [CrossRef]

67. Nesbitt, H.W.; Young, G.M. Prediction of some weathering trends of plutonic and volcanic rocks based on thermodynamic and kinetic considerations. *Geochim. Cosmochim. Acta* **1984**, *48*, 1523–1534. [CrossRef]

68. Fedo, C.M.; Nesbitt, H.W.; Young, G.M. Unraveling the effects of potassium me-tasomatism in sedimentary rocks and paleosols, with implications for paleo-weathering conditions and provenance. *Geology* **1995**, *23*, 921–924. [CrossRef]

69. Cox, R.; Lowe, D.R. A conceptual review of regional-scale controls on the com-position of clastic sediment and the co-evolution of continental blocks and their se-dimentary cover. *J. Sediment. Res.* **1995**, *65*, 1–12.

70. Cox, R.; Lowe, D.R.; Cullers, R.L. The influence of sediment recycling and base-ment composition on evolution of mudrock chemistry in the southwestern United States. *Geochim. Cosmochim. Ac.* **1995**, *59*, 2919–2940. [CrossRef]

71. Boynton, W.V. Cosmochemistry of the rare earth elements: Meteorite studies dev. *Geochemistry* **1984**, *2*, 63–114.

72. Haskin, M.A.; Haskin, L.A. Rare earths in European shales redetermination. *Science* **1966**, *154*, 507–509. [CrossRef]

73. Taylor, S.R.; Mclennan, S.M. *The Continental Crust: Its Composition and Evolution: An Examination of the Geochemical Record Preserved in Sedimentary Rocks*; Blackwell Scientific Publication: Oxford London, UK, 1985.

74. Wu, F.Y.; Zhao, G.C.; Wilde, S.A.; Sun, D.Y. Nd isotopic constraints on crustal formation in the North China Craton. *J. Asian Earth Sci.* **2005**, *24*, 523–545. [CrossRef]

75. Wu, F.Y.; Zhang, Y.B.; Yang, J.H.; Xie, L.W.; Yang, Y.H. Zircon U-Pb and Hf isotopic constraints on the Early Archean crustal evolution in Anshan of the North China Craton. *Precambrian Res.* **2008**, *167*, 339–362. [CrossRef]

76. Diwu, C.R.; Sun, Y.; Lin, C.L.; Liu, X.M.; Wang, H.L. Zircon U-Pb ages and Hf isotopes and their geological significance of Yiyang TTG gneisses from Henan province. *China. Acta Petrol. Sin.* **2007**, *23*, 253–262, (in Chinese with English abstract).

77. Diwu, C.R.; Sun, Y.; Yuan, H.L.; Wang, H.L.; Zhong, X.P.; Liu, X.M. U-Pb ages and Hf isotopes for detrital zircons from quartzite in the Paleoproterozoic Songshan Group on the southwestern margin of the North China Craton. *Chin. Sci. Bull.* **2008**, *53*, 2828–2839. [CrossRef]

78. Dong, M.M.; Wang, C.M.; Santosh, M.; Shi, K.X.; Du, B.; Chen, Q.; Zhu, J.X.; Liu, X.J. Geochronology and petrogenesis of the Neoarchean-Paleoproterozoic Taihua Complex, NE China: Implications for the evolution of the North China Craton. *Precambrian Res.* **2020**, *346*, 105792. [CrossRef]

79. Zhao, G.C.; Wilde, S.A.; Sun, M.; Guo, J.H.; Kröner, A.; Li, S.Z.; Li, X.P.; Wu, C.M. SHRIMP U-Pb zircon geochronology of the Huaian Complex: Constraints on Late Archean to Paleoproterozoic crustal accretion and collision of the Trans-North China Orogen. *Am. J. Sci.* **2008**, *308*, 270–303. [CrossRef]

80. Wu, M.L.; Zhao, G.C.; Sun, M.; Yin, C.Q.; Li, S.Z.; Tam, P.Y. Petrology and P-T path of the Yishui mafic granulites: Implications for tectonothermal evolution of the Western Shandong Complex in the Eastern Block of the North China Craton. *Precambrian Res.* **2012**, *222*, 312–324. [CrossRef]

81. Wu, M.L.; Zhao, G.C.; Sun, M.; Li, S.Z.; He, Y.H.; Bao, Z. Zircon U-Pb geochro-nology and Hf isotopes of major lithologies from the Yishui Terrane: Implications for the crustal evolution of the Eastern Block, North China Craton. *Lithos* **2013**, *170*, 164–178. [CrossRef]

82. Wu, S.J.; Hu, J.M.; Ren, M.H.; Gong, W.B.; Liu, Y.; Yan, J.Y. Petrography and zircon U-Pb isotopic study of the Bayanwulashan Complex: Constrains on the Paleoproterozoic evolution of the Alxa Block, westernmost North China Craton. *J. Asian Earth Sci.* **2014**, *94*, 226–239. [CrossRef]

83. Zhao, T.P.; Zhang, Z.H.; Zhou, Y.Y.; Wang, S.Y.; Liu, C.S.; Liang, H.J.; Zhang, B.C.; Hu, G.H. *Precambrian Geology of the Songshan Area, Henan Province*; Geological Publishing House: Beijing, China, 2012; pp. 1–206, (In Chinese with English abstract).

84. Zhai, M.G.; Bian, A.G. Amalgamation of the supercontinent of the North China Craton and its break up during late-middle Proterozoic. *Sci. China (D)* **2001**, *43*, 219–232. [CrossRef]
85. Kusky, T.M. Geophysical and geological tests of tectonic models of the North China Craton. *Gondwana Res.* **2011**, *20*, 26–35. [CrossRef]
86. Diwu, C.R.; Sun, Y.; Guo, A.L.; Wang, H.L.; Liu, X.M. Crustal growth in the North China Craton at ~2.5 Ga: Evidence from in situ zircon U-Pb ages, Hf isotopes and whole-rock geochemistry of the Dengfeng complex. *Gondwana Res.* **2011**, *20*, 149–170. [CrossRef]
87. Zhou, Y.Y.; Zhao, T.P.; Wang, C.Y.; Hu, G.H. Geochronology and geochemistry of 2.5 to 2.4 Ga granitic plutons from the southern margin of the North China Craton: Implications for a tectonic transition from arc to post-collisional setting. *Gondwana Res.* **2011**, *20*, 171–183. [CrossRef]
88. Wang, X.; Huang, X.L.; Yang, F.; Luo, Z.X. Late Neoarchean magmatism and tectonic evolution recorded in the Dengfeng Complex in the southern segment of the Trans-North China Orogen. *Precambrian Res.* **2017**, *302*, 180–197. [CrossRef]
89. Zhang, J.; Zhang, H.F.; Li, L.; Wang, J.L. Neoarchean-Paleoproterozoic tectonic evolution of the southern margin of the North China Craton: Insights from geochemical and zircon U-Pb-Hf-O isotopic study of metavolcanic rocks in the Dengfeng complex. *Precambrian Res.* **2018**, *318*, 103–121. [CrossRef]
90. Liu, C.H.; Zhao, G.C.; Sun, M.; Zhang, J.; Yin, C.Q.; He, Y.H. Detrital zircon U-Pb dating, Hf isotopes and whole-rock geochemistry from the Songshan Group in the Dengfeng Complex: Constraints on the tectonic evolution of the Trans-North China Orogen. *Precambrian Res.* **2012**, *192–195*, 1–15. [CrossRef]
91. Guo, J.H.; Sun, M.; Zhai, M.G. Sm-Nd and SHRIMP U-Pb zircon geochronology of high-pressure granulites in the Sanggan area, North China Craton: Timing of Paleoproterozoic continental collision. *J. Asian Earth Sci.* **2005**, *24*, 629–642. [CrossRef]
92. Zhang, J.; Zhao, G.C.; Li, S.Z.; Sun, M.; Wilde, S.A.; Liu, S.W.; Yin, C.Q. Polyphase deformation of the Fuping Complex, Trans-North China Orogen: Structures, SHRIMP U-Pb zircon ages and tectonic implications. *J. Struct. Geol.* **2009**, *31*, 177–193. [CrossRef]
93. Lu, J.S.; Wang, G.D.; Wang, H.; Chen, H.X.; Wu, C.M. Metamorphic P-T-t paths retrieved from the amphibolites, Lushan terrane, Henan Province and reappraisal of the Paleoproterozoic tectonic evolution of the Trans-North China Orogen. *Precambrian Research* **2013**, *238*, 61–77. [CrossRef]
94. Yang, Q.Y.; Santosh, M. Paleoproterozoic arc magmatism in the North China Craton: No Siderian global plate tectonic shutdown. *Gondwana Res.* **2015**, *28*, 82–105. [CrossRef]
95. Yang, Q.Y.; Santosh, M. Charnockite magmatism during a transitional phase: Implications for Late Paleoproterozoic ridge subduction in the North China Craton. *Precambrian Res.* **2015**, *261*, 188–216. [CrossRef]
96. Herron, M.M. Geochemical classification of terrigenous sands and shales from core or log data. *J. Sediment. Res.* **1988**, *58*, 820–829.
97. Nance, W.B.; Taylor, S.R. Rare earth element patterns and crustal evolution-I. Australian post—Archean sedimentary rocks. *Geochim. Et Cosmochim. Acta* **1976**, *40*, 1539–1551. [CrossRef]
98. Bhatia, M.R. Rare earth element geochemistry of Australian Paleozoic graywackes and mudrocks: Provenance and tectonic control. *Sediment. Geol.* **1985**, *45*, 97–113. [CrossRef]
99. Murray, R.W. Chemical criteria to identify the depositional envionment of chert: General principles and applications. *Sendimentary Geol.* **1994**, *90*, 213–232. [CrossRef]
100. Mao, G.Z.; Liu, C.Y. Application of Geochemistry in Provenance and Depositional Setting Analysis. *J. Earch Sci. Environ.* **2011**, *33*, 337–348, (In Chinese with English abstract).
101. McLennan, S.M.; Hemming, S.; McDaniel, D.K.; Hanson, G.N. Geochemical approaches to sedimentation, provenance, and tectonics. *Geol. Soc. Am. Spec. Pap.* **1993**, *284*, 21–40.
102. Allegre, C.J.; Minster, J.F. Quantitative models of trace element behavior in magmatic processes. *Earth Planet. Sci. Lett.* **1978**, *38*, 1–25. [CrossRef]
103. Floyd, P.A.; Leveridge, B.E. Tectonic environment of the Devonian Gramscatho basin, south Cornwall: Framework mode and geochemical evidence from turbiditic sandstones. *J. Geol. Soc.* **1987**, *144*, 531–542. [CrossRef]
104. Condie, K.C. Chemical composition and evolution of the upper continental crust: Contrasting results from surface samples and shales. *Chemical geology* **1993**, *104*, 1–37. [CrossRef]
105. Zhao, T.P.; Jin, C.W. Rewiew on the study of the Xiong'er Group in past 40 years. *North China J. Geol. Miner. Resour.* **1999**, *14*, 18–26, (In Chinese with English abstract).
106. Zhao, T.P.; Wang, J.P.; Zhang, Z.H. *Proterozoic Geology of Mt. Wangwushan Mountain and Adjacent Areas*; China Land Press: Beijing, China , 2005. (In Chinese)
107. Bhatia, M.R.; Crook, K.A.W. Trace element characteristics of graywackes and tectonic setting discrimination of sedimentary basins. *Contrib. Mineral. Petrol.* **1986**, *92*, 181–193. [CrossRef]
108. Bhatia, M.R. Plate tectonics and geochemical composition of sandstones. *J. Geol.* **1983**, *91*, 611–627. [CrossRef]
109. Bhatia, M.R.; Taylor, S.R. Trace-element geochemistry and sedimentary provinces: A study from the Tasman Geosyncline, Australia. *Chem. Geol.* **1981**, *33*, 115–125. [CrossRef]
110. Xu, Y.J.; Du, Y.S.; Yang, J.H.; Huang, H. Sedimentary geochemistry and prove-nance of the lower and middle Devonian Laojunshan Formation, the north Qilian orogenic belt. *Sci. China Earth Sci.* **2010**, *53*, 356–367. [CrossRef]

111. Wang, W.; Zhou, M.F.; Yan, D.P.; Li, J.W. Depositional age, provenance, and tectonic setting of the Neoproterozoic Sibao Group, Southeastern Yangtze Block, South China. *Precambrian Res.* **2012**, *192–195*, 107–124. [CrossRef]
112. Zhang, M.; Liu, Z. Geochemistry of pelitic rocks from the Middle Permian Lucaogou Formation, Sangonghe area, Junggar basin, Northwest China: Implications for source weathering, recycling, provenance and tectonic setting. *Geol. J.* **2015**, *50*, 552. [CrossRef]

Article

Geochronology and Geochemistry of Paleoproterozoic Mafic Rocks in Northern Liaoning and Their Geological Significance

Jingsheng Chen [1,2], Yi Tian [3], Zhonghui Gao [4], Bin Li [1,2,*], Chen Zhao [1,2], Weiwei Li [4], Chao Zhang [1,2] and Yan Wang [1,2]

[1] Shenyang Geological Survey Center of China Geological Survey, Shenyang 110034, China; jschen0712@126.com (J.C.); aaron198807@163.com (C.Z.); congray@163.com (C.Z.); wangy68413@163.com (Y.W.)
[2] Northeast Geological S&T Innovation Center of China Geological Survey, Shenyang 110034, China
[3] Liaoning Geological and Mineral Survey Institute Co., Ltd., Shenyang 110031, China; lnsytianyi@163.com
[4] Institute of Geology and Mineral Resources of Liaoning Co., Ltd., Shenyang 110029, China; 13897938707@163.com (Z.G.); 15040028747@163.com (W.L.)
* Correspondence: libin@mail.cgs.gov.cn

Abstract: Petrological, geochronological, and geochemical analyses of mafic rocks in northern Liaoning were conducted to constrain the formation age of the Proterozoic strata, and to further study the source characteristics, genesis, and tectonic setting. The mafic rocks in northern Liaoning primarily consist of basalt, diabase, gabbro, and amphibolite. Results of zircon U-Pb chronology reveal four stages of mafic magma activities in northern Liaoning: the first stage of basalt (2209 ± 12 Ma), the second stage of diabase (2154 ± 15 Ma), the third stage of gabbro (2063 ± 7 Ma), and the fourth stage of magmatic protolith of amphibolite (2018 ± 13 Ma). Combined with the unconformity overlying Neoproterozoic granite, the formation age of the Proterozoic strata in northern Liaoning was found to be Paleoproterozoic rather than Middle Neoproterozoic by the geochronology of these mafic rocks. A chronological framework of mafic magmatic activities in the eastern segment of the North China Craton (NCC) is proposed. The mafic rocks in northern Liaoning exhibit compositional ranges of 46.39–50.33 wt% for SiO_2, 2.95–5.08 wt% for total alkalis ($K_2O + Na_2O$), 6.17–7.50 wt% for MgO, and 43.32–52.02 for the Mg number. TiO_2 contents lie between 1.61 and 2.39 wt%, and those of MnO between 0.17 and 0.21 wt%. The first basalt and the fourth amphibolite show low total rare earth element contents. Normalized against primitive mantle, they are enriched in large ion lithophile elements (Rb, Ba, K), depleted in high field strength elements (Th, U, Nb, Ta, Zr, Ti), and exhibit negative anomalies in Sr and P, as well as slight positive anomalies in Zr and Hf. The second diabase and the third gabbro have similar average total rare earth element contents. The diabase shows slight negative Eu anomalies (Eu/Eu* = 0.72–0.88), enrichment in large ion lithophile elements (Ba), depletion in Rb, and slight positive anomalies in high field strength elements (Th, U, Nb, Ta, Zr, Hf, Ti), with negative anomalies in K, Sr, and P. The gabbro is enriched in large ion lithophile elements (Rb, Ba, K), depleted in high field strength elements (Th, U, Nb, Ta, Zr, Hf), and exhibits positive anomalies in Eu (Eu/Eu* = 1.31–1.37). The contents of Cr, Co, and Ni of these four stages of mafic rocks are higher than those of N-MORB. The characteristics of trace element ratios indicate that the mafic rocks belong to the calc-alkaline series and originate from the transitional mantle. During the process of magma ascent and emplacement, it is contaminated by continental crustal materials. There are residual hornblende and spinel in the magma source of the first basalt. The other three magma sources contain residual garnet and spinel. The third gabbro was formed in an island arc environment, and the other three stages of mafic rocks originated from the Dupal OIB and were formed in an oceanic island environment. The discovery of mafic rocks in northern Liaoning suggests that the Longgang Block underwent oceanic subduction and extinction in both the north and south in the Paleoproterozoic, indicating the possibility of being in two different tectonic domains.

Citation: Chen, J.; Tian, Y.; Gao, Z.; Li, B.; Zhao, C.; Li, W.; Zhang, C.; Wang, Y. Geochronology and Geochemistry of Paleoproterozoic Mafic Rocks in Northern Liaoning and Their Geological Significance. *Minerals* **2024**, *14*, 717. https://doi.org/10.3390/min14070717

Academic Editors: Federica Zaccarini and Giorgio Garuti

Received: 5 May 2024
Revised: 14 July 2024
Accepted: 15 July 2024
Published: 16 July 2024

Keywords: Paleoproterozoic mafic rocks; zircon U-Pb chronology; geochemistry; feature of source; tectonic setting

1. Introduction

As one of the oldest cratons in the world, the North China Craton (NCC) is a product of the amalgamation of multiple micro landmasses during the Late Neoarchean to Paleoproterozoic [1–7]. The researchers have different explanations on the tectonic evolution of the NCC. Four Archean landmasses and three Paleoproterozoic orogenic belts have been identified by some researchers (Figure 1c; [1,2,7]), and the aggregation process is summarized as follows: (1) In the western part of the NCC, the Yinshan Block and the Ordos Block converged along the Khondalite Belt at 1.95 Ga, forming a unified land block known as the Western Block. (2) In the eastern part, the Longgang Block and the Liaonan–Rangnim Block converged along the Jiao-Liao-Ji Belt (JLJB) at 1.92 Ga resulting in the formation of the Eastern Block. (3) The final collision occurred at 1.85 Ga between the Eastern and Western Blocks along the Trans-Central Orogenic Belt, leading to the final amalgamation of the NCC into a unified Precambrian continent [1,2,7]. However, other scholars have a different understanding of the convergence process of the NCC, with regard to the amalgamation of smaller tectonic units into larger continental landmasses at the end of the Archean and into the Paleoproterozoic with the formation of the Columbia Continent [3,4]. The Western Block collided with the arc-modified margin of the composite Eastern Block at 2.43 Ga leading to the formation of the Central Orogenic Belt with the imbricated arc and fore-arc ophiolitic mélanges. The northern margin of the craton was modified to become an Andean-style arc from 2.3 Ga to 1.9 Ga soon after this collision, and numerous magmatic rocks, volcanic and volcaniclastic rocks, and thick clastic sediments occurred in the continental-margin arc and retro-arc foreland basins. From about 1.88 to 1.79 Ga, the Columbia/Nuna Continent collided with the NCC along the northern margin of the craton resulting in the formation of the Inner Mongolia–Northern Hebei Orogen (IMNHO) [3,4,8,9]. Thus, the Paleoproterozoic geological bodies exposed within the orogenic belt are the key to reconstructing the tectonic evolution of the NCC.

In the eastern segment of the NCC, the well-known Paleoproterozoic JLJB is composed of a large amount of Paleoproterozoic magmatic, sedimentary, and metamorphic rocks, and has been studied extensively [1,2,10–20]. As the focus of debate, four primary tectonic evolution models of the JLJB have been proposed by researchers including: (1) the opening and closing of a Paleoproterozoic intracontinental rift [21–23]; (2) the collision of a continent–arc–continent system [24–27]; (3) a complete Wilson cycle encompassing Paleoproterozoic rifting–extension–ocean basin–subduction–collision [2,6,7,28,29]; and (4) the opening and closing of a back-arc basin [10,13–15] or retro-arc foreland basin [4,5,8]. Large-scale Paleoproterozoic mafic dykes developed in the JLJB [13,14], and they are direct carriers reflecting the stages of magmatic activity, the characteristics of magma source, and the tectonic setting. The study of these rocks provides valuable insights into the composition and behavior of the Earth's interior. Thus, the mafic rocks in the JLJB are the key to discussing the evolution of the tectonic belt.

In addition to the JLJB, many Paleoproterozoic magmatic and metamorphic events have been identified in the Qingyuan terrane of the northern margin of the NCC [30–32]. The Paleoproterozoic Inner Mongolia–Northern Hebei Orogen, as defined by Kusky et al. (2007), has been extended to the northern Liaoning but lacks direct geochronological evidence [4,5,9]. A suite of low-grade metamorphic rocks consisting of dolomite, sandstone, and siltstone has been exposed in northern Liaoning. Due to a lack of direct evidence of geochronology, its deposition age has been controversial, and it is temporarily classified as Mesoproterozoic based solely on lithological comparisons [33]. Large-scale mafic rocks intrude into this stratigraphic sequence. According to whole-rock K–Ar dating results, these mafic rocks are considered to have formed in the Early Triassic [33]. However, during the field investigation, the authors found that these mafic rocks intrude into the dolomite, or are enveloped by marble, and have undergone high erosion, alteration, and metamorphism, similar to the large-scale Paleoproterozoic mafic dykes developed in the Liaoyang–Haicheng area in the south of the Longgang Block, rather than the Mesozoic mafic dykes [13,14]. Meanwhile, a 2.12 Ga metabasic dyke has recently been reported in Qingyuan [30]. Therefore, are the mafic rocks in northern Liaoning also products of the Paleoproterozoic magmatic events? What is the relationship with the mafic rocks in the JLJB and the tectonic setting in which they were formed? The composition of mafic–ultramafic rocks can directly reflect their sources and tectonic settings. Thus, the mafic rocks in northern Liaoning were selected for petrographic, geochronological, and geochemical studies to constrain the deposition age of the Proterozoic sediments and reconstruct the Paleoproterozoic tectonic evolution in this area.

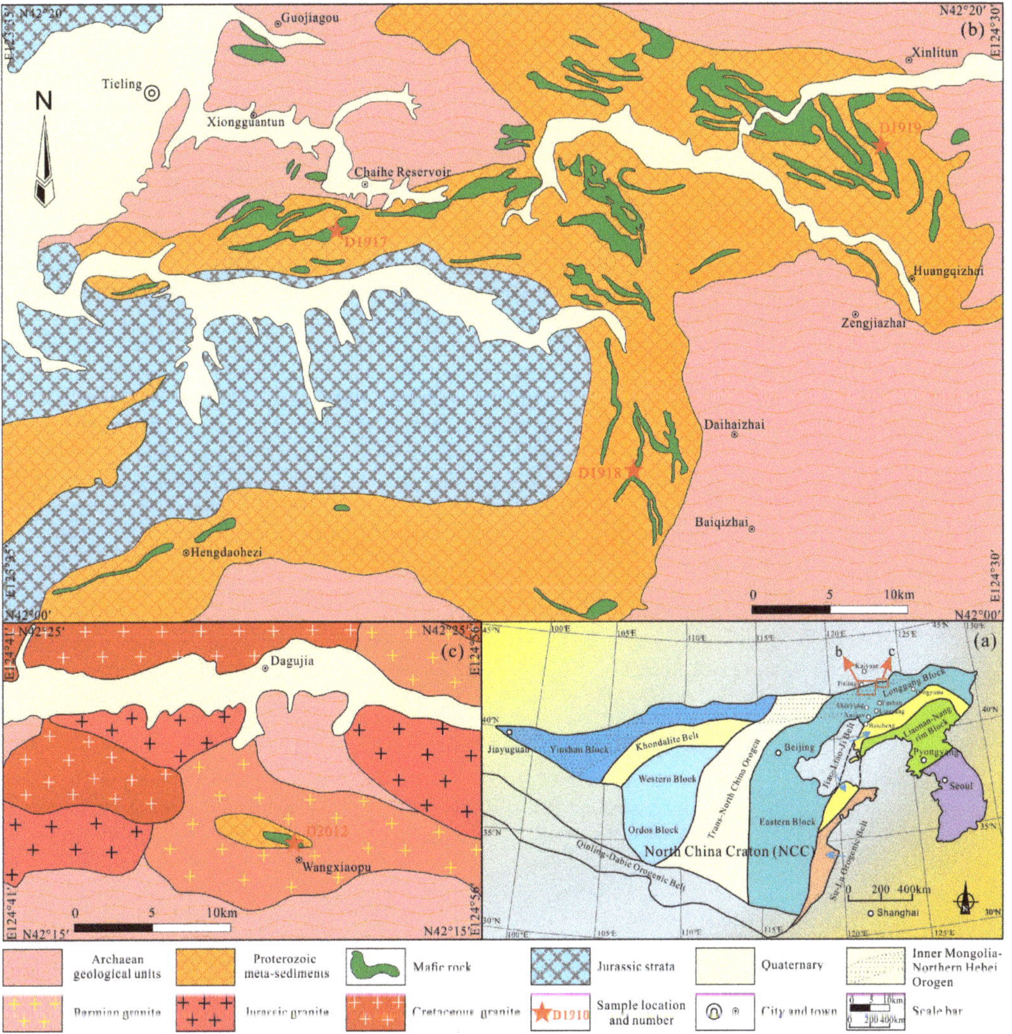

Figure 1. The tectonic location (**a**) (modified after Zhao, 2005 [1]); simplified geological map and sample location (**b**,**c**).

2. Geological Setting and Sample Descriptions

2.1. Geological Setting

The Precambrian basement of Liaodong Peninsula, located in the northeastern segment of the NCC, is composed of the Archean crystalline basement including the Liaonan–Nangrim Block in the southeast and the Longgang Block in the northwest, and the Paleoproterozoic JLJB in the center (Figure 1a; [1]). Archean to Paleoproterozoic supracrustal and granitoid rocks are exposed in the Liaonan–Nangrim Block in the southern Liaoning [34]. The Paleoproterozoic JLJB between these two blocks mainly contains metavolcano-sedimentary successions and granitic to mafic intrusions that were metamorphosed to greenschist–amphibolite facies [13–20]. The Longgang Block in the northern Liaoning exhibits extensive occurrences of tonalite–trondhjemite–granodiorite rocks (TTGs) [35–37] and basic volcanic rock [38]. These rocks, which date back approximately 3.8 Ga, have been discovered in the Anshan area. Furthermore, various zircons of magmatic, xenocrystal, and detrital origins, with ages $\geq$ 3.0 Ga, have been identified in this region [39].

The Proterozoic geological units are mainly exposed in the northern Liaoning, located from north of Shenyang–Fushun to Kaiyuan, and from Tieling to Wangxiaopu in the east. Based on rock composition, the stratigraphy is compared with the Changcheng and Jixian Systems' strata in western Liaoning, and is placed in the Mesoproterozoic. According to the current classification scheme, it is divided from bottom to top into the Daposhan Formation, Kangzhuangzi Formation, and Guanmenshan Formation of the Changcheng System, the Tongjiajie Formation, Hutouling Formation, Erdaogou Formation, Shimen Formation, and Yangshitun Formation of the Jixian System, the Yubeigou Formation and Yaomotaichong Formation of Neoproterozoic, and the Yintun Formation of the Nanhua System. These strata have undergone low-grade metamorphism and overlaid Archean gneiss with angular unconformity [33].

Large-scale mafic rocks as sills or dykes are widely distributed in northern Liaoning, generally trending east–west, and intruding into the Kangzhuangzi Formation, Guanmenshan Formation, Tongjiajie Formation, and Hutouling Formation with minor mafic intrusions into Neoarchean gneiss (Figure 1b). These mafic rocks can be subdivided into two belts: the northern belt extends from Guojia Gou to Xinlitun, and the southern belt extends from Chaihe Reservoir to Zengjiazhai. Additionally, sporadic outcrops of mafic rocks can be observed in the Xiongguantun area of Tieling City. They exhibit obvious contact metamorphic belt and chilled margins, with varying widths and abundant country rock xenoliths. The Precambrian basement in this area is covered by the Jurassic Qianwanling Formation, Nankangzhuang Formation, and Yingshugou Formation with unconformity (Figure 1b). Furthermore, through the 1:50,000 regional geological survey, a suite of two-mica schist, marble, and basalt (partially pillow-shaped) rock assemblage was identified in Wangxiaopu Village, which is in fault contact with the surrounding Permian granite and is intruded by later quartz veins (Figure 1c).

2.2. Petrological Characteristics of Mafic Rocks

In this study, 4 geochronological samples and 28 geochemical samples were collected from the intrusions of diabase, gabbro sills, and basalt associated with marble in the Proterozoic strata for analysis, respectively. The sampling locations are shown in Figure 1 and Table 1.

Sample D1917, basalt, was collected from 500 m south of Yunpangou Village, Xiongguantun Town (Figure 1b). It crops out as a pillow structure, with a fine-grained contact margin between the pillows (Figure 2a), and intrudes into the Guanmenshan Formation. It is black to grey in color with a porphyritic texture, massive structure, and a matrix of microcrystalline texture (Figure 2b,c). The phenocrysts are composed of plagioclase (~5%) and pyroxene (~5%). Plagioclase, subhedral to euhedral columnar crystals ranging from 0.5–2 mm in diameter, has undergone chloritization, epidotization, and calcitization. Pyroxene, euhedral columnar crystals, with a particle size of 0.5–2 mm, has undergone alteration into amphibole and carbonation. The matrix consists of microcrystalline plagioclase (~55%)

and pyroxene (~30%). Plagioclase is crystallized in subhedral columns with a particle size of 0.2–0.5 mm. Pyroxenes exhibit xenomorphic granular textures with a particle size of less than 0.2 mm. The accessory minerals are magnetite, apatite, etc., and the alteration minerals include chlorite, epidote, calcite, etc. (Figure 2c).

Figure 2. Occurrence and micro-pictures of mafic rocks from northern Liaoning. (**a,b,c**): D1917 basalt; (**a**): pillow-shaped basalt; (**b**): edge and center phases of pillow-shaped basalt; (**c**): microscopic characteristics of basalt; (**d–f**): D1918 diabase; (**d**): field occurrence of diabase dyke; (**e**): specimen of diabase; (**f**): microscopic characteristics of diabase; (**g–i**): D1919 gabbro; (**g**): gabbro intruded into marble; (**h**): spherical weathering of gabbro; (**i**): microscopic characteristics of gabbro; (**j–l**): D2012 amphibolite, (**j**). field occurrence of amphibolite; (**k**). amphibolite wrapped in marble in a pillow shape; (**l**): microscopic characteristics of amphibolite; Pl: plagioclase; Px: pyroxene; Ol: olivine; Hbl: hornblende; Spn: sphene; Ep: epidote.

Table 1. Location and lithology for the magmatic rocks from the Kaiyuan Area, North Liaoning.

Sample	GPS Location	Lithology	Results (Ma)
D1917	125°57′05.27″, 41°21′16.88″	Basalt	2154 ± 15
D1918	124°12′36.40″, 42°05′01.27″	Diabase	2209 ± 12
D1919	124°13′57.47″, 42°14′18.33″	Gabbro	2063 ± 7
D2012	124°51′44.51″, 42°19′09.73″	Amphibolite	2018 ± 13

Sample D1918, diabase, collected from northwest of Baiqizhai Town (Figure 1b), intrudes into the dolomite of the Guanmenshan Formation (Figure 2d) with a subhedral granular texture and massive structure (Figure 2e). The mineral components of the diabase are plagioclase (~55%), hornblende (~35%), biotite (~5%), and quartz (~3%) (Figure 2f) Plagioclase, with a subhedral columnar texture, and a polysynthetic twin, with a particle size of 0.5–2 mm and some particles of 2–3 mm, exhibits varying degrees of epidotization and zoidization. Hornblende occurs as brown–green, subhedral columnar crystals ranging from 0.2 to 2 mm in diameter, fading to light green hornblende in varying degrees. Biotite is brown, idiomorphic flaky, with a particle size of 0.2–2 mm, and exhibits varying degrees of chloritization showing a pseudomorphic or residual structure. Xenomorphic crystals of quartz fill interstices between plagioclase and hornblende grains with a particle size of <0.2 mm. Accessory minerals include magnetite, apatite, and sphene (Spn), with a content of about 2%. Alteration includes sericitization, chloritization, and epidotization (Figure 2f)

Sample D1919, gabbro, collected from north of Huangqizhai Town (Figure 1b), intrudes into dolomite (Figure 2g,h). It shows a gabbroic texture, an embedded olivine texture, and a massive structure (Figure 2i), and consists of plagioclase (~55%), pyroxene (~30%), olivine (~8%), hornblende (~5%), and biotite (~2%). Plagioclase shows a subhedral columnar texture and a polysynthetic twin, with a main particle size of 0.2–2 mm and some particles of 2–5 mm and 5–8 mm, and exhibits zoisitization. Pyroxene, subhedral columnar crystals ranging from 0.2 to 2 mm in diameter, shows an embedded olivine texture containing granular olivine, and partially shows brown amphibole reaction edges. Olivine, with a xenomorphic granular texture and a particle size of 0.1–0.5 mm, is strongly altered to serpentine, microscale biotite, magnetite, etc., retaining its structure. Colorless to green hornblende, xenomorphic to granular in shape, shows amphibole cleavage with a particle size of 0.2–0.5 mm, and has undergone varying degrees of chloritization. Biotite is brown, idiomorphic flaky, with a particle size of 0.2–0.5 mm. The accessory minerals are magnetite, ilmenite, etc. (Figure 2i).

Sample D2012, amphibolite, collected from the north of Wangxiaobao Village (Figure 1c), intruded by later felsic veins (Figure 2j), was wrapped in marble in a pillow shape in the field (Figure 2k), which displayed as the rock assemblage of oceanic islands. It is composed of primary minerals such as hornblende (~77%), plagioclase (~12%), and quartz (~6%) exhibiting a fine granular texture, columnar recrystallization texture, and gneissic structure (Figure 2l). Gray to green hornblende crystallizes as long columns with a particle size ranging from 0.05 to 85 mm. Plagioclase, subhedral-xenomorphic plate-columnar-shaped, with a particle size of 0.12–0.50 mm, shows obvious sericitization and clayification with polysynthetic twin partially. Quartz, with a xenomorphic granular texture and a particle size of 0.04–0.30 mm, shows a wavy extinction. A large amount of opaque dark minerals are distributed in ribbons and clumps, with a total content of about 5%. Alteration includes chloritization and epidotization (Figure 2l).

2.3. Zircons in These Mafic Rocks

Commonly, some researchers contend that zircons are exclusively observed in felsic rocks due to the gradual saturation of Zr (and Si) during magma evolution. Conversely, the origin of zircons from mafic rocks, particularly fine-grained basalt and diabase, is considered highly improbable [40,41]. However, in recent research, some researchers believe that zircon crystallization in low-Zr mafic magmas is possible [42]. They explored that possibility using 2D finite elements to model the crystallization of MORB melts confined in pores, and found that zircon-saturated volumes may form locally at the growing mineral–melt interfaces if the growth rate of a low Kd^{Zr} mineral (<0.2) is much faster than the diffusion rate of the rejected Zr^{4+} away into the melt, thus leading to the precipitation of zircon in low-Zr mafic magmas. Thus, zircon crystallization in low-Zr mafic magmas is perfectly possible under confined crystallization [42]. Through microscopic identification, zircons were identified in thin sections of these four mafic rocks. The micrographs are as follows (Figure 3).

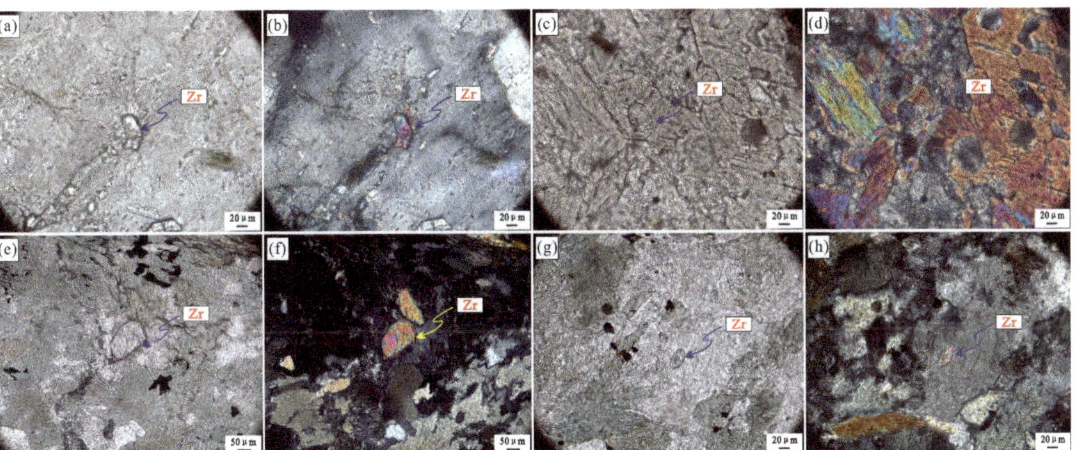

Figure 3. Micro-pictures of zircons in the mafic rocks from northern Liaoning. (**a**,**b**): D1917 zircon in pillow-shaped basalt; (**c**,**d**): D1918 zircon in diabase; (**e**,**f**): D1919 zircon in gabbro; (**g**,**h**): D2012 zircon in amphibolite; Zr: zircon.

3. Analytical Methods

3.1. Sample Preparation

Samples for geochronological analyses were first cleaned, crushed, and ground after being collected from the field. The zircon crystals were then separated from these samples using conventional heavy liquid and magnetic techniques at Langfang Yuneng Mineral Separation Co., Ltd. in Langfang, China. The separated zircons were carefully hand-picked under a binocular microscope. To examine their internal structures, the selected zircons were embedded in epoxy resin, polished, and imaged using a scanning electron microscope with cathodoluminescence (CL) at the Beijing Gaonian Navigation Technology Limited Company in Beijing, China. CL images of four samples were obtained using a CAMECA SX51 microprobe (CAMRCA, Gennevilliers, France), operating at 50 kv and 15 nA.

3.2. Zircon LA-ICP-MS U-Pb Isotope Dating

The LA-ICP-MS U-Pb isotope analyses were performed using an Agilent 7500a (Agilent Technologies Co., Ltd, Santa Clara, CA, USA) quadrupole ICP-MS with a UP-193 Solid-State Laser (193 nm, New Wave Research Inc., Shanghai, China). The laser spot size was set to 32 μm for most of the analyses, with a laser energy density of 10 J/cm^2 and a repetition rate of 8 Hz. The laser sampling procedure consisted of a 30 s blank, followed by a 30 s sampling ablation, and a 2 min sample-chamber flushing after the ablation. The ablated material was carried into the ICP-MS by a high-purity Helium gas stream with a flux of 1.15 L/min. The entire laser path was fluxed with Ar (600 mL/min) to ensure energy stability. The counting times were 20 ms for ^{204}Pb, ^{206}Pb, ^{207}Pb, and ^{208}Pb, 15 ms for ^{232}Th and ^{238}U, 20 ms for ^{49}Ti, and 6 ms for other elements. NIST 610 glass was used as an external standard and Si as an internal standard for calibrations in the zircon analyses. U-Pb isotope fractionation effects were corrected using zircon 91500 [43] as an external standard. Isotopic ratios and element concentrations of zircons were calculated using Glitter [44]. Concordia ages and diagrams were obtained using Isoplot/Ex (3.0) [44]. The common lead was corrected using LA-ICP-MS Common Lead Correction (ver. 3.15), following the method of Andersen (2002) [45]. The analytical data are presented on U-Pb Concordia diagrams with 1σ errors. The mean ages are weighted means at 95% confidence levels [44]. The analyses of four samples were conducted at the Key laboratory of Mineral Resources Evaluation in Northeast Asia, Ministry of Natural Resources. The data are listed in Table 2.

Table 2. LA-ICP-MS zircon U-Pb dating data of mafic rocks from northern Liaoning.

Sample No.	Pb (μg/g)	Th (μg/g)	U (μg/g)	Th/U	^{207}Pb/^{206}Pb Ratio	1σ	^{207}Pb/^{235}U Ratio	1σ	^{206}Pb/^{238}U Ratio	1σ	^{207}Pb/^{206}Pb Ages	1σ	^{207}Pb/^{235}U Ages	1σ	^{206}Pb/^{238}U Ages	1σ
D1917-1	147	204	204	1.00	0.16593	0.00205	11.11433	0.14256	0.48580	0.00550	2517	10	2533	12	2552	24
D1917-2	113	152	158	0.97	0.16924	0.00222	11.30379	0.15295	0.48440	0.00559	2550	10	2549	13	2546	24
D1917-3	165	151	248	0.61	0.16773	0.00201	11.23557	0.14037	0.48583	0.00546	2535	9	2543	12	2553	24
D1917-4	80	96	115	0.84	0.15909	0.00245	10.67661	0.16715	0.48672	0.00591	2446	12	2495	15	2556	26
D1917-5	104	126	150	0.84	0.16642	0.00228	11.13507	0.15676	0.48528	0.00568	2522	11	2534	13	2550	25
D1917-6	136	173	256	0.67	0.13358	0.00185	7.24344	0.10270	0.39328	0.00453	2146	11	2142	13	2138	21
D1917-7	248	208	383	0.54	0.15805	0.00210	10.60423	0.14510	0.48660	0.00562	2435	10	2489	13	2556	24
D1917-8	55	166	181	0.92	0.16471	0.00251	11.02476	0.17113	0.48546	0.00590	2505	12	2525	14	2551	26
D1917-9	107	142	175	0.81	0.17198	0.00301	11.50736	0.20313	0.48529	0.00629	2577	14	2565	16	2550	27
D1917-10	34	53	48	1.11	0.16709	0.00336	11.17697	0.22481	0.48516	0.00669	2529	17	2538	19	2550	29
D1917-11	181	241	263	0.92	0.16475	0.00252	11.02789	0.17202	0.48548	0.00593	2505	12	2525	15	2551	26
D1917-12	110	16	190	0.08	0.11216	0.00165	5.06862	0.07596	0.32775	0.00379	1835	13	1831	13	1827	18
D1917-13	56	63	84	0.75	0.16491	0.00264	11.03467	0.17975	0.48530	0.00604	2507	13	2526	15	2550	26
D1917-14	126	108	198	0.54	0.17321	0.00275	11.59110	0.18697	0.48534	0.00605	2589	13	2572	15	2550	26
D1917-15	81	85	132	0.64	0.16791	0.00233	11.23826	0.16055	0.48543	0.00575	2537	11	2543	13	2551	25
D1917-16	86	100	168	0.60	0.13479	0.00178	7.34147	0.10054	0.39503	0.00453	2161	11	2154	12	2146	21
D1917-17	40	52	67	0.78	0.16737	0.00265	11.20165	0.18090	0.48541	0.00604	2532	13	2540	15	2551	26
D1917-18	156	204	233	0.87	0.16325	0.00198	10.92303	0.13888	0.48530	0.00553	2490	10	2517	12	2550	24
D1918-1	433	579	329	1.76	0.12674	0.00165	4.73294	0.06320	0.27084	0.00302	2053	11	1773	11	1545	15
D1918-2	66	192	346	0.55	0.06855	0.00126	1.27825	0.02361	0.10524	0.00155	885	20	836	11	618	9
D1918-3	34	199	325	0.61	0.06050	0.00145	0.56447	0.01342	0.06766	0.00081	622	31	454	9	422	5
D1918-4	11	47	23	2.07	0.13647	0.00693	5.34881	0.26801	0.28956	0.00662	2183	54	1853	42	1639	33
D1918-5	105	156	295	0.53	0.13855	0.00200	7.80004	0.11473	0.40831	0.00472	2209	12	2208	13	2207	22
D1918-6	19	353	401	0.88	0.05312	0.00189	0.29153	0.01019	0.05181	0.00053	334	55	260	8	352	3
D1919-1	551	998	1166	0.86	0.12831	0.00145	6.72157	0.07992	0.37990	0.00419	2075	9	2075	11	2076	20
D1919-2	372	565	819	0.69	0.11940	0.00128	4.47229	0.05086	0.27164	0.00296	1947	9	1726	9	1549	15
D1919-3	328	1128	1005	1.12	0.11497	0.00187	3.19372	0.05222	0.20145	0.00237	1879	14	1456	13	1183	13
D1919-4	187	322	396	0.81	0.12081	0.00168	5.02826	0.07156	0.30184	0.00345	1968	12	1824	12	1700	17
D1919-5	432	597	962	0.62	0.11532	0.00150	4.79112	0.06421	0.30130	0.00338	1885	11	1783	11	1698	17
D1919-6	714	1159	871	1.33	0.12667	0.00198	6.55743	0.10378	0.37542	0.00445	2052	13	2054	14	2055	21
D1919-7	799	1533	843	1.82	0.11443	0.00129	3.09836	0.03670	0.19636	0.00214	1871	10	1432	9	1156	12
D1919-8	694	1404	1140	1.23	0.11692	0.00131	3.17513	0.03728	0.19694	0.00214	1910	9	1451	9	1159	12
D1919-9	806	181	254	0.71	0.12653	0.00137	5.71060	0.06523	0.32730	0.00355	2050	9	1933	10	1825	17
D1919-10	611	1042	1350	0.77	0.12085	0.00134	4.63869	0.05410	0.27837	0.00303	1969	9	1756	10	1583	15
D1919-11	537	698	1196	0.58	0.12378	0.00151	5.18714	0.06559	0.30390	0.00336	2011	10	1851	11	1711	17
D1919-12	399	435	930	0.47	0.11951	0.00171	4.96315	0.07215	0.30119	0.00343	1949	12	1813	12	1697	17
D1919-13	578	767	1327	0.58	0.12760	0.00138	6.64670	0.07600	0.37777	0.00409	2065	9	2066	10	2066	19
D1919-14	597	1330	1279	1.04	0.11898	0.00138	4.61516	0.05564	0.28131	0.00307	1941	10	1752	10	1598	15
D1919-15	628	1179	1345	0.88	0.12776	0.00147	6.66303	0.08020	0.37825	0.00413	2067	9	2068	11	2068	19

Table 2. *Cont.*

| Sample No. | Pb | Th | U | Th/U | Isotopic Ratios | | | | | | Ages (Ma) | | | | | |
| | | | | | $^{207}Pb/^{206}Pb$ | | $^{207}Pb/^{235}U$ | | $^{206}Pb/^{238}U$ | | $^{207}Pb/^{206}Pb$ | | $^{207}Pb/^{235}U$ | | $^{206}Pb/^{238}U$ | |
	µg/g	µg/g	µg/g		Ratio	1σ	Ratio	1σ	Ratio	1σ	Ages	1σ	Ages	1σ	Ages	1σ
D1919-16	758	1279	867	1.48	0.12795	0.00148	6.64009	0.08005	0.37640	0.00411	2070	9	2065	11	2059	19
D1919-17	247	373	539	0.69	0.12699	0.00204	6.45484	0.10483	0.36867	0.00437	2057	14	2040	14	2023	21
D1919-18	268	222	426	0.52	0.12721	0.00198	6.48693	0.10213	0.36985	0.00433	2060	13	2044	14	2029	20
D1919-19	610	836	1352	0.62	0.11767	0.00132	4.44471	0.05225	0.27399	0.00296	1921	9	1721	10	1561	15
D1919-20	763	921	791	1.16	0.12337	0.00164	4.68489	0.06379	0.27543	0.00307	2006	11	1765	11	1568	16
D2012-1	223	612	385	1.59	0.12430	0.00309	6.28771	0.16997	0.36747	0.00890	2019	21	2017	24	2017	42
D2012-2	78	97	102	0.95	0.15723	0.00568	9.88062	0.36881	0.45644	0.01234	2426	30	2424	34	2424	55
D2012-3	79	104	89	1.17	0.12421	0.01113	7.42792	0.65583	0.43433	0.01872	2018	36	2164	79	2325	84
D2012-4	53	141	144	0.98	0.12424	0.01331	4.55272	0.46856	0.26612	0.01318	2018	23	1741	86	1521	67
D2012-5	322	647	422	1.53	0.12418	0.00274	5.65294	0.13968	0.33053	0.00789	2017	20	1924	21	1841	38
D2012-6	83	117	178	0.66	0.12429	0.00296	6.17418	0.16200	0.36066	0.00872	2019	21	2001	23	1985	41
D2012-7	338	376	628	0.60	0.15747	0.00470	9.93961	0.31371	0.45823	0.01188	2429	24	2429	29	2432	53
D2012-8	98	151	155	0.98	0.15584	0.00733	9.78661	0.46316	0.45576	0.01418	2411	41	2415	44	2421	63
D2012-9	262	786	490	1.60	0.12434	0.00329	6.02196	0.17227	0.35136	0.00873	2019	23	1979	25	1941	42
D2012-10	490	1059	936	1.13	0.12424	0.00285	4.68979	0.12025	0.27383	0.00664	2018	20	1765	21	1560	34
D2012-11	301	263	553	0.48	0.15727	0.00436	9.87122	0.29528	0.45526	0.01168	2427	23	2423	28	2419	52
D2012-12	169	155	319	0.48	0.15791	0.00332	9.89645	0.23891	0.45448	0.01100	2433	19	2425	22	2415	49
D2012-13	586	999	675	1.48	0.12436	0.00279	5.90907	0.14947	0.34455	0.00839	2020	20	1963	22	1909	40
D2012-14	87	136	211	0.64	0.12432	0.01120	5.79273	0.51744	0.33781	0.01512	2019	16	1945	77	1876	73
D2012-15	313	1264	735	1.72	0.12418	0.00345	6.76699	0.20284	0.39505	0.01003	2017	24	2081	27	2146	46
D2012-16	166	178	302	0.59	0.15738	0.00401	9.93189	0.27757	0.45748	0.01156	2428	21	2428	26	2428	51
D2012-17	87	89	111	0.80	0.15522	0.01408	9.77099	0.86913	0.45628	0.02187	2404	38	2413	82	2423	97

3.3. Major and Trace Element Analyses

After conducting major, trace, and rare earth element (REE) analyses for 28 samples, the results were obtained from the Northeast China Supervision and Inspection Center of Mineral Resources, Ministry of Natural Resources (Shenyang, China). The samples underwent petrographic examination and the altered rock surfaces were removed before being crushed and ground to 74 µm using an agate mill. The major element contents of the whole rock were determined using X-ray fluorescence spectrometry (XRF), with analytical precisions exceeding 2%. The trace element and REE contents were determined using X Series II ICP-MS from American Thermal Power Company in Salt Lake City Utah State. A total of 0.1 g of the sample was placed in a digestion tank, with 1 mL of concentrated nitric acid and 1 mL of hydrofluoric acid added. The digestion tank was placed in an oven and was heated up to 180 degrees for 10–12 h. After removal, it was heated at 120 degrees on an electric heating plate in an open environment. When about 2–3 mL of the digestion solution was left, it was heated up to 240 degrees. After redissolving, 0.5% dilute nitric acid was added to the mark for measurement. Twelve measurements were conducted on samples, with analytical precisions exceeding 5% for elements with contents of >10 µg/g, exceeding 8% for elements with contents of <10 µg/g, and 10% for the transition metals. The results are listed in Table 3.

Table 3. Major element (%), rare element, and trace element ($\mu g/g$) composition of mafic rocks from northern Liaoning.

Sample No.	D1918-1	D1918-2	D1918-3	D1918-4	D1918-5	D1918-6	D1917-1	D1917-2	D1917-3	D1917-4	D1917-5	D1917-6	D1917-7	D1917-8
Lithology				Diabase							Basalt			
SiO_2	51.11	52.86	50.22	50.13	49.04	48.60	47.51	49.81	48.59	50.38	49.71	50.42	47.03	51.18
TiO_2	1.60	1.63	1.59	1.61	1.65	1.61	2.35	2.36	2.51	2.49	2.36	2.36	2.37	2.33
Al_2O_3	13.34	12.95	13.41	13.13	13.14	13.33	14.36	14.08	14.90	14.41	14.13	13.86	14.20	13.56
Fe_2O_3	1.85	1.39	2.96	3.22	3.31	3.69	3.72	2.18	2.86	2.38	2.54	1.79	2.84	0.40
FeO	11.82	11.40	11.54	11.24	12.06	11.60	12.08	11.99	12.18	11.39	11.77	11.90	12.28	12.13
MnO	0.19	0.19	0.22	0.22	0.24	0.22	0.22	0.20	0.20	0.18	0.20	0.20	0.21	0.18
MgO	6.30	6.23	6.28	6.52	6.73	6.91	6.80	6.80	6.67	7.05	6.87	6.90	6.30	6.76
CaO	9.55	9.18	9.34	9.61	9.48	9.98	6.69	6.23	6.02	5.40	6.41	6.55	8.79	7.59
Na_2O	2.01	1.97	2.18	1.96	2.08	1.91	4.24	4.45	4.31	4.29	4.42	4.40	3.38	3.60
K_2O	0.80	1.04	0.99	1.12	0.76	0.92	0.39	0.34	0.37	0.33	0.33	0.31	0.60	0.44
P_2O_5	0.18	0.17	0.18	0.18	0.20	0.20	0.27	0.25	0.27	0.24	0.25	0.24	0.26	0.22
LOI	1.23	1.01	1.11	1.05	1.30	1.08	1.34	1.24	1.15	1.44	1.11	1.15	1.72	1.65
SUM	99.97	100.01	100.01	99.99	99.98	100.03	99.96	99.93	100.03	99.98	100.09	100.07	99.97	100.04
$Mg^{\#}$	45.69	46.99	44.32	45.35	44.63	45.45	44.25	46.74	44.86	48.41	46.78	47.91	43.33	49.36
A/NK	3.2	2.97	2.88	2.95	3.1	3.23	1.94	1.83	1.99	1.95	1.85	1.83	2.29	2.12
A/CNK	0.74	0.71	0.72	0.69	0.73	0.86	0.91	0.9	0.99	1.08	0.88	0.85	0.81	0.86
R1	2412	2503	2226	2257	2227	2212	1417	1544	1483	1635	1540	1616	1663	1959
R2	1593	1545	1574	1609	1606	1671	1336	1282	1266	1211	1302	1314	1532	1413
La	13.1	13.1	13.4	12.2	12.3	12.5	19.0	18.2	20.3	20.0	18.9	20.3	19.5	18.81
Ce	26.9	26.6	27.1	25.3	27.3	25.4	39.8	41.1	42.9	41.1	39.9	42.5	40.7	40.45
Pr	3.8	3.7	3.80	3.69	3.6	3.57	5.96	5.66	6.44	6.17	5.85	6.19	6.04	6.01
Nd	17.3	16.7	17.2	16.6	16.2	16.4	26.4	25.4	28.5	27.8	25.8	27.8	27.1	27.2
Sm	4.56	4.30	4.42	4.34	4.17	4.20	6.04	5.91	6.50	6.44	5.94	6.38	6.28	6.30
Eu	1.57	1.44	1.42	1.40	1.35	1.39	1.56	1.41	1.53	1.43	1.56	1.69	1.71	1.72
Gd	4.69	4.62	4.64	4.56	4.37	4.56	5.54	5.29	5.88	5.66	5.32	5.72	5.70	5.68
Tb	1.01	0.97	0.98	0.97	0.92	0.93	1.04	0.98	1.05	1.03	1.02	1.06	1.04	1.02
Dy	6.10	6.01	6.07	6.05	5.83	5.87	5.76	5.41	5.81	5.58	5.54	5.76	5.66	5.62
Ho	1.26	1.21	1.24	1.24	1.20	1.19	1.04	1.02	1.07	1.01	1.04	1.06	1.06	1.06
Er	3.51	3.45	3.63	3.54	3.35	3.36	2.72	2.64	2.73	2.69	2.81	2.80	2.71	2.74
Tm	0.55	0.54	0.54	0.55	0.51	0.53	0.40	0.39	0.42	0.40	0.42	0.41	0.41	0.41
Yb	3.71	3.60	3.71	3.61	3.42	3.51	2.45	2.50	2.62	2.55	2.62	2.54	2.57	2.56
Lu	0.55	0.53	0.54	0.54	0.52	0.52	0.35	0.35	0.37	0.37	0.37	0.38	0.37	0.38
ΣREE	88.6	86.9	88.6	84.6	85.0	84.0	118.1	116.2	126.1	122.2	117.1	124.5	120.8	119.9
LREE	67.3	66.0	67.3	63.6	64.9	63.5	98.8	97.6	106.1	102.9	97.9	104.8	101.3	100.5
HREE	21.4	20.9	21.3	21.0	20.1	20.5	19.3	18.6	20.0	19.3	19.1	19.7	19.5	19.5
LREE/HREE	3.15	3.15	3.15	3.02	3.23	3.10	5.12	5.26	5.32	5.34	5.12	5.31	5.19	5.16
La_N/Yb_N	2.54	2.62	2.60	2.43	2.58	2.56	5.55	5.22	5.56	5.63	5.17	5.72	5.43	5.26
Eu/Eu^*	1.04	0.99	0.96	0.96	0.96	0.97	0.82	0.77	0.76	0.72	0.85	0.86	0.87	0.88
Ce/Ce^*	0.93	0.93	0.93	0.92	1.01	0.93	0.92	0.99	0.92	0.91	0.93	0.93	0.92	0.93

Table 3. *Cont.*

Sample No.	D1918-1	D1918-2	D1918-3	D1918-4	D1918-5	D1918-6	D1917-1	D1917-2	D1917-3	D1917-4	D1917-5	D1917-6	D1917-7	D1917-8
Lithology				Diabase							Basalt			
Y	26.9	27.6	26.7	27.6	26.3	26.7	24.7	23.6	24.6	24.4	25.0	25.5	25.2	25.02
Sc	43	43	42	44	44	44	32	33	34	34	33	33	33	33
V	311	317	307	308	324	330	338	349	376	373	353	356	349	348
Cr	108	111	94	103	116	106	112	117	125	130	123	126	124	123
Co	49.8	49.9	50.5	51.7	52.2	53.1	42.3	41.3	41.5	42.3	51.2	50.7	51.6	49.1
Ni	71.3	73.6	71.9	75.5	75.1	74.2	75.1	67.7	72.0	75.1	68.9	70.1	71.9	73.0
Be	0.8	0.8	0.8	0.8	0.8	0.9	1.6	1.5	1.4	1.5	1.6	1.5	1.0	1.11
Rb	20	30	21	31	14	23	4	5	5	4	4	4	7	6
Sr	188	165	162	164	160	176	210	209	272	273	215	232	284	263
Ba	152	218	276	250	193	138	144	128	185	200	140	151	334	286
Zr	109	104	106	102	106	106	144	149	154	162	150	162	151	148
Nb	13.5	13.0	13.7	13.3	12.2	12.0	20.3	21.1	21.7	21.5	19.8	21.7	21.8	21.3
Hf	3.58	3.43	3.60	3.44	3.50	3.50	4.50	4.69	4.84	5.06	4.91	4.99	4.70	4.65
Ta	0.82	0.79	0.82	0.84	0.86	0.85	1.23	1.30	1.35	1.41	1.30	1.24	1.21	1.21
Th	1.86	1.67	1.73	1.53	1.68	3.62	1.31	1.36	1.73	1.71	1.35	3.47	1.86	1.67
U	0.4	0.4	0.4	0.4	0.4	0.4	0.4	0.3	0.4	0.4	0.3	0.8	0.6	0.39
$(La/Sm)_N$	1.86	1.97	1.96	1.81	1.90	1.93	2.03	1.99	2.01	2.01	2.05	2.05	2.00	1.93
La/Sm	2.88	3.05	3.04	2.81	2.94	2.98	3.14	3.08	3.12	3.11	3.18	3.17	3.10	2.99
Ta/Yb	0.22	0.22	0.22	0.23	0.25	0.24	0.50	0.52	0.51	0.55	0.50	0.49	0.47	0.47
Th/Yb	0.50	0.46	0.47	0.42	0.49	1.03	0.53	0.54	0.66	0.67	0.52	1.37	0.72	0.65
Ce/Yb	7.26	7.40	7.30	7.02	8.00	7.22	16.2	16.4	16.4	16.1	15.2	16.8	15.9	15.8
La/Yb	3.54	3.65	3.62	3.38	3.60	3.57	7.74	7.28	7.74	7.84	7.21	7.98	7.57	7.34
Sc/Ni	0.50	0.58	0.59	0.58	0.59	0.59	0.42	0.48	0.48	0.46	0.48	0.47	0.46	0.45
Zr/Y	4.04	3.77	3.99	3.70	4.05	3.96	5.84	6.33	6.28	6.66	5.99	6.38	6.00	5.91
Nb/Y	0.50	0.47	0.51	0.48	0.46	0.45	0.82	0.89	0.88	0.88	0.79	0.85	0.87	0.85
Nb/Th	7.24	7.79	7.90	8.71	7.29	3.33	15.5	15.5	12.6	12.5	14.6	6.2	11.8	12.8
La/Nb	0.98	1.01	0.98	0.92	1.01	1.04	0.94	0.86	0.93	0.93	0.95	0.93	0.89	0.88
Ba/Nb	11.30	16.72	20.16	18.81	15.76	11.48	7.12	6.07	8.52	9.30	7.06	6.96	15.3	13.4
Th/Yb	0.50	0.46	0.47	0.42	0.49	1.03	0.53	0.54	0.66	0.67	0.52	1.37	0.72	0.65
Zr/Nb	8.06	7.99	7.78	7.68	8.71	8.78	7.11	7.08	7.11	7.56	7.56	7.50	6.91	6.94
Th/Zr	0.02	0.02	0.02	0.01	0.02	0.03	0.01	0.01	0.01	0.01	0.01	0.02	0.01	0.01
Ba/La	11.59	16.57	20.53	20.54	15.68	11.02	7.61	7.04	9.12	9.97	7.41	7.45	17.2	15.2
La/Yb	3.54	3.65	3.62	3.38	3.60	3.57	7.74	7.28	7.74	7.84	7.21	7.98	7.57	7.34
Zr/Hf	30.36	30.25	29.55	29.69	30.39	30.20	32.05	31.86	31.82	32.06	30.55	32.58	32.11	31.83
Hf/Ta	4.34	4.34	4.40	4.11	4.06	4.13	3.65	3.61	3.60	3.59	3.77	4.02	3.87	3.84
La/Ta	15.95	16.59	16.43	14.56	14.24	14.79	15.41	14.01	15.06	14.21	14.51	16.33	16.02	15.52
Th/Ta	2.26	2.11	2.12	1.83	1.94	4.27	1.06	1.05	1.28	1.21	1.04	2.80	1.53	1.38
Ta/Hf	0.23	0.23	0.23	0.24	0.25	0.24	0.27	0.28	0.28	0.28	0.27	0.25	0.26	0.26
Th/Hf	0.52	0.49	0.48	0.44	0.48	1.03	0.29	0.29	0.36	0.34	0.28	0.70	0.40	0.36

Table 3. *Cont.*

Sample No.	D1917-9	D1917-10	D1919-1	D1919-2	D1919-3	D1919-4	D1919-5	D1919-6	D2012-1	D2012-2	D2012-3	D2012-4	D2012-5	D2012-6
Lithology	Basalt				Gabbro							Amphibolite		
SiO_2	47.72	50.34	48.83	51.17	47.03	51.02	48.93	51.44	45.29	46.10	48.37	46.63	45.46	46.48
TiO_2	2.36	2.38	2.15	2.16	2.21	2.14	2.19	2.14	1.90	2.00	1.73	1.99	2.29	1.70
Al_2O_3	14.33	13.97	15.03	14.55	15.10	14.70	14.98	14.83	14.46	14.82	14.26	14.07	14.05	15.00
Fe_2O_3	2.99	1.71	3.61	2.39	5.67	2.36	4.11	2.87	2.36	2.53	1.88	1.14	2.34	2.52
FeO	11.76	11.52	9.24	9.65	9.15	9.86	9.35	9.20	11.35	11.94	11.57	13.10	13.29	9.24
MnO	0.20	0.18	0.19	0.19	0.17	0.16	0.15	0.15	0.19	0.24	0.21	0.21	0.22	0.16
MgO	6.54	6.30	6.18	6.38	6.31	6.17	6.12	5.87	7.17	7.65	7.11	8.51	7.85	6.72
CaO	7.80	7.55	7.35	7.12	7.45	7.02	7.34	6.75	11.45	9.32	8.35	8.74	8.19	13.98
Na_2O	3.40	3.77	3.30	3.38	3.22	3.33	3.44	3.45	2.05	3.02	3.29	2.62	3.07	1.61
K_2O	0.64	0.48	1.68	1.57	1.89	1.78	1.68	1.80	1.78	1.14	1.60	1.09	1.17	1.15
P_2O_5	0.26	0.24	0.39	0.36	0.43	0.38	0.41	0.38	0.19	0.20	0.17	0.19	0.22	0.16
LOI	1.97	1.59	2.02	1.02	1.36	1.11	1.22	1.05	1.76	1.08	1.40	1.71	1.81	1.34
SUM	99.99	100.05	99.96	99.94	99.99	100.04	99.93	99.94	99.93	100.03	99.94	100.01	99.97	100.05
$Mg^{\#}$	44.87	46.47	47.11	49.33	44.34	48.10	45.78	47.26	48.91	49.22	49.11	52.03	47.84	51.23
A/NK	2.28	2.08	2.08	2.00	2.06	1.98	2.01	1.94	2.73	2.39	2.00	2.56	2.22	3.85
A/CNK	0.94	0.86	1.00	0.86	0.90	0.88	0.88	0.90	0.67	0.74	0.76	0.83	0.85	0.58
R1	1687	1796	1546	1727	1371	1684	1488	1655	1765	1617	1591	1840	1547	2148
R2	1441	1394	1389	1364	1407	1345	1384	1305	1865	1667	1527	1633	1542	2122
La	19.87	20.17	22.94	19.91	22.20	21.69	21.20	23.28	12.58	12.56	11.95	13.95	13.42	11.02
Ce	41.42	42.64	43.30	38.81	43.12	41.29	41.12	43.74	28.70	28.83	26.09	32.84	30.75	23.93
Pr	6.19	6.35	6.06	5.45	6.05	5.78	5.70	6.19	4.22	4.45	3.84	4.57	4.39	3.50
Nd	27.55	28.67	26.64	24.08	26.32	25.13	24.91	27.15	17.63	19.56	15.88	19.33	19.15	15.07
Sm	6.32	6.54	5.57	5.15	5.77	5.27	5.23	5.61	4.59	4.94	4.30	4.73	5.30	3.70
Eu	1.72	1.64	2.31	2.11	2.38	2.15	2.24	2.28	1.50	1.60	1.42	1.72	1.64	1.42
Gd	5.84	5.88	5.13	4.71	5.04	4.82	4.75	5.08	3.86	4.73	4.25	4.56	4.78	3.98
Tb	1.06	1.09	0.88	0.82	0.91	0.84	0.84	0.89	0.60	0.76	0.66	0.78	0.90	0.55
Dy	5.74	5.87	4.62	4.40	4.82	4.55	4.34	4.64	3.82	4.08	3.38	4.00	4.89	3.64
Ho	1.06	1.10	0.88	0.86	0.90	0.86	0.82	0.89	0.73	0.71	0.70	0.75	0.75	0.74
Er	2.72	2.90	2.28	2.23	2.34	2.27	2.15	2.28	1.88	2.13	1.75	2.17	2.48	1.94
Tm	0.41	0.42	0.35	0.33	0.34	0.34	0.33	0.34	0.32	0.30	0.25	0.32	0.36	0.26
Yb	2.60	2.69	2.24	2.09	2.19	2.15	2.01	2.19	1.95	2.19	1.64	1.81	2.03	1.79
Lu	0.37	0.38	0.31	0.30	0.32	0.31	0.29	0.31	0.23	0.31	0.28	0.30	0.33	0.24
ΣREE	122.9	126.3	123.5	111.3	122.7	117.4	115.9	124.9	82.6	87.1	76.4	91.8	91.2	71.8
LREE	103.1	106.0	106.8	95.5	105.8	101.3	100.4	108.2	69.2	71.9	63.5	77.1	74.7	58.6
HREE	19.8	20.3	16.7	15.8	16.9	16.1	15.5	16.6	13.4	15.2	12.9	14.7	16.5	13.1
LREE/HREE	5.21	5.21	6.40	6.06	6.28	6.28	6.46	6.51	5.17	4.73	4.92	5.25	4.52	4.47
La_N/Yb_N	5.48	5.39	7.36	6.82	7.28	7.25	7.55	7.63	4.64	4.12	5.24	5.52	4.74	4.42
Eu/Eu*	0.87	0.81	1.32	1.31	1.35	1.30	1.37	1.31	1.09	1.01	1.01	1.13	0.99	1.13
Ce/Ce*	0.92	0.92	0.90	0.91	0.91	0.90	0.92	0.89	0.97	0.94	0.94	1.01	0.98	0.95

Table 3. *Cont.*

Sample No.	D1917-9	D1917-10	D1919-1	D1919-2	D1919-3	D1919-4	D1919-5	D1919-6	D2012-1	D2012-2	D2012-3	D2012-4	D2012-5	D2012-6
Lithology	Basalt				Gabbro							Amphibolite		
Y	25.78	26.19	22.19	20.70	22.47	21.39	21.24	22.17	18.01	19.70	17.93	19.16	22.90	17.96
Sc	32	32	27	27	27	27	27	27	24	25	25	26	30	22
V	346	337	293	297	310	303	299	294	250	264	248	267	306	228
Cr	122	127	99	99	107	167	105	101	140	123	137	127	125	126
Co	51.3	46.4	68.5	60.0	61.6	56.7	66.9	62.7	50.3	45.8	46.8	50.7	53.2	43.8
Ni	70.8	69.4	82.3	75.2	66.7	72.4	69.7	77.6	105.3	96.6	97.1	107.7	91.5	91.8
Be	1.09	1.09	0.95	0.98	0.94	1.03	0.94	0.97	1.07	1.19	1.23	1.20	1.13	1.58
Rb	8.36	6.16	38.05	35.93	40.14	40.37	38.06	42.54	41.35	30.96	51.02	33.16	28.29	27.77
Sr	251	235	439	429	372	406	412	423	154	327	413	272	370	138
Ba	362	344	715	737	682	839	650	733	59	179	99	175	115	47
Zr	157	158	117	101	111	111	100	117	112	116	102	124	135	106
Nb	21.1	20.2	9.0	9.0	9.2	8.6	10.7	10.2	11.3	11.4	10.0	11.1	13.1	10.2
Hf	4.83	5.00	3.65	3.36	3.61	3.57	3.11	3.64	3.25	3.73	3.13	3.83	3.89	2.96
Ta	1.25	1.26	0.63	0.62	0.60	0.61	0.64	0.67	0.87	0.97	0.63	0.96	0.67	0.44
Th	1.68	1.88	2.45	2.04	2.13	1.65	1.67	2.05	0.94	1.16	0.89	1.19	0.94	0.76
U	0.36	0.33	0.37	0.42	0.42	0.29	0.32	0.34	0.25	0.25	0.16	0.23	0.24	0.16
$(La/Sm)_N$	2.03	1.99	2.66	2.50	2.48	2.66	2.62	2.68	1.77	1.64	1.80	1.90	1.63	1.92
La/Sm	3.14	3.08	4.12	3.87	3.85	4.12	4.06	4.15	2.74	2.54	2.78	2.95	2.53	2.98
Ta/Yb	0.48	0.47	0.28	0.29	0.27	0.28	0.32	0.30	0.45	0.45	0.39	0.53	0.33	0.25
Th/Yb	0.65	0.70	1.10	0.97	0.97	0.77	0.83	0.94	0.48	0.53	0.54	0.66	0.46	0.43
Ce/Yb	15.9	15.9	19.4	18.5	19.7	19.2	20.4	20.0	14.7	13.2	16.0	18.1	15.1	13.4
La/Yb	7.64	7.51	10.26	9.51	10.15	10.11	10.53	10.64	6.46	5.74	7.31	7.70	6.61	6.16
Sc/Ni	0.46	0.46	0.33	0.36	0.41	0.37	0.39	0.34	0.23	0.26	0.25	0.24	0.33	0.24
Zr/Y	6.07	6.02	5.26	4.90	4.95	5.20	4.72	5.26	6.23	5.91	5.68	6.49	5.90	5.89
Nb/Y	0.82	0.77	0.40	0.43	0.41	0.40	0.50	0.46	0.63	0.58	0.56	0.58	0.57	0.57
Nb/Th	12.55	10.74	3.66	4.40	4.30	5.22	6.41	4.96	12.03	9.86	11.27	9.35	13.97	13.37
La/Nb	0.94	1.00	2.56	2.22	2.43	2.52	1.98	2.28	1.12	1.10	1.20	1.25	1.03	1.08
Ba/Nb	17.1	17.0	79.8	82.3	74.5	97.6	60.8	71.9	5.2	15.7	10.0	15.7	8.8	4.6
Th/Yb	0.55	0.70	1.10	0.97	0.97	0.77	0.83	0.94	0.48	0.53	0.54	0.66	0.46	0.43
Zr/Nb	7.42	7.80	13.01	11.31	12.16	12.93	9.39	11.45	9.96	10.21	10.20	11.17	10.34	10.35
Th/Zr	0.01	0.01	0.02	0.02	0.02	0.01	0.02	0.02	0.01	0.01	0.01	0.01	0.01	0.01
Ba/La	18.2	17.1	31.2	37.0	30.7	38.7	30.7	31.5	4.7	14.2	8.33	12.54	8.56	4.26
La/Yb	7.54	7.51	10.26	9.51	10.15	10.11	10.53	10.64	6.46	5.74	7.31	7.70	6.61	6.16
Zr/Hf	32.42	31.54	31.94	30.12	30.83	31.17	32.23	32.09	34.54	31.26	32.57	32.44	34.72	35.68
Hf/Ta	3.37	3.97	5.80	5.46	6.02	5.85	4.83	5.46	3.72	3.83	4.93	3.98	5.79	6.68
La/Ta	15.92	16.02	36.47	32.33	37.02	35.60	32.90	34.92	14.43	12.90	18.85	14.49	19.97	24.84
Th/Ta	1.35	1.50	3.90	3.31	3.55	2.70	2.59	3.08	1.07	1.19	1.40	1.24	1.39	1.72
Ta/Hf	0.26	0.25	0.17	0.18	0.17	0.17	0.21	0.18	0.27	0.26	0.20	0.25	0.17	0.15
Th/Hf	0.35	0.38	0.67	0.61	0.59	0.46	0.54	0.57	0.29	0.31	0.28	0.31	0.24	0.26

4. Analytical Results

4.1. Zircon U-Pb Geochronology

Sample D1917 is a basalt collected from Xiongguantun Town Yunpangou Village in Tieling City. Eighteen isotopic analyses were conducted on 18 zircons from this sample, and can be divided into three groups. Dark edges with different widths occurred around most zircons from the first group, which may be the proliferative edges formed by recrystallization during metamorphism. They are mostly self-shaped rhombohedral with diameters of 60–140 μm and length–width ratios of 1.5:1 to 3:1, generally black, clear oscillatory zoning (Figure 4), and the combination with their higher Th/U ratios (0.54–1.11), higher contents of total rare earth elements (REEs) (average contents of 585.65 μg/g), Nb, and Ta (average contents of 1.94 μg/g and 0.58 μg/g), and a significant positive Ce anomaly and negative Eu anomaly (Figure 5a), indicates a magmatic origin [46–49]. Fifteen analyses yielded an upper intercept age of 2550 ± 16 Ma (n = 15, MSWD = 0.1), which is consistent with the ^{207}Pb/^{206}Pb weighted average age of 2544 ± 18 Ma (n = 9, MSWD = 3.5) (Figure 6a). The age of the first group is interpreted as the captured zircons from the surrounding Archean rocks. The second group is composed of two zircons (6, 16) with diameters of 60–70 μm and length–width ratios around 2:1 (Figure 4), which combined with their evident internal structure and Th/U ratios (0.60, 0.67), higher REEs (average contents of 840.5 μg/g), Nb, and Ta (average contents of 1.65 μg/g and 0.53 μg/g), and a significant positive Ce anomaly and negative Eu anomaly (Figure 5a), indicate a magmatic origin [46–49]. Two analyses yield a concordant weighted average ^{207}Pb/^{206}Pb age of 2154 ± 15 Ma (n = 2, MSWD = 0.93) (Figure 6a). Thus, this represents the magmatic crystallization age. Only one zircon (data 12) belongs to the third group. It exhibits rounded and embayed boundaries, largely homogeneous central regions with an unzoned internal structure (Figure 4) and low ratio of Th/U (0.08), lower contents of REEs (127.03 μg/g), Nb, and Ta (0.04 μg/g and 0.02 μg/g), and a flat Ce and Eu distribution (Figure 5a), which indicate a metamorphic zircon [46–49]. This zircon has a concordant ^{207}Pb/^{206}Pb age of 1831 ± 13 Ma (Figure 6a) and is interpreted as the time of late metamorphism.

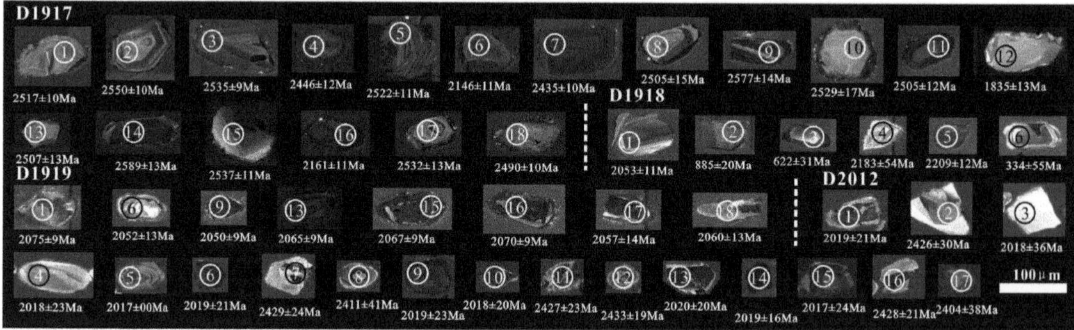

Figure 4. Cathodoluminescence (CL) images of the selected zircons from the mafic rocks in northern Liaoning. The circles on zircons represent analyzed spots.

Only six zircons have been separated from the D1918 diabase sample, and they are euhedral stubby prisms (length/width ratios 1.5:1–3:1), clear and colorless to light-brown, ranging in size from 50 to 120 μm, with larger grains being fragments presumably broken during processing. In cathodoluminescence (CL) imaging, the internal structures vary from high-CL to low-CL contrast fine oscillatory zoning in both CL-bright and CL-dark grains (Figure 4). Combined with high Th/U ratios (0.60, 0.67), higher contents of REEs (average contents of 1271.34 μg/g), Nb, and Ta (average contents of 1.96 μg/g and 0.76 μg/g), and a significant positive Ce anomaly and negative Eu anomaly (Figure 5b), this indicates that the zircons in this sample are magmatic-derived zircon [46–49]. Six analyses yielded an upper intercept age of 2203 ± 50 Ma (n = 6, MSWD = 5.7), which is consistent with the ^{207}Pb/^{206}Pb age (2209 ± 12 Ma) in the concordant line (Figure 6b). Thus, this age is interpreted as the time of magmatic emplacement.

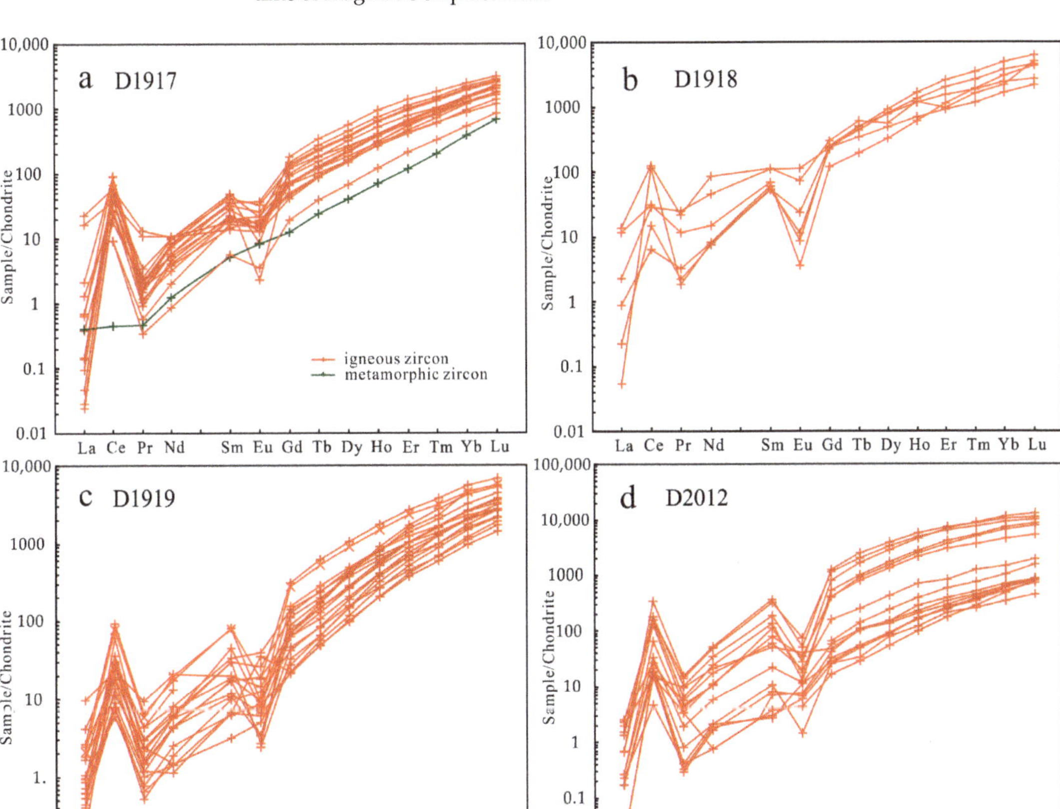

Figure 5. Chondrite-normalized REE distribution diagrams for different zircons from mafic rocks in northern Liaoning.

Zircons from the gabbro sample D1919 are euhedral prisms ranging in size from 50 to 130 μm with length/width ratios from 2:1 to 4:1 with some rounding of their apices. Although exhibiting a low contrast, CL imaging of the internal structure reveals dominantly fine oscillatory zoning (Figure 4). With the Th/U ratios ranging from 0.47 to 1.82, higher contents of REEs (average contents of 933.35 μg/g), Nb, and Ta (average contents of 2.97 μg/g and 1.19 μg/g), and a significant positive Ce anomaly and negative Eu anomaly (Figure 5c), these zircons were considered to be of magmatic origin [46–49]. A total of 20 U-Pb isotopic analyses were conducted on 20 zircons from this sample, yielding an upper intercept ^{207}Pb/^{206}Pb age of 2055 ± 23 Ma (n = 20, MSWD = 3.0), which is consistent with the weighted average ^{207}Pb/^{206}Pb age of 2063 ± 7 Ma (n = 8, MSWD = 0.18) from eight zircons on the concordant line (Figure 6c). This age is interpreted as the crystallization age of the gabbro.

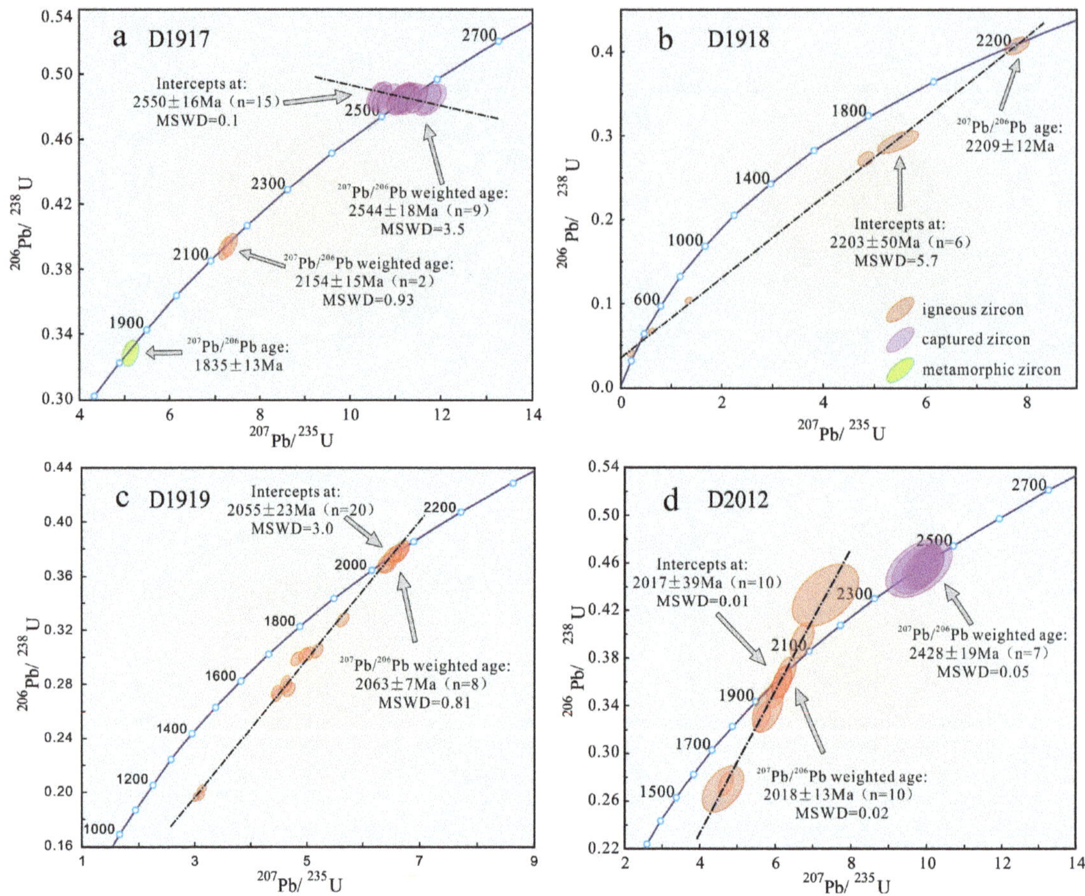

Figure 6. Concordia diagrams for zircons analysed from mafic rocks in the northern Liaoning.

Sample D2012, which was collected from Wangxiaopu Village, yielded prismatic and sub-angular zircon ranging in size from 35 μm to 120 μm with length/width ratios from 1:1 to 4:1. The zircon grains contain numerous inclusions and are divided into two groups according to the CL images (Figure 4). The rounded or irregular cores and white rims are present in these grains from the first group and are CL-dark with no obvious zoning. The ratios of Th/U range from 0.48 to 0.98, with high contents of REEs (average contents of 232.59 μg/g), Nb, and Ta (average contents of 0.95 μg/g and 0.48 μg/g), and a significant positive Ce anomaly and negative Eu anomaly (Figure 5d) indicate a magmatic origin [46–49]. Seven analyses were obtained from seven zircons, yielding a concordant $^{207}Pb/^{206}Pb$ age of 2428 ± 19 Ma (n = 7, MSWD = 0.05) (Figure 6d). Due to the small size of zircon, both the core and rim were collected during sampling, so this age represents a mixed age, and these zircons are interpreted as the captured zircons from the surrounding Archean rocks. The other groups of zircons generally have an evident internal structure with fine, oscillatory zoning and Th/U ratios (0.59–1.72), higher contents of REEs (average contents of 2736.25 μg/g), Nb and Ta (average contents of 13.40 μg/g and 1.81 μg/g), and a significant positive Ce anomaly and negative Eu anomaly (Figure 5d), which show the characteristics of magmatic origin zircon [46–49]. Ten analyses were concordant to near-concordant and yielded an upper intercept $^{207}Pb/^{206}Pb$ age of 2017 ± 39 Ma (n = 10, MSWD = 0.01), and a weighted average $^{207}Pb/^{206}Pb$ age of 2018 ± 13 Ma (n = 10, MSWD = 0.02) was obtained from these same ten grains (Figure 6d). This age is interpreted as the magmatic protolith age.

In summary, all the zircons separated from mafic rock samples in northern Liaoning of the northern margin of the NCC record four distinct age groups including the first basalt (~2209 Ma (D1918)), the second diabase (~2154 Ma (D1917)), the third gabbro (~2063 Ma (D1919), and the forth amphibolite (~2018 Ma (D2012)), which represent the emplacement ages for the mafic rocks, respectively.

4.2. Major and Trace Element Geochemistry

The average SiO_2 content of the first three periods of mafic rocks is 49.27–50.33 wt%, while the fourth amphibolite has a lower average SiO_2 content (46.39 wt%). The Al_2O_3 content is moderate (13.22–14.86 wt%). The total alkali content ($K_2O + Na_2O$) is moderate, with the lowest being 2.95 wt% in the first basalt and ranging from 3.93 to 5.08 wt% in the other three stages of mafic rocks. On the TAS diagram (Figure 7a), most of the samples belong to the gabbro, and on the R1-R2 diagram (Figure 7b), they fall within the gabbro and olivine-gabbro area. The K_2O content of samples is very variable, with average values of 0.94 wt%, 0.42 wt%, 1.73 wt%, and 1.32 wt%, respectively. On the SiO_2-K_2O diagram, the samples belong to the calc-alkaline to high-K calc-alkaline series (Figure 7c), and Ta/Yb vs. Ce/Yb ratios allow an allocation to the calc-alkaline series (Figure 7d). The average Fe_2O_3 content of the samples is 2.13–3.50 wt%, with a high average FeO content of 9.41–11.90 wt%, an average MgO content of 6.17–7.50 wt%, and a low average content of TiO_2 (1.61–2.39 wt%) and MnO (0.17–0.21 wt%). $Mg^{\#}$ is low, ranging from 43.32 to 52.02, which is much lower than that of primary basaltic rocks ($Mg^{\#}$ = 70 [50]). The first basalt and the fourth amphibolite have a relatively high CaO content (9.52–10.00 wt%) and low P_2O_5 content (0.18–0.19 wt%), while the other two periods of mafic rocks have a relatively low CaO content (6.90–7.17 wt%) and high P_2O_5 content (0.25–0.39 wt%).

There are differences in the characteristics of rare earth elements and trace elements in these mafic rocks. The total contents of rare earth elements (REEs) in the first basalt and the fourth amphibolite are low, with average contents of 83 µg/g and 86 µg/g, respectively. On the chondrite-normalized rare earth element (REE) diagram, they exhibit similar characteristics to ocean island basalt (OIB), with a right-skewed smooth curve (Figure 8a), indicating an enrichment of light rare earth elements (LREEs) and a flat distribution pattern of heavy rare earth elements (HREEs). The differentiation between LREEs and HREEs is not significant, with an average $(La/Yb)_N$ value of 2.55–4.78 and a slight positive europium anomaly (Eu/Eu* = 0.98–1.06) (Figure 8a). In the primitive mantle-normalized trace element spider diagram, these two mafic rocks demonstrate similar characteristics, with an enrichment of large ion lithophile elements (LILEs) (Rb, Ba, K, etc.), depletion of high field strength elements (HFSEs) (Th, U, Nb, Ta, Zr, Ti, etc.), and negative anomalies of Sr and P, as well as slight positive anomalies of Zr and Hf (Figure 8b). The contents of Cr (with average contents of 106 µg/g and 129 µg/g), Co (with average contents of 51.2 µg/g and 48.4 µg/g), and Ni (with average contents of 73.6 µg/g and 98 µg/g) in these two mafic rocks are higher than normal mid-ocean ridge basalt (N-MORB).

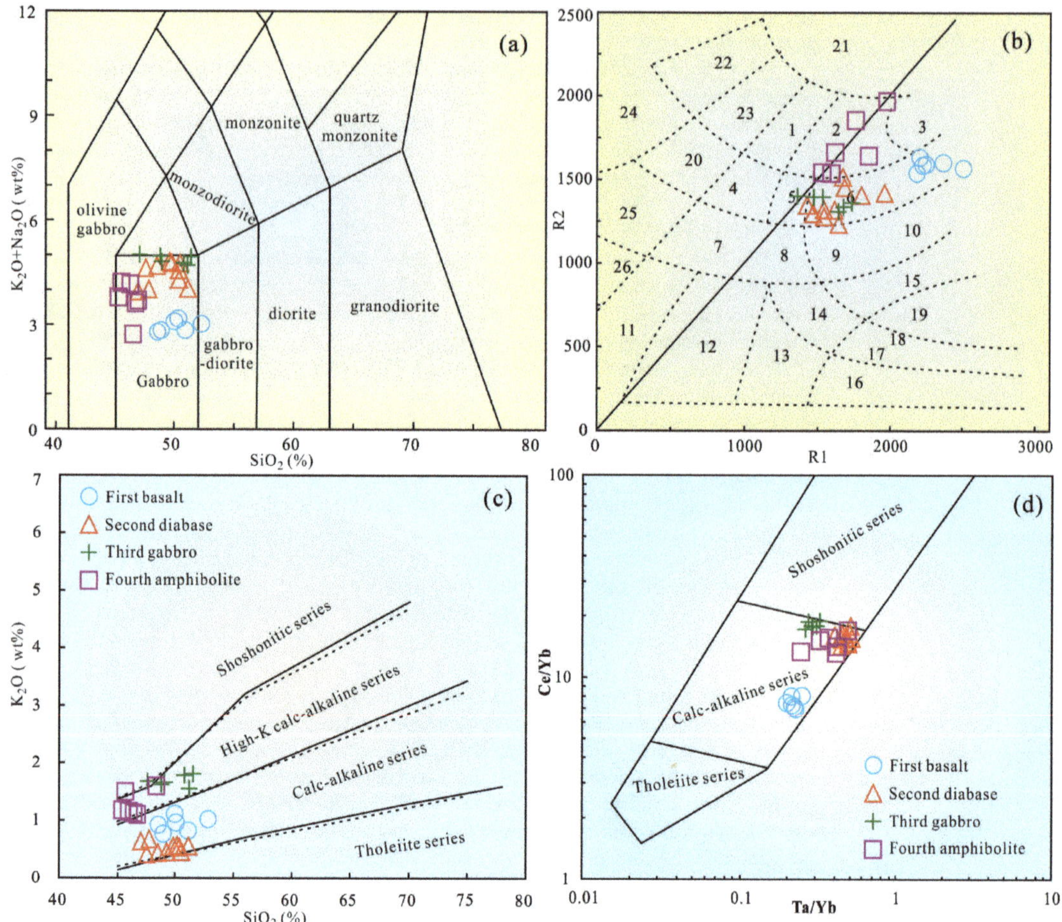

Figure 7. SiO$_2$ vs. total alkali (Na$_2$O + K$_2$O) ((**a**), after [51]), R1 vs. R2 ((**b**), after [52]), SiO$_2$-K$_2$O ((**c**), after [53]) and Ta/Yb vs. Ce/Yb ((**d**), after [54]) diagrams for mafic rocks from northern Liaoning. (**b**): 1—alkaline gabbro (alkaline basalt); 2—olivine gabbro (olivine basalt); 3—gabbro norite (tholeiite); 4—syenite gabbro (trachyte basalt); 5—monzonite gabbro (andesite coarse basalt); 6—gabbro (basalt); 7—trachyandesite (syenite); 8—monzonite (andesite); 9—monzodiorite (trachyte); 10—diorite (andesite); 11—nepheline syenite (trachyte phonolite); 12—syenite (trachyte); 13—quartz syenite (quartz trachyte); 14—quartz monzonite (quartz andesite); 15—tonalite (dacite); 16—alkaline granite (alkaline rhyolite); 17—syenogranite (rhyolite); 18—monzogranite (dacite rhyolite); 19—granodiorite (rhyolite dacite); 20—essenite aegirine gabbro; 21—peridotite (picrite); 22—nepheline (picrite nepheline); 23—qilieyan (basanite); 24—neonite (nepheline); 25—essenite; 26—nepheline syenite (phonolite).

The second diabase and the third gabbro show similar average REE contents, which are 119.28 μg/g and 121.41 μg/g, respectively, and are between OIB and N-MORB [55]. On the chondrite-normalized REE diagram, the samples show a right-skewed curve (Figure 8a), indicating an enrichment of LREEs and a flat pattern of HREEs. There is a slight differentiation between them, with average (La/Yb)$_N$ values of 5.44 and 7.32, respectively. The diabase shows a slight negative europium anomaly (Eu/Eu* = 0.72–0.88), while the gabbro exhibits a positive europium anomaly (Eu/Eu* = 1.31–1.37) (Figure 8a). In the primitive mantle-normalized trace element spider diagram, the diabase is enriched in LILEs (Ba, etc.), relatively depleted in Rb and HFSEs (Th, U, Nb, Ta, Zr, Hf, Ti, etc.), and shows slight positive anomalies of K, Sr, and P, and Eu shows negative anomalies (Figure 8b); the gabbro shows the enrichment of LILEs (Rb, Ba, K), depletion of HFSEs (Th, U, Nb, Ta, Zr, Hf, etc.), and a positive anomaly of Eu (Figure 8b). The contents of Cr (with average contents of 123 µg/g and 113 µg/g), Co (with average contents of 46.8 µg/g and 62.7 µg/g), and Ni (with average contents of 71.4 µg/g and 74.0 µg/g) in these two mafic rocks are slightly higher than N-MORB.

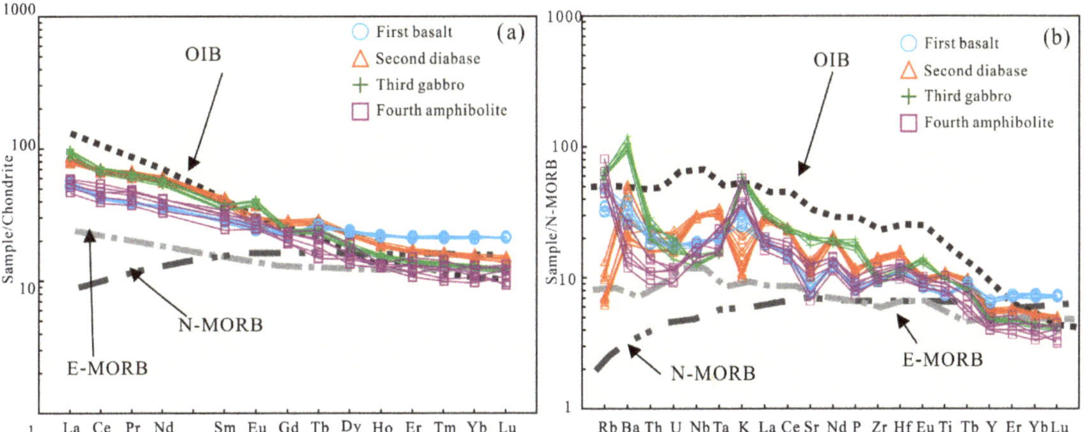

Figure 8. Chondrite-normalized rare earth element patterns (**a**) and primitive mantle-normalized trace element spider diagram (**b**) for the mafic rocks from northern Liaoning. (The values of chondrite and primitive mantle are from [55]).

5. Discussion

5.1. The Geochronological Significance of Mafic Rocks

5.1.1. Constraints on the Age of Proterozoic Strata in Northern Liaoning

The Proterozoic strata, exposed in northern Liaoning, are placed in the Changcheng, Jixian, and Nanhua System of the Meso-Neoproterozoic, according to the comparison of rock assemblage with the Proterozoic strata in western Liaoning [33]. However, after field investigation, this suite of strata has certain differences from the strata of Mesoproterozoic in western Liaoning. The Proterozoic strata in northern Liaoning has generally undergone metamorphism, with the lithology consisting of dolomite, meta-sandstone, meta-siltstone, and slate. These characteristics of rock composition and low-grade metamorphism are similar to those of the North Liaohe Groups and Laoling Groups in JLJB [56]. The mafic rocks studied in this paper intrude into the Kangzhuangzi Formation, Guanmenshan Formation, Tongjiajie Formation, and Hutouling Formation in northern Liaoning. They are considered to have formed in the Early Triassic according to whole-rock K–Ar dating results of 238.2 Ma [33].

However, the credibility of this chronology result is doubted: (1) In northern Liaoning, these mafic rocks have intruded into the earlier strata with objective contact metamorphic belts, rather than the Neoproterozoic strata (Figure 1b). (2) Partial pillow-shaped mafic rocks, associated with marble and dolomite, are in tectonic contact with the surrounding Permian granite (Figure 1c). (3) Metamorphism is widely developed in these mafic rocks. Pyroxene has undergone metamorphism into amphibole and carbonation or with brown amphibole reaction edges partially. Plagioclase has undergone epidotization, zoisitization, etc. Olivine has intensely metamorphosed into microscale biotite, magnetite, etc. Hornblende has undergone chloritization. Some mafic rocks have metamorphosed into amphibolite (Figure 2k,l).

Four zircon U-Pb ages have been obtained from the mafic rocks in northern Liaoning, ~2209 Ma, ~2154 Ma, ~2055 Ma, and ~2018 Ma, recording four periods of mafic magmatic activity. These results indicate that the emplacement age of the mafic rocks is the Paleoproterozoic rather than the Early Triassic, which means the intruded strata by, or associated with, the mafic rocks were formed in the Paleoproterozoic or earlier. Combined with the unconformity of overlying the Neoproterozoic granite, the formation age of these strata should be corrected to Paleoproterozoic.

5.1.2. Paleoproterozoic Mafic Magmatic Events in the Eastern Segment of the NCC

In this study, four episodes of mafic magmatic activities have been identified in northern Liaoning. Meanwhile, a 2118 Ma metamorphic basic dyke has recently been reported in the Qingyuan area [30]. The magmatic activities of mafic rocks in the eastern segment of the NCC are primarily concentrated in the JLBT and persist from approximately 2209 Ma to around 1820 Ma. The rock assemblage with this activity includes basalt, diabase, gabbro, and amphibolite, which are found as intrusions, sills, dykes, or strata interlayers coexisting with Paleoproterozoic meta-sedimentary and metamorphic rocks. Based on previous studies and new data obtained in this study, a chronological framework for these Paleoproterozoic mafic magmatic activities in the eastern segment of the NCC has been established. This framework provides a timeline of when these events occurred, helping scientists understand the sequence of geological processes during this time (Figure 8, Table 4).

Table 4. Summary of geochronological data of Paleoproterozoic mafic rocks in the eastern segment of the NCC.

No.	Sample	Lithology	U-Pb Age (Ma)	Location	Analytical Method	References
1	D021	Amphibolite	1952 ± 38	Zhujiagou	Zircon (LA-ICPMS)	[17]
2	D019	Amphibolite	2024 ± 33	124°58′14.9″, 40°47′22.5″	Zircon (LA-ICPMS)	[18]
3	D014	Amphibolite	2053 ± 34	124°55′53″, 40°39′13″	Zircon (LA-ICPMS)	[18]
4	D018	Amphibolite	2130 ± 19	124°58′43.8″, 40°47′12.8″	Zircon (LA-ICPMS)	[18]
5	HLY-3	Pillow basalt	1928 ± 16	Helan Town	Zircon (LA-ICPMS)	[19]
6	D15054	Amphibolite	1985 ± 31	Helan Town	Zircon (LA-ICPMS)	[19]
7	D1423-1	Amphibolite	2079 ± 21		Zircon (LA-ICPMS)	[19]
8	D1488-1	Amphibolite	2145 ± 19		Zircon (LA-ICPMS)	[19]
9	TWD15003	Meta-diabase	1821 ± 78	Lianshanguan Town	Zircon (LA-ICPMS)	[20]
10	TWD15008	Meta-diabase	2010 ± 27	Helan Town	Zircon (LA-ICPMS)	[20]
11	A1102	Meta-gabbro	2110 ± 31	Qianshan Town	Zircon (SHRIMP)	[57]
12	D1002-B1	Amphibolite	1995 ± 13	Huanghuadianzi Town	Zircon (LA-ICPMS)	[58]
13	D4034-B1	Amphibolite	2150 ± 21	Huanghuadianzi Town	Zircon (LA-ICPMS)	[58]
14		Gabbro	1880 ± 6		LA-ICP-MS	[59]
15		Amphibolite	1886 ± 26	Helan Town	LA-MC-ICP-MS	[59]
16		Gabbro	1914 ± 40	Helan Town	LA-MC-ICP-MS	[59]
17	SJZ07-2.1	Amphibolite	2167 ± 31	Sanjiazi Town	Zircon (LA-ICPMS)	[60]
18	DZ74-1	Meta-gabbro	2144 ± 16	Bahui Town	Zircon (LA-ICPMS)	[10]
19	DZ85-1	Meta-gabbro	2157 ± 17	Helan Town	Zircon (LA-ICPMS)	[10]
20	DZ73-1	Meta-gabbro	2159 ± 12	Helan Town	Zircon (LA-ICPMS)	[10]
21	DZ91-1	Meta-diabase	2161 ± 12	Mafeng Town	Zircon (LA-ICPMS)	[10]
22	DZ78-1	Amphibolite	2161 ± 45	Helan Town	Zircon (LA-ICPMS)	[10]
23	DZ40-2	Amphibolite	2159 ± 28	Fengcheng	Zircon (LA-ICPMS)	[11]
24	NLX02-4	Amphibolite	2163 ± 22	Helan Town	Zircon (SHRIMP)	[61]
25	09LG29	Gabbro	1828 ± 13	Helan Town	Zircon (SHRIMP)	[62]

Table 4. *Cont.*

No.	Sample	Lithology	U-Pb Age (Ma)	Location	Analytical Method	References
26	09LG28	Meta-mafic rock	1875 ± 28	Helan Town	Zircon (SHRIMP)	[62]
27	598XLLZ2	Meta-gabbro	2115 ± 13	Qianshan Town	Zircon (CAMECA)	[63]
28	598XLLZ2	Meta-gabbro	2115 ± 3	Qianshan Town	Baddeleyite (CAMECA)	[63]
29	SJZ07-5	Amphibolite	2054–2061	Sanjiazi Town	Zircon (LA-ICPMS)	[13]
30	16KD55-1-1	Amphibolite	2063 ± 23	Sanjiazi Town	Zircon (LA-ICPMS)	[13]
31	D2066-11	Amphibolite	2083 ± 13	Helan Town	Zircon (LA-ICPMS)	[13]
32	SJZ11-1	Amphibolite	2119 ± 19	Sanjiazi Town	Zircon (LA-ICPMS)	[13]
33	D1009-5	Meta-diabase	2100 ± 12	Helan Town	Zircon (LA-ICPMS)	[14]
34	D1009-7	Meta-diabase	2110 ± 23	Helan Town	Zircon (LA-ICPMS)	[14]
35	D1002-2	Meta-diabase	2133 ± 14	Helan Town	Zircon (LA-ICPMS)	[14]
36	D5048-4	Amphibolite	2164 ± 6	Helan Town	Zircon (LA-ICPMS)	[14]
37	D9001-1	Meta-gabbro	2118.6 ± 6.3	Helan Town	Zircon (LA-ICPMS)	[13]
38	16KD68-1	Meta-gabbro	2188.2 ± 8.5	Helan Town	Zircon (LA-ICPMS)	[13]
39	HP-9	Meta-mafic vein	2157 ± 21	Hupiyu Pluton	Zircon (LA-ICPMS)	[64]
40	DD24-1	Amphibolite	2059 ± 22	Mafeng Town	Zircon (LA-ICPMS)	[65]
41	YK12-1-4	Gabbro	2125 ± 6	Pailou Town	Zircon (LA-ICPMS)	[27]
42		Gabbro	2113 ± 15	Longchang Town	Zircon (LA-ICPMS)	[66]
43	15Q18	Metabasic dykes	2118 ± 18	124°56.57′, 42°12.66′	Zircon (LA-ICPMS)	[30]
44	D1917	basalt	2154 ± 15	125°57′05.27″, 41°21′16.88″	Zircon (LA-ICPMS)	This study
45	D1918	Diabase	2209 ± 12	124°12′36.40″, 42°05′01.27″	Zircon (LA-ICPMS)	This study
46	D1919	Gabbro	2063 ± 7	124°13′57.47″, 42°14′18.33″	Zircon (LA-ICPMS)	This study
47	D2012	Amphibolite	2018 ± 13	124°51′44.51″, 42°19′09.73″	Zircon (LA-ICPMS)	This study

In the eastern segment of the NCC, the Paleoproterozoic mafic rocks underwent four events of magmatism: The period from 2210 to 2100 Ma was marked by an especially intense activity of mafic magmatism. This mafic magmatic event started around 2209 Ma and persisted until 2100 Ma, with two significant peaks occurring at approximately 2162 Ma and 2108 Ma, respectively. The geological implications of this phenomenon suggest the potential onset and continuous development of ocean opening processes, commencing around 2209 Ma and extending until 2100 Ma when the ocean basin reached its maximum extent. From 2100 to 2000 Ma, the mafic magmatic activity gradually decreased, possibly indicating the onset of subduction and reduced basaltic magma activity in a compressional tectonic setting. Occasional mafic magmatic activity was observed from 2000 to 1900 Ma, which may be associated with a continent–continent collision and crustal thickening after 2000 Ma, resulting in reduced magmatic activity. After 1900 Ma, there was an increase in mafic magmatic activities. This could be attributed to the post-orogenic extensional phase. In this tensional tectonic setting, a certain scale of mafic magmatism occurred. However, due to the thicker crust at this time compared to earlier periods, the intensity of mafic magmatism during this stage was not as strong as that observed in the initial phase (Figure 9, Table 4).

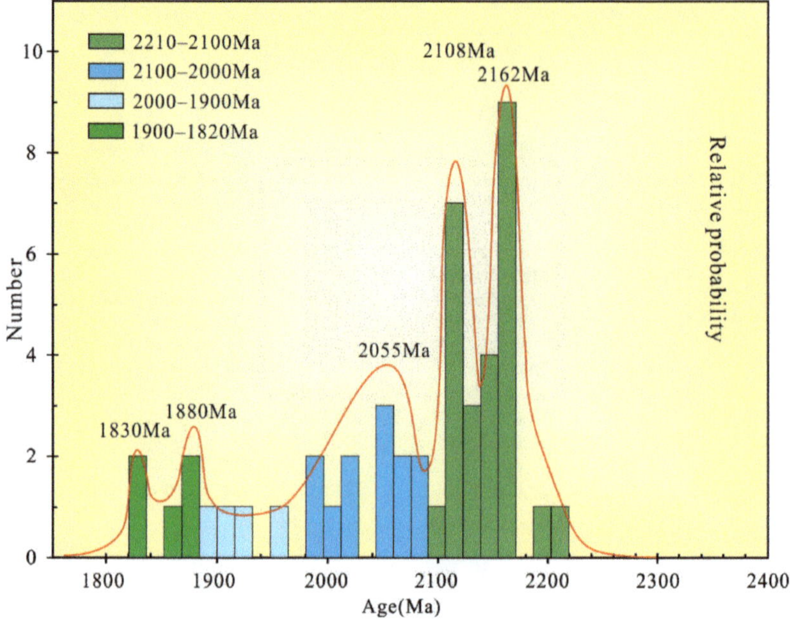

Figure 9. Age histogram of mafic rocks from the eastern segment of the NCC.

5.2. Source Characteristics and Genesis of the Paleoproterozoic Mafic Rocks

The ratio of some incompatible elements is stable during partial melting and fractional crystallization processes, which can largely reflect the characteristics of the source. For example, since Nb and Ta have similar valence states and ionic radii, their ratios are also comparable in igneous rocks from the same source. Similarly, Zr and Hf have similar valence states, ionic radii, and distribution coefficients in various minerals, so similar ratios and trends can be observed in rocks derived from the same origin. The ratios of Nb/Ta, Nb/La, Ta/Th, Ce/Pb, La/Pb, Nb/U, Zr/Nb, and Zr/Hf in the four periods of mafic rocks all show differences (Table 3), indicating that they come from different sources. These mafic rocks have undergone medium- to high-grade metamorphism and later experienced alteration processes such as sericitization, chloritization, and epidotization. During the process, highly active elements may migrate due to changing conditions, while REEs and HFSEs remain stable [67]. These elements are used to analyze and discuss the magma series, genesis, and source characteristics of metamorphic rocks [67].

Generally, mafic rocks originate from the mantle source [68]. The Paleoproterozoic mafic rocks are characterized by the enrichment of LREEs and LILEs, depletion of HREEs and HFSEs such as Nb, Ta, Ti, and Zr, and a slight negative or positive Eu anomaly, indicating characteristics of a mantle source [68]. The REE and trace element patterns of the Paleoproterozoic mafic rocks are different from N-MORB and E-MORB, but similar to OIB (Figure 8a,b), suggesting their origin from the lithospheric mantle rather than the asthenospheric mantle [68]. On the Nb/Th-Zr/Nb diagram, the first basalt is located between the primitive mantle, enriched mantle, and recycled slab, while the second diabase and fourth amphibolite are located at the transitional zone between the primitive mantle and recycled subduction slab, and the third gabbro is relatively close to the enriched mantle (Figure 10a). This indicates that the sources of these mafic rocks are transitional mantle.

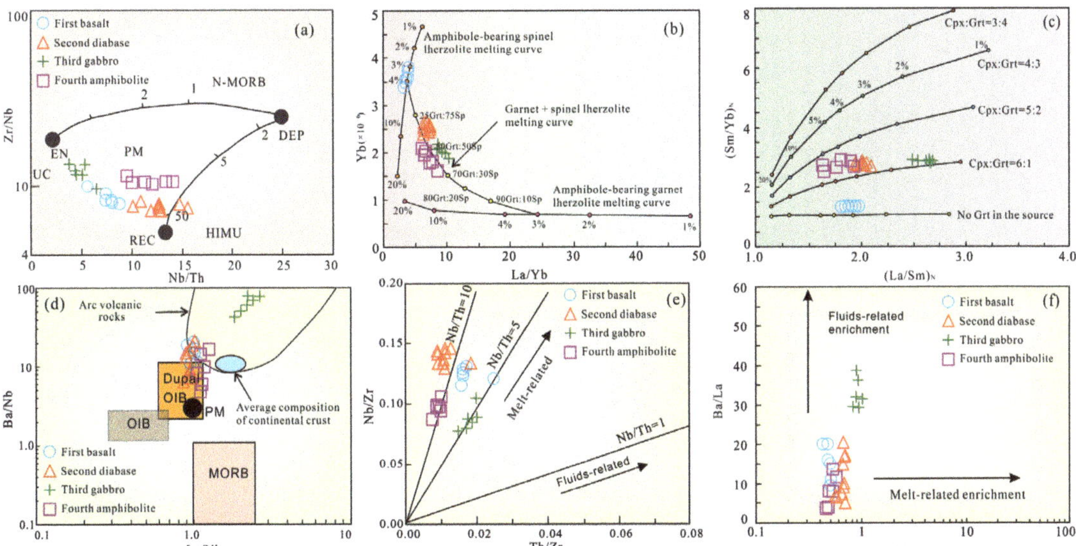

Figure 10. Source characteristics of the Triassic gabbro from the Kaiyuan Area. (**a**) After [69], DEP—depleted mantle, EN—enriched mantle, N-MORB—normal mid-ocean ridge basalt, PM—primitive mantle, REC—recycled plate, UC—upper crust. (**b**,**c**) After [70], Grt—garnet, SP—spinel. (**c**) Cpx—clinopyroxene. (**d**) After [71]. (**e**,**f**) After [72].

Due to the similar distribution coefficients, trace elements are difficult to fractionate during partial melting or fractional crystallization; they are usually used to reflect the nature of the parent magma in the magma source [70]. The first basalt shows flat patterns for REEs, with a low LREE/HREE ratio, indicating that there may be no residual garnet in the source area (Figure 10b,c). On the La/Yb-Yb diagram, it is located at the 3–4% melting position of the Amphibole-bearing spinel lherzolite melting curve (Figure 10b), indicating the presence of residual amphibole and spinel in the magma source. The other stages of mafic rocks show similar patterns of enrichment in LREEs, relatively flat patterns of HREEs, high LREE/HREE ratios, and low Yb contents, indicating the presence of residual garnet in the source (Figure 10b,c). On the La/Yb-Yb diagram, they are located near the Garnet spinel lherzolite melting curve (Figure 10b), indicating the residual garnet and spinel in the magma source. The garnet to spinel ratio of the diabase is between 50:50 and 25:75, and the ratio of the gabbro and amphibolite is between 50:50 and 70:30 (Figure 10b). Differing from the partial melting of spinel lherzolite, during the partial melting process of different proportions of clinopyroxene and garnet, the properties of La/Sm and Sm/Yb show different characteristics, with Sm/Yb not changing with the decrease in La/Sm [73]. On the $(La/Sm)_N$-$(Sm/Yb)_N$ diagram (Figure 10c), consistent with the above conclusion, there is no garnet in the source, and the first basalt is the product of 3%–4% melting of amphibole-bearing spinel lherzolite. The other three stages of mafic rocks are products of partial melting of garnet–spinel lherzolite with a ratio of clinopyroxene to garnet of 6:1 to 5:2, only with different degrees of partial melting (Figure 10c [74]). The diabase, gabbro, and amphibolite show partial melting degrees of about 3%, 1%–2%, and 3%–5%, respectively (Figure 10c). All mafic rocks show negative Sr anomalies (Figure 10b), indicating the possible residual plagioclase in the source.

The mafic rocks exhibit similar enrichments in LREEs and LILEs such as Rb, Ba, and K, as well as depletions in HFSEs such as Nb, Ta, Zr, Hf, Ti, and P. In the La/Nb-Ba/Nb diagram (Figure 10d), the third gabbro falls within the area of island arc volcanic rocks, while the other mafic rocks fall within the transitional zone between Dupal OIB and arc volcanic rocks. This could be related to the introduction of fluids or involvement of mantle components, indicating that metasomatic processes occurred in the source [75]. Thus, is the metasomatism caused by melts or fluids? The mafic rocks show high contents of Cr (average content of 106 μg/g, 123 μg/g, 113 μg/g, and 129 μg/g), Co (average content of 51.2 μg/g, 46.8 μg/g, 62.7 μg/g, and 48.4 μg/g), and Ni (average content of 73.6 μg/g, 71.4 μg/g, 74.0 μg/g, and 98 μg/g) compared to N-MORB, indicating that their source has undergone melt metasomatism, which is consistent with the trends shown in the Th/Zr-Nb/Zr diagram (Figure 10e). When the mantle-derived magma interacts with crustal material on the subducting plate and subducted oceanic crust, the magma generally exhibits low Na_2O, P_2O_5, and TiO_2 contents, positive anomalies of Nb, Ta, and Ti [76], or low Ce/Th ($\approx$8), low Ba/Th ($\approx$111), and obvious negative Ce anomaly [77]. The geochemical characteristics of these four episodes of mafic rocks are as follows: The first basalt exhibits a low Ce/Th (14.32), low Ba/Th (114.74), low Na_2O, P_2O_5, and TiO_2 contents, negative anomalies of Nb, Ta, and Ti (Figure 8b), and a slight negative Ce anomaly (average Ce/Ce* = 0.94). The second diabase exhibits high Na_2O, P_2O_5, and TiO_2 contents and positive anomalies of Nb, Ta, and Ti (Figure 8b), with low Ba/Th (114.74) but high Ce/Th (24.47), and a slightly negative Ce anomaly (average Ce/Ce* = 0.93). The third gabbro exhibits high Na_2O, P_2O_5, and TiO_2 contents, low Ce/Th (14.32), negative anomalies of Nb and Ta, no anomaly in Ti (Figure 8b), high Ba/Th (371.73), high Ce/Th (21.33), and a slightly negative Ce anomaly (average Ce/Ce* = 0.91). The fourth amphibolite exhibits a low Ba/Th (110.08) and a positive anomaly in Ti, but low Na_2O, P_2O_5, and TiO_2 contents, negative anomalies of Nb and Ta (Figure 8b), high Ce/Th (29.47), and a slightly negative Ce anomaly (average Ce/Ce* = 0.97). These characteristics indicate that all four stages of mafic rocks have undergone dual metasomatism by both fluid and melt, which is consistent with the trends shown in the Th/Zr-Nb/Zr and Th/Yb-Ba/La diagrams (Figure 10e,f). The third gabbro has undergone greater fluid metasomatism, while the others are mainly influenced by melt metasomatism.

Generally, mantle-derived magma may undergo assimilation and contamination with crustal materials during its ascent and emplacement. The geochemical data of the mafic rocks show moderate potassium and alkali contents (Figure 7c), low $Mg^{\#}$ ($Mg^{\#}$ = 43.32–52.02), enrichment in LILEs (Rb, Ba, K, etc.), and depletion in HFSEs (Nb, Ta, Zr, etc.). These characteristics suggest that the magma may have undergone contamination by crustal materials [78]. The Nb/Ta (average values of 15.63, 16.53, 15.01, and 15.74) and Zr/Hf (average values of 30.07, 31.88, 31.04, and 33.53) of the mafic rocks are similar to the values of the continental crust in eastern China (Nb/Ta = 15.38, Zr/Hf = 35.56, according to [79]), and lower than those of mid-ocean ridge basalts (MORB) and primitive mantle (Nb/Ta = 17.7, Zr/Hf = 36.1 [55]), indicating the influence of crustal contamination on the mafic magmas. The La/Sm is often used to reflect the degree of crustal contamination [80]. A higher La/Sm ratio often suggests a greater influence from crustal components, while a lower ratio may indicate a more pristine mantle signature [80]. The average La/Sm ratios of the four episodes of mafic rocks are 2.95, 3.11, 4.03, and 2.75, respectively, indicating the incorporation of continental crustal material during magma ascent. The intensity of crustal contamination is highest in the third gabbro, followed by the first basalt and second diabase, and weakest in the fourth amphibolite.

In conclusion, the Paleoproterozoic mafic rocks originated from a transitional mantle source. The first basalt, containing residual hornblende and spinel in the source, is the product of 3%–4% partial melting of amphibole-bearing spinel lherzolite. The source of the other three stages of mafic rocks contains residual garnet and spinel, and experienced partial melting of garnet–spinel lherzolite with a clinopyroxene to garnet ratio ranging from 6:1 to 5:2. The garnet to spinel ratios in these three mafic rocks range from 50:50 to 25:75 and 50:50 to 70:30, with partial melting degrees of 3%, 1%–2%, and 3%–5%, respectively. During the ascent and emplacement of all mafic magmas, they were contaminated by crustal material and incorporated continental crustal material.

5.3. Tectonic Setting of Paleoproterozoic Mafic Rocks

Mafic rocks formed in various tectonic settings show different TiO_2 contents [81]. Island arc basalt (IAB) typically contains the lowest TiO_2 content, around 0.98 wt% [81]. MORB contains a TiO_2 content of 1.5 wt%, while OIB contains the highest TiO_2 content at 2.63 wt% [81]. Within-plate basalts (WPB), on the other hand, show higher TiO_2 contents ranging from 2.23 wt% to 2.9 wt% [53]. In terms of trace element abundances, WPB generally show higher Nb and Ta contents, ranging from 13 to 84 µg/g and 0.73 to 5.9 µg/g, respectively. IAB, in contrast, exhibits very low Nb and Ta contents, ranging from 1.7 to 2.7 µg/g and 0.1 to 0.18 µg/g, respectively [53]. According to Condie's (1989) study on element ratios in different tectonic settings of basaltic rocks, WPB and MORB are enriched in TiO_2 and HFSEs. The element ratio shows the following characteristics: Nb/La > 0.8, Ti/Y > 350, Ti/V > 30, Hf/Ta < 5, La/Ta < 15, and Th/Ta < 3 [53,82]. On the contrary, it is similar to IAB on the active continental margin.

In this study, the third gabbro exhibits high TiO_2, Nb, and Ta contents (averages of 2.17 wt%, 9.42 µg/g, and 0.63 µg/g, respectively), as well as high Ti/Y and Ti/V ratios (averages of 587 and 42.5, respectively). This may be attributed to the influence of melt contamination, resulting in excessively high TiO_2 contents. The average trace element ratios of Nb/La = 0.43, Hf/Ta = 5.57, La/Ta = 34.87, and Th/Ta = 3.19 indicate that it differs from WPB and MORB but shares similarities with continental arc basalts (CAB) [53,82]. The gabbro also exhibits characteristics of island arc volcanic rocks in the La/Nb-Ba/Nb diagram (Figure 10d). Furthermore, in the Hf/3-Th-Nb/16 triangular diagram and Nb/Yb-Th/Yb diagram, the gabbro falls within the continental arc region (Figure 11a,b). These characteristics suggest that the gabbro was formed in an island arc environment.

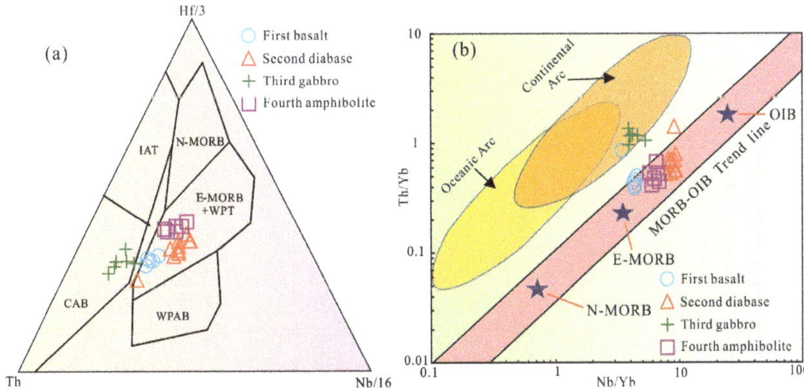

Figure 11. Identification diagram of tectonic setting for the mafic rocks from northern Liaoning. (**a**) Hf/3 versus Th versus Nb/16 (after [83]); (**b**) Nb/Yb versus Th/Yb diagram (after [83]).

The other three phases of mafic rocks show high TiO_2 (averages of 1.61 wt%, 2.39 wt%, 1.94 wt%), Nb (averages of 13 µg/g, 21 µg/g, 11.2 µg/g), and Ta (averages of 0.83 µg/g, 1.28 µg/g, 0.76 µg/g) contents. The average trace element ratios of Nb/La (averages of 1.01, 1.08, 0.89), Ti/Y (averages of 351, 558, 587), Ti/V (averages of 30.01, 39.5, 43.4), Hf/Ta (averages of 4.23, 3.78, 4.82), La/Ta (averages of 15.4, 15.3, 16.6), and Th/Ta (averages of 2.42, 1.42, 1.33) indicate that they are similar with WPB and MORB. They also share similar characteristics in the Hf/3-Th-Nb/16 triangular diagram (Figure 11a). The first basalt and the second diabase intrude into the limestone as sills, while the fourth amphibolite is surrounded by marble. These field features are similar to the rock assemblage of oceanic islands or seamounts. The geochemical characteristics of these three phases of mafic rocks are similar to OIB. In the La/Nb-Ba/Nb diagram, they mainly originate from the Dupal OIB and transition to arc volcanic rocks (Figure 10d), and in the Nb/Yb-Th/Yb diagram, they fall on the MORB-OIB trend line, indicating that these three phases of mafic rocks were formed in an oceanic island environment (Figure 11b).

The formation mechanism of the Paleoproterozoic orogenic belt (JLJB) in the eastern segment of the NCC has been a subject of ongoing debate and differing interpretations among geologists [4,5,8,9,13–20,30,84]. One view suggests that it resulted from subduction and continental collision from the northern plate of the Longgang block, with the JLJB forming in an arc-back basin, and the northern margin of the NCC belonging to a continental margin sedimentary environment [4,5]. Previous studies on high-pressure granulite metamorphic rocks in the northern margin of the NCC proposed the existence of an east-west striking Paleoproterozoic orogenic belt (IMNHB), which may have formed during the assembly of the Columbia supercontinent, influenced by the closure, disappearance, and collision of the northern ocean of the NCC [4,5,8,9]. It is believed that this orogenic belt extended from Hebei to northern Liaoning [9]. The Paleoproterozoic mafic magmatic events and metamorphism in the Hebei–Northern Liaoning area also support the existence of this orogenic belt [30,84]. The mafic rocks identified in this study were formed in an oceanic island and island arc tectonic setting, indicating the existence of subduction, and the extinction of ocean and continent–continent collision in the north of the Longgang Block during the Paleoproterozoic, which is consistent with Kusky's viewpoint.

However, in recent years, according to regional geological investigations and studies on magmatic rocks, and metamorphism and sedimentary rocks, more and more scholars believe that the JLJB is a locally ordered and globally disordered mélange, and the formation of JLJB is related to northward subduction of the Nangrim Block in the south towards to the Longgang Block in the north [1,2,13–20]. The genetic models of JLJB in the subduction environment can be divided into two types: the continent–arc–continent collision model [13–15,24–27] and the rift–subduction–collision cycle model [2,6,7,18–20,28,29].

The debate between these two views lies in whether the subduction occurred in the south or north of the Longgang Block. The geochronological framework of mafic magmatic activity in the eastern segment of the NCC shows that there were similar mafic magmatic activities in the northern and southern parts of the Longgang Block during the Paleoproterozoic. Therefore, the northern and southern parts of the Longgang Block may have been in two different tectonic domains in the Paleoproterozoic. This issue requires further research in the future.

6. Conclusions

(1) Zircon U-Pb ages (2209 ± 12 Ma, 2154 ± 15 Ma, 2063 ± 7 Ma, 2018 ± 13 Ma) and metamorphic ages (1835 ± 13 Ma) have been obtained from mafic rocks in northern Liaoning, which constrained the formation age of the Proterozoic strata to be Paleoproterozoic. Based on previous data, the Paleoproterozoic mafic magmatic activities in the eastern segment of the NCC can be classified into four stages: the most frequent activity occurred between 2210 and 2100 Ma, gradually decreased between 2100 and 2000 Ma, with occasional activities between 2000 and 1900 Ma, and increased activities after 1900 Ma.

(2) Geochemical characteristics reveal that the four stages of mafic rocks belong to the calc-alkaline series and originated from transitional mantle. During the process of magma ascent and emplacement, the magma underwent contamination by crustal materials. The first basalt contains residual hornblende and spinel in the source, and is a product of 3–4% partial melting of amphibole-bearing spinel lherzolite. The other mafic rocks contain residual garnet and spinel in the source, and experienced partial melting of garnet–spinel lherzolite with a clinopyroxene to garnet ratio ranging from 6:1 to 5:2. The proportions of garnet to spinel range from 25:75 to 70:30, and the degrees of partial melting are 3%, 1%–2%, and 3%–5%, respectively.

(3) Trace element data indicate that the third gabbro exhibits characteristics of continental island arc basalts, suggesting it was formed in an island arc environment. The other three stages of mafic rocks originated from the Dupal OIB and formed in an oceanic island environment. The identification of Paleoproterozoic magmatic and subsequent

metamorphic events in the mafic rocks from northern Liaoning, as well as the verification that the northern orogenic belt of the NCC extends to this region, suggests that the north of the Longgang Block may be under an oceanic subduction tectonic setting. JLJB was formed under the background of a northward subduction of the Nangrim Block or island arc in the south of the Longgang Block. Therefore, it is possible that the northern and southern parts of the Longgang block were located at different tectonic domains in the Paleoproterozoic.

Author Contributions: Conceptualization, J.C.; formal analysis, J.C., Y.T., B.L. and W.L.; investigation, J.C., Y.T., Z.G., B.L., C.Z. (Chen Zhao), W.L., C.Z. (Chao Zhang). and Y.W.; writing—original draft preparation, J.C. and B.L.; writing—review and editing, J.C. and B.L.; projection administration, Y.W. All authors have read and agreed to the published version of the manuscript.

Funding: This research was funded by the National Natural Science Foundation of China (U2244213) and the China Geological Survey (Grants DD20242929, DD20190042).

Data Availability Statement: The authors confirm that the data generated or analyzed during this study are provided in full within the published article.

Acknowledgments: We thank the editors and anonymous reviewers for their critical reviews and excellent suggestions that helped to improve this manuscript. We thank the staff of the Key laboratory of Mineral Resources Evaluation in Northeast Asia, Ministry of Natural Resources, for their advice and assistance during the zircon U-Pb dating by LA-ICP-MS. We also thank the Northeast China Supervision and Inspection Center of Mineral Resources, Ministry of Natural Resources, Shenyang, China, for their assistance in the major and trace element analysis.

Conflicts of Interest: Yi Tian, Zhonghui Gao, Weiwei Li were employed by the company Institute of Geology and Mineral resources of Liaoning Co., Ltd. The remaining authors declare that the research was conducted in the absence of any commercial or financial relationships that could be construed as a potential conflict of interest.

References

1. Zhao, G.C.; Sun, M.; Wilde, S.A.; Li, S.Z. Late Archean to Paleoproterozoic evolution of the North China Craton: Key issues revisited. *Precambrian Res.* **2005**, *136*, 177–202. [CrossRef]
2. Zhao, G.C.; Cawood, P.A.; Li, S.Z.; Wilde, S.A.; Sun, M.; Zhang, J.; He, Y.; Yin, C. Amalgamation of the North China Craton: Key issues and discussion. *Precambrian Res.* **2012**, *222–223*, 56–76. [CrossRef]
3. Faure, M.; Trap, P.; Lin, W.; Monié, P.; Bruguier, O. Polyorogenic Evolution of the Paleo-Proterozoic Trans-North China Belt: New Insights from the Lvliangshan-Hengshan-Wutaishan and Fuping Massifs. *Episodes* **2007**, *30*, 96–107. [CrossRef]
4. Kusky, T.M. Geophysical and geological tests of tectonic models of the North China Craton. *Gondwana Res.* **2011**, *20*, 26–35. [CrossRef]
5. Kusky, T.M.; Polat, A.; Windley, B.F.; Burke, K.C.; Dewey, J.F.; Kidd, W.S.F.; Maruyama, S.; Wang, J.P.; Deng, H.; Wang, Z.S.; et al. Insights into the tectonic evolution of the North China Craton through comparative tectonic analysis: A record of outward growth of Precambrian continents. *Earth-Sci. Rev.* **2016**, *162*, 387–432. [CrossRef]
6. Zhai, M.G.; Santosh, M. The early Precambrian odyssey of North China Craton: A synoptic overview. *Gondwana Res.* **2011**, *20*, 6–25. [CrossRef]
7. Zhai, M.G.; Santosh, M. Metallogeny of the North China Craton: Link with secular changes in the evolving Earth. *Gondwana Res.* **2013**, *24*, 275–297. [CrossRef]
8. Kusky, T.M.; Li, J.H. Paleoproterozoic tectonic evolution of the North China Craton. *J. Asian Earth Sci.* **2003**, *22*, 383–397. [CrossRef]
9. Kusky, T.M.; Li, J.H.; Santosh, M. The Paleoproterozoic North Hebei orogen: North China Craton's collisional suture with the Columbia supercontinent. *Gondwana Res.* **2007**, *12*, 4–28. [CrossRef]
10. Meng, E.; Liu, F.L.; Liu, P.H.; Liu, C.H.; Yang, H.; Wang, F.; Shi, J.R.; Cai, J. Petrogenesis and tectonic significance of Paleoproterozoic meta-mafic rocks from central Liaodong Peninsula, northeast China: Evidence from zircon U-Pb dating and in situ Lu-Hf isotopes, and whole-rock geochemistry. *Precambrian Res.* **2014**, *247*, 92–109. [CrossRef]
11. Meng, E.; Wang, C.Y.; Li, Y.G.; Li, Z.; Yang, H.; Cai, J.; Ji, L.; Jin, M.Q. Zircon U-Pb-Hf isotopic and whole-rock geochemical studies of Paleoproterozoic metasedimentary rocks in the northern segment of the Jiao-Liao-Ji Belt, China: Implications for provenance and regional tectonic evolution. *Precambrian Res.* **2017**, *298*, 472–489. [CrossRef]
12. Liu, J.H.; Liu, F.L.; Ding, Z.J.; Liu, P.H.; Guo, C.L.; Wang, F. Geochronology, petrogenesis and tectonic implications of Paleoproterozoic granitoid rocks in the Jiaobei Terrane, North China Craton. *Precambrian Res.* **2014**, *255*, 685–698. [CrossRef]
13. Xu, W.; Liu, F.L.; Santosh, M.; Liu, P.H.; Tian, Z.H.; Dong, Y.S. Constraints of mafic rocks on a Paleoproterozoic back-arc in the Jiao-Liao-Ji Belt, North China Craton. *J. Asian Earth Sci.* **2018**, *166*, 195–209. [CrossRef]

14. Xu, W.; Liu, F.L.; Tian, Z.H.; Liu, L.S.; Ji, L.; Dong, Y.S. Source and petrogenesis of Paleoproterozoic meta-mafic rocks intruding into the North Liaohe Group: Implications for back-arc extension prior to the formation of the Jiao-Liao-Ji Belt, North China Craton. *Precambrian Res.* **2018**, *307*, 66–81. [CrossRef]

15. Xu, W.; Liu, F.L. Geochronological and geochemical insights into the tectonic evolution of the Paleoproterozoic Jiao-Liao-Ji Belt, Sino-Korean Craton. *Earth-Sci. Rev.* **2019**, *193*, 162–198. [CrossRef]

16. Chen, J.S.; Xing, D.H.; Liu, M.; Li, B.; Yang, H.; Tian, D.X.; Yang, F.; Wang, Y. Zircon U–Pb chronology and geological significance of felsic volcanic rocks in the Liaohe Group from the Liaoyang area, Liaoning Province. *Acta Petrol. Sin.* **2017**, *33*, 2792–2810. (In Chinese with English Abstract)

17. Chen, J.S.; Jiang, Z.Q.; Li, W.W.; Li, B.; Liu, M.; Yang, F.; Xing, D.H.; Wang, Y.; Tan, H.Y. The formation ages of the Langzishan and Lieryu formations in Lianshanguan area, Benxi, Liaoning province, and its geological significance. *Geol. Bull. China.* **2018**, *37*, 1693–1703. (In Chinese with English Abstract)

18. Chen, J.S.; Tian, D.X.; Xing, D.H.; Li, B.; Liu, M.; Yang, F.; Yang, Z.Z. Zircon U-Pb Geochronology and Its Geological Significance of the Basic Volcanic Rocks from the Li'eryu Formation, Liaohe Group in Kuandian Area. *Earth Sci.* **2020**, *45*, 3282–3294. (In Chinese with English Abstract)

19. Chen, J.S.; Li, W.W.; Xing, D.H.; Yang, Z.Z.; Tian, D.X.; Zhang, L.D.; Li, B.; Liu, M.; Yang, F. Zircon U-Pb geochronology of volcanic rocks from Gaojiayu Formation, Liaohe Group, Liaoning Province and its geological significance. *Earth Sci.* **2020**, *45*, 3934–3949. (In Chinese with English Abstract)

20. Chen, J.S.; Yang, Z.Z.; Tian, D.X.; Xing, D.H.; Zhang, L.D.; Yang, F.; Li, B.; Liu, M.; Shi, Y.; Zhang, C. Geochronological Framework of Paleoproterozoic Intrusive Rocks and Its Constraints on Tectonic Evolution of the Liao-Ji Belt, Sino-Korean Craton. *J. Earth Sci.* **2021**, *32*, 8–24. [CrossRef]

21. Zhang, Q.S. Early proterozoic tectonic styles and associated mineral deposits of the North China platform. *Precambrian Res.* **1988**, *39*, 1–29. [CrossRef]

22. Li, S.Z.; Zhao, G.C. SHRIMP U-Pb zircon geochronology of the Liaoji granitoids:constraints on the evolution of the Paleoproterozoic Jiao-Liao-Ji belt in the eastern block of the North China Craton. *Precambrian Res.* **2007**, *158*, 1–16. [CrossRef]

23. Wang, F.; Liu, F.L.; Liu, P.H.; Cai, J.; Schertl, H.P.; Ji, L.; Liu, L.S.; Tian, Z.H. In situ zircon U-Pb dating and whole-rock geochemistry of metasedimentary rocks from South Liaohe Group, Jiao-Liao-Ji orogenic belt: Constraints on the depositional and metamorphic ages, and implications for tectonic setting. *Precambrian Res.* **2017**, *303*, 764–780. [CrossRef]

24. Bai, J. *The Precambrian Geology and Pb-Zn Mineralization in the Northern Margin of North China Platform*; Geological Publishing House: Beijing, China, 1993; pp. 1–136. (In Chinese with English Abstract)

25. Faure, M.; Lin, W.; Monie, P.; Bruguier, O. Palaeoproterozoic arc magmatism and collision in Liaodong Peninsula (north-east China). *Terra Nova* **2004**, *16*, 75–80. [CrossRef]

26. Li, Z.; Chen, B. Geochronology and geochemistry of the Paleoproterozoic metabasalts from the Jiao-Liao-Ji Belt, North China Craton: Implications for petrogenesis and tectonic setting. *Precambrian Res.* **2014**, *255*, 653–667. [CrossRef]

27. Yuan, L.L.; Zhang, X.H.; Xue, F.H.; Han, C.M.; Chen, H.H.; Zhai, M.G. Two episodes of Paleoproterozoic mafic intrusions from Liaoning province, North China Craton: Petrogenesis and tectonic implications. *Precambrian Res.* **2015**, *264*, 119–139. [CrossRef]

28. Li, S.Z.; Zhao, G.C.; Santosh, M.; Liu, X.; Dai, L.M.; Suo, Y.H.; Tam, P.Y.; Song, M.C.; Wang, P.C. Paleoproterozoic structural evolution of the southern segment of the Jiao-Liao-Ji Belt, North China Craton. *Precambrian Res.* **2012**, *200–203*, 59–73. [CrossRef]

29. Zhao, G.C.; Zhai, M.G. Lithotectonic elements of Precambrian basement in the North China Cration: Review and tectonic implications. *Gondwana Res.* **2013**, *23*, 1207–1240. [CrossRef]

30. Duan, Z.Z.; Wei, C.J.; Li, Z. Metamorphic P-T paths and zircon U-Pb ages of Paleoproterozoic metabasic dykes in eastern Hebei and northern Liaoning: Implications for the tectonic evolution of the North China Craton. *Precambrian Res.* **2019**, *326*, 124–141. [CrossRef]

31. Wu, D.; Wei, C.J. Metamorphic evolution of two types of garnet amphibolite from the Qingyuan terrane, North China Craton: Insights from phase equilibria modelling and zircon dating. *Precambrian Res.* **2021**, *355*, 106091. [CrossRef]

32. Cui, R.Z.; Wei, C.J.; Duan, Z.Z. Two phases of granulite facies metamorphism during the Neoarchean and Paleoproterozoic in the Qingyuan terrane, North China Craton. *Acta Petrol. Sin.* **2023**, *39*, 2257–2278. (In Chinese with English Abstract) [CrossRef]

33. Bureau of Geology and Mineral Resource of Liaoning Province. *Regional Geology of Liaoning Province*; Geological Publishing House: Beijing, China, 2014; pp. 595–763. (In Chinese)

34. Zhao, G.C.; Cao, L.; Wilde, S.; Sun, M.; Choe, W.; Li, S.Z. Implications based on the first SHRIMP U–Pb zircon dating on Precambrian granitoid rocks in North Korea. *Earth Planet. Sci. Lett.* **2006**, *251*, 365–379. [CrossRef]

35. Li, Z.; Meng, E.; Wang, C.Y.; Li, Y.G. Early Precambrian tectonothermal events in the Southern Jilin Province, China: Implications for Neoarchean crustal evolution of the northeastern North China Craton. *Mineral. Petrol.* **2019**, *113*, 185–205. [CrossRef]

36. Li, Z.; Wei, C.J.; Chen, B.; Fu, B.; Gong, M.Y. Late Neoarchean reworking of the Mesoarchean crustal remnant in northern Liaoning, North China Craton: A U-Pb-Hf-O-Nd perspective. *Gondwana Res.* **2020**, *80*, 350–369. [CrossRef]

37. Peng, P.; Wang, C.; Wang, X.P.; Yang, S.Y. Qingyuan high-grade granite-greenstone terrain in the Eastern North China Craton: Root of a Neoarchaean arc. *Tectonophysics* **2015**, *662*, 7–21. [CrossRef]

38. Li, Z.; Wei, C.J. Two Types of Neoarchean basalts from Qingyuan greenstone belt, North China Craton: Petrogenesis and tectonic implications. *Precambrian Res.* **2017**, *292*, 175–193. [CrossRef]

39. Wan, Y.S.; Song, B.; Yang, C.; Liu, D.Y. Zircon SHRIMP U-Pb geochronology of Archaean rocks from the Fushun-Qingyuan area, Liaoning province and its geological significance. *Acta Geol. Sin.* **2005**, *79*, 78–87. (In Chinese with English Abstract)
40. Dickinson, J.E., Jr.; Hess, P.C. Zircon saturation in lunar basalts and granites. *Earth Planet. Sci. Lett.* **1982**, *57*, 336–344. [CrossRef]
41. Shao, T.; Xia, Y.; Ding, X.; Cai, Y.; Song, M. Zircon saturation in terrestrial basaltic melts and its geological implications. *Solid Earth Sci.* **2019**, *4*, 27–42. [CrossRef]
42. Bea, F.; Bortnikov, N.; Cambese, A.; Chakraborty, S.; Molina, J.F.; Montero, P.; Morales, I.; Silantiev, S.; Zinger, T. Zircon crystallization in low-Zr mafic magmas: Possible or impossible? *Chem. Geol.* **2022**, *602*, 120898. [CrossRef]
43. Wiedenbeck, M.; Allé, P.; Corfu, F.; Griffin, W.L.; Meier, M.; Oberli, F.; Quadt, A.; Roddick, J.C.; Spiegel, W. Three natural zircon standards for U-Th-Pb, Lu-Hf, trace element and REE analyses. *Geostand. Newsletter.* **1995**, *19*, 1–23. [CrossRef]
44. Ludwig, K.R. User's manual for isoplot 3.0: A geochronological toolkit for microsoft excel. *Berkeley Geochronol. Cent. Spec. Publ.* **2003**, *4*, 75.
45. Andersen, T. Correction of common Lead in U-Pb analyses that do not report ^{204}Pb. *Chem. Geol.* **2002**, *192*, 59–79. [CrossRef]
46. Belousova, E.; Griffin, W.; O'Reilly, S.Y.; Fisher, N. Igneous zircon: Trace element composition as an indicator of source rock type. *Contrib. Mineral. Petrol.* **2002**, *143*, 602–622. [CrossRef]
47. Hoskin, P.W.O.; Ireland, T.R. Rare earth element chemistry of zircon and its use as a provenance indicator. *Geology* **2000**, *28*, 627–630. [CrossRef]
48. Li, C.M. A review on the minerageny and situ microanalytical dating techniques of zircons. *Geol. Surv. Res.* **2009**, *33*, 161–174. (In Chinese with English Abstract)
49. Liu, D.Y.; Wilde, S.A.; Wan, Y.S.; Shiyan Wang, S.Y.; Valley, J.W.; Kita, N.; Dong, C.Y.; Xie, H.Q.; Yang, C.X.; Zhang, Y.X.; et al. Combined U-Pb, hafnium and oxygen isotope analysis of zircons from meta-igneous rocks in the southern North China Craton: Reveal multiple events in the Late Mesoarchean-Early Neoarchean. *Chem. Geol.* **2009**, *261*, 139–153. [CrossRef]
50. Dupuy, C.; Dostal, J. Trace element geochemistry of some continental tholeiites. *Earth Planet. Sci. Lett.* **1984**, *67*, 61–69. [CrossRef]
51. Middlemost, E.A.K. Naming materials in the magma/igneous rock system. *Earth-Sci. Rev.* **1994**, *37*, 215–224. [CrossRef]
52. De la Roche, H.; Leterrier, J.; Grandclaude, P.; Marchal, M. A classification of volcanic and plutonic rocks using R1R2-diagram and major-element analyses—Its relationships with current nomenclature. *Chem. Geol.* **1980**, *29*, 183–210. [CrossRef]
53. Peccerillo, A.; Taylor, S.R. Geochemistry of Eocene calc-alkaline volcanic rocks from the Kastamonu area, northern Turkey. *Contrib. Mineral. Petrol.* **1976**, *58*, 63–81. [CrossRef]
54. Pearce, J.A. Trace elements characteristic of lavas from destructive plate boundaries. Andesites. In *Thorpe R S. Orogenic Andesites and Related Rocks*; John Wiley & Sons: Chichester, UK, 1982; pp. 525–548.
55. Sun, S.S.; McDonough, W.F. Chemical and isotopic systematics of oceanic basalts: Implications for mantle composition and processes. *Geol. Soc. Lond. Spec. Publ.* **1989**, *42*, 313–345. [CrossRef]
56. Liu, F.L.; Liu, P.H.; Wang, F.; Liu, C.H.; Cai, J. Progresses and overviews of voluminous meta-sedimentary series within the Paleoproterozoic Jiao-Liao-Ji orogenic/mobile belt, North China Craton. *Acta Petrol. Sin.* **2015**, *31*, 2816–2846. (In Chinese with English Abstract)
57. Dong, C.Y.; Ma, M.Z.; Liu, S.J.; Xie, H.Q.; Liu, D.Y.; Li, X.M.; Wan, Y.S. Middle Paleoproterozoic crustal extensional regime in the North China Craton: New evidence from SHRIMP zircon U-Pb dating and whole-rock geochemistry of meta-gabbro in the Anshan-Gongchangling area. *Acta Petrol. Sin.* **2012**, *28*, 2785–2792. (In Chinese with English Abstract)
58. Gao, B.S.; Dong, Y.S.; Li, F.Q.; Wang, P.S.; Gan, Y.C.; Chen, M.S.; Tian, Z.H. Petrogenesis of the Li'eryu Formation of the South Liaohe Group in the Huanghuadian area, Liaodong Peninsula. *Acta Petrol. Sin.* **2017**, *33*, 2725–2742. (In Chinese with English Abstract)
59. Bureau of Geology and Mineral Resource of Liaoning Province. *Liaoning 1:50,000 Erpendianzi Regional Geological Survey*; Geological Publishing House: Beijing, China, 2019; 276p. (In Chinese)
60. Liu, P.H.; Cai, J.; Zou, L. Metamorphic P-T-t path and its geological implication of the Sanjiazi garnet amphibolites from the northern Liaodong Penisula, Jiao-Liao-Ji belt:Constraints on phase equilibria and zircon U-Pb dating. *Acta Petrol. Sin.* **2017**, *33*, 2649–2674. (In Chinese with English Abstract)
61. Qin, Y. Geochronological Constraints in the Tectonic Evolution of the Liao-Ji Paleoproterozoic Rift Zone. Ph.D. Thesis, Jilin University, Changchun, China, 2013; pp. 1–167. (In Chinese with English Abstract)
62. Wang, H.C.; Lu, S.N.; Chu, H.; Xiang, Z.Q.; Zhang, C.J.; Liu, H. Zircon U-Pb age and tectonic setting of meta-basalts of Liaohe Group in Helan area, Liaoyang, Liaoning Province. *J. Jilin Univ.* **2011**, *41*, 1322–1334. (In Chinese with English Abstract)
63. Wang, X.P.; Peng, P.; Wang, C.; Yang, S.Y. Petrogenesis of the 2115 Ma Haicheng mafic sills from the Eastern North China Craton: Implications for an intra-continental rifting. *Gondwana Res.* **2016**, *39*, 347–364. [CrossRef]
64. Yang, M.C.; Chen, B.; Yan, C. Petrogenesis of Paleoproterozoic gneissic granites from Jiao-Liao-Ji Belt of North China Craton and their tectonic implications. *J. Earth Sci. Environ.* **2015**, *37*, 31–51. (In Chinese with English Abstract)
65. Yu, J.J.; Yang, D.B.; Feng, H.; Lan, X. Chronology of amphibolite protolith in Haicheng of southern Liaoning: Evidence from LA-ICP-MS zircon U-Pb dating. *Glob. Geol.* **2007**, *26*, 391–396, 408. (In Chinese with English Abstract)
66. Zhao, Y.; Kou, L.L.; Zhang, P.; Bi, Z.W.; Li, D.T.; Chen, C. Characteristics of geochemistry and Hf isotope from meta-gabbro in Longchang area, Liaodong Peninsula: Implications on evolution of the Jiao-Liao-Ji Paleoproterozoic Orogenic Belt. *Earth Sci.* **2019**, *44*, 3333–3345. (In Chinese with English Abstract)

67. Kerrich, R.; Polat, A.; Wyman, D.; Hollings, P. Trace element systematics of Mg-, to Fe-tholeiitic basalt suites of the Superior Province: Implications for Archean mantle reservoirs and greenstone belt genesis. *Lithos* **1999**, *46*, 163–187. [CrossRef]
68. Sklyarov, E.V.; Gladkochub, D.P.; Mazukabzov, A.M.; Menshagin, Y.V.; Watanabe, T.; Pisarevsky, S.A. Neoproterozoic mafic dike swarms of the Sharyzhalgai metamorphic massif, southern Siberian Craton. *Precambrian Res.* **2003**, *122*, 359–376. [CrossRef]
69. Condie, K.C.; Frey, B.A.; Kerrich, R. The 1.75 Ga Iron King Volcanics in West Central Arizona: A remnant of an accreted Oceanic Plateau derived from a mantle plume with a deep depleted component. *Lithos* **2002**, *64*, 49–62. [CrossRef]
70. McKenzie, D.; O'Nions, R.K. Partial melt distribution from inversion of rare earth element concentrations. *J. Petrol.* **1991**, *32*, 1021–1091. [CrossRef]
71. Zhang, X.H.; Yuan, L.L.; Xue, F.H.; Zhang, Y.B. Contrasting Triassic ferroan granitoids from northwestern Liaoning, North China: Magmatic monitor of Mesozoic decratonization and a craton–orogen boundary. *Lithos* **2012**, *144–145*, 12–23. [CrossRef]
72. Woodhead, J.D.; Hergt, J.M.; Davidson, J.P.; Eggins, S.M. Hafnium isotope evidence for conservative element mobility during subduction zone process. *Earth Planet. Sci. Lett.* **2001**, *192*, 331–346. [CrossRef]
73. Aldanmaz, E.; Pearce, J.A.; Thirlwall, M.F.; Mitchell, J.G. Petrogenetic evolution of Late Cenozoic, post-collision volcanism in Western Anatolia, Turkey. *J. Volcanol. Geotherm. Res.* **2000**, *102*, 67–95. [CrossRef]
74. Jourdan, F.; Bertrand, H.; Schaerer, U.; Blichert-Toft, J.; Feraud, G.; Kampunzu, A.B. Major and trace element and Sr, Nd, Hf and Pb isotope compositions of the Karoo large igneous province, Botswana-Zimbabwe: Lithosphere vs mantle plume contribution. *Petrology* **2007**, *48*, 1043–1077. [CrossRef]
75. Sun, S.S.; Nesbitt, R.W. Geochemical regularities and genetic significance of ophiolitic basalts. *Geology* **1978**, *6*, 689–693. [CrossRef]
76. Sajona, F.G.; Maury, R.C.; Pubellier, M.; Leterrier, J.; Bellon, H.; Cotton, J. Magmatic source enrichment by slab-derived melts in a young post-collision setting, Central Mindanao (Philippines). *Lithos* **2000**, *54*, 173–206. [CrossRef]
77. Plank, T.; Langmuir, C.H. The chemical composition of subducting sediment and its consequences for the crust and mantle. *Chem. Geol.* **1998**, *145*, 325–394. [CrossRef]
78. Wu, F.Y.; Li, X.H.; Yang, J.H.; Zheng, Y.F. Discussions on the petrogenesis of granites. *Acta Petrol. Sin.* **2007**, *23*, 1217–1238. (In Chinese with English Abstract)
79. Chi, Q.H.; Yan, M.C. *Handbook of Elemental Abundance for Applied Geochemistry*; Geological Publishing House: Beijing, China, 2007; pp. 1–148. (In Chinese)
80. Zhang, Y.M.; Pei, X.Z.; Li, Z.C.; Li, R.B.; Liu, C.J.; Pei, L.; Chen, Y.X.; Wang, M. Zircon U-Pb geochronology, geochemistry and its geological implication of the early indosinian basic complex in the Qinghai nanshan tectonic belt. *Earth Sci.* **2019**, *44*, 2461–2477. (In Chinese with English Abstract)
81. Wilson, M. *Igneous Petrogenesis: A global Tectonic Approach*; Unwin Hyman: London, UK, 1989; pp. 1–466.
82. Condie, K.C. Geochemical changes in baslts and andesites across the Archean? Proterozoic Boundary: Identification and significance. *Lithos* **1989**, *23*, 1–18. [CrossRef]
83. Pearce, J.A. A user's guide to basalt discrimination diagrams. In *Trace Element Geochemistry of Volcanic Rocks: Applications for Massive Sulphide Exploration*; Wyman, D.A., Ed.; Short Course Notes; Geological Association of Canada: St. John's, NF, Canada 1996; Volume 12, pp. 79–113.
84. Duan, Z.Z.; Wei, C.J.; Qian, J.H. Metamorphic P-T paths and zircon U–Pb age data for the Paleoproterozoic metabasic dykes of high-pressure granulite facies from Eastern Hebei, North China Craton. *Precambrian Res.* **2015**, *271*, 295–310. [CrossRef]

MDPI AG
Grosspeteranlage 5
4052 Basel
Switzerland
Tel.: +41 61 683 77 34

Minerals Editorial Office
E-mail: minerals@mdpi.com
www.mdpi.com/journal/minerals

www.ingramcontent.com/pod-product-compliance
Lightning Source LLC
LaVergne TN
LVHW070113170726
843515LV00007B/1684